Neurobiology of
Glycoconjugates

Neurobiology of Glycoconjugates

Edited by
Richard U. Margolis
New York University Medical Center
New York, New York

and
Renée K. Margolis
State University of New York
Health Science Center at Brooklyn
Brooklyn, New York

Springer Science+Business Media, LLC

Library of Congress Cataloging in Publication Data

Neurobiology of glycoconjugates / edited by Richard U. Margolis and Renée K. Margolis.
    p.   cm.
    Includes bibliographies and index.
    ISBN 978-1-4757-5957-0   ISBN 978-1-4757-5955-6 (eBook)
    DOI 10.1007/978-1-4757-5955-6
    1. Glycoproteins. 2. Glycolipids. 3. Neurochemistry. I. Margolis, Richard U. II. Margolis, Renée K.
    [DNLM: 1. Glycoconjugates — metabolism. 2. Neurobiology. 3. Neurochemistry. QU 75 N494]
    QP552.G59N48   1989
    599′.01924 — dc20
    DNLM/DLC   89-8487
    for Library of Congress   CIP

© 1989 Springer Science+Business Media New York
Originally published by Plenum Press, New York in 1989
Softcover reprint of the hardcover 1st edition 1989

Contributors

Salvatore Carbonetto • Neurosciences Unit, Montreal General Hospital Research Institute, McGill University, Montreal, Quebec H3G 1A4, Canada

Steven S. Carlson • Department of Physiology and Biophysics, University of Washington, Seattle, Washington 98195

Glyn Dawson • Departments of Pediatrics and Biochemistry and Molecular Biology, and the Kennedy Mental Retardation Research Center, University of Chicago, Chicago, Illinois 60637

J. Dodd • Department of Physiology, Columbia University, and College of Physicians and Surgeons, New York, New York 10032

Philippe Douville • Center for Neuroscience Research, McGill University, Montreal General Hospital Research Institute, Montreal, Quebec H3G 1A4, Canada

Jeffry F. Goodrum • Biological Sciences Research Center and Department of Pathology, University of North Carolina, Chapel Hill, North Carolina 27599

James W. Gurd • Department of Biochemistry, Scarborough Campus, University of Toronto, West Hill, Ontario M1C 1A4, Canada

Larry W. Hancock • Departments of Pediatrics and Biochemistry and Molecular Biology, and the Kennedy Mental Retardation Research Center, University of Chicago, Chicago, Illinois 60637

M. A. Hynes • Howard Hughes Medical Institute, Center for Neurobiology and Behavior, Columbia University, and College of Physicians and Surgeons, New York, New York 10032

T. M. Jessell • Howard Hughes Medical Institute, Center for Neurobiology and Behavior, Columbia University, and Department of Biochemistry and Molecular Biophysics, College of Physicians and Surgeons, New York, New York 10032

Robert W. Ledeen • Departments of Neurology and Biochemistry, Albert Einstein College of Medicine, Bronx, New York 10461

Renée K. Margolis • Department of Pharmacology, State University of New York, Health Science Center at Brooklyn, Brooklyn, New York 11203

Richard U. Margolis • Department of Pharmacology, New York University Medical Center, New York, New York 10016

Pierre Morell • Biological Sciences Research Center and Department of Biochemistry, University of North Carolina, Chapel Hill, North Carolina 27599

Richard H. Quarles • Section on Myelin and Brain Development, Laboratory of Molecular and Cellular Neurobiology, National Institute of Neurological Disorders and Stroke, National Institutes of Health, Bethesda, Maryland 20892

Urs Rutishauser • Department of Genetics and Center for Neuroscience, Case Western Reserve University, School of Medicine, Cleveland, Ohio 44106

Megumi Saito • Department of Biochemistry and Molecular Biophysics, Medical College of Virginia, Virginia Commonwealth University, Richmond, Virginia 23298

Nancy B. Schwartz • Departments of Pediatrics and Biochemistry and Molecular Biology, and the Kennedy Mental Retardation Research Center, University of Chicago, Chicago, Illinois 60637

Neil R. Smalheiser • Department of Pediatrics, and the Kennedy Mental Retardation Research Center, University of Chicago, Chicago, Illinois 60637

George C. Stone • Division of Molecular Biology, Nathan Kline Institute, Orangeburg, New York 10962

Bryan P. Toole • Department of Anatomy and Cellular Biology, Tufts University Health Sciences Center, Boston, Massachusetts 02111

Charles J. Waechter • Department of Biochemistry, University of Kentucky College of Medicine, A. B. Chandler Medical Center, Lexington, Kentucky 40536

Robert K. Yu • Department of Biochemistry and Molecular Biophysics, Medical College of Virginia, Virginia Commonwealth University, Richmond, Virginia 23298

Preface

Ten years ago the general area covered in this volume was comprehensively surveyed for the first time (*Complex Carbohydrates of Nervous Tissue,* Plenum Press, 1979). The revised title of the present volume reflects a number of important changes in content and emphasis, which have been made possible by the accelerating pace of research in this field. A third of the chapters did not appear in the predecessor volume, and over half of the remainder have new authors. Many advances relating to the functional aspects of nervous tissue glycoconjugates have resulted from the application of sensitive and specific immunocytochemical techniques to studies on their localization and developmental changes. In the past ten years, the importance of various extracellular matrix macromolecules such as laminin, fibronectin, tenascin, and heparan sulfate proteoglycans has also become generally recognized, as has the role of specialized oligosaccharide structures in cell interactions and developmental processes. In certain cases, connecting links have been discovered between seemingly unrelated research areas, such as the finding that a number of nervous tissue glycoproteins with probable roles in cell interactions (e.g., Thy-1, NCAM, and the myelin-associated glycoprotein) all belong to the immunoglobulin superfamily.

The next decade of research will undoubtedly prove to be particularly exciting. The application of molecular biological approaches should provide information concerning functionally significant protein domains and amino acid sequence homologies, as well as on the role of modifications such as alternative splicing and of different glycoforms. Perhaps most importantly, the ability to manipulate the expression of these glycoconjugates will allow us to better understand their developmental regulation and functional roles. We therefore hope that this volume will again serve as a convenient and reliable survey of current knowledge in the area of nervous tissue glycoconjugates, and will also provide both direction and impetus for a deeper understanding of their biological relevance.

Richard U. Margolis and Renée K. Margolis

New York

Contents

Chapter 3

Structure and Localization of Glycoproteins and Proteoglycans

Renée K. Margolis and Richard U. Margolis

Chapter 4

Biosynthesis of Glycoproteins

Charles J. Waechter

Chapter 5

Biosynthesis of Glycosaminoglycans and Proteoglycans

Nancy B. Schwartz and Neil R. Smalheiser

Chapter 6

Lysosomal Degradation of Glycoproteins and Glycosaminoglycans

Larry W. Hancock and Glyn Dawson

Chapter 7

Glycoproteins of the Synapse

   James W. Gurd

Chapter 8

Glycoproteins of Myelin and Myelin-Forming Cells

   Richard H. Quarles

Chapter 9

Axonal Transport and Intracellular Sorting of Glycoconjugates

Jeffry F. Goodrum, George C. Stone, and Pierre Morell

Chapter 10

Synaptic Vesicle Glycoproteins and Proteoglycans

Steven S. Carlson

Chapter 11

Carbohydrate Recognition, Cell Interactions, and Vertebrate Neural Development

M. A. Hynes, J. Dodd, and T. M. Jessell

Chapter 12

Polysialic Acid as a Regulator of Cell Interactions

Urs Rutishauser

Chapter 13

Extracellular Matrix Adhesive Glycoproteins and Their Receptors in the Nervous System

Philippe Douville and Salvatore Carbonetto

Chapter 14

Hyaluronate and Hyaluronate-Binding Proteins of Brain

Bryan P. Toole

Chapter 15

Inborn Errors of Complex Carbohydrate Catabolism

Glyn Dawson and Larry W. Hancock

Structure and Localization of Gangliosides

Robert K. Yu and Megumi Saito

1. INTRODUCTION

Gangliosides are sialic acid-containing glycosphingolipids (GSLs) found primarily in the plasma membrane of virtually all vertebrate tissues and are particularly abundant in the nervous system. They constitute part of the glycocalyx network surrounding the cell surface and are crucial in determining the properties and functions of cells. Structurally, they contain a hydrophobic ceramide chain to which a hydrophilic oligosaccharide is glycosidically linked. Some representative brain gangliosides are shown in Figure 1.

Heterogeneity of the oligosaccharide molecular structure is a hallmark of gangliosides, as of GSLs in general. Presently, nearly 90 different gangliosides have been isolated from various sources (see Table 2). Approximately half of them have been found in the nervous system. The number of new gangliosides is expected to increase with the advent of improved isolation and analytical methods (Ledeen and Yu, 1982; Ando, 1983; Egge *et al.*, 1985; Yu *et al.*, 1986; Ginsburg, 1987).

2. STRUCTURES

GSLs are generally classified according to the carbohydrate structures linked to the ceramide moiety. Approximately 70 different oligosaccharide structures of gangliosides have been identified to date, and when variations in sialic acid structures are taken into account, the total number of ganglioside structures is about 90. Of these, a

---

Robert K. Yu and Megumi Saito • Department of Biochemistry and Molecular Biophysics, Medical College of Virginia, Virginia Commonwealth University, Richmond, Virginia 23298.

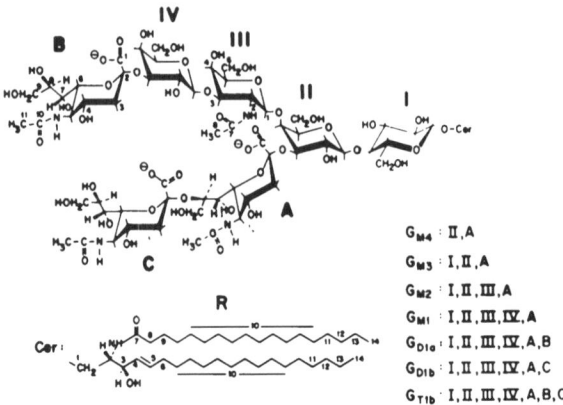

G_{M4} : II,A

G_{M3} : I,II,A

G_{M2} : I,II,III,A

G_{M1} : I,II,III,IV,A

G_{D1a} : I,II,III,IV,A,B

G_{D1b} : I,II,III,IV,A,C

G_{T1b} : I,II,III,IV,A,B,C

Figure 1. Structures of some representative brain gangliosides.

considerable number of new ganglioside structures have been identified since the predecessor volume appeared in 1979. Based on the oligosaccharide structures, gangliosides can be categorized into five major families: gala, hemato, ganglio, lacto, and globo series. As shown in Table 1, the gala series is derived from galactosylceramide and only one ganglioside, GM4, is included in this family. All other series of gangliosides originate from lactosylceramide, and are divided into four families according to the type of sugars linked to the galactose moiety of lactosylceramide. The lacto and globo series are further classified into two subgroups, respectively, according to the linkage between the third and fourth sugars. Table 2 shows representative oligosaccharide core structures of gangliosides and their abbreviations.

Variations in sialic acid structures also contribute to the diversity in ganglioside structures. The sialic acid residues in gangliosides are present either as *N*-acetyl-

Table 1. Families of Gangliosides

Family	Abbreviation	Structure
Gala	Gal	Galβ1-1'Cer
Hemato	Lac	Galβ1-4Glc1-1'Cer
Gangliotriaose	GgOse$_3$	GalNAcβ1-4Galβ1-4Glc1-1'Cer
Gangliotetraose	GgOse$_4$	Galβ1-3GalNAcβ1-4Galβ1-4Glc1-1'Cer
Lactotetraose	LcOse$_4$	Galβ1-3GlcNAcβ1-3Galβ1-4Glc1-1'Cer
Neolactotetraose	nLcOse$_4$	Galβ1-4GlcNAcβ1-3Galβ1-4Glc1-1'Cer
Neolactohexaose	nLcOse$_6$	Galβ1-4GlcNAcβ1-3Galβ1-4GlcNAcβ1-3Galβ1-4Glc1-1'Cer
Neolactooctaose	nLcOse$_8$	Galβ1-4GlcNAcβ1-3Galβ1-4GlcNAcβ1-3Galβ1-4GlcNAcβ1-3Galβ1-4Glc1-1'Cer
Globotetraose	GbOse$_4$	GalNAcβ1-3Galα1-4Galβ1-4Glcβ1-1'Cer
Globopentaose	GbOse$_5$	Galβ1-3GalNAcβ1-3Galα1-4Galβ1-4Glc1-1'Cer
Isoglobopentaose	iGbOse$_5$	Galβ1-3GalNAcβ1-3Galα1-3Galβ1-4Glcβ1-1'Cer

Table 2. Carbohydrate Structures of Gangliosides

Structure	Abbreviation	Source	Reference
Gala series			
Gal- 3 | NeuAcα2	I³NeuAc-Gal	Human brain Human myelin Avian myelin Egg yolk	Kuhn and Wiegandt (1964) Ledeen et al. (1973) Cochran et al. (1981) Li et al. (1978)
Hemato series			
Galβ1-4Glc- 3 | NeuAcα2	II³NeuAc-Lac GM3	Human brain Human liver Bovine liver, spleen, kidney Bovine adrenal medulla Rabbit tissues	Kuhn and Wiegandt (1964) Seyfried et al. (1978) Wiegandt (1973) Ledeen et al. (1968) Iwamori and Nagai (1981)

Galβ1-4Glc-
3
|
NeuGcα2
	II³NeuGc-Lac
	GM3 (NeuGc)
		Horse erythrocytes
		
		Bovine adrenal mudulla
		Bovine liver,
		  spleen, kidney
			Yamakawa and Suzuki
			  (1951), Klenk and
			  Lauenstein (1953)
			Ledeen et al. (1968)
			Wiegandt (1973)

Galβ1-4Glc- 3 | (4-OAc) NeuGcα2	II³(4-OAcNeuGc)-Lac	Horse erythrocytes	Hakomori and Saito (1969)
Galβ1-4Glc- 3 | (8-OMe)NeuGcα2	II³(8-OMeNeuGc)-Lac	Starfish hepatopancreas	Smirnova et al. (1987)

(continued)

Table 2. (Continued)

Structure	Abbreviation	Source	Reference
Galβ1-4Glc- 3 | NeuAcα2-8NeuAcα2	II³(NeuAc)₂-Lac GD3	Human brain Bovine liver, spleen, kidney Bovine retina	Kuhn and Wiegandt (1964) Wiegandt (1973) Handa and Burton (1969)
Galβ1-4Glc- 3 | NeuAcα2-8NeuGcα2	II³(NeuAcα2-8NeuGcα2)-Lac GD3(NeuAc/NeuGc)	Bovine liver, spleen, kidney	Wiegandt (1973)
Galβ1-4Glc- 3 | NeuGcα2-8NeuAcα2	II³(NeuGcα2-8NeuAcα2)-Lac GD3(NeuGc/NeuAc)	Bovine liver, spleen, kidney Rabbit thymus	Wiegandt (1973) Iwamori and Nagai (1978)
Galβ1-4Glc- 3 | NeuGcα2-8NeuGcα2	II³(NeuGc)₂-Lac GD3(NeuGc)₂	Bovine liver, spleen, kidney Cat erythrocytes	Wiegandt (1973) Handa and Handa (1965)
Galβ1-4Glc- 3 | 9-OAcNeuAcα2-8NeuAcα2	II³(9-OAcNeuAcα2-8NeuAc)-Lac 9-OAcGD3	Melanoma	Cheresh et al. (1984)
Galβ1-4Glc- 3 | NeuAcα2-8NeuAcα2-8NeuAcα2	II³(NeuAc)₃-Lac GT3	Fish brain Pig kidney	Price et al. (1975); Yu and Ando (1980), Avrova et al. (1979) Murakami-Murofushi et al. (1981)

Monosialo

Ganglio series

Structure	Abbreviation	Localization	Reference
GalNAcβ1-4Galβ1-4Glc- 　　　　　3 　　　　　| 　　　　NeuAcα2	II^3NeuAc-GgOse$_3$ GM2	Human brain Tay–Sachs brain	Kuhn and Wiegandt (1964) Ledden and Salsman (1965)
GalNAcβ1-4Galβ1-4Glc- 　　　　　3 　　　　　| 　　　　NeuAc2	Lyso-GM2	Tay–Sachs brain	Neuenhofer *et al.* (1985)
GalNAcβ1-4Galβ1-4Glc- 　　　　　3 　　　　　| 　　　　NeuGcα2	II^3NeuGc-GgOse$_3$ GM2(NeuGc)	Bovine spleen	Wiegandt (1973)
GalNAcβ1-4Galβ1-4Glc- 　　　　　3 　　　　　| (8-O-SO$_4$)NeuAcα2	II^3(8-O-SO$_4$)NeuAc-GgOse$_3$	Bovine gastric mucosa	Slomicny *et al.* (1981)
GalNAcβ1-4Galβ1-4Glc- 　　　　　3 　　　　　| (8-O-SO$_4$)NeuGcα2	II^3(8-O-SO$_4$)NeuGc- GgOse$_3$	Bovine gastric mucosa	Slomiany *et al.* (1981)
Galβ1-3GalNAcβ1-4Galβ1-4Glc- 　　　　　　　　　3 　　　　　　　　　| 　　　　　　　NeuAcα2	II^3NeuAc-GgOse$_4$ GMl(GMla)	Human brain GM1 gangliosidosis Bovine spleen	Kuhn and Wiegandt (1963a) Ledeen *et al.* (1965) Wiegandt (1973)
Galβ1-3GalNAcβ1-4Galβ1-4Glc- 　　　　　　　　　3 　　　　　　　　　| 　　　　　　　NeuGcα2	II^3NeuGc-GgOse$_4$ GMl(NeuGc)	Bovine spleen, kidney, liver	Wiegandt (1973)

(continued)

Table 2. (Continued)

Structure	Abbreviation	Source	Reference

Galβ1-3GalNAcβ1-4Galβ1-4Glc-
4
|
GalNAcβ1
  3
  |
  NeuAcα2
	$IV^4\beta GalNAc,II^3NeuAc$-GgOse$_4$
	GM1-GalNAc
		Human brain	Iwamori and Nagai (1978)

Galβ1-3GalNAcβ1-4Galβ1-4Glc- 3 3 | | Galα1 NeuAcα2	$IV^3\alpha Gal,II^3NeuAc$-GgOse$_4$ GM1-Gal	Frog fat body	Ohashi (1980)
Galβ1-3GalNAcβ1-4Galβ1-4Glc- 3 3 | | Galα1 NeuAcα2 3 | Galβ1	$IV^3(Gal\beta 1$-$3Gal\alpha),II^3$NeuAc-GgOse$_4$ GM1-Gal$_2$	Frog fat body	Ohashi (1980)
Galβ1-3GalNAcβ1-4Galβ1-4Glc- 3 3 | | Galα1 NeuAcα2 3 | Galβ1 3 | Galα1	$IV^3(Gal\alpha 1$-$3Gal\beta 1$-$3Gal\alpha),II^3$NeuAc-GgOse$_4$ GM1-Gal$_3$	Frog fat body	Oshashi (1980)
Galβ1-3GalNAcβ1-4Galβ1-4Glc- 2 3 | | Fucα1 NeuAcα2	$IV^2\alpha Fuc,II^3NeuAc$-GgOse$_4$ GM1-Fuc	Bovine brain Bovine thyroid Pig adipose Mini-pig brain Boar testis	Ghidoni et al. (1976) Macher et al. (1979), Van Dessel et al. (1979) Ohashi and Yamakawa (1977) Fredman et al. (1981) Suzuki et al. (1975)

Structure	Designation	Source	Reference
Galβ1-3GalNAcβ1-4Galβ1-4Glc- 　　　　　　2　　　　　3 　　　　　　|　　　　　| 　　　　Fucα1　　NeuGcα2	IV²αFuc,II³NeuGc-GgOse₄ GM1-Fuc(NeuGc)	Bovine liver Pig adipose Mini-pig brain	Wiegandt (1973) Ohashi and Yamakawa (1977) Fredman et al. (1981)
Galβ1-3GalNAcβ1-4Galβ1-4Glc- 　　　　　　3　　　　　3 　　　　　　|　　　　　| 　　　　Fucα1　　NeuAcα2	IV³αFuc,II³NeuAc-GgOse₄	Pig adipose	Ohashi and Yamakawa (1977)
Galα1 　| 　2 Galβ1-3GalNAcβ1-4Galβ1-4Glc- 　　　　　　3　　　　　3 　　　　　　|　　　　　| 　　　　Fucα1　　NeuAcα2	IV³αFuc,IV²αGal, II³NeuAc-GgOse₄	Rat hepatoma Rat stomach PC12 cells	Holmes and Hakomori (1982) Bouhours et al. (1987) Ariga et al. (1987)
GalNAcα1-3GalNAcβ1 　　　　　　　| 　　　　　　　3 Galβ1-3GalNAcβ1-4Galβ1-4Glc- 　　　　　　　3 　　　　　　　| 　　　　　NeuAcα2	IV³(GalNAcα1-3GalNAcβ1), II³NeuAc-GgOse₄	English sole liver	Ostrander et al. (1988)
Galβ1-3GalNAcβ1-4GAlβ1-4Glc- 　　　　　　　3 　　　　　　　| 　　　　　NeuAcα2	IV³NeuAc-GgOse₄ GMlb	Human erythrocytes Rat tumor Human brain	Watanabe et al. (1979a) Hirabayashi et al. (1979) Ariga and Yu (1987)

(continued)

Table 2. (Continued)

Structure	Abbreviation	Source	Reference
GalNAcβ1 | 4 Galβ1-3GalNAcβ1-4Galβ1-4Glc- 3 | NeuAcα2	IV⁴βGalNAc,IV³NeuAc-GgOse₄ GMlb-GalNac	Tay–Sachs brain	Itoh et al. (1981)
GalNAcβ1 | 4 Galβ1-3GalNAcβ1-4Galβ1-4Glc- 3 | NeuGcα2	IV⁴βGalNAc,IV³NeuGc-GgOse₄	Mouse T lymphocytes	Muthing et al. (1987)
Galβ1-3GalNAcβ1 | 4 Galβ1-3GalNAcβ1-4Galβ1-4Glc- 3 | NeuAcα2	IV⁴(Galβ1-3GalNAcβ1), IV³NeuAc-GgOse₄	Mouse spleen	Nakamura et al. (1987)
Disialo GalNAcβ1-4Galβ1-4Glc- 3 | NeuAcα2 8 | NeuAcα2	II³(NeuAc)₂-GgOse₃ GD2	Human brain	Kuhn and Wiegandt (1964), Klenk and Naomi (1968)
Galβ1-3GalNAcβ1-4Galβ1-4Glc- 3 | NeuAcα2 NeuAcα2	IV³NeuAc,II³NeuAc-GgOse₄ GDla	Human brain Bovine adrenal medulla	Kuhn and Wiegandt (1963b), Klenk and Gielen (1963) Price et al. (1975)

Structure	Abbreviation	Source	Reference
Galβ1-3GalNAcβ1-4Galβ1-4Glc- 3 | NeuAcα2 NeuGcα2	IV³NeuAc,II³NeuGc-GgOse₄ GDIa(NeuAc/NeuGc)	Bovine brain Bovine adrenal medulla	Ghidoni et al. (1976) Price et al. (1975)
Galβ1-3GalNAcβ1-4Galβ1-4Glc- 3 | NeuGcα2 NeuAcα2	IV³NeuGc,II³NeuAc-GgOse₄ GDIa(NeuGc/NeuAc)	Bovine brain	Ghidoni et al. (1976)
Galβ1-3GalNAcβ1-4Galβ1-4Glc- 3 | NeuGcα2 NeuGcα2	IV³NeuGc,II³NeuGc-GgOse₄ GDIa(NeuGc)₂	Bovine spleen, kidney	Wiegandt (1973)
Galβ1-3GalNAcβ1-4Galβ1-4Glc- 3 | (9-OAc)NeuAcα2 NeuAcα2	IV³NeuAc(9-OAc),II³NeuAc-GgOse₄	Rat erythrocytes	Gowda et al. (1984)
GalNAcβ1 4 | Galβ1-3GalNAcβ1-4Galβ1-4Glc- 3 | NeuAcα2 NeuAcα2	IV⁴βGalNAc,IV³NeuAc, II³NeuAc-GgOse₄ GDIa-GalNAc	Human brain	Svennerholm et al. (1973)
NeuAcα2 | 6 Galβ1-3GalNAcβ1-4Galβ1-4Glc- 3 | NeuAcα2	IV³NeuAc,III⁶NeuAc-GgOse₄ GDIα	Frog brain Rat ascites hepatoma cells Mouse lymphoma cell line	Ohashi (1979) Taki et al. (1986) Murayama et al. (1986)

(continued)

Table 2. (Continued)

Structure	Abbreviation	Source	Reference
Galβ1-3GalNAcβ1-4Galβ1-4Glc- 3 | NeuAcα2-8NeuAcα2	$II^3(NeuAc)_2$-GgOse$_4$ GDlb	Human brain	Kuhn and Wiegandt (1963b), Klenk et al. (1967)
	GDlb-L(GDlb inner ester)	Human brain	Riboni et al. (1986)
Galβ1-3GalNAcβ1-4Galβ1-4Glc- 3 | NeuGcα2-8NeuGcα2	$II^3(NeuGC)_2$-GgOse$_4$ GDlb(NeuGc)$_2$	Bovine adrenal medulla	Ariga et al. (1984)
Fucα1 | 2 Galβ1-3GalNAcβ1-4Galβ1-4Glc- 3 | NeuAcα2-8NeuAcα2	$IV^2αFuc,II^3(NeuAc)_2$-GgOse$_4$ GDlb-Fuc	Human brain Pig cerebellum Mini-pig brain	Ando and Yu (1979) Sonnino et al. (1978) Fredman et al. (1981)
Fucα1 | 2 Galβ1-3GalNAcβ1-4Galβ1-4Glc- 3 | Galα1 NeuAcα2-8NeuAcα2	$IV^2αFuc,IV^3αGal,$ $II^3(NeuAc)_2$-GgOse$_4$	PC12 cells	Ariga et al. (1987)
Galβ1-3GalNAcβ1-4Galβ1-4Glc- 3 | NeuAcα2-8NeuAcα2	$IV^3(NeuAc)_2$-GgOse$_4$ GDlc	Mouse thymoma	Bartoszewicz et al. (1986)
Trisialo GalNAcβ1-4Galβ1-4Glc- 3 | NeuAcα2-8NeuAcα2-8NeuAcα2	$II^3(NeuAc)_3$-GgOse$_3$ GT2	Fish brain	Yu and Ando (1980), Ishizuka and Wiegandt (1972)

Galβ1-3GalNAcβ1-4Galβ1-4Glc-
    3
    |
    NeuAcα2
    |
NeuAcα2-8NeuAcα2

$IV^3(NeuAc)_2,II^3NeuAc\text{-}GgOse_4$ GT1a — Human brain — Ando and Yu (1977)

Galβ1-3GalNAcβ1-4Galβ1-4Glc-
    3
    |
NeuAcα2 NeuAcα2-8NeuAcα2

$IV^3NeuAc,II^3(NeuAc)_2\text{-}GgOse_4$ GT1b — Human brain — Kuhn and Wiegandt (1963b), Klenk et al. (1967)

Galβ1-3GalNAcβ1-4Galβ1-4Glc-
    3
    |
    NeuAcα
(9-OAc)NeuAcα2-8NeuAcα2

$IV^3NeuAc,II^3(9\text{-}OAcNeuAcα2\text{-}8NeuAc)\text{-}GgOse_4$ GT1L — Mouse brain — Ghidoni et al. (1980)

Galβ1-3GalNAcβ1-4Galβ1-4Glc-
    3
    |
NeuAcα2-8NeuAcα2-8NeuAcα2

$II^3(NeuAc)_3\text{-}GgOse_4$ GT1c — Fish brain — Yu and Ando (1980)

Galβ1-3GalNAcβ1-4Galβ1-4Glc-
    3
    |
NeuAcα2
    6
    |
NeuAcα2

$IV^3NeuAc,III^6(NeuAc)_2\text{-}GgOse_4$ — Ohashi (1980)

Galβ1-3GalNAcβ1-4Galβ1-4Glc-
    3
    |
NeuAcα2-8NeuAcα2

$IV^3(NeuAc)_2,III^6NeuAc\text{-}GgOse_4$ — Ohashi (1980)

(continued)

Table 2. (Continued)

Structure	Abbreviation	Source	Reference
Tetrasialo			
Galβ1-3GalNAcβ1-4Galβ1-4Glc- 　　　　3 　　　　| 　　NeuAcα2 NeuAcα2-8NeuAcα2	$IV^3NeuAc,II^3(NeuAc)_3$- GgOse$_4$ GQ1c	Fish brain	Ando and Yu (1979), Ishizuka and Wiegandt (1972)
Galβ1-3GalNAcβ1-4Galβ1-4Glc- 　　　　3 　　　　| NeuAcα2-8NeuAcα2 　　　NeuAcα2-8NeuAcα2	$IV^3(NeuAc)_2,II^3(NeuAc)_2$- GgOse$_4$ GQ1b	Human, bovine, and chicken brain	Ando and Yu (1979), Fredman et al. (1981)
Galβ1-3GalNAcβ1-4Galβ1-4Glc- 　　　　3 　　　　| NeuAcα2-8NeuAcα2 (9-OAc)NeuAcα2-8NeuAcα2	$IV^3(NeuAc)_2,II^3$ (9-OAcNeuAc,NeuAc) GgOse$_4$ 9-OAcGQ1b	Mouse brain	Chigorno et al. (1982)
Galβ1-3GalNAcβ1-4Galβ1-4Glc- 　　3　　　6 　　|　　　| NeuAcα2-8NeuAcα2 　　　NeuAcα2	$IV^3(NeuAc)_2,III^6(NeuAc)_2$- GgOse$_4$		Wiegandt (1985)
Pentasialo			
Galβ1-3GalNAcβ1-4Galβ1-4Glc- 　　　　3 　　　　| NeuAcα2-8NeuAcα2 　　NeuAcα2-8NeuAcα2-8NeuAcα2	$IV^3(NeuAc)_2,II^3(NeuAc)_3$- GgOse$_4$ GP1c	Fish brain	Ishizuka and Wiegandt (1972)

Lacto series

$IV^3NeuAc\text{-}LcOse_4$

Galβ1-3GlcNAcβ1-3Galβ1-4Glc-
3
|
NeuAcα2

Human meconium
Human carcinomas

Prieto and Smith (1986)
Nilsson et al. (1985)

$IV^3NeuAc,III^6NeuAc\text{-}LcOse_4$

Galβ1-3GlcNAcβ1-3Galβ1-4Glc-
3   6
| |
NeuAcα2   NeuAcα2
            Fucα1
            |
            4

Human colonic
adenocarcinoma
Human glioma
cell line

Fukushi et al. (1986)
Mansson et al. (1986)

$IV^3NeuAc,III^4Fuc\text{-}LcOse_4$

Galβ1-3GlcNAcβ1-3Galβ1-4Glc-
3
|
NeuAcα2
Fucα1
|
4

Human carcinomas

Magnani et al. (1982)

$IV^3NeuAc,III^6NeuAc,III^4\alpha Fuc\text{-}LcOse_4$

Galβ1-3Glc NAcβ1-3Galβ1-4Glc-
3   6
| |
NeuAcα2   NeuAcα2

Human colonic
adenocarcinoma

Nudelman et al. (1986)

(continued)

Table 2. (Continued)

Structure	Abbreviation	Source	Reference
Neolacto series			

Galβ1-4GlcNAcβ1-3Galβ1-4Glc-
  3
  $\mid$
NeuAcα2
	IV^3NeuAc-nLcOse$_4$
	LM1
		Human erythrocytes
		
		Human peripheral nerve
		Pig adipose
		
		Rat sciatic nerve
			Ando et al. (1973), Wherrett (1973)
			Li et al. (1973)
			Ohashi and Yamakawa (1977)
			Chou et al. (1985)

Galβ1-4GlcNAcβ1-3Galβ1-4Glc-
  3
  $\mid$
NeuGcα2
	IV^3NeuGc-nLcOse$_4$	Bovine spleen, kidney	Wiegandt (1973)

Galβ1-4GlcNAcβ1-3Galβ1-4Glc-
  3   3
  $\mid$ $\mid$
NeuAcα2   Fucα1
	IV^3NeuAc,III^{3}αFuc-nLcOse$_4$	Human kidney
		Human pancreatic adenocarcinoma
			Rauvala (1976)
			Mansson et al. (1985)

Galβ1-4GlcNAcβ1-3Galβ1-4Glc-
  3   4
  $\mid$ $\mid$
NeuAcα2   GalNAcβ1
	IV^3NeuAc,II^{4}βGalNAc-
	nLcOse$_4$
		Mullet roe	DeGasperi et al. (1987)

GalNAcβ1
  $\mid$
  4
Galβ1-4GlcNAcβ1-3Galβ1-4Glc-
  3
  $\mid$
NeuAcα2
	IV^4βGalNAc,IV^{3}NeuAc-
	nLcOse$_4$
		Mullet roe
		Human erythrocytes
			DeGasperi et al. (1987)
			Gillard et al. (1988)

Structure	Abbreviation	Source	Reference

GalNAcβ1
|
4
Galβ1-4GlcNAcβ1-3Galβ1-4Glc-
3
|
NeuAcα2   GalNAcβ1
           |
           4
	$IV^4\beta GalNAc,IV^3NeuAc$
	$II^4\beta GalNAc$-$nLcOse_4$
		Mullet roe	DeGasperi et al. (1987)

Galβ1-4GlcNAcβ1-3Galβ1-4Glc- 6 | NeuAcα2	IV^6NeuAc-$nLcOse_4$	Bovine spleen, kidney Human erythrocytes Human pancreatic adenocarcinoma	Wiegandt (1973) Watanabe et al. (1979a) Mansson et al. (1985)
Galβ1-4GlcNAcβ1 | 6 Galβ1-4GlcNAcβ1-3Galβ1-4Glc- 6 | NeuAcα2	$IV^6NeuAc,II^6(Gal\beta1$-$4GlcNAc\beta1)$-$nLcOse_4$	Bovine buttermilk	Takamizawa et al. (1986)
Galβ1-4GlcNAcβ1-3Galβ1-4Glc- 3 | NeuAcα2-3GalNAcβ1	$IV^3(NeuAc\alpha2$-$3GalNAc\beta1)$-$nLcOse_4$	Human erythrocytes	Watanabe and Hakomori (1979)
Galβ1-4GlcNAcβ1-3Galβ1-4Glc- 3 | NeuAcα2-8NeuAcα2	$IV^3(NeuAc)_2$-$nLcOse_4$	Human kidney	Rauvala et al. (1978)

(continued)

Table 2. (Continued)

Structure	Abbreviation	Source	Reference
Galβ1-4GlcNAcβ1-3Galβ1-Glc- \ 3 \	\ NeuAcα2-8NeuAcα2-8NeuAcα2	$IV^3(NeuAc)_3$-nLcOse$_4$	Hog kidney cortex	Murakami-Murofushi et al. (1983)
Galβ1-4GlcNAcβ1-3Galβ1-4Glc- \ 3 \	\ NeuAcα2	VI^3NeuAc-nLcOse$_6$	Human spleen \ Bovine erythrocytes \ Human placenta	Wiegandt (1974) \ Chien et al. (1978) \ Taki et al. (1988)
Galβ1-4GlcNAcβ1-3Galβ1-4Glc- \ 3 \	\ NeuGcα2	VI^3NeuGc-nLcOse$_6$	Bovine erythrocytes	Chien et al. (1978)
Galβ1-4GlcNAcβ1-3Galβ1-4Glc- \ 3 \	\ Fucα1 \ 3 \	\ NeuAcα2	$VI^3NeuAc,III^3\alpha Fuc$-nLcOse$_6$	Human leukemia cells	Fukuda et al. (1986)
Galβ1-4GlcNAcβ1-3Galβ1-4Glc- \ 3 3 \		\ NeuAcα2 Fucα1	$VI^3NeuAc,V^3\alpha Fuc$-$III^3\alpha Fuc$-nLcOse$_6$	Human colonic adenocarcinoma	Fukushi et al. (1984)
Galβ1-4GlcNAcβ1 \	\ 6 \ Galβ1-4GlcNAcβ1-3Galβ1-4Glc- \ 3 \	\ NeuAcα2	$VI^3NeuAc,IV^6(Gal\beta1$-$4GlcNAc\beta1)$-nLcOse$_6$	Human erythrocytes	Watanabe et al. (1979b)

```
Galα1-3Galβ1-4GlcNAcβ1
                      |
                      6
Galβ1-4GlcNAcβ1-3Galβ1-4GlcNAcβ1-3Galβ1-4Glc-
3
|
Siaα2
```

$VI^3Sia,IV^6(Gal\alpha1-3\ Gal\beta1-4GlcNAc\beta1)-nLcOse_6$

Bovine erythrocytes

Watanabe et al. (1979a)

```
Fucα1-2Galβ1-4GlcNAcβ1
                      |
                      6
Galβ1-4GlcNAcβ1-3Galβ1-4GlcNAcβ1-3Galβ1-4Glc-
3
|
NeuAcα2
```

$VI^3NeuAc,IV^6(Fuc\alpha1-2Gal\beta1-4GlcNAc\beta1)-nLcOse_6$

Human erythrocytes

Watanabe et al. (1978), Kannagi et al. (1983)

```
GalNAcα1
       |
       3
Fucα1-2Galβ1-4GlcNAcβ1
                      |
                      6
Galβ1-4GlcNAcβ1-3Galβ1-4GlcNAcβ1-3Galβ1-4Glc-
3
|
NeuAcα2
```

$VI^3NeuAc,IV^6[Fuc\alpha1-2(GalNAc\alpha1-3)Gal\beta1-4GlcNAc\beta1]-nLcOse_6$

Human erythrocytes

Kannagi et al. (1983)

```
Galα1
     |
     3
Fucα1-2Galβ1-4GlcNAcβ1
                      |
                      6
Galβ1-4GlcNAcβ1-3Galβ1-4GlcNAcβ1-3Galβ1-4Glc-
3
|
NeuAcα2
```

$VI^3NeuAc,IV^6[Fuc\alpha1-2(Gal\alpha1-3)Gal\beta1-4GlcNAc\beta1]-nLcOse_6$

Human erythrocytes

Kannagi et al. (1983)

(continued)

Table 2. (Continued)

Structure	Abbreviation	Source	Reference
Galβ1-4GlcNAcβ1-3Galβ1-4GlcNAcβ1-3Galβ1-4Glc- 6 | NeuAcα2	IV^6NeuAc-nLcOse$_6$	Human erythrocytes	Watanabe et al. (1979b)
NeuAcα2-3Galβ1-4GlcNAcβ1 | 6 Galβ1-4GlcNAcβ1-3Galβ1-4GlcNAcβ1-3Galβ1-4Glc- 3 | NeuAcα2	VI^3NeuAc,IV^6(NeuAcα2-3Galβ1-4GlcNAcβ1)-nLcOse$_6$	Human erythrocytes Human placenta	Kundu et al. (1983) Taki et al. (1988)
Galβ1-4GlcNAcβ1-3Galβ1-4GlcNAcβ1-3Galβ1-4Glc- 3 | NeuAcα2	VIII^3NeuAc-nLcOse$_8$	Human erythrocytes	Kundu et al. (1981), Iwamori and Nagai (1981)
Galβ1-4GlcNAcβ1-3Galβ1-4GlcNAcβ1-3Galβ1-4Glc- 3 | NeuGcα2	VIII^3NeuGc-nLcOse$_8$	Bovine erythrocytes	Dasgupta et al. (1985)
	Globo series		
GalNAcβ1-3Galα1-4Galβ1-4Glc- 3 | NeuAcα2	IV^3NeuAc-GbOse$_4$	Human teratocarcinoma cell line	Schwarting et al. (1983)

Structure	Shorthand designation	Localization	Reference
Galβ1-3GalNAcβ1-3Galα1-4Galβ1-4Glc- 3 | NeuAcα2	V^{3}NeuAc-GbOse$_5$	Chicken muscle	Chien and Hogan (1983), Kannagi et al. (1983)
Galβ1-3GalNAcβ1-3Galα1-4Galβ1-4Glc- 3 | NeuAcα2-8NeuAcα2	V^3(NeuAc)$_2$-GbOse$_5$	Chicken muscle	Hogan et al. (1982)
Galβ1-3GalNAcβ1-3Galα1-4Galβ1-4Glc- 6 3 | | NeuAcα2 NeuAcα2	V^{3}NeuAc, V^{6}NeuAc-GbOse$_5$	Human erythrocytes	Kundu et al. (1983)
Isoglobo series			
Galβ1-3GalNAcβ1-3Galα1-3Galβ1-4Glc- 3 | NeuAcα2	V^{3}NeuAc-iGbOse$_5$		Breimer et al. (1982)
Unclassified			
Galβ1-3GalNAcβ1-4Glc- 3 | NeuAcα2			Watanabe et al. (1979a)

neuraminic acid (NeuAc) or *N*-glycolylneuraminic acid (NeuGc). Occasionally, both types of sialic acid can be present in the same structure. The brain gangliosides of vertebrates usually contain NeuAc although there are exceptions such as those of bovine neurohypophysis, which contain a high proportion of NeuGc (Clarke, 1975). Sialic acids can also be present in forms *O*-acetylated at the 4- or 9-hydroxyl group (Table 2). The natural occurrence of a lactone form of gangliosides, GD1b inner ester, has recently been demonstrated. These structures are unstable under alkaline conditions and are easily destroyed during isolation. Gangliosides which are sulfated or methylated at the 8-hydroxyl group of the sialic acid have also been recognized.

2.1. Ganglio Series

The majority of brain gangliosides belong to this family. Structurally, they are characterized by the presence of *N*-acetylgalactosamine (GalNAc). Since both GalNAc and sialic acid are linked to the same galactose, all gangliosides of this family (except GMlb) have branched structures.

Most adult mammalian brains contain four major gangliosides: GM1, GD1a, GD1b, and GT1b. Lower vertebrates such as fish have relatively little monosialogangliosides, while tetra- and pentasialogangliosides predominate (Ishizuka *et al.*, 1970; Ando and Yu, 1979; Hilbig and Rahmann, 1987; Rosner and Rahmann, 1987). However, such variations do not necessarily conform to a simple phylogenetic progression, as shown by the finding that ray and lamprey brains contain higher proportions of monosialogangliosides compared to mammals (Avrova, 1971).

2.2. Hemato and Gala Series

Hematosides have a core structure of Gal(β1-4)Glc(1-1')Cer with one to three sialic acid residues. GM3 and GD3 are usually minor components in adult brain, but are abundant in embryonic brain (Yu *et al.*, 1988). Their concentrations also tend to increase in pathological conditions characterized by gliosis (Seyfried and Yu, 1985). Since GM3 is rather abundant in extraneural tissues, it has sometimes been regarded as a characteristic ganglioside of these tissues. However, this is not invariably the case. GD3 is known to be abundant in retina of certain species (Dreyfus *et al.*, 1976). GT3, the trisialo form in this series, was first demonstrated in fish brain (Ishizuka and Wiegandt, 1972; Yu and Ando, 1980). Sialosylgalactosylceramide (GM4), the only ganglioside in the gala series, is the third most abundant ganglioside in human white matter and is known to be concentrated in myelin of primate and avian species.

2.3. Lacto Series

This series of gangliosides is characterized by the presence of *N*-acetylglucosamine (GlcNAc) linked to lactosylceramide. Based on the linkage of the GlcNAc residue with a fourth galactose, Galβ1—3GlcNAc and Galβ1-4GlcNAc, this family is further divided into lacto and neolacto series, respectively. A considerable number of gangliosides, which are usually present in extraneural tissues, are included in the neolacto series. Glycolipids classified as belonging to lacto series had not been identified until demonstrated by Magnani *et al.* (1982).

Recently, gangliosides with hybrid carbohydrate structures of the ganglio and neolacto series have been identified. These glycolipids have the same core oligosaccharide structures as those of the neolacto series but with branching carbohydrate structures comprised of GalNAc residues (Table 2). No consensus has been reached as to whether these gangliosides should be classified as an independent family.

2.4. Globo and Isoglobo Series

Several species of gangliosides belonging to the globo series have recently been identified in human and chicken tissues (Table 2), whereas only one ganglioside is known in the isoglobo series.

3. LOCALIZATION

3.1. Whole Nerve Tissues

Table 3 lists the ganglioside concentrations and distributions of a variety of nerve tissues. In general, cerebral gray matter contains considerably higher concentrations of gangliosides than does white matter or peripheral nervous tissues. This is consistent with the earlier concept that the bulk of gangliosides in brain reside in neurons and neuronal subfractions (see below). The brains of higher vertebrates contain at least four major ganglio series gangliosides—GM1, GD1a, GD1b, and GT1b—which account for 80–90% of the total gangliosides. Primate and avian species contain an additional ganglioside, GM4, which is particularly abundant in white matter, having been shown to be specifically localized in CNS myelin and oligodendroglia (Ledeen *et al.*, 1973; Yu and Iqbal, 1979; Ueno *et al.*, 1980; Yu *et al.*, 1988).

Brain ganglioside concentrations and distributions change dramatically during development (Suzuki, 1965; Vanier *et al.*, 1971; Merat and Dickerson, 1973; Irwin *et al.*, 1980; Rosner, 1982; Seyfried and Yu, 1985; Rosner and Rahmann, 1987; Yu *et al.*, 1988). Two well-described features during development are the general increase in total ganglioside concentration and the predominance of GM3 and GD3 during early embryonic ages (Table 2). At later ages, the ganglio series gangliosides increase in concentration. These changes are apparently caused by a shift from the synthesis of simple, hemato series gangliosides to the synthesis of the more complex gangliosides during development.

3.2. Nerve Cells

To understand the biological function of gangliosides in the nervous system, it is essential to know their cellular localization. Gangliosides of neurons, astroglia, and oligodendroglia have been studied by many investigators (see Table 4). The range of ganglioside concentrations in these cell types varies considerably from one report to

Table 3. Ganglioside Composition of Nervous Tissues

	Tissue	Concentration (μg sialic acid per g wet wt.)	GM4	GM3	GM2	GM1	GD1a	GD3	GT1a	GD2	GD1b	GT1b	GQ1b	GT3	GT2	GT1c	GQ1c	GP+GH	Reference
Human	Cerebral GM	875	1.5	2.7	4.1	14.9	21.7	1.8	5.4	8.0	18.2	16.3	5.0						Ando et al. (1978)
	Cerebral WM	275	8.6	4.8	2.5	21.6	16.6	2.2	8.8	3.1	16.9	11.1	2.7						Ando et al. (1978)
	Spinal cord	87	12.8	14.0	3.7	16.8	4.2	0.5	16.4	0.9	18.7	9.1	2.9						Ueno et al. (1978)
	Spinal cord GM	232																	Ledeen and Yu (1982)
	Spinal cord WM	133																	Ledeen and Yu (1982)
	Sciatic nerve	≈80																	MacMillan and Wherrett (1969)
	Femoral nerve	34																	Svennerholm et al. (1972)
	Retina	113						≈50											Holm et al. (1972)
Chimpanzee	Cerebral GM	782																	Yu et al. (1974)
	Cerebral WM	374	6.3	1.6	1.3	26.2	20.6			3.5	15.4	22.1	3.3						Yu et al. (1974)
Bovine	Cerebral GM	992	0.4	1.0	1.8	13.9	27.8	3.0	2.9		15.5	22.0	6.6						Yu and Ledeen (1970)
	Cerebral WM	205	3.4	1.4	2.2	21.2	25.2	4.8	1.7		14.8	17.1	5.4						Yu and Ledeen (1970)

Animal	Tissue												Reference
	Neurohypophysis	453								25.9			Clarke (1975)
	Intradural root	93			8.6	15.0	50.6						Fong et al. (1976)
	Retina	171				18.7	32.3			22.6	26.4		Handa and Burton (1969)
	Optic nerve	155											Holm and Mansson (1974)
Sheep	Cerebral GM	1073	0.6	1.0	2.7	32.8	40.3		2.0	8.5	9.6	2.5	Yu and Ledeen (1970)
	Cerebral WM	342											Yu and Ledeen (1970)
Pig	Cerebral GM	987	0.6	0.9	0.2	24.7	44.1		1.6	13.2	11.6	3.2	Yu and Ledeen (1970)
	Cerebral WM	406											Yu and Ledeen (1970)
Rabbit	Whole brain	1156			12.0	44.0				11.4	28.6	4.0	Tettamanti et al. (1973)
	Cerebral GM	916	1.1	1.5	26.5	35.6	3.9	2.1	1.3	9.8	14.0	2.9	Ueno et al. (1978)
	Cerebral WM	436		1.6	34.6	16.4	5.4	1.8	1.1	18.6	13.3	5.5	Ueno et al. (1978)
	Spinal cord	195	2.9	1.3	29.0	14.7	10.3	2.0	1.2	14.3	15.6	6.1	Ueno et al. (1978)
	Sciatic nerve	83	4.1	7.1	42.3					16.7	23.8	6.1	Yates and Wherrett (1974)
	Retina	178											Holm et al. (1972)

(continued)

Table 3. (Continued)

Tissue	Concentration (μg sialic acid per g wet wt.)	GM4	GM3	GM2	GM1	GD1a	GD3	GT1a	GD2	GD1b	GT1b	GQ1b	GT3	GT2	GT1c	GQ1c	GP+GH	Reference
Rat Whole brain	783			1.2	14.0	30.2			10.6	16.0	22.8	5.3						Avrova (1971)
	1150			10.2	10.2	34.5				18.5	28.5	7.4						Tettamanti et al. (1973)
Embryonic day 14	43	19.7	2.2			1.8	44.3	2.3	4.5	8.7	10.3	3.7						Avrova (1971)
Embryonic day 16	50	16.8	2.2		0.5	4.9	37.6		5.6	8.3	15.8	6.9						Yu et al. (1988)
Embryonic day 18	116	4.5			1.9	12.6	21.1		8.3	13.3	26.8	11.5						Yu et al. (1988)
Embryonic day 20	160	3.8			3.2	19.0	12.6		6.7	13.3	28.8	12.8						Yu et al. (1988)
Embryonic day 22	169	2.9	1.2		4.1	25.2	7.9		6.1	12.0	29.5	11.1						Yu et al. (1988)
Newborn	286	1.5	1.0		5.4	29.0	3.3		3.4	11.0	34.4	11.0						Yu et al. (1988)
Adult	768	1.7	0.4		12.9	27.8	0.6		3.8	15.8	25.4	11.5						Yu et al. (1988)
Cerebral GM	877	2.3	0.2		13.2	34.8	1.9	.18	0.2	13.0	23.7	5.9						Ando et al. (1978)
Cerebellum	406	2.3	2.3		12.4	40.6			5.8	8.8	18.3	16.4						Abe and Norton (1974)
Sciatic nerve	11																	Klein and Mandel (1975)
Superior cervical ganglion	93																	Harris and Klingman (1972)
Retina	149	6.1			2.7	18.1	36.5			17.7	15.3	9.9						Edel-Harth et al. (1973)

Animal	Region	Total	1	2	3	4	5	6	7	8	9	10	11	12	13	Reference
Mouse	Cerebrum	964							9.6	36.0	4.0	8.2	13.5	16.6	8.0	Seyfried et al. (1979)
	Cerebellum	784						5.0	20.0	8.0	8.0	2.8	15.8	21.3	11.9	Seyfried et al. (1979)
	Brain stem	650						8.2	13.9	5.0	5.7	5.1	18.7	19.1	14.4	Seyfried et al. (1979)
Chicken	Whole brain	767							1.6	23.6	36.8	6.4	7.1	18.7	5.8	Avrova (1971)
	Whole brain	390						4.8	0.3	15.8	27.0	16.6	10.9	15.9	5.4	Dreyfus et al. (1975)
	Cerebral GM	—						2.2	5.9	10.0	31.1	18.3	15.5	8.7	4.7	Cochran et al. (1981)
	Cerebral WM	—						16.0	4.6	21.5	16.5	13.5	10.2	9.3	6.6	Cochran et al. (1981)
	Spinal cord	245														Cochran et al. (1981)
	Retina	131						12.5	0.7	8.5	32.7	15.4	9.1	14.0	5.3	Dreyfus et al. (1975)
Pigeon	Whole brain	693							6.0	15.5	31.7	6.9	8.9	24.1	7.0	Avrova (1971)
	Spinal cord	253														Cochran et al. (1981)
	Spinal cord	196														Cochran et al. (1981)
Snake	Whole brain	320														Eldredge et al. (1963)
Turtle	Whole brain	220														Eldredge et al. (1963)
Frog	Whole brain	315								0.7	3.9	15.4	17.5	16.2	36.3	Avrova (1971)
Cod	Whole brain	193	0.4	0.4	0.1	10.1	3.5	1.5	0.9	2.5	5.3	4.6	7.4	44.0	14.4	Yu and Ando (1980)

Table 4. Ganglioside Composition of Isolated Brain Cells

	Cell	Concentration (μg sialic acid per mg protein)	GM4	GM3	GM2	GM1	GD3	GD1a	GT1a	GD2	GD1b	GT1b	GQ1b	GT3	GT2	GT1c	GQ1c	GP+GH	Reference
Neuron	Human brain	1.3	0.9	4.1		22.6	4.5	21.5	7.2	3.9	20.7	11.3	2.2						Yu and Iqbal (1979)
	Ox Dieter's nucleus	3050[a]																	Derry and Wolfe (1967)
	Pig brain stem	1100–1600[a]																	Tamai et al. (1971)
	Rabbit cortex	2.9				24.9		42.3		0.2	13.6	17.4	1.7						Hamberger and Svennerholm (1971)
	Rat cortex (15–20 day)	1900[a]		8.2	10.0	23.3		28.1	4.9		9.3	8.4	4.2						Abe and Norton (1974)
	Rat brain	0.77																	Skrivanek et al. (1978)
		2.0																	Maccioni et al. (1978)
	(20 day)	0.93	10.0		1.8	15.2		36.6	5.5	4.6	11.7	14.3							Byrne et al. (1989)
	Hamster	—																	Sbaschnig-Agler et al. (1988)
	Chicken embryo (cultured)	5.9				12.2		41.2	5.7		9.3	26.3	5.3						
Astroglia	Rabbit cortex	5.6				23.3		48.2		0.6	15.0	11.8	1.2						Hamberger and Svennerholm (1971)

Sample												Reference
Rat cortex (15–20 day)	4300[a]		8.6	8.4	24.7	29.6		4.8	9.1	7.2	3.8	Abe and Norton (1974)
Rat brain	2.1											Skrivanek et al. (1978)
(20 day)	2.8	11.0	1.0	13.7	34.8		1.6	4.2	9.8	18.1	7.0	Byrne et al. (1989)
Rat cortex (10–30 day)	3.2											Norton and Poduslo (1971)
Rat astrocytoma	—	9.1			9							Dawson and Sweeley (1971)
Hamster cortex	—				15.4	38.6	3.8		17.9	17.9	6.4	Robert et al. (1975)
Chicken embryo (cultured)	3.2	49.1					33.4					Sbaschnig-Agler et al. (1988)
Oligodendroglia: Human brain	0.34	5.9	8.1	5.7	20.1	16.4	11.9	3.3	15.0	9.3	1.9	Yu and Iqbal (1979)
Bovine brain	0.15	1.5	11.2	4.6	16.2	10.3	36.3	3.5	5.3	7.5	1.5	Yu et al. (1989)
Calf WM	1.3											Poduslo and Norton (1972)
Rat brain (30 day)	0.92	9.3	1.1	6.5	29.0	12.5	3.1	1.5	10.8	19.8	6.6	Yu et al. (1989)
(60 day)	0.54	5.3	1.5	3.3	24.4	13.2	5.1	2.8	12.2	21.4	8.4	Yu et al. (1989)

[a] Expressed as μg sialic acid per g dry wt.

Table 5. Ganglioside Composition of Neuronal Subfractions

	Subfraction	Concentration (µg sialic acid per mg protein)	GM4	GM3	GM2	GM1	GD1a	GD3	GD2	GD1b	GT1b	GQ1b	GT3	GT2	GT1c	GQ1c	GP+GH	Reference
Axon	Ox	0.38		13.1	16.1	32.0	19.7	0.9			12.3						3.6	DeVries and Norton (1974)
Axolemma	Human	13.9																DeVries et al. (1981)
	Bovine	15.1																DeVries et al. (1981)
	Rat	24.1																DeVries et al. (1981)
Synaptic vesicle	Rat	5.1																Lapetina et al. (1968)
		2.9																Breckenridge et al. (1973)
Torpedo californica		78.6																Ledeen et al. (1988)
Torpedo marmorata		57.6																Ledeen et al. (1988)
Synaptosomes	Human	7.1																Kornguth et al. (1974)
	Ox	16.4																Wiegandt (1967)
	Calf	8.0																Tettamanti et al. (1980)
	Rabbit	13.9				18.3	45.9		2.2	12.7	17.6						3.3	Hamberger and Svennerholm (1971)

Preparation	Species											Reference
		12.0			10.0	45.7			10.6	29.5		4.2	Tettamanti et al. (1973)
	Rat	9.6											Dekirmenjian and Brunngraber (1969)
		9.3		1.5	20.0	32.7			7.8	18.5	16.6	3.2	Avrova et al. (1973)
		7.4											Maccioni et al. (1978)
		10.8	0.5	0.1	11.4	38.8	4.6	6.5	12.6	19.7		6.6	Yohe et al. (1980)
	(20 day)	10.0											Byrne et al. (1989)
	Guinea pig	8.3											Eichberg et al. (1964)
	Calf	27.8											Tettamanti et al. (1980)
Synaptic plasma membranes	Rat, cholinergic	16.7											Lapetina et al. (1968)
	Rat, non-cholinergic	7.3											Lapetina et al. (1968)
	Rat, cholinergic	29.7		0.7	10.6	35.5			8.0	19.1	23.0	3.1	Avrova et al. (1973)
	Rat, non-cholinergic	18.5		2.0	13.6	32.8			10.7	18.7	20.0	2.2	Avrova et al. (1973)
	Rat	19.3											Brunngraber et al. (1967)
		44.6			7.8	34.4			21.4	29.3		7.2	Breckenridge et al. (1972)

(continued)

Table 5. (Continued)

Subfraction	Concentration (μg sialic acid per mg protein)	Ganglioside composition (%)																Reference
		GM4	GM3	GM2	GM1	GD1a	GD3	GT1a	GD2	GD1b	GT1b	GQ1b	GT3	GT2	GT1c	GQ1c	GP+GH	
	8.4																	Ledeen et al. (1976)
	15.8																	Cruz and Gurd (1981)
	14.2		3.3	0.8	19.9	29.6	1.0	3.6		16.8	17.8	6.7						Skrivanek et al. (1982)
Guinea pig	17.0																	Whittaker (1969)
Growth cone membranes — Rat, fetal (17–19 days)	24.0		5.5		6.2	18.0	13.4	8.7		10.9	24.4	11.9						Sbaschnig-Agler et al. (1988)
Soluble — Bovine cortex	1.3				28.5	40.2	4.0			10.8	13.4	3.1						Sonnino et al. (1979)
Bovine cortex/nerve ending	1.5, 0.34				27.0	41.2	4.1			11.9	12.8	3.0						Sonnino et al. (1979)
Rat synaptosomes	0.56																	Lapetina et al. (1968)
	0.34																	Ledeen et al. (1976)
	1.5																	Sonnino et al. (1979)
Rat neuron and glia	1.3																	Sonnino et al. (1979)

another, probably owing to the differences in isolation and analytical procedures. In general, bulk-isolated astroglia and neurons tend to have similar levels of gangliosides. Since astroglia have a much higher surface area to volume ratio, hence more plasma membrane per cell, it is presumed that neurons still contain the highest concentrations of gangliosides (Norton and Poduslo, 1971). Oligodendroglia, on the other hand, contain significantly less gangliosides (Yu and Iqbal, 1979; Yu et al., 1988).

Although the ganglioside patterns of bulk-isolated neurons and astroglia are similar (Byrne et al., 1989), specific subpopulations of nerve cells may exhibit preferential enrichment of certain ganglioside species. For example, using a series of neurological mouse mutants which lose specific populations of cerebellar neurons at various stages of development, Seyfried and Yu (1984) found GD1a to be preferentially enriched in granule cells and GT1a in Purkinje cells. Their studies also confirmed earlier observations that GD3 may be more closely associated with fibrous astroglia than with resting astroglia (Yu and Manuelidis, 1978; Yu et al., 1982).

Whereas the bulk-isolated astroglia from mammalian brains contain predominantly gangliosides of the ganglio series (Table 4), primary cultures of astrocytes appear to contain an abundance of GM3 and GD3 (Robert et al., 1975; Mandel et al., 1980) with practically no GM1 (Asou and Brunngraber, 1983, 1984). In a more recent study, GM3 and GD3 comprised 75–85% of the total gangliosisde in cultured chick astroglial cells, the remainder being mainly structures other than those of the gangliotetraose series (Sbaschnig-Agler et al., 1988). These cells also do not appear to possess the ability to synthesize ganglio series gangliosides, as indicated by a very low level of UDP-N-acetylgalactosaminyl:GM3 N-acetylgalactosaminyltransferase activity. The source of the ganglio series gangliosides in bulk-isolated astroglia thus remains unclear.

During development, neurons may also transiently express specific gangliosides. For example, GD3 and 9-O-acetyl GD3 have been found to be associated with immature neuroectodermal cells (Goldman et al., 1984; Levine et al., 1984, 1986; Schlosshauer et al., 1988). The expression of ganglioside GQ1c is apparently developmentally regulated in neuronal (Eisenbarth et al., 1979; Kasai and Yu, 1983; Fredman et al., 1984; Kim et al., 1986) as well as in glial precursor cells (Raff et al., 1983). Knowledge of the dynamic changes in ganglioside composition which occur in cellular proliferation and maturation will be critical for understanding their functions.

In contrast to myelin, which has a relatively simple ganglioside pattern (see below), oligodendroglia have a very complex ganglioside composition, including the presence of high levels of ganglio and hemato series gangliosides (Table 4). The most conspicuous feature, however, is the presence of GM4, a unique marker for CNS myelin and oligodendroglia (Ledeen et al., 1973; Yu and Iqbal, 1979; Yu et al., 1988).

3.3. Subcellular Fractions

The question as to how gangliosides are distributed over the surface of neuronal cells remains an open one. It has been postulated that gangliosides are probably not localized only at the nerve ending, but rather are evenly distributed over the entire cell surface (Ledeen, 1978). For example, ganglioside levels in a number of mammalian

Table 6. Ganglioside Composition of Myelin

Source	Concentration (μg sialic acid per 100 mg dry myelin)	Ganglioside composition (%)																Reference
		GM4	GM3	GM2	GM1	GD1a	GD3	GT1a	GD2	GD1b	GT1b	GQ1b	GT3	GT2	GT1c	GQ1c	GP+GH	
Human Cerebrum	63	26.6	1.2	3.7	34.7	6.9	0.5	2.7	0.3	16.1	6.3	1.0						Ueno et al. (1978)
	62	20.3	3.5	3.2	31.7	7.8	1.8	1.2		19.5	8.9	2.1						Cochran et al. (1982)
Sciatic nerve	29																	Fong et al. (1976)
Chimpanzee Brain	102	7.3		4.3	39.6	10.6				21.3	13.6	2.5						Cochran et al. (1982)
Bovine Brain	60	1.0		5.0	49.0	7.9		3.0		20.5	9.1	4.4						Cochran et al. (1982)
Sciatic nerve	22																	Fong et al. (1976)
Intradural root	27				15.4	38.6	3.8			17.9	17.9	6.4						Fong et al. (1976)
Rabbit Brain	82			2.2	59.0	6.5				16.8	10.8	4.7						Cochran et al. (1982)
Cerebrum	97			3.2	55.6	8.1	2.7	0.5	0.2	14.9	10.8	3.2						Ueno et al. (1978)
Spinal cord	47			3.5	44.6	7.8	5.3	2.4	1.0	15.3	12.5	5.3						Ueno et al. (1978)
Rat Brain	106	4.6	1.4	2.0	63.7	5.8	3.0	2.6		6.6	6.2	4.1						Cochran et al. (1982)
15 day	49				55.8	18.7				13.7	11.8							Suzuki (1967)
30 day	40				57.8	9.8				11.0	8.4							Suzuki (1967)
60 day	40				82.6	6.7				6.3	4.4							Suzuki (1967)
190 day	52				92.0	4.5				2.6	0.9							Suzuki (1967)
425 day	68				89.0	5.3				3.8	1.9							Suzuki (1967)
Guinea pig Brain	91	4.8		2.7	51.8	7.6	1.1	1.4		14.2	11.5	4.6						Cochran et al. (1982)

Species	Tissue/Age																Reference
Mouse	Brain	114	4.7	0.3	3.6	59.6	8.0	4.2	4.8	8.0	4.2	2.6					Cochran et al. (1982)
	23 day	56															Yu and Yen (1975)
	48 day	74															Yu and Yen (1975)
	183 day	84															Yu and Yen (1975)
	270 day	114															Yu and Yen (1975)
	490 day	118															Yu and Yen (1975)
Chicken	Brain, embryonic (17 day)	266															Cochran et al. (1983)
	1 day	243	34.1	7.4	1.7	30.0	14.6	4.0		2.4	4.4	1.4					Cochran et al. (1983)
	11 day	251	40.0	5.9	1.4	30.3	10.2	3.7		1.5	4.3	1.4					Cochran et al. (1983)
	49 day	245	38.0	3.2	1.4	33.9	9.8	4.7		2.7	3.8	1.4					Cochran et al. (1983)
	63 day	251	34.4	6.5	1.5	25.6	15.0	4.7		3.8	5.1	2.0					Cochran et al. (1983)
	135 day	225	31.8	2.9	2.0	33.8	11.4	5.8		5.7	5.0	1.6					Cochran et al. (1983)
	Spinal cord	245	31.4	2.9	3.4	34.0	11.1	7.3		5.4	5.0	1.4					Cochran et al. (1983)
Pigeon	Brain	253	24.3	1.5	1.2	27.3	13.8	7.2		10.8	11.3	2.5					Cochran et al. (1981)
	Spinal cord	196	32.2	2.6	3.4	32.4	8.5	7.2		5.4	6.5	1.8					Cochran et al. (1981)
Frog	Brain	44			2.8	16.5	5.4	3.2	24.2	2.0			3.4	12.6	14.4	15.0	Cochran et al. (1982)
Cod	Brain	31			5.6	3.5	6.2	2.2	3.9				3.5	17.3	28.5		Cochran et al. (1982)

brain synaptosomal fractions have an average value of about 10 μg sialic acid/mg protein. The average value for the synaptic plasma membrane preparations is higher, ranging from about 10 to 30 μg sialic acid/mg protein (Table 5). This is similar to that reported for several axolemmal fractions (DeVries *et al.*, 1981). The growth cone membranes, prepared from fetal rat brain, contain higher levels of gangliosides than do synaptic plasma membranes (Sbaschnig-Agler *et al.*, 1988). Since the growth cone membranes are rich in overall lipid content, they may represent highly specialized membranes during neuronal differentiation. Interestingly, the ganglioside composition of growth cones is generally similar to that of mature synaptic membranes except for the presence of more GD3 and less GD1a (Sbaschnig-Agler *et al.*, 1988), a phenomenon also observed for the whole fetal brain of the same age (Yu *et al.*, 1988).

Earlier reports on the presence of gangliosides in synaptic vesicles have generally been received with skepticism owing to the difficulty in obtaining contamination-free preparations. A recent report (Ledeen *et al.*, 1988) provides convincing evidence that synaptic vesicles isolated from the electric organs of *Torpedinidae* contain relatively high concentrations of gangliosides (Table 5). The ganglioside composition of these vesicular preparations is complex, with about 50% of the gangliotetraose structure. It would be interesting to reexamine the synaptic vesicles from mammalian species for their ganglioside composition.

There have been several reports on the soluble, cytosolic pool of gangliosides in mammalian brain (Table 5). The levels are generally extremely low, ranging from 0.3 to 1.5 μg sialic acid/mg protein. The gangliosides are primarily of the gangliotetraose series (Sonnino *et al.*, 1979). CSF also contains low concentrations (0.1–0.3 μg sialic acid/ml) of gangliosides (Ledeen and Yu, 1972). Lumbar fluid has a pattern somewhat similar to that of brain, whereas the pattern for the ventricular fluid resembles that of plasma.

The ganglioside composition of myelin has been studied extensively in different species and during development (Table 6), and found to be quite different from that of whole brain, neurons, astrocytes, or other brain membranes. In general, myelin from mature brain contain high concentrations of GM1. In primate and avian species, there is an additional ganglioside, GM4, which appears to be a useful marker for myelin and oligodendroglia (Ledeen *et al.*, 1973; Yu and Iqbal, 1979; Yu *et al.*, 1988). Since myelin is an extension of the oligodendroglial plasma membrane and the latter has an extremely complex ganglioside pattern compared to myelin, it would be interesting to elucidate the metabolic relationships between these two compartments.

ACKNOWLEDGMENT

This work was supported by USPHS Grants NS-11853, NS-23102, and NS-26994.

4. REFERENCES

Abe, T., and Norton, W. T., 1974, The characterization of sphingolipids from neurons and astroglia of immature rat brain, *J. Neurochem.* **23**:1025–1036.

Ando, S., 1983, Gangliosides in the nervous system, *Neurochem. Int.* **5**:507–537.

Ando, S., and Yu, R. K., 1977, Isolation and characterization of a novel trisialoganglioside, GTla, from human brain, *J. Biol. Chem.* **252**:6247–6250.

Ando, S., and Yu, R. K., 1979, Isolation and characterization of two isomers of brain tetrasialogangliosides, *J. Biol. Chem.* **254**:12224–12229.

Ando, S., Kon, K., Isobe, M., and Yamakawa, T., 1973, Structural study on tetraglycosylceramide and gangliosides isolated from human red blood cells, *J. Biochem.* **73**:893–895.

Ando, S., Chang, N.-C., and Yu, R. K., 1978, High-performance thin-layer chromatography and densitometric determination of brain ganglioside compositions of several species, *Anal. Biochem.* **89**:437–450.

Ariga, T., and Yu, R. K., 1987, Isolation and characterization of ganglioside GM1b from normal human brain, *J. Lipid Res.* **28**:285–291.

Ariga, T., Sekine, M., Yu, R. K., and Miyatake, T., 1984, Isolation and characterization of a novel disialoganglioside from bovine adrenal medulla, *Arch. Biochem. Biophys.* **232**:305–309.

Ariga, T., Kobayashi, K., Kuroda, Y., Yu, R. K., Suzuki, M., Kitagawa, H., Inagaki, F., and Miyatake, T., 1987, Characterization of tumor-associated fucogangliosides from PC12 pheochromocytoma cells, *J. Biol. Chem.* **262**:14146–14153.

Asou, H., and Brunngraber, E. G., 1983, Absence of ganglioside GM1 in astroglial cells from 21-day old rat brain. Immunohistochemical, histochemical and biochemical studies, *Neurochem. Res.* **8**:1045–1057.

Asou, H., and Brunngraber, E. G., 1984, Absence of ganglioside GM1 in astroglial cells from newborn rat brain, *Neurochem. Int.* **6**:81–89.

Avrova, N. F., 1971, Brain ganglioside patterns of vertebrates, *J. Neurochem.* **18**:667–674.

Avrova, N. F., Chenykaeva, E. Y., and Obukhova, E. L., 1973, Ganglioside composition and content of rat brain subcellular fractions, *J. Neurochem.* **20**:997–1004.

Avrova, N. F., Li, Y.-T., and Obukhova, E. L., 1979, On the composition and structure of individual gangliosides from the brain of elasmo-branches, *J. Neurochem.* **32**:1807–1815.

Bartoszewicz, Z., Koscielak, J., and Pacuszka, T., 1986, Structure of a new disialoganglioside GD1c from spontaneous murine thymoma, *Carbohydr. Res.* **151**:77–88.

Bouhours, J.-F., Bouhours, D., and Hansson, G. C., 1987, Developmental changes of gangliosides of the rat stomach, *J. Biol. Chem.* **262**:16370–16375.

Breckenridge, W. C., Gombos, G., and Morgan, I. G., 1972, The lipid composition of adult rat brain synaptosomal plasma membranes, *Biochim. Biophys. Acta* **266**:695–707.

Breckenridge, W. C., Morgan, I. G., Zanta, J. P., and Vincendonn, G., 1973, Adult rat brain synaptic vesicles. II. Lipid composition, *Biochim. Biophys. Acta* **320**:681–686.

Breimer, M. E., Hansson, G. C., Karlsson, K. A., and Leffler, H., 1982, Glycosphingolipids of rat tissues. Different composition of epithelial and nonepithelial cells of rat small intestine, *J. Biol. Chem.* **257**:557–568.

Brunngraber, E. G., Dekirmerjian, H., and Brown, B. D., 1967, The distribution of protein-bound N-acetylneuraminic acid in subcellular fractions of rat brain, *Biochem. J.* **103**:73–78.

Byrne, M. C., Farooq, M., Sbaschnig-Agler, M., Norton, W. T., and Ledeen, R. W., 1989, Gangliosides content of astroglia and neurons isolated from maturing rat brain: Consideration of the source of astroglial gangliosides, *J. Neurochem.*, in press.

Cheresh, D. A., Varki, A. P., Varki, N., Stallcup, W. B., Levine, J., and Reisfeld, R. A., 1984, A monoclonal antibody recognizes an O-acetylated sialic acid in a human melanoma-associated ganglioside, *J. Biol. Chem.* **259**:7453–7459.

Chien, J.-L., and Hogan, E. L., 1983, Novel pentahexosyl ganglioside of the globo series purified from chicken muscle, *J. Biol. Chem.* **258**:10727–10730.

Chien, J.-L., Li, S.-C., Laine, R. A., and Li, Y.-T., 1978, Characterization of gangliosides from bovine erythrocyte membranes, *J. Biol. Chem.* **253**:4031–4035.

Chigorno, V., Sonnino, S., Ghidoni, R., and Tettamanti, G., 1982, Isolation and characterization of a tetrasialoganglioside from mouse brain, containing 9-O-acetyl, N-acetylneuraminic acid, *Neurochem. Int.* **4**:531–539.

Chou, K. H., Nolan, C. E., and Jungalwara, F. B., 1985, Subcellular fractionation of rat sciatic nerve and specific localization of ganglioside LM1 in rat nerve myelin, *J. Neurochem.* **44**:1898–1912.

Clarke, J. T. R., 1975, Gangliosides of the bovine neurohypophysis, *J. Neurochem.* **24**:533–538.

Cochran, F. B., Jr., Yu, R. K., Ando, S., and Ledeen, R. W., 1981, Myelin gangliosides: An unusual pattern in the avian central nervous system, *J. Neurochem.* **36**:696–702.

Cochran, F. B., Yu, R. K., and Ledeen, R. W., 1982, Myelin gangliosides in vertebrates, *J. Neurochem.* **39**:773–779.

Cochran, F. B., Ledeen, R. W., and Yu, R. K., 1983, Gangliosides and proteins in developing chicken brain myelin, *Dev. Brain Res.* **6**:27–32.

Cruz, T. F., and Gurd, J. W., 1981, The effects of development on activity, specificity and endogenous substrates of synaptic membrane sialidase, *Biochim. Biophys. Acta* **675**:201–208.

Dasgupta, S., Chien, J.-L., and Hogan, E. L., 1985, N-Acetyl- and N-glycolyl neuraminosyl lacto-N-noroctaosylceramides from bovine erythrocytes, *Fed. Proc.* **44**:1089.

Dawson, G., and Sweeley, C. C., 1971, Mass spectrometry of neutral, mono- and disialoglycosphingolipids, *J. Lipid Res.* **12**:56–64.

DeGasperi, R., Koerner, T. A. W., Quarles, R. H., Ilyas, A. A., Ishikawa, Y., Li, S.-C., and Li, Y.-T., 1987, Isolation and characterization of gangliosides with hybrid neolacto-ganglio-type sugar chains, *J. Biol. Chem.* **262**:17149–17155.

Dekirmenjian, H., and Brunngraber, E. G., 1969, Distribution of protein-bound N-acetylneuraminic acid in subcellular particulate fractions prepared from rat whole brain, *Biochim. Biophys. Acta* **177**:1–10.

Derry, D. M., and Wolfe, L. S., 1967, Gangliosides in isolated neurons and glial cells, *Science* **158**:1450–1452.

DeVries, G. H., and Norton, W. T., 1974, The lipid composition of axons from bovine brain, *J. Neurochem.* **22**:259–264.

DeVries, G. H., Payne, W., and Saul, R. G., 1981, Axolemma-enriched fractions isolated from bovine CNS myelinated axons, *Neurochem. Res.* **6**:521–537.

Dreyfus, H., Urban, P. F., Edel-Harth, S., and Mandel, P., 1975, Developmental patterns of gangliosides and of phospholipids in chicken retina and brain, *J. Neurochem.* **25**:245–250.

Dreyfus, H., Urban, P. F., Harth, S., Preti, A., and Mandel, P., 1976, Retinal gangliosides: Composition, evolution with age. Biosynthetic and metabolic approaches, *Adv. Exp. Med. Biol.* **71**:163–188.

Edel-Harth, S., Dreyfus, H., Bosch, P., Rebel, G., Urban, P. F., and Mandel, P., 1973, Gangliosides of whole retina and rod outer segments, *FEBS Lett.* **35**:284–288.

Egge, H., Peter-Katalinie, J., Reuter, G., Schuar, R., Ghidoni, R., Sonnino, S., and Tettamanti, G., 1985, Analyses of gangliosides using fast atom bombardment mass spectrometry, *Chem. Phys. Lipids* **37**:127–141.

Eichberg, J., Whittaker, V. P., and Dawson, R. M. C., 1964, Distribution of lipids in subcellular particles of guinea-pig brain, *Biochem. J.* **92**:91–100.

Eisenbarth, G. S., Walsh, F. S., and Nirenberg, M., 1979, Monoclonal antibody to a plasma membrane antigen of neurons, *Proc. Natl. Acad. Sci. USA* **76**:4913–4917.

Eldredge, N. T., Read, G., and Cutting, W., 1963, Sialic acids in the brain and tissues of various animals: Analytical and physiological data, *Med. Exp. (Basel)* **8**:265–277.

Fong, J. W., Ledeen, R. W., Kundu, S. K., and Brostoff, S., 1976, Gangliosides of peripheral nerve myelin, *J. Neurochem.* **26**:157–162.

Fredman, P., Mansson, J.-E., Svennerholm, L., Samuelsson, B., Pascher, I., Pimlott, W., Karlsson, K.-A., and Klinghardt, G. W., 1981, Chemical structures of three fucogangliosides isolated from nervous tissue of mini-pig, *Eur. J. Biochem.* **116**:553–564.

Fredman, P., Magnani, J. L., Nirenberg, M., and Ginsberg, V., 1984, Monoclonal antibody A2B5 reacts with many gangliosides in neuronal tissue, *Arch. Biochem. Biophys.* **233**:661–666.

Fukuda, N. M., Dell, A., Tiller, P. R., Vorki, A., Klock, J. C., and Fukuda, M., 1986, Structure of a novel sialylated fucosyl lacto-N-norhexaosylceramide isolated from chronic myelogenous leukemia cells, *J. Biol. Chem.* **261**:2376–2383.

Fukushi, Y., Nudelman, E., Levery, S. B., and Hakomori, S., 1984, Novel fucolipids accumulating in human adenocarcinoma. III. A hybridoma antibody (FH6) defining a human cancer associated difuco-ganglioside (VI^{3}NeuAcV^{2}III^{3}Fuc$_2$nLc$_6$), *J. Biol. Chem.* **259**:10511–10517.

Fukushi, Y., Nudelman, E., Levery, S. B., Higuchi, T., and Hakomori, S., 1986, A novel disialoganglioside (IV^{3}NeuAcIII^{6}NeuAcLc$_4$) of human adenocarcinoma and the monoclonal antibody (FH9) defining this disialosyl structure, *Biochemistry* **25**:2859–2866.

Ghidoni, R., Sonnino, S., Tettamanti, G., Wiegandt, H., and Zambotti, V., 1976, On the structure of two new gangliosides from beef brain, *J. Neurochem.* **27**:511–515.

Ghidoni, R., Sonnino, S., Tettamanti, G., Baumann, N., Reuter, G., and Schauer, R., 1980, Isolation and characterization of trisialoganglioside from mouse brain containing 9-O-acetyl-neuraminic acid, *J. Biol. Chem.* **255**:6990–6995.

Gillard, B. K., Blanchard, D., Bouhours, J.-F., Cartron, J.-P., van Kuik, J. A., Kamerling, J. P., Vliegenthart, J. F. G., and Marcus, D. M., 1988, Structure of a ganglioside with Cad blood group antigen activity, *Biochemistry* **27**:4601–4606.

Ginsburg, V. (ed.), 1987, Complex Carbohydrates, Part E, *Methods Enzymol.* **138**.

Goldman, J. E., Hiramo, M., Yu, R. K., and Seyfried, T. N., 1984, GD3 ganglioside is a glycolipid characteristic of immature neuroectodermal cells, *J. Neuroimmunol.* **7**:179–192.

Gowda, D. C., Reuter, G., Shukla, A. K., and Schauer, R., 1984, Identification of a disialoganglioside (GDla) containing terminal N-acetyl-9-O-acetyl neuraminic acid in rat erythrocytes, *Hoppe-Seyler's Z. Physiol. Chem.* **365**:1247–1253.

Hakomori, S., and Saito, T., 1969, Isolation and characterization of a glycosphingolipid having a new sialic acid, *Biochemistry* **8**:5082–5088.

Hamberger, A., and Svennerholm, L., 1971, Composition of gangliosides and phospholipids of neuronal and glial cell enriched fractions, *J. Neurochem.* **18**:1821–1829.

Handa, N., and Handa, S., 1965, The chemistry of lipids of posthemolytic residue or stroma of erythrocytes. XVI. Chemical structure of glycolipid of cat erythrocyte stroma, *Jpn. J. Exp. Med.* **35**:331–341.

Handa, S., and Burton, R. M., 1969, Lipid of retina. I. Analysis of gangliosides in beef retina by thin layer chromatography, *Lipids* **4**:205–208.

Harris, J. V., and Klingman, J. D., 1972, Detection, determination and metabolism *in vitro* of gangliosides in mammalian sympathetic ganglia, *J. Neurochem.* **19**:1267–1278.

Hilbig, R., and Rahmann, H., 1987, Phylogeny of vertebrate brain gangliosides, in: *Gangliosides and Modulation of Neuronal Functions* (H. Rahmann, ed.), NATO ASI Series, Vol. H7, pp. 333–350, Springer-Verlag, Berlin.

Hirabayashi, Y., Taki, T., and Matsumoto, M., 1979, Tumor ganglioside—Natural occurrence of GM1b, *FEBS Lett.* **100**:253–257.

Hogan, E. L., Gappel, R. D., and Chien, J.-L., 1982, Membrane glycosphingolipids in chicken muscular dystrophy, *Adv. Exp. Med. Biol.* **152**:273–278.

Holm, M., and Mansson, J.-E., 1974, Gangliosides of the bovine optic nerve, *FEBS Lett.* **45**:159–161.

Holm, M., Mansson, J.-E., Vanier, M.-T., and Svennerholm, L., 1972, Gangliosides of human, bovine and rabbit retina, *Biochim. Biophys. Acta* **280**:356–364.

Holmes, E., and Hakomori, S., 1982, Isolation and characterization of a new fucoganglioside accumulated in precancerous rat liver and in rat hepatoma induced by N-2-acetylaminofluorene, *J. Biol. Chem.* **257**:7698–7703.

Irwin, L. N., Michael, D. B., and Irwin, C. C., 1980, Ganglioside patterns of fetal rat and mouse brain, *J. Neurochem.* **34**:1527–1530.

Ishizuka, I., and Wiegandt, H., 1972, An isomer of trisialoganglioside and the structure of tetra and pentasialogangliosides from fish brain, *Biochim. Biophys. Acta* **260**:279–289.

Ishizuka, I., Kloppenburg, M., and Wiegandt, H., 1970, Characterization of gangliosides from fish brain, *Biochim. Biophys. Acta* **210**:299–305.

Itoh, T., Li, Y.-T., Li, S.-C., and Yu, R. K., 1981, Isolation and characterization of a novel monosialosyl-pentahexosyl ceramide from Tay-Sachs brain, *J. Biol. Chem.* **256**:165–169.

Iwamori, M., and Nagai, Y., 1978, Isolation and characterization of a novel ganglioside, monosialosyl pentahexosyl ceramide from human brain, *J. Biochem.* **84**:1601–1608.

Iwamori, M., and Nagai, Y., 1981, Monosialogangliosides of rabbit skeletal muscle. Characterization of N-acetylneuraminosyl lacto-N-noroctaosyl ceramide, *J. Biochem.* **89**:1253–1264.

Kannagi, R., Roelbe, D., Peterson, K. A., Okada, Y., Levery, S. B., and Hakomori, S., 1983, Characterization of an epitope (determinant) structure in a developmentally regulated glycolipid antigent defined by a cold agglutinin fi, recognition of α-sialosyl and α-L-fucosyl groups in a branched structure, *Carbohydr. Res.* **120**:143–157.

Kasai, N., and Yu, R. K., 1983, The monoclonal antibody A2B5 is specific to ganglioside GQ1c, *Brain Res.* **277**:155–158.

Kim, S. U., Moretto, G., Lee, V., and Yu, R. K., 1986, Neuroimmunology of gangliosides in human neurons and glial cells in culture, *J. Neurosci. Res.* **15**:303–321.

Klein, F., and Mandel, P., 1975, Gangliosides of the peripheral nervous system of the rat, *Life Sci.* **16**:751–758.

Klenk, E., and Gielen, W., 1963, Über ein zweites hexosamin haltiges gangliosid aus menschengehirn, *Hoppe-Seyler's Z. Physiol. Chem.* **330**:218–226.

Klenk, E., and Lauenstein, I., 1953, Über die Glykolipoide und Sphingomyeline des Stromas der Pferdeerythrocyten, *Hoppe-Seyler's Z. Physiol. Chem.* **295**:164–173.

Klenk, E., and Naomi, M., 1968, Über eine komponente des Gemisches der Gehirnganglioside, die durch Neuraminidaseein-Wirkung in das Tay-Sachs-Gangliosid ubergeht, *Hoppe-Seyler's Z. Physiol. Chem.* **349**:288–292.

Klenk, E., Hof, L., and Georgias, L., 1967, Zur kenntnis der Gehirnglioside, *Hoppe-Seyler's Z. Physiol. Chem.* **348**:149–166.

Kornguth, S., Wannamaker, B., Kolodny, E., Geison, R., Scott, G., and O'Brien, J. F., 1974, Subcellular fractions from Tay-Sachs brains: Ganglioside, lipid, and protein composition and hexosaminidase activities, *J. Neurol. Sci.* **22**:383–406.

Krawn, I., Rosner, H., Cosovic, C., and Stavljenic, A., 1984, Topographic atlas of the gangliosides of the adult human brain, *J. Neurochem.* **43**:979–989.

Kuhn, R., and Wiegandt, H., 1963a, Die Konstitution der Ganglio-N-tetraose und des Gangliosides G_1, *Chem. Ber.* **96**:866–880.

Kuhn, R., and Wiegandt, H., 1963b, Die Konstitution der Ganglioside, *Z. Naturforsch.* **18b**:541–543.

Kuhn, R., and Wiegandt, H., 1964, Weitere Ganglioside aus Menschenhirn, *Z. Naturforsch.* **19b**:256–257.

Kundu, S. K., Marcus, D. M., Pascher, I., and Samuelsson, B. E., 1981, New gangliosides from human erythrocytes, *Fed. Proc.* **40**:1545.

Kundu, S. K., Samuelsson, B. E., Pascher, I., and Marcus, D. M., 1983, New gangliosides from human erythrocytes, *J. Biol. Chem.* **258**:13857–13866.

Lapetina, E. G., Soto, E. F., and DeRobertis, E., 1968, Lipid and proteolipids in isolated subcellular membranes of rat brain cortex, *J. Neurochem.* **15**:437–445.

Ledeen, R. W., 1978, Ganglioside structures and distribution: Are they localized at the nerve ending? *J. Supramol. Struct.* **8**:1–17.

Ledeen, R. W., and Salsman, K., 1965, Structure of the Tay-Sachs' ganglioside, *Biochemistry* **4**:2225–2233.

Ledeen, R. W., and Yu, R. K., 1972, Gangliosides of CSF and plasma: Their relation to the nervous system, *Adv. Exp. Med. Biol.* **19**:77–93.

Ledeen, R. W., and Yu, R. K., 1982, Gangliosides: Structure, isolation, and analysis, *Methods Enzymol.* **83**:139–191.

Ledeen, R. W., Salsman, K., Gonatas, J., and Taghavy, A., 1965, Structure comparison of the major monosialogangliosides from brains of normal human, gargoylism and late infantile systematic lipidosis. I, *J. Neuropathol. Exp. Neurol.* **24**:341–351.

Ledeen, R. W., Salsman, K., and Cabrera, M., 1968, Gangliosides of bovine adrenal medulla, *Biochemistry* **7**:2287–2295.

Ledeen, R. W., Yu, R. K., and Eng, L. F., 1973, Gangliosides of human myelin: Sialosylgalactosylceramide (G_7) as a major component, *J. Neurochem.* **21**:829–839.

Ledeen, R. W., Skrivanek, J. A., Tirri, L. J., Margolis, R. K., and Margolis, R. U., 1976, Gangliosides of the neuron: Localization and origin, *Adv. Exp. Med. Biol.* **71**:83–103.

Ledeen, R. W., Parsons, S. M., Diebler, M. F., Sbaschnig-Agler, M., and Lazereg, S., 1988, Ganglioside composition of synaptic vesicles from *Torpedo* electric organ, *J. Neurochem.* **51**:1465–1469.

Levine, J. M., Beasley, L., and Stallcup, W. B., 1984, The D1.1 antigen, a cell surface marker for germinal cells of the central nervous system, *J. Neurosci.* **4**:820–831.

Levine, J. M., Beasley, L., and Stallcup, W. B., 1986, Localization of a neuroectoderm-associated cell surface antigen in the developing and adult rat, *Dev. Biol. Res.* **27**:211–222.

Li, S.-C., Chien, J.-L., Wan, C. C., and Li, Y.-T., 1978, Occurrence of glycosphingolipids in chicken egg yolk, *Biochem. J.* **173**:697–699.

Li, Y.-T., Mansson, J. E., Vanier, M. T., and Svennerholm, L., 1973, Structure of the major glucosamine-containing ganglioside of human tissues, *J. Biol. Chem.* **248**:2634–2636.

Maccioni, H. J. F., Defilpo, S. S., Landa, C. A., and Caputto, R., 1978, The biosynthesis of brain gangliosides. Ganglioside-glycosylating activity in rat brain neuronal perikarya fraction, *Biochem, J.* **174**:673–680.

Macher, M. A., Pacuszka, T., Mullin, B. R., Sweeley, C. C., Brady, R. O., and Fishman, P. H., 1979, Isolation and identification of a fucose-containing ganglioside from bovine thyroid gland, *Biochim. Biophys. Acta* **588**:35–43.

MacMillan, V. H., and Wherrett, J. R., 1969, A modified procedure for the analysis of mixtures of tissue gangliosides, *J. Neurochem.* **16**:1621–1624.

Magnani, J. L., Nilsson, B., Brockhaus, M., Sopf, D., Steplewski, S., Koprowski, H., and Ginsburg, V., 1982, A monoclonal antibody-defined antigen associated with gastrointestinal cancer is a ganglioside containing sialylated lacto-*N*-fucopertaose II, *J. Biol. Chem.* **257**:14365–14369.

Mandel, P., Dreyfus, H., Yusufi, A. N. K., Sarlieve, L., Robert, J., Neskovic, N., Harth, S., and Rebel, G., 1980, Neuronal and glial cell cultures, a tool for investigation of ganglioside function, *Adv. Exp. Med. Biol.* **125**:515–531.

Mansson, J.-E., Fredman, P., Nilsson, O., Lindholm, L., Holmgren, J., and Svennerholm, L., 1985, Chemical structure of carcinoma ganglioside antigens defined by monoclonal antibody C-50 and some allied gangliosides of human pancreatic adenocarcinoma, *Biochim. Biophys. Acta* **834**:110–117.

Mansson, J.-E., Fredman, P., Bigner, D. D., Molin, K., Rosengren, B., Friedman, H. S., and Svennerholm, L., 1986, Characterization of a new ganglioside of the lactotetraose series in murine xenografts of a human glioma cell line, *FEBS Lett.* **201**:109–113.

Merat, A., and Dickerson, J. W. T., 1973, The effect of development on the gangliosides of rat and pig brain, *J. Neurochem.* **20**:873–880.

Murakami-Murofushi, K., Tadano, K., Koyama, I., and Ishizuka, I., 1981, A trisialosyl ganglioside GT_3 of hog kidney. Structure and biosynthesis in vitro, *J. Biochem.* **90**:1817–1820.

Murakami-Murofushi, K., Tadano, K., and Ishizuka, I., 1983, A novel trisialosyl ganglioside. $IV^3(NeuAc)_3nLcOse_4$ (er) from hog kidney cortex, *J. Biochem.* **93**:621–629.

Murayama, K., Levery, S. B., Schirrmacher, V., and Hakomori, S., 1986, Quantitative differences in position of sialylation and surface expression of glycolipids between murine lymphomas with low metastatic (Eb) and high metastatic (ESb) potentials and isolation of a novel disialoganglioside (GDIα) from EB cells, *Cancer Res.* **46**:1395–1402.

Muthing, J., Egge, H., Kniep, B., and Muhlradt, P. F., 1987, Structural characterization of gangliosides from murine T lymphocytes, *Eur. J. Biochem.* **163**:407–416.

Nakamura, K., Suzuki, M., Inagaki, F., Yamakawa, T., and Akemi, S., 1987, A new ganglioside showing choleragenoid-binding activity in mouse spleen, *J. Biochem.* **101**:825–835.

Neuenhofer, S., Conzelmann, E., Schwarzmann, G., Egge, H., and Sandhoff, K., 1986, Occurrence of lysoganglioside lyso-GM2 (II^3-Neu5Ac-gangliotriaosyl-sphingosine) in GM2 gangliosidosis brain, *Biol. Chem. Hoppe-Seyler* **367**:241–244.

Nilsson, O., Mansson, J.-E., Lindholm, L., Holmgren, J., and Svennerholm, L., 1985, Sialosyllac-tosyltetraosylceramide, a novel ganglioside antigen detected in human carcinomas by a monoclonal antibody, *FEBS Lett.* **182**:398–402.

Norton, W. T., and Poduslo, S. E., 1971, Neuronal perikarya and astroglia of rat brain: Chemical composition during myelination, *J. Lipid Res.* **12**:84–90.

Nudelman, E., Fukushi, Y., Levery, S. B., Higuchi, T., and Hakomori, S., 1986, Novel fucolipids of human adenocarcinoma: Disialosyl Le^a antigen ($III^4FucIII^6NeuAcIV^3NeuAcLc_4$) of human colonic adenocarcinoma and the monoclonal antibody (FH7) defining this structure, *J. Biol. Chem.* **261**:5487–5495.

Ohashi, M., 1979, A comparison of the ganglioside distributions of fat tissues in various animals by two dimensional thin layer chromatography, *Lipids* **14**:52–57.

Ohashi, M., 1980, A new type of ganglioside. The structures of three novel gangliosides from the fat body of the frog, *J. Biochem.* **88**:583–589.

Ohashi, M., and Yamakawa, T., 1977, Isolation and characterization of glycosphingolipids in pig adipose tissue, *J. Biochem.* **81**:1675–1690.

Ostrander, G. K., Levery, S. B., Hakomori, S., and Holmes, E. H., 1988, Isolation and characterization of the major acidic glycosphingolipids from the liver of the English sole (Parophys vetulus), *J. Biol. Chem.* **263**:3103–3110.

Poduslo, S. E., and Norton, W. T., 1972, Isolation and some chemical properties of oligodendroglia from calf brain, *J. Neurochem.* **19**:727–736.

Price, H., Kundu, S., and Ledeen, R. W., 1975, Structures of gangliosides from bovine adrenal medulla, *Biochemistry* **14**:1512–1518.

Prieto, P. A., and Smith, D. F., 1986, A new ganglioside in human meconium detected with antiserum against human milk sialyl-tetrasacharide A, *Arch. Biochem. Biophys.* **249**:243–253.

Raff, M. C., Abney, E. R., Cohen, J., Lindsay, R., and Nobel, M., 1983, Two types of astrocytes in cultures of developing rat white matter, differences in morphology, surface gangliosides and growth characteristics, *J. Neurosci.* **3**:1289–1300.

Rauvala, H., 1976, The fucoganglioside of human kidney, *FEBS Lett.* **62**:161–164.

Rauvala, H., Krusius, T., and Finne, J., 1978, Disialosyl paragloboside—A novel ganglioside isolated from human kidney, *Biochim. Biophys. Acta* **531**:266–274.

Riboni, L., Sonnino, S., Acquotti, D., Malesci, A., Ghidoni, R., Egge, H., Mingrino, S., and Tettamanti, G., 1986, Natural occurrence of ganglioside lactogens. Isolation and characterization of GD1b inner ester from adult human brain, *J. Biol. Chem.* **261**:8514–8519.

Robert, J., Freysz, L., Sensenbrenner, M., Mandel, P., and Rebel, G., 1975, Gangliosides of glial cells: A comparative study of normal astroblasts in tissue culture and glial cell isolated on sucrose—Ficoll gradients, *FEBS Lett.* **50**:144–146.

Rosner, H., 1982, Ganglioside changes in the chicken optic lobes as biochemical indicators of brain development and maturation, *Brain Res.* **239**:49–61.

Rosner, H., 1982, Uniform distribution and similar turnover rates of individual gangliosides along axons of retinal ganglion cells in the chicken, *Brain Res.* **236**:63–75.

Rosner, H., and Rahmann, H., 1987, Ontogeny of vertebrate brain gangliosides, in: *Gangliosides and Modulation of Neuronal Functions* (H. Rahmann, ed.), NATO ASI Series, Vol. H7, pp. 373–390, Springer-Verlag, Berlin.

Sbaschnig-Agler, M., Dreyfus, H., Norton, W. T., Sensenbrenner, M., Farooq, M., Byrne, M. C., and Ledeen, R. W., 1988, Gangliosides of cultured astroglia, *Brain Res.* **461**:98–106.

Schlosshauer, B., Blum, A., Mandez-Otero, R., Barnstable, C., and Constantine-Paton, M., 1988, Developmental regulation of ganglioside antigens recognized by the JONES antibody, *J. Neurosci.* **8**:580–592.

Schwarting, G. A., Carroll, P. G., and DeWolf, W. C., 1983, Fucosyl-globoside and sialosylgloboside are new glycolipids isolated from human tetracarcinoma cells, *Biochem. Biophys. Res. Commun.* **112**:935–940.

Seyfried, T. N., and Yu, R. K., 1984, Cellular localization of gangliosides in the mouse cerebellum: Analysis using neurological mutants *Adv. Exp. Med. Biol.* **174**:169–181.

Seyfried, T. N., and Yu, R. K., 1985, Ganglioside GD3 structure, distribution and possible function, *Mol. Cell. Biochem.* **68**:3–10.

Seyfried, T. N., Ando, S., and Yu, R. K., 1978, Isolation and purification of human liver hematoside, *J. Lipid Res.* **19**:538–543.

Seyfried, T. N., Glaser, G. H., and Yu, R. K., 1979, Genetic variability for regional brain gangliosides in five strains of young mice, *Biochem. Genet.* **17**:43–55.

Skrivanek, J. A., Ledeen, R. W., Norton, W. T., and Farooq, M., 1978, Ganglioside distribution in rat cortex, *Trans. Am. Soc. Neurochem.* **9**:133.

Skrivanek, J. A., Ledeen, R. W., Margolis, R. U., and Margolis, R. K., 1982, Gangliosides associated with microsomal subfractions of brain: Comparison with synaptic plasma membranes, *J. Neurobiol.* **13**:95–106.

Slomiany, B. L., Kojima, K., Banas-Gruszka, Z., Murty, V. L. N., Galicki, N. I., and Slomiany, A., 1981, Characterization of the sulfated monosialosyl triglycosylceramide from bovine gastric mucosa, *Eur. J. Biochem.* **119**:647–650.

Smirnova, G. P., Kochetkov, N. K., and Sadovskaya, V. L., 1987, Gangliosides of the starfish Aphelasterias japonica, evidence for a new linkage between two N-glycolylneuraminic acid residues through the hydroxy group of the glycolic acid residue, *Biochim. Biophys. Acta* **920:**47–55.

Sonnino, S., Ghidoni, R., Galli, G., and Tettamanti, G., 1978, On the structure of a new fucose containing ganglioside from pig cerebellum, *J. Neurochem.* **31:**947–956.

Sonnino, S., Ghidoni, R., Marchesini, S., and Tettamanti, G., 1979, Cytosolic gangliosides: Occurrence in calf brain as ganglioside–protein complexes, *J. Neurochem.* **33:**117–121.

Suzuki, A., Ishizuka, J., and Yamakawa, T., 1975, Isolation and characterization of a ganglioside containing fucose from boar testis, *J. Biochem.* **78:**947–954.

Suzuki, K., 1965, The pattern of mammalian brain gangliosides. III. Regional and developmental differences, *J. Neurochem.* **12:**969–979.

Suzuki, K., 1967, Ganglioside pattern of normal and pathological brains, in: *Inborn Disorders of Sphingolipid Metabolism* (S. M. Aronson and B. W. Volk, eds.), pp. 215–230, Pergamon Press, New York.

Svennerholm, L., Bruce, A., Mansson, J.-E., Rynmark, B.-M., and Varnier, M.-T., 1972, Spingolipids of human skeletal muscle, *Biochim. Biophys. Acta* **280:**626–636.

Svennerholm, L., Mansson, J.-E., and Li, Y.-T., 1973, Isolation and structural determination of a novel ganglioside, a disialosylpentahexosyl-ceramide from human brain, *J. Biol. Chem.* **248:**740–742.

Takamizawa, K., Iwamori, M., Mutai, M., and Nagai, Y., 1986, Gangliosides of bovine buttermilk—Isolation and characterization of a novel monosialoganglioside with a new branching structure, *J. Biol. Chem.* **261:**5625–5630.

Taki, T., Hirabayashi, Y., Ishikawa, H., Ando, S., Kon, K., Tanaka, Y., and Matsumoto, M., 1986, A ganglioside of rat ascites hepatoma AH 7974F cells—Occurrence of a novel disialoganglioside (GDIα) with a unique N-acetylneuraminosyl(α2-6)-N-acetylgalactosamine structure, *J. Biol. Chem.* **261:**3075–3078.

Taki, T., Matsuo, K., Yamamoto, K., Matsubara, T., Hayashi, A., Abe, T., and Matsumoto, M., 1988, Human placenta gangliosides, *Lipids* **23:**192–198.

Tamai, Y., Matsukawa, S., and Satake, M., 1971, Gangliosides in neuron, *J. Biochem.* **69:**235–238.

Tettamanti, G., Bonali, F., Marchesini, S., and Zambotti, V., 1973, A new procedure for the extraction and fractionation of brain gangliosides, *Biochim. Biophys. Acta* **296:**160–170.

Tettamanti, G., Preti, A., Cestaro, B., Venerando, B., Lombardo, A., Ghidoni, R., and Sonnino, S., 1980, Gangliosides, neuraminidase and sialyl transferase at the nerve endings, *Adv. Exp. Med. Biol.* **125:** 263–280.

Ueno, K., Ando, S., and Yu, R. K., 1978, Gangliosides of human, cat and rabbit spinal cords and cord myelin, *J. Lipid Res.* **19:**863–871.

Van Dessel, G. A. F., Lagrou, A. R., Hilderson, H. J. J., Dierick, W. S. H., and Lauwers, W. J., 1979, Structure of the major gangliosides from bovine thyroid, *J. Biol. Chem.* **254:**9305–9310.

Vanier, M.-T., Holm, M., Ohman, R., and Svennerholm, L., 1971, Developmental profiles of gangliosides in human and rat brain, *J. Neurochem.* **18:**581–592.

Watanabe, K., and Hakomori, S., 1979, Gangliosides of human erythrocytes. A novel ganglioside with a unique N-acetylneuraminosyl-(2-3)-N-acetylgalactosamine structure, *Biochemistry* **18:**5502–5504.

Watanabe, K., Powell, M., and Hakomori, S., 1978, Isolation and characterization of a novel fucoganglioside of human erythrocyte membranes, *J. Biol. Chem.* **253:**8962–8967.

Watanabe, K., Hakomori, S., Childs, R. A., and Feizi, T., 1979a, Characterization of a blood group I-active ganglioside, *J. Biol. Chem.* **254:**3221–3228.

Watanabe, K., Powell, M. E., and Hakomori, S., 1979b, Isolation and characterization of gangliosides with a new sialosyl linkage and core structures, *J. Biol. Chem.* **254:**8223–8229.

Wherrett, J. R., 1973, Characterization of the major ganglioside in human red cells and of a related tetrahexosyl ceramide in white cells, *Biochim. Biophys. Acta* **326:**63–73.

Whittaker, V. P., 1969, The synaptosome, in: *Handbook of Neurochemistry,* Vol. 2 (A. Lajtha, ed.), pp. 327–364, Plenum Press, New York.

Wiegandt, H., 1967, The subcellular localization of gangliosides in the brain, *J. Neurochem.* **14:**671–674.

Wiegandt, H., 1973, Gangliosides of extraneural organs, *Hoppe-Seyler's Z. Physiol. Chem.* **354:**1049–1054.

Wiegandt, H., 1974, Monosialo-lactoisohexaosyl-ceramide: A ganglioside from human spleen, *Eur. J. Biochem.* **45**:367–369.

Wiegandt, H., 1985, Gangliosides, in: *Glycolipids* (H. Wiegandt, ed.), Vol. 10, pp. 199–260, Elsevier, Amsterdam.

Wiegandt, H., and Schulze, B., 1969, Spleen gangliosides: The structure of ganglioside GLNn T1 (NGNA), *Z. Naturforsch.* **24b**:945–946.

Yamakawa, T., and Suzuki, S., 1951, The chemistry of posthemolytic residue or stroma of erythrocytes. I. Concerning the ether-insoluble lipids of lyophilized horse blood stroma, *J. Biochem.* **38**:199–212.

Yates, A. J., and Wherrett, J. R., 1974, Changes in the sciatic nerve of the rabbit and its tissue constituents during development, *J. Neurochem.* **23**:993–1003.

Yohe, H. C., Ueno, K., Chang, N.-C., Glaser, G. H., and Yu, R. K., 1980, Incorporation of N-acetylmannosamine into rat brain subcellular gangliosides: Effect of pentylenetetrazol-induced convulsions on brain gangliosides, *J. Neurochem.* **34**:560–568.

Yu, R. K., and Ando, S., 1980, Structures of some new complex gangliosides of fish brain, *Adv. Exp. Med. Biol.* **125**:33–45.

Yu, R. K., and Iqbal, K., 1979, Sialosylgalactosyl ceramide as a specific marker for human myelin and oligodendroglia perikarya: Gangliosides of human myelin, oligodendroglia and neurons, *J. Neurochem.* **32**:293–300.

Yu, R. K., and Ledeen, R. W., 1970, Gas–liquid chromatographic assay of lipid-bound sialic acids: Measurement of gangliosides in brain of several species, *J. Lipid Res.* **11**:506–516.

Yu, R. K., and Manuelidis, E. E., 1978, Ganglioside alterations in guinea pig brains at end stages of experimental Creutzfeldt–Jakob disease, *J. Neurol. Sci.* **35**:15–23.

Yu, R. K., and Yen, S. I., 1975, Gangliosides in developing mouse brain myelin, *J. Neurochem.* **25**:229–232.

Yu, R. K., Ledeen, R. W., Gajdusek, D. C., and Gibbs, C. J., 1974, Ganglioside changes in slow virus diseases: Analyses of chimpanzee brains infected with Kuru and Creutzfeldt–Jakob agents, *Brain. Res.* **70**:103–113.

Yu, R. K., Ueno, K., Glaser, G. H., and Tourtellotte, W. W., 1982, Lipid and protein alterations of spinal cord and cord myelin of multiple sclerosis, *J. Neurochem.* **39**:464–477.

Yu, R. K., Koerner, T. A. W., Scarsdale, J. N., and Prestegard, J. H., 1986, Elucidation of glycolipid structure by proton nuclear magnetic resonance spectroscopy, *Chem. Phys. Lipids* **42**:27–48.

Yu, R. K., Macala, L. J., Taki, T., Weinfeld, H. M., and Yu, F. S., 1988, Developmental changes in ganglioside composition and synthesis in embryonic rat brain, *J. Neurochem.* **50**:1825–1829.

Yu, R. K., Macala, L. J., Farooq, M., Sbaschnig-Agler, M., Norton, W. T., and Ledeen, R. W., 1989, Ganglioside and lipid composition of bulk-isolated rat and bovine oligodendroglia, *J. Neurosci. Res.*, in press.

Biosynthesis, Metabolism, and Biological Effects of Gangliosides

Robert W. Ledeen

1. INTRODUCTION AND OVERVIEW

Early speculation on ganglioside function was fashioned in the belief that these substances are localized primarily in the neuron. Current awareness of their widespread distribution has considerably broadened the range of potential functions receiving consideration, although neuronal aspects continue to command special interest owing to the relatively high content and molecular complexity of gangliosides in these cells. Further interest derives from the fact that neurons contain approximately two-thirds of their total glycoconjugate sialic acid in these lipid-bound forms, in contrast to extraneural cells which contain only minor proportions of ganglioside. However, it should be emphasized that despite such relative enrichment, gangliosides comprise only an estimated 2–3% of the neuronal plasma membrane phospholipid content on a molar basis. Even considering that this figure doubles when focusing on the outer half of the bilayer where they are localized, they remain relatively minor components of the membrane. In terms of cell surface carbohydrate, they are viewed as making an important contribution to the glycocalyx surrounding the neuronal membrane and determining many of its surface characteristics. This general perception, while probably correct, has been lacking in details concerning the specific roles of the myriad structural forms known to occur in this and other membranes.

Determining how this complex constellation of oligosaccharide structures fulfills the special requirements of the neuron will undoubtedly occupy the efforts of many investigators for a long time to come. However, some useful clues have begun to emerge, particularly in the realm of specific proteins with which gangliosides interact in a manner that leads to modulated activity. These include a number of receptors, enzymes, and perhaps certain structural components as well.

Robert W. Ledeen • Departments of Neurology and Biochemistry, Albert Einstein College of Medicine, Bronx, New York 10461

On the basis of this admittedly limited sampling, it can be postulated that gangliosides (and perhaps glycolipids in general) function primarily through association with specific proteins. The variety of glycolipid structures might then be understood in relation to the variety of cell surface proteins with which each is specifically designed to interact. Several examples of this kind, in which a protein's functional activity is "fine-tuned" by the presence of a ganglioside or other glycolipid, are now on record. If this model is found to have general validity, discovery of new glycolipids (which continues apace) would then raise the automatic question: with which protein does it associate in the membrane and in what manner does it modulate that protein's function? Certain of the topics covered here will be considered from this point of view. Progress in the biosynthesis, transport, and metabolism of gangliosides and the manner in which these relate to function will be considered in the first portion of this chapter. For further information the reader is referred to several comprehensive reviews that appeared recently on this subject (Schachter and Roseman, 1980; Ando, 1983; Ledeen, 1983; Burczak et al., 1984; Wiegandt, 1985; Basu et al., 1988).

2. GANGLIOSIDE BIOSYNTHESIS

The principal pathways for ganglioside synthesis and degradation have been elucidated through a combination of *in vivo* and *in vitro* paradigms. Synthesis is known to involve stepwise transfer of individual sugars from their nucleotide conjugates onto the growing oligosaccharide chain. The concept of "cooperative sequential specificity" was originally proposed (Roseman, 1970) as a system of coordinated glycosyltransferases. Subsequent work (Caputto et al., 1976) demonstrated two distinct ganglioside pools, a small one of transient precursors in which biosynthesis occurs, and a large pool of end products in membranes. According to this model, each ganglioside is synthesized by a separate multienzyme complex with no mixing of pools. Studies with neuroblastoma cells (Miller-Podraza et al., 1982) supported such synthesis occurring with a very small pool of intermediates and indicated the end product pool to be the plasma membrane which contained approximately three-fourths of the total cell gangliosides. There is now good evidence to indicate that the site of ganglioside synthesis suggested in those experiments is the Golgi apparatus (see below). It has been determined (Miller-Podraza and Fishman, 1982) that transfer of gangliosides from the site of synthesis to the cell surface requires approximately 20 min, consistent with other membrane constituents. This ganglioside transfer may involve vesicular transport in keeping with the generally accepted membrane flow model (Morre et al., 1979), but details of the mechanism remain to be elucidated.

2.1. In Vitro Studies Establishing the Major Pathways

The major routes for biosynthesis of GM3, GM2, and the members of the gangliotetraose family, as established primarily with *in vitro* preparations, are depicted in Figure 1. The principal enzyme source employed in the early studies (Kaufman et al., 1967; Roseman, 1970) was embryonic chick brain, which served to establish the *a* and

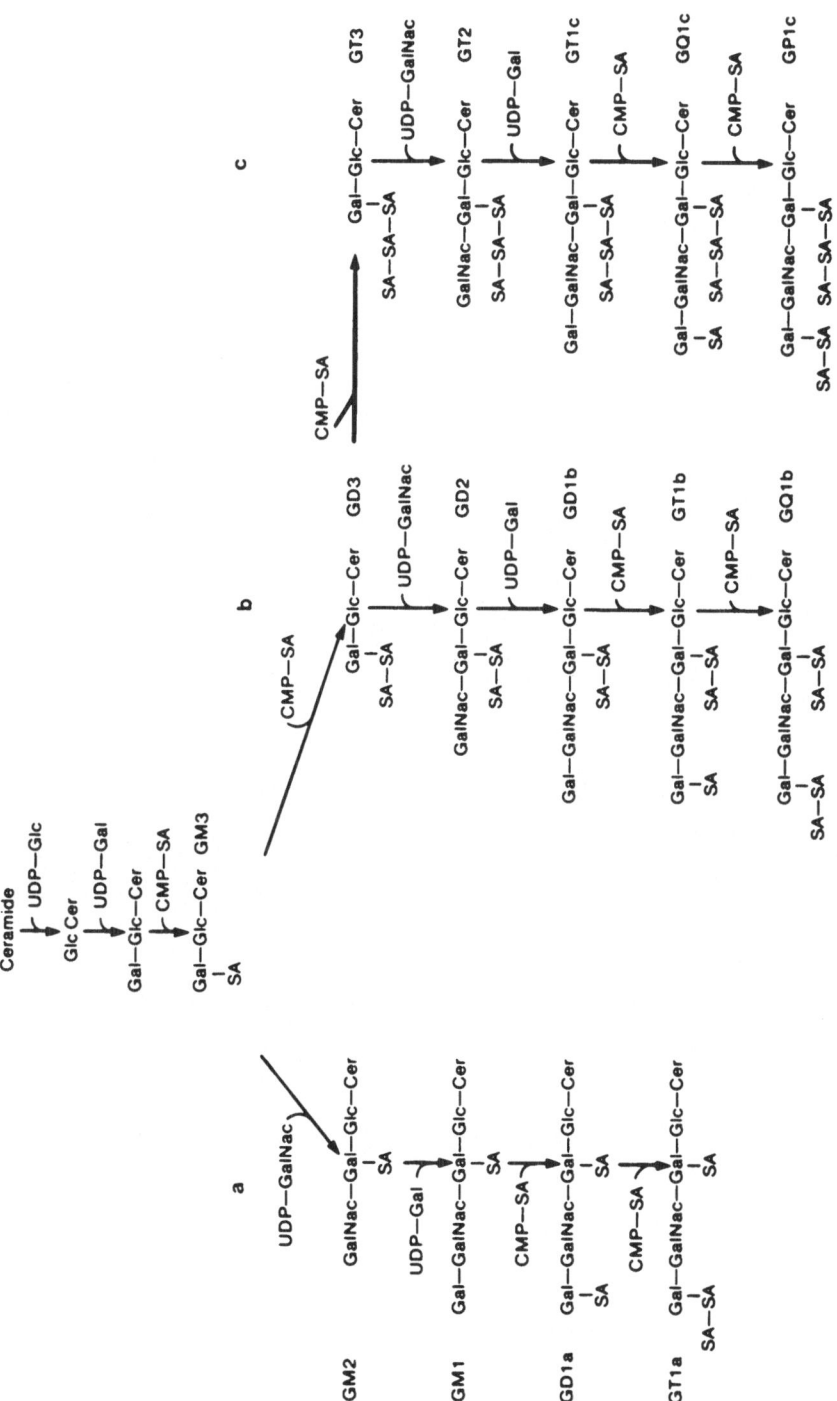

Figure 1. Pathways for biosynthesis of gangliosides of the ganglio series.

b pathways (Figure 1). Many of the same reactions were later observed in other species and tissues. These reactions have been further elucidated using solubilized, and in some cases purified, enzymes (Basu *et al.*, 1988).

The *c* pathway (Figure 1) was conceived more recently as a result of structural studies on cod fish brain gangliosides (Yu and Ando, 1980). This was suggested to represent the major pathway in certain fishes and thus to constitute a phylogenetically older route. Essential links in this pathway were provided by isolation of three new trisialogangliosides from fish brain: GT3, GT2, and GT1c (Yu and Ando, 1980). The sialosyltransferase which converts GD3 to GT3 was detected in hog kidney (Mura-kami-Murofushi *et al.*, 1981) and GT3 itself was isolated from the same source.

Various branch points, important as control sites, are evident in these pathways. One of these is GM3, which can react with CMP-NeuAc to form GD3 or UDP-GalNAc to form GM2. The *N*-acetylgalactosaminyltransferase catalyzing the latter reaction provides the committed step for synthesis of GM2 and components of the *a* pathway of the gangliotetraose family. Diminished activity accompanies altered ganglioside patterns in transformed cells (Hakomori, 1981). There is evidence that activity of this enzyme may be controlled in part by cAMP (see below). Competition and inhibition experiments employing GM3, GD3, and LacCer as substrates with Golgi vesicles from rat liver pointed to a single GalNAc-transferase catalyzing the formation of GM2, GD2, and GA2 (GgOse$_3$Cer), respectively (Pohlentz *et al.*, 1988).

In vitro studies of the above type have indicated several different sialosyltransferases involved in the synthesis of the ganglio, globo, and neolacto series of gan-

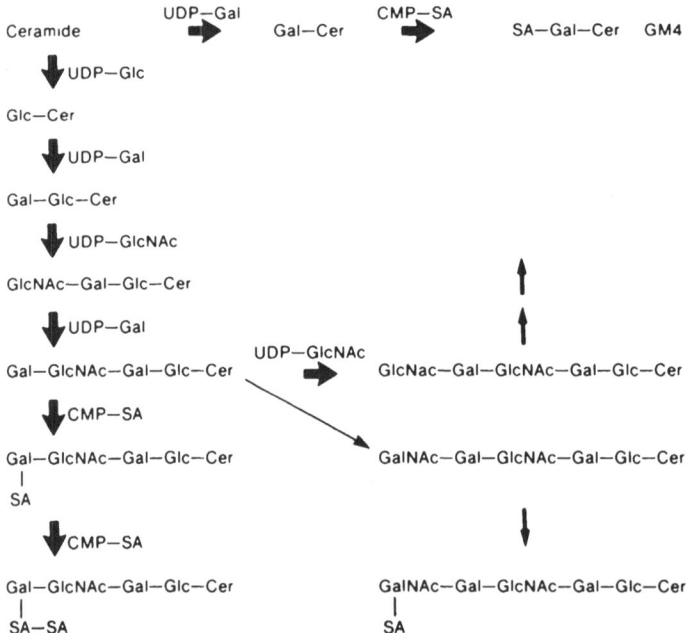

Figure 2. Pathways for biosynthesis of gangliosides of the neolacto series and GM4. Heavy arrows indicate reactions demonstrated *in vitro;* thin arrows indicate proposed reactions not yet observed.

gliosides in a variety of tissues. The enzyme CMP-NeuAc:GM1 sialosyltransferase which converts GM1 to GD1a was shown to differ from that forming GM3 from lactosylceramide in embryonic chick (Kaufman *et al.*, 1967, 1968) and neonatal rat (Ng and Dain, 1977) brain. Biosynthesis of GD3 from GM3 involves another sialosyl-transferase which differs from the above two in regard to subcellular localization (Pacuszka *et al.*, 1978) and enzymatic properties (Kaufman *et al.*, 1967, 1968). This is believed to be the same enzyme which converts IV^3NeuAc-nLcOse$_4$Cer (sia-losylparagloboside) to its disialo analogue (Higashi *et al.*, 1985). Competition and inhibition experiments analogous to those described above for GalNac-transferase were used to demonstrate a single sialosyltransferase as responsible for conversion of GM1, GD1b, and GA1 (GgOse$_4$Cer) to GD1a, GT1b, and GM1b, respectively (Pohlentz *et al.*, 1988).

Several aspects of the neolacto pathway (Figure 2) have been elucidated and virtually all the reactions leading to IV^3NeuAc-nLcOse$_4$Cer have been observed *in vitro* (Basu and Basu, 1982). There was previous indication that the sialosyltransferase involved in synthesis of the latter substance may be the same one forming GM1b via the so-called aminoglycolipid pathway (Yip and Dain, 1969; Handa and Burton, 1969), but recent evidence suggests these are different enzymes (S. Basu, personal communication). The existence of the aminoglycolipid pathway was brought into question recently with the observation (Yanagisawa *et al.*, 1987) that LacCer could not be converted to GA2, but this was attributed as possibly due to poor availability of LacCer to GalNAc-transferase in detergent-containing assays (Pohlentz *et al.*, 1988).

2.2. Site of Biosynthesis

Early subcellular localization studies suggested the nerve ending as the site of ganglioside biosynthesis (for review, see Brunngraber, 1979), but subsequent work has tended to discount that theory. Such studies have indicated that most of the glycosyl-transferase activities detected in synaptosomes and related subfractions resulted from contamination with membranes of the Golgi apparatus and/or endoplasmic reticulum (Reith *et al.*, 1972; Landa *et al.*, 1979). The Golgi apparatus has been shown to play a prominent role in ganglioside biosynthesis in extraneural tissues such as liver (Keenan *et al.*, 1974; Richardson *et al.*, 1977; Yusuf *et al.*, 1983a), spleen (Basu *et al.*, 1976), kidney (Fleischer, 1977), and thyroid (Pacuszka *et al.*, 1978). The fact that gan-gliosides (as well as glycoproteins) were synthesized *in vitro* by a subcellular mem-brane fraction from the neuronal perikarya which resembled the Golgi apparatus both morphologically and biochemically suggested the same may apply to these cells (Landa *et al.*, 1981). Suggestions of compartmentation within the Golgi membranes resulted from the use of monensin to block transport between compartments (Saito *et al.*, 1984).

Definitive evidence that the main locus of ganglioside synthesis is the neuronal perikaryon was provided in studies with the chick optic system, in which the cell bodies are well-separated anatomically from the nerve endings (Landa *et al.*, 1979). The latter structures receive gangliosides through axonal transport (see below). While the primary site of synthesis in the neuron appears to be the Golgi apparatus of the cell body, the possibility of limited synthesis in the nerve ending must still be considered.

For example, a report (Preti *et al.*, 1980) has claimed that two sialosyltransferases exist in synaptic membranes, one acting on LacCer and the other on glycoprotein acceptor(s). Other workers, however, did not find that activity in the nerve ending (Maccioni *et al.*, 1978; Ng and Dain, 1977; Van den Eijnden and van Dijk, 1974) and none of the other glycosyltransferases have been observed there. The possibility of a "sialylation-desialylation cycle" was suggested (Preti *et al.*, 1980), taking account of the presence of sialidase in the same membrane (see below). Arguments against such a cycle have been presented (see below). Ganglioside sialosyltransferase along with other glycosyltransferases have been detected on the cell surface of neurons in culture (Dreyfus *et al.*, 1981; Matsui *et al.*, 1983), but the participation of such cell surface glycosyltransferases in membrane biogenesis is questionable (Keenan and Morre, 1975).

Comparison of cell types revealed neurons to have much greater capability for synthesis of glucosylceramide, the first step in forming ganglioside oligosaccharide chains (Radin *et al.*, 1972; Jones *et al.*, 1972). On that basis it was suggested (Radin *et al.*, 1972) that neurons are the primary site of ganglioside synthesis in brain. Support for that idea has come from the demonstration that two key enzymes required for ganglioside synthesis are virtually lacking in isolated glia: CMP-NeuAc:LacCer sialosyltransferase (Stoffyn *et al.*, 1981) and UDP-GalNac:GM3 *N*-acetylgalactosaminyltransferase (Byrne *et al.*, 1988). The latter enzyme was also found to be deficient in cultured astrocytes (Sbaschnig-Agler *et al.*, 1988). The fact that glial cells, particularly astrocytes, have a relatively high content of ganglioside suggests either that the low levels of synthesis suffice for the needs of the cell or that gangliosides synthesized in the neuron are transferred to the glia (Byrne *et al.*, 1988).

2.3. Regulation of Biosynthesis

Owing to the marked alterations in ganglioside content and composition which accompany various developmental, pathological, and functional changes, efforts are increasingly directed toward elucidation of regulatory processes which govern such changes. Enzymes considered to have key regulatory roles are those at branch points, e.g., those which complete the synthesis of GM3, GM2, and GD3. The activities of these as well as other Golgi-localized enzymes were shown to depend on such factors as lipid environment (Yusuf *et al.*, 1983a), availability of substrates and divalent cations (Yusuf *et al.*, 1983a,b; Sommers and Hirschberg, 1982), phosphorylation-dephosphorylation (Burczak *et al.*, 1984), and feedback control (see below).

Regulatory enzyme activities are often influenced by the biological state of the cell or tissue. Thus, GM3 synthetase, or CMP-sialic acid:LacCer sialosyltransferase, was shown to be elevated severalfold in two cell lines on treatment with retinoic acid (Burczak *et al.*, 1984) and sodium butyrate (Fishman *et al.*, 1974; Macher *et al.*, 1978). Retinoids have the property of causing transformed cells to regain normal growth behavior and show density-dependent inhibition (Patt *et al.*, 1978). The same enzyme was shown to be activated by phorbol ester (Burczak *et al.*, 1983), a treatment known to enhance protein kinase C and subsequent phosphorylation. cAMP, activator of a different protein kinase, was also found to increase GM3 content of cells (Moskal *et al.*, 1974). These results along with additional findings (Burczak *et al.*, 1984) point to regulation of this enzyme by the well-known mechanism of phosphorylation–dephosphorylation.

A parallel situation appears to exist for UDP-GalNAc:GM3 N-acetylgalactosaminyltransferase, and possibly for UDP-Gal:GM2 galactosyltransferase, enzymes which complete the synthesis of GM2 and GM1, respectively. Both of these enzymes were inhibited in opiate-receptor-positive mouse neuroblastoma cells on treatment with β-endorphin or [D-Ala^2,D-Leu^5]enkephalin, with resultant reduction in ganglioside biosynthesis (Dawson et al., 1980; McLawhon et al., 1981). The effect was attributed to opiate-induced reduction in cAMP. Supporting this interpretation was the recent demonstration (Scheideler and Dawson, 1986) of direct activation of the above GalNAc-transferase by cAMP addition to microsomes from neonatal rat brain. Whether the rapid increase observed in the activity of this enzyme (Maccioni et al., 1984a,b) and perhaps others (Dreyfus et al., 1980a), during development, can be attributed to the phosphorylation mechanism or to more basic changes at the nuclear level cannot be answered at present.

Feedback control as a mechanism operating in Golgi-localized synthesis was suggested in the finding (Nores and Caputto, 1984) that the end products GT1b and GD1a were potent inhibitors of GM2-synthetase in detergent-free microsomes from chick retina. Those results were confirmed and extended in a study with pure, intact Golgi vesicles from rat liver, both with and without detergent (Yusuf et al., 1987). Although some cross inhibition was observed, GM2-synthetase was most strongly inhibited by GD1a, and GD3-synthetase (CMP-sialic acid:GM3 sialosyltransferase) by GQ1b. This indicated pathway specificity in that the regulatory steps for the a- and b-series pathways were preferentially inhibited by their respective end products. It was pointed out recently by the authors of that study that it would have been more appropriate to consider the effects of end-product gangliosides on GM3-synthetase (CMP-sialic acid:LacCer sialosyltransferase) rather than GM2-synthetase, since the former is considered the starting enzyme of the a series (Pohlentz et al., 1988).

Another manifestation of feedback control was seen in experiments carried out with the chicken optic system, which indicated that the rate of ganglioside synthesis in the retina and the amount axonally transported to the tectum varies with the physiological state of the system (Caputto et al., 1982). Higher labeling occurred in chickens exposed to light than in those maintained in darkness, while the information causing this regulation in the retina originated in the nerve terminal (Caputto and Caputto, 1986).

Attention has turned in recent years to gene-controlled mechanisms of cellular expression of gangliosides and other glycosphingolipids. This has involved a number of different approaches. Applying genetic linkage analysis to inbred strains of mice possessing distinct ganglioside profiles in erythrocytes and liver, the gene locus that regulates the expression of GM1(NeuGc) was shown to occupy a segment of chromosome 17 in the vicinity of the H-2K locus (Hashimoto et al., 1983). The T/t complex located in proximity to the H-2 locus regulates galactosyltransferase activity (Shur and Bennette, 1978). A very interesting but somewhat speculative approach has viewed the glycosyltransferase gene family as an evolutionary precursor of the immunoglobulin (Ig) gene family, and considers the polymorphism and phenotypic character of glycosyltransferases as analogous to that seen in the immunoglobulins (Roth, 1985). This theory would predict glycosyltransferase activity in some MHC antigens, a phenomenon that was recently observed (Furukawa *et al.*, 1985). Various glycosyltransferases are now being successfully cloned.

Gene transfer studies are providing another useful approach. The first of these (Hakomori and Kannagi, 1983) employed relatively large oncogene-containing DNAs isolated from human lung and bladder carcinoma cell lines which altered the glycosphingolipids of recipient NIH 3T3 cells. More recently, DNA tumor virus- and RNA tumor retrovirus oncogenes transfected into the rat fibroblastic cell line 3Y1 caused pronounced and characteristic changes in ganglioside pattern (Nagai *et al.*, 1986). As an example, the adenovirus transforming gene, E1, and its transcriptional subfragment, EIA, caused neoexpression of GD3 coinciding with enhanced activity of CMP-NeuAc:GM3(2-8)sialosyltransferase. Similar results were obtained with transfection of c-*myc* DNA (Nakaishi *et al.*, 1988a); in both cases the products are expressed intranuclearly. In contrast, transfection with "extranuclear type" oncogenes (e.g., *ras, src, fes, fps*) in which the products are expressed in the cytoplasm or on the cell surface membrane, invariably induced neosynthesis of IV^3NeuAc-nLcOse$_4$Cer with concomitant decrease in GM3 (Nakaishi *et al.*, 1988b). A correlation was observed between levels of GD3 expression and colony-forming activity of *myc*-transformed 3Y1 cells, suggesting a specific role for GD3 in *myc*-induced transformation (Nakaishi *et al.*, 1988a).

3. METABOLISM

3.1. In Vivo Studies

Turnover studies of gangliosides have given quite variable results, due in part to intrinsic differences among systems as well as failure to correct for such factors as precursor reutilization. Younger animals have generally shown faster turnover rates than older ones, and glucose a shorter half-life than other precursors. For example, rat brain studies using glucosamine gave a half-life of 24 days for total gangliosides (Burton *et al.*, 1964; Suzuki, 1967) whereas galactose (Burton *et al.*, 1964) and glucose (Suzuki, 1967) gave values of 20 and 10 days, respectively. The shorter half lives most likely reflected less reutilization of metabolized precursor and are therefore considered closer to the true values. The half-life of 60 days obtained with acetate (Holm and Svennerholm, 1972) indicated extensive reutilization of hydrophobic components. Whole brain studies have generally revealed similar turnover rates for individual gangliosides, consistent with the absence of precursor–product relationships required by the existence of a separate synthesizing complex for each ganglioside (see above). However, use of whole brain can obscure important compartmental differences, as revealed in a study of myelin GM1 which was found to turn over more slowly than GM1 of whole rat brain (Suzuki, 1970).

Focusing on the sialic acid moiety, double-label experiments revealed similar turnover rates for sialidase-sensitive and -resistant NeuAc in brain gangliosides (Maccioni *et al.*, 1971), suggesting a single pool of CMP-NeuAc for synthesis. However, an important consideration came to light in the finding (Ferwerda *et al.*, 1981) that the NeuAc components of free NeuAc, CMP-NeuAc, lipid-bound NeuAc, and protein-bound NeuAc all had similar specific radioactivities a few days after intraventricular injection of labeled ManNAc in the rat. That and the further observation that all four pools lost specific radioactivity at the same rate were interpreted as evidence for active recycling of sialic acid molecules or precursors. It was concluded that the calculated half-life of 3.5 weeks was not a true half-life of brain sialoglycoconjugates but merely the rate of leakage out of the brain of labeled NeuAc and/or precursors. Using the rate of incorporation of labeled ManNAc, the authors calculated approximate half lives of 6–8 and 2–3 days for sialic acid residues of gangliosides and glycoproteins, respectively. The problem of precursor reutilization was also dealt with through use of deuterium labeling and mass fragmentography (Ando *et al.*, 1981).

In vivo metabolic studies with exogenously administered gangliosides have become of interest due to the growing use of gangliosides as therapeutic agents. Gangliosides injected into the circulation bind first to serum albumin (Tettamanti *et al.*, 1981) and then find their way into many tissues. The half-life of gangliosides in blood, measured in hours, was considerably longer than that of neutral glycosphingolipids (Barkai and Di Cesare, 1975). Brain took up small though significant quantities of labeled GM1 from the circulation, much of it occurring in the soluble fraction at 4 hr after injection (Tettamanti *et al.*, 1981; Lang, 1981; Ghidoni *et al.*, 1986a). This was in contrast to liver in which most of the radiolabel was in the particulate fraction.

The metabolic fate of exogenous GM1 entering the liver was shown to involve both degradation and synthesis (Ghidoni *et al.*, 1986b). Detection of radiolabeled GM2, GM3, and neutral glycosphingolipids in the lysosomal fraction reflected degradation, while labeled GD1a and GD1b in the Golgi apparatus fraction indicated synthetic modification. The latter involved both direct sialylation of GM1 and neosynthesis with reutilization of degradation products. Similar conclusions were reached in studies with fibroblasts treated with radiolabeled (Sonderfeld *et al.*, 1985) or spin-labeled (Klein *et al.*, 1988) GM2. It was shown that the exogenous ganglioside inserted itself into the plasma membrane and, following endocytosis, was transported to the lysosomal and Golgi compartments for degradative and synthetic conversions, respectively.

3.2. Catabolism: Localization

Ganglioside degradation proceeds in stepwise fashion with liberation of individual carbohydrates in reverse order to biosynthesis. Most of the enzymes in the catabolic pathway have features that identify them as lysosomal in origin: sedimentation in particulate fractions between 800 and 15,000g, enhanced activity with detergents, and acidic pH optima. Degradative enzymes appear to be present in all cell types of the brain, though opinions differ as to relative activity. Thus, β-galactosidase was described as considerably higher in neuronal cell bodies than in astroglial and oligo-dendroglial fractions of adult rabbit and bovine brain (Freysz *et al.*, 1979). This supported an earlier claim that β-galactosidase is a neuronal marker (Sinha and Rose,

1973). Other workers reported this enzyme to occur prominently in both neurons and glia (Abe *et al.*, 1979; Ragahavan *et al.*, 1972). The discrepancy was attributed to age differences (Freysz *et al.*, 1979) following an earlier demonstration (Arbogast and Arsenis, 1974) of marked developmental changes among several lysosomal hydrolases.

3.2.1. Sialidase

This enzyme has been reported to occur in both lysosomal and nonlysosomal compartments. The latter includes the synaptic membrane of the neuron (Schengrund and Rosenberg, 1970; Tettamanti *et al.*, 1972) and the plasma membranes of cells generally (Schengrund *et al.*, 1976, 1979). Sialidase in synaptic membranes has been shown to catalyze hydrolysis of endogenous gangliosides in the same membrane, as well as exogenous gangliosides (Schengrund and Rosenberg, 1970; Tettamanti *et al.*, 1972). However, hydrolysis of exogenous GD1a by rat brain microsomes was shown to require prior insertion of the ganglioside into the membrane; the loosely attached trypsin-removable pool was not affected (Scheel *et al.*, 1985). Considering the coexistence of sialidase and its natural substrate (ganglioside) in the same plasma membrane, the question arises as to how the latter avoids hydrolysis while in the membrane. One possibility is protective association of gangliosisde with other membrane proteins, an example being GM4 which, in the presence of myelin basic protein, was protected against sialidase (Yohe *et al.*, 1983).

Catabolism of the major oligosialogangliosides of brain commences with sialidase-catalyzed cleavage of terminal sialic acid(s). A particulate enzyme from human brain degraded GT1b preferentially to GD1b rather than GD1a, indicating the sialosylgalactosyl grouping to be more susceptible than sialosylsialosyl (Ohman *et al.*, 1970). However, the fact that the disialosyl grouping of GT1a reacted more rapidly than GD1b suggested that the sluggishness of the latter may result from steric hindrance (Ando and Yu, 1977). Even more pronounced steric hindrance is responsible for the nearly complete resistance of GM1 and GM2 to most animal and bacterial sialidases. Such hindrance has been depicted as resulting from an "oxygen cage" surrounding the ketosidic linkage of NeuAc (Schauer *et al.*, 1980; Harris and Thornton, 1978).

3.2.2. Catabolism of GM1

After sialidase cleavage to GM1, the major catabolic pathway appears to be GM1 → GM2 → GM3. The latter product is fully susceptible to sialidase, the steric barriers having been removed. The first step in the metabolic degradation of GM1 involves β-galactosidase. There are at least two genetically distinct lysosomal β-galactosidases in mammalian brain, one of which hydrolyzes GM1 (and GgOse$_4$Cer) and the other GalCer (Suzuki *et al.*, 1980). Both enzymes can hydrolyze LacCer, although the existence of a separate enzyme remains a possibility. A protein activator of GM1 β-galactosidase was isolated from human liver (S.-C. Li *et al.*, 1979) and subsequently shown to be a nonspecific sphingolipid activator (see below).

Following the action of β-galactosidase, the product (GM2) is further catabolized by *N*-acetyl-β-hexosaminidase to GM3. This lysosomal enzyme catalyzes the hydrolysis of β-GalNAc or β-GlcNac residues from the nonreducing ends of gangliosides, neutral glycolipids, and glycoproteins. It exists as two major isozymes, formed by association of two peptide subunits. Hexosaminidase A is a dimer of two subunits, α and β, which are encoded on different chromosomes (15 and 5, respectively) but are structurally very similar. Hexosaminidase B is a homodimer of two β subunits (for recent reviews, see Mahuran *et al.*, 1985; Sandhoff *et al.*, 1989). A dimer of α subunits ("hexosaminidase S") is usually also present in trace amounts but is of unknown physiologic relevance. Each of the two subunits carries an active site but with different substrate specificities (Kytzia and Sandhoff, 1985). Both hexosaminidase A and B act on glycoproteins whereas only hexosaminidase A is able to degrade GM2.

The latter reaction requires an activator protein (Y.-T. Li *et al.*, 1973) which is different from that activating β-galactosidase (S.-C. Li *et al.*, 1979). These activator proteins, situated within the lysosome, appear to function by removing single sphingolipid molecules from the lysosomal membrane and presenting them to the hydrolytic enzyme as a 1 : 1 complex (Conzelmann *et al.*, 1982). Two such protein cofactors are known to play a physiologic role: one is specific for the hydrolysis of GM2 and related glycolipids by hexosaminidase A ("GM2 activator"), the other is less specific with respect to enzyme as well as substrate and promotes—at least *in vitro* —a number of reactions. It is generally referred to as the "sulfatide activator" or "SAP-1"; it is also the activator of GM1 β-galactosidase. (For a recent review, see Furst *et al.*, 1986.) Recent results indicate that this latter, nonspecific activator is encoded as a large precursor (Fujibayashi and Wenger, 1986) which after biosynthesis is proteolytically cleaved to yield three different proteins (Furst *et al.*, 1988). An activator exhibiting the properties of a nonspecific natural detergent has been described (S.-C. Li *et al.*, 1988).

Recently a unique type of β-hexosaminidase was isolated from roe of striped mullet which cleaves GalNAc from GM2 without the assistance of either activator protein or detergent (DeGasperi *et al.*, 1988). It was reactive toward the oligosaccharide of GM2 but not neutral glycolipids such as GgOse$_3$Cer or GbOse$_4$Cer.

3.2.3. Gangliosidoses

Ganglioside storage diseases are a group of metabolic disorders caused by deficiencies in activity of one of the above hydrolases (see also Chapter 15). Profound deficiency of GM1 β-galactosidase was shown to be the genetic origin of "generalized gangliosidosis" (Okada and O'Brien, 1968), later named GM1 gangliosidosis. Accumulation of GM1 could have different origins, e.g., a structural gene mutation involving the synthesis of a protein which has greatly decreased enzyme activity although being present at near-normal levels (Norden and O'Brien, 1975).

The several forms of GM2 gangliosidosis result from whole or partial failure to express β-hexosaminidase. These have been attributed to a variety of mutations (for reviews, see Sandhoff and Christomanou, 1979; Sandhoff and Conzelmann, 1984; Dawson *et al.*, 1986). Classical Tay–Sachs disease (GM2 gangliosidosis, Type B) involves an α-locus defect with loss of hexosaminidase A activity. Since the unaffected hexosaminidase B is unable to metabolize GM2 or GA2, these progressively accumulate in the nervous system. Type O (Sandhoff disease) involves a β-locus defect with resulting loss of both hexosaminidase A and B activities. In addition to neuronal storage of GM2 and GA2, patients with this variant store globoside in extraneuronal tissues. Residual hexosaminidase is attributed to hexosaminidase S. Another category of mutation is seen in Type AB wherein the defect lies in the GM2 activator protein (Conzelmann and Sandhoff, 1978). This prevents metabolism of GM2 and GA2 despite the full presence of hexosaminidase A and B activities.

Many variations of the Type B condition are now known which involve defects at the α locus. These include failure to synthesize α chains (Myerowitz and Proia, 1984; Myerowitz *et al.*, 1985), synthesis of insoluble α chains (Proia and Neufeld, 1982), and synthesis of α chains which fail to associate with β chains (d'Azzo *et al.*, 1984). Synthesis of labile α chains is another possibility (d'Azzo *et al.*, 1984). An interesting mutation is the B1 variant in which hexosaminidase A appears normal catalytically when tested with conventional chromogenic or fluorogenic substrates, but which is unable to hydrolyze GM2 or the artificial substrate 4-methylumbelliferyl *N*-acetylglucosamine 6-sulfate (Kytzia *et al.*, 1983). The nucleotide sequence of a cDNA clone of the α chain was found to be completely normal except for a single base substitution at No. 533 (Ohno and Suzuki, 1988a). Gene analysis of the classical Ashkenazi Jewish form of the disease revealed a single nucleotide transversion at the 5′ donor site of intron 12 from the normal G to C (Ohno and Suzuki, 1988b; Arpaia *et al.*, 1988). Hence, this junctional mutation was shown to result in functional abnormality.

While the above examples are generally lethal infantile forms of α-locus defects, a large number of more benign juvenile and adult-onset forms are known (Sandhoff and Christomanou, 1979; Navon *et al.*, 1986). These are characterized by partial deficiency of hexosaminidase A or a catalytically defective enzyme. A number of these variants are expressed clinically as motor neuron disease, which has led to the suggestion (Dawson *et al.*, 1986) that motoneurons might be particularly susceptible to this type of impaired ganglioside metabolism.

3.3. Transport and Transfer

3.3.1. Axonal Transport: CNS

The problem of logistics is particularly acute for the neuron with its enormous network of processes. Gangliosides share a common mechanism with other membrane components in being synthesized in the cell body and conveyed by fast axonal transport to axonal and nerve-ending membranes (see also Chapter 9). This rate has been estimated at 70–100 mm/day in the goldfish and approximately four times that in mammals. Dendritic gangliosides are presumed to arrive by a similar mechanism, although this has not been directly demonstrated. Virtually all the ganglioside transport studies of the CNS have employed the optic system of one species or another. Beginning with the goldfish (Forman and Ledeen, 1972), it was shown that intraocular injection of radiolabeled ganglioside precursors, such as glucosamine and *N*-acetylmannosamine, gave rise to radiolabeled gangliosides in the retina which were translocated to the contralateral optic tectum. Similar studies were subsequently carried out with the optic systems of fish (Rösner et al., 1973), chick (Rösner, 1975; Landa et al., 1979), rabbit (Ledeen et al., 1981), and rat (Gammon et al., 1985). The possibility of local modification of ganglioside oligosaccharide chains within the axon or nerve ending has been considered (Rösner et al., 1973; Ledeen et al., 1976), although no firm evidence for this has been presented. Such a mechanism would require axonal transport of precursor(s) (e.g., nucleotide sugars), and a recent study (Igarashi et al., 1985) has claimed that CMP-sialic acid undergoes such transport at an intermediate rate. However, the possibility of extraaxonal diffusion (Haley et al., 1979) as an alternative explanation was not ruled out.

In a study employing the rabbit optic system (Ledeen et al., 1981), it was shown that all gangliosides undergo transport simultaneously and that the ganglioside pattern characterizing this one type of CNS neuron is similar to that for brain as a whole. The results further showed that axons do not simply serve as channels for the flow of gangliosides to the nerve endings, but are themselves targets of transport. The axonal and nerve-ending membranes thus appeared to behave as a unit in the uptake and turnover of gangliosides. Similar conclusions were reached in studies employing the chick (Rösner and Merz, 1982) and rat (Gammon et al., 1985) visual systems. The latter study demonstrated that this behavior of gangliosides was in contrast to that of glycoproteins and proteoglycans, which were transported primarily to the nerve terminal-containing structures.

3.3.2. Axonal Transport: PNS

Recent work has demonstrated a basic similarity between CNS and PNS in regard to ganglioside flow. Fast anterograde transport of these substances was shown to occur in rat sciatic nerve (Yates et al., 1984; Aquino et al., 1985, 1987; Harry et al., 1987); the velocity was estimated at approximately 300–400 mm/day. As with the CNS, all molecular species appeared to migrate simultaneously. An interesting difference in the outflow patterns of gangliosides and glycoproteins, following labeling of both by injection of glucosamine into the dorsal root ganglia, was that gangliosides did not show the well-defined crest of radioactivity characteristic of glycoproteins, but rather a pattern with an attenuated crest in a series of rather flat curves representing different times (Harry et al., 1987). This was interpreted to indicate extensive exchange of gangliosides between mobile and stationary axonal structures, in contrast to glycoproteins which were targeted primarily to the nerve ending.

3.3.3. Retrograde Axonal Transport

One advantage of the PNS is the possibility presented by long nerves to study retrograde as well as anterograde transport. Using the double-ligation model (Bisby and Bulger, 1977), this phenomenon was demonstrated for gangliosides in both motoneurons (Aquino et al., 1985) and sensory neurons (Aquino et al., 1987) of rat sciatic nerve. The velocity of such transport could not be calculated directly, but the early return of labeled gangliosides seemed consistent with the relatively rapid velocities (equivalent to one-half anterograde flow) previously estimated for other substances.

Comparison of the pools migrating in the two directions by fractionation of isolated gangliosides according to sialic acid number revealed generally similar patterns. Mono-, di-, tri-, and tetrasialogangliosides were present in approximately the same proportions, not only in terms of direction but also in regard to motor versus sensory axons (Aquino et al., 1985, 1987). These patterns were not grossly different from those of brain. However, more detailed investigation revealed some differences in relation to individual structures (Ledeen et al., 1987). Using sialidase to distinguish between gangliotetraose structures (which give rise to GM1) and other ganglioside families (which produce mainly neutral glycolipids), it was found that motoneurons contain predominantly gangliotetraose species while sensory neurons have less of this type and more belonging to the other families. Sensory neurons also contained neutral glycosphingolipids, undetectable in motoneurons, and these were transported in synchrony with the gangliosides.

The fact that the pattern of anterograde gangliosides was essentially the same as that of the retrograde pool—in both motor and sensory neurons—suggested that there were no major metabolic alterations of gangliosides during their sojourn in the axon and nerve ending. If true, this raises the question of the functional role of such enzymes as sialidase and sialosyltransferase, both of which were reported to be present in the synaptic membrane (see above). The question of possible metabolic or synthetic processing of gangliosides in the axon or nerve ending must accordingly remain open for the present; if it occurs at all, it very likely affects only a minor component.

3.3.4. Ganglioside Transfer Proteins

Transfer proteins, capable of catalyzing exchange and/or transfer of gangliosides between membranes, have been discovered in brain (K. Yamada et al., 1985; Brown et al., 1985; Gammon et al., 1987). These are low-molecular-weight proteins ($M_r \sim$ 20,000) which clearly differ from the GM2 activator protein (see above); the latter was also shown to have transfer activity in vitro (Conzelmann et al., 1982). They are capable of transferring neutral glycolipids as well as gangliosides, but showed no activity toward phospholipids.

The function of such proteins in brain poses an intriguing question. If one assumes this role to be intracellular, a potential problem arises in that glycoconjugates are thought to be sequestered within the luminal portion of those intracellular organelles which contain them, e.g., transport vesicles and the Golgi apparatus. Since transmembrane movement (''flip-flop'') of glycolipids is considered unlikely, gangliosides so situated would not be accessible to catalyzed transfer within the cytosol. (However, see Chapter 3 concerning cytoplasmic glycoconjugates.) One cannot preclude the existence of vesicles or organelles with gangliosides on the cytoplasmic surface, nor the possibility of a role within organelles, e.g., the Golgi matrix. Such possibilities are better assessed once the localization of transfer proteins becomes known. Should they turn out to be extracellular, a possible role in intercellular transfer would need to be considered.

4. FUNCTION

Most of the early attempts to elucidate ganglioside function in brain focused on the excitability of nervous tissue *in vitro* (McIlwain, 1960, 1961). Brain slices which responded to electrical pulses or potassium depolarization with increased oxygen uptake lost this response in the cold or after addition of protamine or histones, but regained it after addition of exogenous gangliosides to the medium. It was proposed that native gangliosides offer acidic sites that function in active cation transport. Although subsequent findings cast some doubt on that interpretation (Evans and McIlwain, 1967; Yogeeswaran *et al.*, 1973), the basic idea of a role for gangliosides in ion transport still warrants consideration. One proposal in that vein postulates interaction with calcium in a manner which modulates neurotransmitter release (Rahmann, 1983, 1987) while another invokes ganglioside facilitation of electrogenic Na$^+$-pump activity (Vyskocill *et al.*, 1985).

The search for a functional role of gangliosides, particularly in the neuron, has given rise to several modes of investigation. One is to study in a correlative manner the changes which occur in ganglioside content, composition, and turnover as a living system matures or alters its functional state. For example, the pronounced increases in ganglioside content and structural complexity which correlate with neuronal differentiation and synaptogenesis (Dreyfus *et al.*, 1980b; Panzetta *et al.*, 1980; Yates, 1986) have suggested a special role for gangliotetraose-type gangliosides in that phase of neuronal development. A widely used paradigm is addition of exogenous gangliosidse to cultured cells or membrane preparations; by this means it has been possible to further examine the role of gangliosides in promoting differentiation of primary neurons and neuroblastoma and to study their effects on specific enzyme and receptor systems. Another approach is use of interventive agents, such as antibodies or toxins, which in some cases have caused revealing changes in cell behavior. In attempting to interpret these diverse findings in molecular terms, attention is increasingly directed to specific proteins which appear to require the presence of one or more specific gangliosides for optimal functioning. This recalls the proposed dynamic annular model wherein gangliosides (and/or other glycolipids) are thought to comprise part of the closely associated ring of annular lipids believed to form microdomains around many membrane proteins (Lee, 1977; Yamakawa and Nagai, 1978). In keeping with this idea, spin-labeled gangliosides incorporated into both artificial and natural membranes, were shown to cluster around and bind to glycoproteins in a reversible manner (Sharom and Grant, 1978). As discussed above, protein modulation might thus be hypothesized as a general function of membrane glycolipids.

4.1. Gangliosides and Receptors

Receptor activity has long been considered a likely function of gangliosides, owing to the example of GM1 as a well-delineated and highly specific receptor for cholera toxin (for a review, see Fishman, 1982). That finding gave rise to a search for other examples with more biological relevance, but aside from some additional toxins, few cases have come to light in which gangliosides act alone as membrane receptors. The bacterial toxins so identified include tetanus (van Heyningen, 1984; Rogers and

Snyder, 1981), botulinum (Kitamura *et al.*, 1980), *E. coli* enterotoxin (Moss *et al.*, 1981), staphylococcal α-toxin (Kato and Naiki, 1976), and *Vibrio parahaemolyticus* hemolysin (Takeda *et al.*, 1975). Tetanus has been employed as a marker for neurons in primary cultures (Dimpfel and Habermann, 1977; Mirsky *et al.*, 1978), but it is not yet clear whether gangliosides or glycoproteins (or a combination of both) constitute the physiological receptor for this toxin (Critchley *et al.*, 1986). Various gangliosides have been implicated as receptors for Sendai virus (Markwell *et al.*, 1981) but here again glycoproteins are also alleged to have a role on the basis of protein blot analysis (Gershoni *et al.*, 1986) and sialoglycoprotein insertion into neuraminidase-treated erythrocytes (T. Suzuki *et al.*, 1984).

4.2. Receptor Modulators

4.2.1. Modulation of Serotonin Receptor

While the evidence argues against gangliosides acting independently as molecular receptors in brain and other tissues, there is a growing indication that they function as cofactors for a number of protein receptors whose activities are thereby modulated. One study (Berry-Kravis and Dawson, 1985) reported that exogenous gangliosides induced a tenfold increase in the affinity of serotonin for the $5HT_1$ receptor of NCB-20 cells and thereby modulated the coupling between this receptor, the G protein, and the adenylate cyclase complex. GQ1b proved the most effective. Using an *in vivo* paradigm, peripherally administered gangliosides caused modulation of serotonin receptor function in normal rat brain (Agnati *et al.*, 1983).

4.2.2. Modulation of Growth Factor Receptors

It has been recognized for some time that gangliosides are likely to have a role in regulating cell proliferation (Hakomori, 1970; Langenbach and Kennedy, 1978). This activity has a bimodal aspect, depending on whether the cells are quiescent or in a proliferative state (Spiegel and Fishman, 1987). Recent evidence suggests that this may be accomplished through modulation of certain growth factor receptors. Exogenous GM3, and to a lesser extent GM1, inhibited growth of BHK cells as stimulated by fibroblast growth factor (Bremer and Hakomori, 1982). The same two gangliosides inhibited growth of Swiss 3T3 cells via the platelet-derived growth factor (PDGF) receptor (Bremer *et al.*, 1984). GM1 was more effective than GM3 while $IV^3NeuAc-nLcOse_4Cer$ and $GbOse_4Cer$ had no effect. In becoming refractory to such stimulation, the cells showed an altered cell surface affinity to PDGF along with greatly reduced tyrosine phosphorylation of the PDGF receptor; however, the number of cell surface receptors remained the same. A similar phenomenon was observed with the epidermal growth factor (EGF) receptor of the KB and A431 human epidermoid carcinoma cell lines which responded to exogenous GM3 (and to a lesser extent GM1) with inhibited growth (Bremer *et al.*, 1986). GM3 caused inhibition of EGF-stimulated phosphorylation of the receptor in membrane preparations of both cells without affecting the binding of EGF to its receptor. It was of interest that both the EGF-dependent receptor phosphorylation (on tyrosine) and its inhibition by GM3 could be observed with partially purified EGF receptor and with membranes isolated from A431 cells that had been cultured in GM3-containing medium. An "allosteric regulator" model was proposed (Bremer and Hakomori, 1984) in which EGF binding to its receptor would be allosterically regulated in relation to tyrosine phosphorylation through binding of gangliosides to specific sites. It was suggested that such binding might inhibit receptor–receptor interaction.

In consideration of the above effects of GM3 and the metabolic behavior of this molecule, in which sialic acid was observed to lose 35% of its label without any measurable loss in the ceramide moiety, it was postulated that sialidase-catalyzed removal of sialic acid from GM3 on the cell surface is part of the mechanism for modulation of cell growth (Usuki *et al.*, 1988a). Such removal was viewed as relieving the inhibition to growth factor-mediated stimulation of cell proliferation caused by proximity of GM3 to the receptor. That *in situ* hydrolysis of cell surface GM3 by extracellular or membrane-bound sialidase might have a role in such regulation was supported by the observation that 2-deoxy-2,3-dehydro-*N*-acetylneuraminic acid, a potent inhibitor of sialidase, caused marked suppression of fibroblast cell growth in a concentration-dependent manner (Usuki *et al.*, 1988b).

4.2.3. Modulation of Cell Adhesion Receptors

Involvement of gangliosides in cell–substratum adhesion was suggested by the ability of GD1a and GT1b to inhibit fibronectin-mediated attachment of cells to collagen-coated substrates and cell adhesion to fibronectin–collagen complexes (Kleinman *et al.*, 1979). Subsequently, the same two gangliosides were shown to inhibit spreading of BHK cells on fironectin-coated tissue culture flasks (Perkins *et al.*, 1982). In experiments with the NCTC 2071A ganglioside-deficient cell line which was able to synthesize but not bind fibronectin, it was shown that such binding could be effected after the cells had taken up exogenous gangliosides (K. H. Yamada *et al.*, 1983). Again, GT1b and GD1a were more active than the others. Using cultures of fibroblasts which had taken up fluorescent-labeled ganglioside from the medium, it was demonstrated that gangliosides codistributed with the fibrillar networks of fibronectin associated with the cells (Spiegel *et al.*, 1984, 1985).

The direct interaction of gangliosides with fibronectin is weak (Perkins *et al.*, 1982), and the fact that variants of BALB/c 3T3 cells, which were believed to lack appreciable complex gangliosides, could organize a fibronectin matrix comparable to that of the parent cells seemed to argue against such a role for gangliosides (Griffiths *et al.*, 1986). However, it was pointed out that very low levels of complex gangliosides in the cell, which may require highly sensitive methods for detection, can suffice to produce the effect (Fishman, 1986).

The matter was recently clarified somewhat by use of a variant of the NCTC 2071A line which has detectable amounts of complex gangliosides such as GD1a and GT1b on the surface (Spiegel *et al.*, 1986). In contrast to the parent cells, this variant was able to retain and organize fibronectin into a fibrillar extracellular matrix. When the variant cells were treated with sialidase and the B subunit of cholera toxin, the surface fibronectin was lost. Hence, membrane gangliosides would appear to be essential for fibronectin organization.

Another approach which indicated a role in adhesion was the use of anti-GD2 and anti-GD3 monoclonal antibodies to inhibit attachment and spreading of human melanoma cells on a number of extracellular matrix proteins including fibronectin, vitronectin, collagen, laminin, and Arg-Gly-Asp-containing synthetic peptides (Cheresh *et al.*, 1986). The latter sequence is common to a variety of adhesive proteins, such as those mentioned above, which are recognized by a class of divalent cation-dependent cell surface receptors. Use of the antibody as an ultrastructural tool revealed that GD2, the major ganglioside of M21 melanoma cells, was concentrated on microprocesses emanating from the surface and making direct contact with a fibronectin substrate (Cheresh and Klier, 1986). In the presence of Ca^{2+}, this ganglioside colocalized with the vitronectin receptor on the surface of the M21 cells and their focal adhesion plaques (Cheresh *et al.*, 1987). It also copurified with this receptor on an affinity column containing a peptide with the Arg-Gly-Asp sequence. Reconstitution experiments indicated a Ca^{2+} requirement for receptor recognition of the peptide; significant binding occurred without GD2 but the latter together with Ca^{2+} produced optimal binding (Cheresh *et al.*, 1988). Enrichment of specific gangliosides in substratum adhesion sites of certain cells, including neuroblastoma, is further indication of a role in cell adhesion (Mugnai *et al.*, 1984).

In the nervous system, laminin is important in cell-adhesion processes and hence it is significant that this protein was reported to bind to gangliosides, especially GD1a (Laitinen *et al.*, 1987). It was concluded that the adhesive and neurite-promoting effect of laminin is dependent on its interaction with gangliosides at the neuronal surface. In studies of retinotectal specificity, a ganglioside or related glycosphingolipid was implicated in the preferential adhesion of chick retinal cells to the surfaces of intact optic tecta (Marchase, 1977). The adhesion was inhibited by liposomes containing GM2, which was proposed as the recognition molecule. The complementary adhesive protein was postulated to be the enzyme UDP-Gal:GM2 galactosyltransferase, and a double-gradient model involving these two molecules was proposed to account for retinotectal specificity (Roth and Marchase, 1976; Marchase, 1977). Ganglioside-specific adhesion to neural retina cells was also observed in another assay (Blackburn *et al.*, 1986) in which GM2 (along with GD3 and GD1a) supported the highest level of adhesion.

4.3. Enzyme Modulation

The ability of exogenous gangliosides to profoundly alter the activities of several membrane-bound enzymes suggests another form of protein modulation by these glycolipids. For example, adenylate cyclase of rat cerebral membranes experienced 50–95% activation by such treatment, whereas membrane-bound cyclic nucleotide phosphodiesterase was not affected (Partington and Daly, 1979; Davis and Daly, 1980). However, the soluble form of the latter was activated. With Na^+,K^+-ATPase of brain membranes, addition of gangliosides enhanced activity by 26–43% (Leon *et al.*, 1981). The four major gangliosides were equally effective, but other amphipathic lipids and free sialic acid were without effect. The process was concentration dependent and appeared to require stable insertion of ganglioside into the membrane bilayer. This in turn required low concentration, perhaps explaining why other studies carried out at higher concentrations led to inhibition (Caputto et al., 1977; Jeserich et al., 1981). A recent study employing spin-labeled derivatives of gangliosides incorporated into Na^+,K^+-ATPase-rich membranes from Squalus acanthias revealed little selectivity in the lipid–protein interaction relative to spin-labeled phosphatidylcholine for derivatives of GM1, GM2, and GM3, but a small selectivity for GD1b (Esmann et al., 1988). The latter selectivity was considerably smaller than that for other negatively charged lipids.

The most extensive studies have been carried out with phosphorylating enzymes, some of which are markedly altered either positively or negatively by gangliosides. The effect on phosphorylation of the EGF receptor was cited above. In a study with microsomal membranes of rat brain, exogenous gangliosides in the presence of Ca^{2+} stimulated phosphorylation of a number of proteins of 45,000 and higher molecular weight (Goldenring et al., 1985). GD1a was the most potent of the gangliosides tested, whereas asialo-GM1, cerebroside, and two acidic phospholipids were without effect. At the same time, this treatment inhibited phosphorylation of other proteins, two of which comigrated with rat myelin basic proteins. A specific study of protein kinase C (Kreutter et al., 1987) revealed that this enzyme was suppressed by individual brain gangliosides and a mixture thereof, such inhibition occurring with myelin basic protein (MBP) as substrate. Essentially the same result was obtained when myelin itself was employed as the source of protein kinase C (Kim et al., 1986). In that case, phosphorylation of MBP was suppressed most effectively by the oligosialogangliosides which, although less abundant than GM1 in myelin, are present in clearly detectable amounts (Cochran et al., 1982). Such species may therefore assert a regulatory role in the endogenous phosphorylation of MBP (Murray and Steck, 1986).

Similar inhibition of MBP phosphorylation was observed on treatment of guinea pig brain myelin with oligosialogangliosides, although phosphorylation of isolated MBP by protein kinase C was claimed to be stimulated by gangliosides provided phosphatidylserine was present (Chan, 1987a). Guinea pig synaptosomal membranes responded to gangliosides with enhanced phosphorylation of a major phosphoprotein of apparent M_r of 62,000 in addition to others (Chan, 1987b). In this system, however, Ca^{2+} was found to have a minimal effect. A novel protein kinase activated directly by gangliosides was partially purified from particulate fractions of brain. This non-Ca^{2+}-requiring preparation could undergo ganglioside-stimulated autophosphorylation of a major phosphoprotein with M_r of 68,000 and also catalyze phosphorylation of exogenous substrates. The same isolation procedure gave rise to another protein kinase that was inhibited by gangliosides (Chan, 1988).

The question naturally arises whether such ganglioside-mediated effects are physiologically significant, considering the possibility of crucial differences in topography. Since protein kinases are generally located intracellularly, one would have to postulate that the relatively small pool of intracellular gangliosides, which includes soluble components (Ledeen et al., 1976; Sonnino et al., 1981), is the active pool, or alternatively, that the relevant kinases are situated on the cell membrane in proximity to the majority of gangliosides. Ecto-protein kinases of this type have been described on the surface of neural cells (Ehrlich *et al.*, 1986) and spermatozoa (Halder and Majumder, 1986). A preliminary report (Nagai and Tsuji, 1988) presented evidence for such a kinase on the plasma membrane of human neuroblastoma (GOTO) cells which is specifically activated by GQ1b. This ganglioside was also selective in triggering differentiation of the same cells (see below). A protein kinase with similar properties, believed to represent a new type (termed Gg kinase), had previously been detected in the plasma membranes of GOTO cells (Tsuji *et al.*, 1985). In that case the substrate specificity was rather broad, in comparison to whole cells in which proteins of M_r of 64,000 and 60,000 were preferred substrates (Nagai and Tsuji, 1988). As to the source of extracellular ATP, it was suggested (Ehrlich *et al.*, 1986) that this could arise from ATP secretion in association with neurotransmitter release at neuromuscular junctions and certain synapses. The idea of cell surface kinases modulated by gangliosides is an attractive model which, if verified, would point to an aspect of protein–ganglioside interaction with far-reaching physiological consequences.

Recent reports (Hanai *et al.*, 1987; Hannun and Bell, 1987) describing modulation of protein kinases by lysogangliosides and other lysophingolipids project another possible mode of ganglioside influence on phosphorylation. Lyso-GM3 inhibited tyrosine phosphorylation of the EGF receptor while de-*N*-acetyl-GM3 strongly promoted the same reaction (Hanai *et al.*, 1987). Inhibition of protein kinase C by lysoglycosphingolipids was suggested to provide the missing biochemical link between sphingolipid accumulation and the pathogenesis of the sphingolipidoses (Hannun and Bell, 1987). However, firm evidence for the presence of lysosphingolipids in normal tissues has yet to be provided, although lyso-GM2 was found in brain tissue of patients with GM2 gangliosidosis (Neuenhofer *et al.*, 1986; Rosengren *et al.*, 1987). Another interesting though still speculative proposal was that gangliosides may be a source of the second messenger-producing fatty acid, arachidonate, based on preliminary evidence for incorporation of this acid into gangliosides and other glycolipids and its extralysosomal release in neural cells (Dawson and Vartanian, 1988).

4.4. Neuritogenic and Neuronotrophic Activity

The pronounced changes in ganglioside content and pattern that occur during brain development have focused interest on the role these substances play in neuritogenesis and synaptogenesis. When these processes were observed *in vivo*, ganglioside expression was found to change in parallel with neuronal differentiation (Willinger and Schachner, 1980; Koulakoff *et al.*, 1983). Essentially the same phenomenon was observed with primary neuronal cultures, such as those from embryonic rat (Yavin and Yavin, 1979; Seifert and Fink, 1984) and chick (Dreyfus *et al.*, 1980b) brain. Certain neuroblastoma cell lines behave similarly when caused to differentiate (Dawson, 1979). Cultured neurons from embryonic chick cerebral hemispheres were found to have two periods of ganglioside accumulation: a first phase corresponding to cell division in which ganglioside content increased slightly, and a second period corresponding to cell maturation in which all the major CNS gangliosides accumulated while GD3 declined (Dreyfus *et al.*, 1980b). The gangliosides which increased belonged primarily to the gangliotetraose family, suggesting a special role for this structural type in differentiation (see below).

4.4.1. Activity of Exogenous Gangliosides: *In Vitro* Effects

An interesting property in relation to neurons came to light with the discovery that exogenously administered gangliosides have neuritogenic and neuronotrophic properties which are manifested *in vivo* as well as *in vitro*. This line of work followed the demonstration (Purpura and Suzuki, 1976; Walkley *et al.*, 1981) that mature neurons in gangliosidosis brains produce aberrant secondary neurites with occasional synapses from regions of the cell (e.g., meganeurites) that are laden with stored ganglioside. The effect of exogenous ganglioside on cell cultures was initially demonstrated with the B104 (Morgan and Seifert, 1979) and neuro-2A (Roisen *et al.*, 1981; Dimpfel *et al.*, 1981) neuroblastoma cell lines, which, under favorable circumstances, develop prolific neurite outgrowth upon addition of single or mixed gangliosides to the culture medium. Several other neurally derived clonal lines behave similarly (Table 1), although the neuritogenic potential of such cells varies considerably. In contrast to the rather restricted structural requirements indicated for endogenous gangliosides (see below), those pertaining to exogenous gangliosides are quite broad for many systems. It was shown, for example, that all of 11 different gangliosides tested had neuritogenic activity toward neuro-2A cells grown in the presence of serum (Byrne *et al.*, 1983). Furthermore, a number of synthetic sialoglycolipids, including some with β-linked sialic acid which cannot be metabolized, were shown to cause the same neuritogenic response as natural gangliosides in various cells (Cannella *et al.*, 1988b,c; Tsuji *et al.*, 1988). Cells showing this kind of broad specificity generally require ganglioside concentrations in the range of 5 to 500 μg/ml, which has been termed "M type" response; this was contrasted with an "N type" response in which the required ganglioside concentration is an order of magnitude lower (Nagai and Tsuji, 1988).

The primary examples of the latter type are the GOTO and NB-1 human neuroblastoma cell lines (Tsuji *et al.*, 1983). In addition to requiring less ganglioside, more stringent structural specificity was shown in that only GQ1b was active. The fact that the oligosaccharide moiety of GQ1b was also active, albeit at considerably lower potency, together with the observation that this oligosaccharide could inhibit the action of GQ1b at very low concentrations, suggested involvement of a receptor-mediated mechanism (Nakajima *et al.*, 1986). This in turn inferred cell–cell interaction as an important determinant in the differentiation and development of these cells. Of possible relevance to this proposal was a recent study indicating the presence of lectinlike receptor(s) for ganglioside(s) on neural membranes (Yasuda *et al.*, 1988).

In regard to the physiological significance of ganglioside-induced neuritogenesis in neuroblastoma cells, it is of considerable interest that in addition to enhancing neurite outgrowth, exogenous gangliosides also induced formation of mature synapse-like contacts in neuro-2A cells that had been pretreated with GABA or sodium bromide (Spoerri, 1983). The morphology of those structures was described as resembling the mature postsynaptic thickening found in Gray Type I synapses.

A number of primary neuronal cultures have similarly responded to gangliosides, beginning with dorsal root ganglia (Roisen *et al.*, 1981) and including a number of preparations from the CNS as well as the PNS (Table 1). GM1 was shown, for example, to promote neurite outgrowth in explant cultures from E8 chicken dorsal root ganglia (Roisen *et al.*, 1981; Leon *et al.*, 1984; Skaper and Varon, 1985). This

Table 1. Studies of Ganglioside-Induced Differentiation in Vitro

Cell source	Reference
Neuroblastoma and PC12 cell lines	
B104	Morgan and Seifert (1979)
Neuro-2A	Roisen et al. (1981)
	Dimpfel et al. (1981)
	Leon et al. (1982)
	Byrne et al. (1983)
	Spoerri (1983)
	Leskawa and Hogan (1985)
	Matta et al. (1986)
	Tsuji et al. (1988)
	Cannella et al. (1988a,b,c)
PC12	Ferrari et al. (1983)
	Katoh-Semba et al. (1984)
	Matta et al. (1986)
	Cannella et al. (1988a,b,c)
GOTO, NB-1	Tsuji et al. (1983)
SB21B1	Rybak et al. (1983)
Primary neuron cultures	
Dorsal root ganglia (chick)	Roisen et al. (1981)
	Leon et al. (1984)
	Doherty et al. (1985a,b)
	Skaper and Varon (1985)
	Skaper et al. (1985)
	Cannella et al. (1988b,c)
Spinal root ganglia (guinea pig)	Hauw et al. (1981)
Ciliary ganglia (chick)	Skaper and Varon (1985)
	Skaper et al. (1985)
Sympathetic ganglia (chick)	Skaper and Varon (1985)
	Skaper et al. (1985)
Cerebrum (chick)	Dreyfus et al. (1984)
	Massarelli et al. (1985)
Hippocampus, forebrain, striatum (chick)	Skaper et al. (1985)

preparation also responded to other natural gangliosides (Doherty et al., 1985a) as well as synthetic sialoglycolipids (Cannella et al., 1988b,c). Nerve growth factor was not added in the latter studies but was believed to be produced by the explant cells. Additional explants giving similar results included chicken E11 sympathetic ganglia and E8 ciliary ganglia, although these required the concurrent presence of nerve growth factor or ciliary neuronotrophic factor, respectively (Skaper et al., 1985).

An extension of the explant technique has been the use of monolayer neuronal cultures following dissociation of PNS or CNS tissue. This was done, for example, with cerebral hemispheres of 8-day-old chick embryos, ganglioside treatment of which caused increases in cell number, cell body area, primary neurite length, and the sprouting of secondary processes (Massarelli *et al.*, 1985). Dissociated neurons from several other sources were similarly studied: chicken E8 ciliary ganglia, E8 and E15 dorsal root ganglia, E8 cerebral cortex, and rat E18 hippocampus (Skaper *et al.*, 1985). All these systems responded to exogenous GM1 with enhanced neurite outgrowth under selected culture conditions, which included the presence of the appropriate neuronotrophic factor. Other studies have reported similar findings with dissociated neurons from E8–E9 chicken dorsal root ganglia (Leon *et al.*, 1984; Doherty *et al.*, 1985b) and the fetal mouse mesencephalon (Leon *et al.*, 1988). These results are analogous to those obtained with PC12 pheochromocytoma cells which failed to respond to GM1 unless they were first primed with nerve growth factor (Ferrari *et al.*, 1983) or unless naive PC12 cells were treated simultaneously with both GM1 and the factor (Ferrari *et al.*, 1983; Katoh-Semba *et al.*, 1984).

4.4.2. Activity of Exogenous Gangliosides: *In Vivo* Effects

Earlier studies with the denervated rat nictitating membrane gave the first indication that exogenously administered gangliosides can accelerate nervous system repair *in vivo* (Ceccarelli *et al.*, 1976). Since then, many additional reports, employing a variety of CNS and PNS animal models, have confirmed that basic finding. Representative examples of these are given in Table 2. The reported ability of exogenous gangliosides to exert neuritogenic and/or neuronotrophic influence in the CNS was somewhat unexpected, considering the presence of the blood–brain barrier and consequent resistance to entry of polar molecules. However, it has been demonstrated that a small but measurable portion of peripherally administered GM1 does enter the brain (Tettamanti *et al.*, 1981; Ghidoni *et al.*, 1986a) and this level has been claimed to correspond to that required for effective interaction with neurons (Leon *et al.*, 1988). The detailed findings of several animal model studies have been described in recent reviews (Ledeen, 1984; Tettamanti *et al.*, 1986; K. Suzuki, 1987; Ledeen *et al.*, 1988) and will not be further considered here.

4.4.3. Mechanistic Considerations

The intriguing question of mechanism has been approached through consideration of both endogenous ganglioside function and exogenous effects. It is an open question whether and to what extent these two aspects are related. As regards exogenous gangliosides, their influence on neuronal cultures and *in vivo* systems has been compared in some respects to that of defined neuritogenic/neuronotrophic agents such as nerve growth factor. Indeed, gangliosides were recently shown to mimic the latter in preventing retrograde degeneration in the nucleus basalis of the rat, caused by cortical lesions (Cuello *et al.*, 1986). In addition, these two agents acted synergistically when given in combination (Cuello *et al.*, 1987). Several other examples of synergism have been reported (Ferrari *et al.*, 1983; Katoh-Semba *et al.*, 1984; Varon *et al.*, 1986; Roisen *et al.*, 1986), giving rise to the hypothesis that exogenous gangliosides are not trophic agents *per se* but serve to potentiate the activities of specific neuronotrophic factors to which the cells respond (Leon *et al.*, 1984; Dal Toso *et al.*, 1986; Skaper *et al.*, 1988;

Table 2. Studies of Ganglioside-Facilitated Recovery in Vivo

Nervous tissue studied	Reference
Peripheral nervous system	
Sup. cerv. ganglion–nictitating	Ceccarelli *et al.* (1976)
Peroneal nerve–EDL muscle	Caccia *et al.* (1979)
	Gorio *et al.* (1980)
Tail ventral nerves	Norido *et al.* (1981)
Sciatic nerve	Sparrow and Grafstein (1982)
	Verghese *et al.* (1982)
	Marini *et al.* (1986)
	Mengs and Stotzem (1987)
Sciatic nerve–EDL muscle	Kalia and DiPalma (1982)
Sciatic nerve–GN muscle	Kleinbeckel (1982)
Sciatic nerve–soleus muscle	Gorio *et al.* (1983)
Sup. gluteal nerve–gluteus maximus muscle	Robb and Keynes (1984)
Central nervous system	
Optic nerve (goldfish)	Grafstein *et al.* (1982)
Hippocampus	Wojcik *et al.* (1982)
	Karpiak (1983)
	Gradkowska *et al.* (1986)
	Ramirez *et al.* (1987)
Nigrostriatal pathway	Toffano *et al.* (1983)
	Agnati *et al.* (1983b)
	Sabel *et al.* (1984,1985)
	Hadjiconstantinou *et al.* (1985)
	Y. S. Li *et al.* (1986)
	Sabel *et al.* (1985)
Spinal cord (transected)	Bose *et al.* (1986)
Nucleus basalis	Cuello *et al.* (1986,1987)
Cholinergic forebrain nuclei	Casamenti *et al.* (1985)
Serotonin and noradrenalin neurons	
(neurotoxins)	Jonsson *et al.* (1984)
	Kojima *et al.* (1984)
Striatum (MPTP lesion)	Hadjiconstantinou and Neff (1988)
Cerebrum (ischemia)	Tanaka *et al.* (1986)
	Karpiak *et al.* (1987)
Limb buds (regenerating, newt)	Maier and Singer (1984)

Leon *et al.*, 1988). Application of this model to neural repair *in vivo* (Consolazione and Toffano, 1988) rests on the current perception that neuronal function and survival in the adult nervous system are crucially dependent on the presence of endogenous neuronotrophic factors (Appel, 1981; Varon *et al.*, 1984).

It has been proposed (Varon *et al.*, 1988) that exogenous gangliosides assert their effects by acting on the cellular machineries charged with execution of the neuronal behavior rather than earlier steps involving neuronotrophic factor interaction with the cell. It was further suggested that exogenous gangliosides accomplish this by supplementing the cell membrane gangliosides, e.g., by inserting themselves into the membrane and raising a suboptimal level of endogenous gangliosides. The exogenous gangliosides would thus act in concert with the endogenous pool to modulate the same cell functions. This concept utilizes the well-demonstrated fact that a portion of the gangliosides which attach to cells from the culture medium are incorporated as fully functional components of the bilayer (Moss et al., 1976; Sharom and Grant, 1978; Toffano et al., 1980; Kanda et al., 1982). An apparent correlation was drawn between the form and extent of associated GM1 and enhancement of neurite outgrowth in neuro-2A cells (Facci et al., 1984), although it still remains to be rigorously established that insertion into the membrane is a necessary concomitant of neuritogenesis.

The broad spectrum of sialoglycolipid structures which show neuritogenic/neuronotrophic effects on neuroblastoma cell lines and primary neurons would seem to argue against such a mechanism. This mechanism would implicitly require at least some structural homology between the exogenous ganglioside and the endogenous one being supplemented, contrary to what was observed. Synthetic compounds with non-metabolizable (β-linked) sialic acid were fully active toward neuroblastoma cells and primary neurons (Cannella et al., 1988b; Tsuji et al., 1988), as were synthetic glyceroganglatosides (Cannella et al., 1988c; Tsuji et al., 1988). The latter compounds demonstrated that a sphingosine-containing ceramide moiety is not necessary, nor is an oligosaccharide chain since sialic acid was the sole carbohydrate attached to the lipid backbone. It was further established that GM1 derivatives with the negative charge removed were still capable of enhancing neurite outgrowth in neuro-2A cells, dorsal root ganglia, and PC12 cells (Cannella et al., 1988a). These results argue for membrane perturbation of a general type which is able to trigger intracellular changes leading to differentiation. The structural requirement for such agents is rather broad: sialic acid (or a close derivative thereof) attached to a lipid moiety. Sialoproteins had no such effect on neuro-2A cells (Ledeen and Cannella, 1987). Such membrane dislocations would be expected to require fairly substantial concentrations of sialolipids (e.g., 0.1 mM), and hence this mechanism might well differ fundamentally from the "N-type" (Nagai and Tsuji, 1988) as well as in vivo systems which require considerably less.

Ganglioside addition to neuroblastoma cells causes rapid surface changes which could conceivably account for such an effect. With neuro-2A cells, exogenous gangliosides stimulated an immediate and dramatic transformation of the smooth somal membranes into microvillus-covered surfaces (Spero and Roisen, 1984a). This resulted in reorganization of the subcortical microfilaments into filamentous bundles within neurites (Spero and Roisen, 1984b). The latter study demonstrated that neurite outgrowth under microfilament-limiting conditions resulted in reduced neurite branching whereas growth under microtubule-limiting conditions allowed initiation but inhibited elongation. Further evidence for microtubule involvement was indicated by significant and selective increase in tubulin mRNA in SB21B1 cells during ganglioside-induced differentiation (Rybak et al., 1983) and in rat brain after nigrostriatal pathway unilateral lesion and treatment with ganglioside (Yavin et al., 1987).

Rapid surface effects of the above type which trigger internal changes raise the possibility of second messenger formation, and this has now been reported. Phosphoinositide breakdown was enhanced by ganglioside addition to neuro-2A cells (Vaswani and Ledeen, 1988) as well as primary neurons in culture (Skaper *et al.*, 1987); an earlier study showing enhanced inositol incorporation (Ferret *et al.*, 1987) suggested a similar effect. With neuro-2A cells the maximal effect was observed 1 hr after ganglioside addition, suggesting prolonged perturbation of the neuronal membrane rather than a rapid receptor-mediated process. The fact that Ca^{2+} influx was enhanced over the same time course pointed to a possible mechanism involving the sequence: membrane perturbation $\rightarrow Ca^{2+}$ influx $\rightarrow$ phospholipase C activation $\rightarrow$ second messenger formation $\rightarrow$ neuritogenesis (Vaswani *et al.*, 1989). It was of some interest that induction of neurite outgrowth in the same cells by retinoic acid or dibutyryl cAMP did not trigger phosphoinositide breakdown, indicating that the phenomenon was not directly associated with neuritogenesis *per se*. Previous work had revealed that ganglioside-induced differentiation of neuro-2A cells resulted in elevated cAMP (Leon *et al.*, 1982), so that a number of second messengers may be involved.

As regards endogenous gangliosides, it is likely that structural requirements for these are more stringent in view of the special role occupied by gangliotetraose species. This is inferred in the elevation of such structures during differentiation of primary neurons (Dreyfus *et al.*, 1980b) and certain transformed cell lines such as PC12 (Margolis *et al.*, 1983). In a study of several neuroblastoma cell lines which show varying capability for neurite outgrowth, some insight into this variability was gained by comparing the endogenous ganglioside patterns (Cannella *et al.*, 1988a). Those cells most responsive to neuritogenic stimuli were found to have the highest levels of gangliotetraose gangliosides. There was no correlation with respect to other structural types or total ganglioside content. More recent work has indicated GM1 to be the most important of the gangliotetraose species (Wu and Ledeen, unpublished observations). The potential for neurite recruitment thus appears to depend on the presence of a threshold level of GM1 (or related species) contributed by the cell to its own membrane.

Additional indication of specific roles for gangliotetraose gangliosides comes from the use of interventive agents. Affinity-purified antibodies to GM1 were reported to block, in a dose-dependent manner, nerve growth factor-induced sprouting of chick embryonic dorsal root ganglia (Schwartz and Spirman, 1982; Roisen *et al.*, 1986). Conditioned media-enhanced neuritogenesis of dorsal root ganglia was blocked by several monoclonal antibodies to GM1, though to varying degrees (Spoerri *et al.*, 1988). On the other hand, GM1 antibodies and B-cholera toxin were claimed not to interfere with the nerve growth factor-mediated fiber outgrowth and survival of neurons in the same system (Doherty and Walsh, 1987). Anti-GM1 antibody had an inhibitory effect on neurite outgrowth in the regenerating optic system of the goldfish when applied to retinal explants (Spirman *et al.*, 1982) or injected into the live animal (Spirman *et al.*, 1984). These somewhat conflicting results point to the need for further study, taking account of intrinsic differences between biological systems as well as the broad differences in reactivity obtained for different antibodies to the same (GM1) antigen (Roisen *et al.*, 1986).

Finally, it should be mentioned that exogenous gangliosides have significant effects on other cells beside neurons. Thus, astroglial cell proliferation was promoted by the four major gangliosides of brain, while $GgOse_4Cer$ and sialic acid were without effect (Katoh-Semba *et al.*, 1986). The same gangliosides produced a block or reversal of the stellation response of these cells to cAMP-inducing agents (Skaper *et al.*, 1986). It is not unreasonable to expect that additional cell types will be discovered that respond in some manner to gangliosides and that such phenomena could help in eventually understanding the biological roles of these substances.

ACKNOWLEDGMENTS

Supported by USPHS NIH Grants NS-04834 and NS-24172. Many helpful suggestions by Professor Konrad Sandhoff, particularly on the subject of ganglioside catabolism, are gratefully acknowledged.

5. REFERENCES

Abe, T., Miyatake, T., Norton, W. T., and Suzuki, K., 1979, Activities of glycolipid hydrolases in neurons and astroglia from rat and calf brains and in oligodendroglia from calf brain, *Brain Res.* **161**:179–182.

Agnati, L. F., Benfenati, F., Battistini, N., Cavicchioli, L., Fuxe, K., and Toffano, G., 1983a, Selective modulation of ^{3}H-spiperone labeled 5-HT receptors by subchronic treatment with the ganglioside GM1 in the rat, *Acta Physiol. Scand.* **117**:311–314.

Agnati, L. F., Fuxe, K., Calza, L., Benfenati, F., Cavicchioli, L., Toffano, G., and Goldstein, M., 1983b, Gangliosides increase the survival of lesioned nigral dopamine neurons and favour the recovery of dopaminergic synaptic function in striatum of rats by collateral sprouting, *Acta Physiol. Scand.* **119**: 347–363.

Ando, A., and Yu, R. K., 1977, Isolation and characterization of a novel trisialoganglioside, GT1a, from human brain, *J. Biol. Chem.* **252**:6247–6250.

Ando, S., 1983, Gangliosides in the nervous system, *Neurochem. Int.* **5**:507–537.

Ando, S., Tanaka, Y., and Ono, Y., 1981, Turnover of glycolipids in mouse brain myelin, in: *Glycoconjugates: Proc. VIth Int. Symp. Glycoconjugates* (T. Yamakawa, T. Osawa, and S. Handa, eds.), pp. 91–92, Japan Sci. Soc. Press, Tokyo.

Appel, S. H., 1981, A unifying hypothesis for the cause of amyotrophic lateral sclerosis, Parkinsonism, and Alzheimer disease, *Ann. Neurol.* **10**:499–505.

Aquino, D. A., Bisby, M. A., and Ledeen, R. W., 1985, Retrograde axonal transport of gangliosides and glycoproteins in the motoneurons of rat sciatic nerve, *J. Neurochem.* **45**:1262–1267.

Aquino, D. A., Bisby, M. A., and Ledeen, R. W., 1987, Bidirectional transport of gangliosides, glycoproteins and neutral glycosphingolipids in the sensory neurons of rat sciatic nerve, *Neuroscience* **20**:1023–1029.

Arbogast, B. W., and Arsenis, C., 1974, The enzymatic ontogeny of neurons and glial cells isolated from postnatal rat cerebral gray matter, *Neurobiology* **4**:21–37.

Arpaia, E., Dumbrille-Ross, A., Maler, T., Maler, K., Neote, K., Tropak, M., Troxel, C., Stirling, J. L., Pitts, J. S., Bapat, P., Lamhonwah, A.-M., Mahuran, D. J., Schuster, S. M., Clarke, J. T. R., Lowden, J. A., and Gravel, R. A., 1988, Identification of an altered splice site in Ashkenazic Tay-Sachs disease, *Nature* **333**:85–86.

Barkai, A., and Di Cesare, J. L., 1975, Influence of sialic acid groups on the retention of glycosphingolipids in blood plasma, *Biochim. Biophys. Acta* **398**:287–293.

Basu, S., Das, K. K., Schaeper, R. J., Banerjee, P., Daussin, F., Basu, M., Khan, F. A., and Zhang, B.-J., 1988, Biosynthesis in vitro of neuronal and non-neuronal gangliosides, in: *New Trends in Ganglioside Research: Neurochemical and Neuroregenerative Aspects*, Vol. 14 (R. W. Ledeen, E. L. Hogan, G. Tettamanti, A. J. Yates, and R. K. Yu, eds.), Liviana Press, Padova, pp. 259–273.

Basu, S., Basu, M., Moskal, J. R., and Chien, J.-L., 1976, Analysis of an A-active nonaglycosylceramide fraction, in: *Glycolipid Methodology* (L. A. Witting, ed.), pp. 123–139, American Oil Chemists Society, Champaign, Ill.

Basu, S., and Basu, M., 1982, Expression of glycosphingolipid glycosyltransferases in development and transformation, in: *The Glycoconjugates*, Vol. 3 (M. Horowitz, ed.), pp. 265–285, Academic Press, New York.

Berry-Kravis, E., and Dawson, G., 1985, Possible role of gangliosides in regulating an adenylate cyclase-linked 5-hydroxytryptamine (5-HT₁) receptor, *J. Neurochem.* **45**:1739–1747.

Bisby, M. A., and Bulger, V. T., 1977, Reversal of axonal transport at a nerve crush, *J. Neurochem.* **29**:313–320.

Blackburn, C. C., Swank-Hill, P., and Schnaar, R. L., 1986, Gangliosides support neural retina cell adhesion, *J. Biol. Chem.* **261**:2873–2881.

Bose, B., Osterholm, J. L., and Kalia, M., 1986, Ganglioside-induced regeneration and reestablishment of axonal continuity in spinal cord-transected rats, *Neurosci. Lett.* **63**:165–169.

Bremer, E. G., and Hakomori, S., 1982, GM₃ ganglioside induces hamster fibroblast growth inhibition in chemically-defined medium: Ganglioside may regulate growth factor receptor function, *Biochem. Biophys. Res. Commun.* **106**:711–718.

Bremer, E. G., and Hakomori, S.-I., 1984, Gangliosides as receptor modulators, in: *Ganglioside Structure, Function, and Biomedical Potential* (R. W. Ledeen, R. K. Yu, M. M. Rapport, and K. Suzuki, eds.), pp. 381–394, Plenum Press, New York.

Bremer, E. G., Hakomori, S.-I., Bowen-Pope, D. F., Raines, E., and Ross, R., 1984, Ganglioside-mediated modulation of cell growth, growth factor binding, and receptor phosphorylation *J. Biol. Chem.* **259**:6818–6825.

Bremer, E. G., Schlessinger, J., and Hakomori, S.-I., 1986, Ganglioside-mediated modulation of cell growth, *J. Biol. Chem.* **261**:2434–2440.

Brown, R. E., Stephenson, F. A., Markello, T., Barenholz, Y., and Thompson, T. E., 1985, Properties of a specific glycolipid transfer protein from bovine brain, *Chem. Phys. Lipids* **38**:79–93.

Brunngraber, E., 1979, *Neurochemistry of Aminosugars: Neurochemistry and Neuropathology of the Complex Carbohydrates*, Thomas, Springfield, Ill.

Burczak, J. D., Moskal, J. R., Trosko, J. E., Fairley, J. L., and Sweeley, C. C., 1983, Phorbol ester-associated changes in ganglioside metabolism, *Exp. Cell Res.* **147**:281–285.

Burczak, J. D., Soltysiak, R. M., and Sweeley, C. C., 1984, Regulation of membrane-bound enzymes of glycosphingolipid biosynthesis, *J. Lipid Res.* **25**:1541–1547.

Burton, R. M., Balfour, Y. M., and Gibbons, J. M., 1964, Gangliosides and cerebrosides turnover rates in rat brain, *Fed. Proc.* **23**:230.

Byrne, M. C., Ledeen, R. W., Roisen, F. J., Yorke, G., and Sclafani, J. R., 1983, Ganglioside-induced neuritogenesis: Verification that gangliosides are the active agents, and comparison of molecular species, *J. Neurochem.* **41**:1214–1222.

Byrne, M. C., Farooq, M., Sbaschnig-Agler, M., Norton, W. T., and Ledeen, R. W., 1988, Ganglioside content of astroglia and neurons isolated from maturing rat brain: Consideration of the source of astroglial gangliosides, *Brain Res.*, **481**:87–97.

Caccia, M. R., Meola, G., Cerri, C., Frattola, L., Scarlato, G., and Aporti, F., 1979, Treatment of denervated muscle by gangliosides, *Muscle Nerve* **2**:382–389.

Cannella, M. S., Wu, G., Vaswani, K. K., and Ledeen, R. W., 1988a, Neuritogenic effects of exogenous gangliosides and synthetic sialoglycolipids: Comparison to endogenous ganglioside requirements, in: *New Trends in Ganglioside Research: Neurochemical and Neuroregenerative Aspects* (R. W. Ledeen, E. L. Hogan, G. Tettamanti, A. J. Yates, and R. K. Yu, eds.), Liviana Press, Padova, pp. 379–390.

Cannella, M. S., Roisen, F. J., Ogawa, T., Sugimoto, M., and Ledeen, R. W., 1988b, Comparison of epi-GM3 with GM3 and GM1 as stimulators of neurite outgrowth, *Dev. Brain Res.* **39**:137–143.

Cannella, M. S., Acher, A. J., and Ledeen, R. W., 1988c, Stimulation of neurite outgrowth in vitro by a glycero-ganglioside, *Int. J. Dev. Neurosci.*, **6**:319–326.

Caputto, B. L., and Caputto, R., 1986, Optic nerve integrity is required for light to affect retina ganglion cell gangliosides, *Neurochem. Res.* **11**:1083–1090.

Caputto, B. L., Nores, G. A., Cemborain, B. N., and Caputto, R., 1982, The effect of light exposure following an intraocular injection of [³H]N-acetylmannosamine on the labeling of gangliosides and glycoproteins of retina ganglion cells and optic tectum of singly caged chicken, *Brain Res.* **245**:231–238.

Caputto, R., Maccioni, H. J., Arce, A., and Cumar, F. A., 1976, Biosynthesis of brain gangliosides, *Adv. Exp. Med. Biol.* **71**:27–44.

Caputto, R., Maccioni, A. H. R., and Caputto, B. L., 1977, Activation of deoxycholate solubilized adenosine triphosphatase by ganglioside and asialoganglioside preparations, *Biochem. Biophys. Res. Commun.* **74**:1046–1052.

Casamenti, F., Bracco, L., Bartolini, L., and Pepeu, G., 1985, Effects of ganglioside treatment in rats with a lesion of the cholinergic forebrain nuclei, *Brain Res.* **338**:45–52.

Ceccarelli, B., Aporti, F., and Finesso, M., 1976, Effects of brain gangliosides on functional recovery in experimental regeneration and reinnervation, in: *Ganglioside Function* (G. Porcellati, B. Ceccarelli, and G. Tettamanti, eds.), pp. 275–293, Plenum Press, New York.

Chan, K.-F. J., 1987a, Ganglioside-modulated protein phosphorylation in myelin, *J. Biol. Chem.* **262**:2415–2422.

Chan, K.-F. J., 1987b, Ganglioside-modulated protein phosphorylation. Partial purification and characterization of a ganglioside-stimulated protein kinase in brain, *J. Biol. Chem.* **262**:5248–5255.

Chan, K.-F. J., 1988, Ganglioside-modulated protein phosphorylation. Partial purification and characterization of a ganglioside-inhibited protein kinase in brain, *J. Biol. Chem.* **263**:568–574.

Cheresh, D. A., and Klier, F. G., 1986, Disialoganglioside GD2 distributes preferentially into substrate-associated microprocesses on human melanoma cells during their attachment to fibronectin, *J. Cell Biol.* **102**:1887–1897.

Cheresh, D. A., Pierschbacher, M. D., Herzig, M. A., and Mujoo, K., 1986, Disialogangliosides GD2 and GD3 are involved in the attachment of human melanoma and neuroblastoma cells to extracellular matrix proteins, *J. Cell Biol.* **102**:688–696.

Cheresh, D. A., Pytela, R., Pierschbacher, M. D., Klier, F. G., Ruoslahti, E., and Reisfeld, R. A., 1987, An Arg-Gly-Asp-directed receptor on the surface of human melanoma cells exists in a divalent cation-dependent functional complex with the disialoganglioside GD2, *J. Cell Biol.* **105**:1163–1173.

Cheresh, D. A., Pytela, R., Pierschbacher, M. D., Ruoslahti, E., and Reisfeld, R. A., 1988, An Arg-Gly-Asp-directed adhesion receptor on human melanoma cells exists in a calcium-dependent functional complex with the disialoganglioside GD2, in: *New Trends in Ganglioside Research: Neurochemical and Neuroregenerative Aspects* (R. W. Ledeen, E. L. Hogan, G. Tettamanti, A. J. Yates, and R. K. Yu, eds.), Liviana Press, Padova, pp. 203–217.

Cochran, F. B., Jr., Yu, R. K., and Ledeen, R. W., 1982, Myelin gangliosides in vertebrates, *J. Neurochem.* **39**:773–779.

Consolazione, A., and Toffano, G., 1988, Ganglioside role in functional recovery of damaged nervous system, in: *New Trends in Ganglioside Research: Neurochemical and Neuroregenerative Aspects* (R. W. Ledeen, E. L. Hogan, G. Tettamanti, A. J. Yates, and R. K. Yu, eds.), Liviana Press, Padova, pp. 523–533.

Conzelmann, E., and Sandhoff, K., 1978, AB variant of infantile GM2 gangliosidosis: Deficiency of a factor necessary for stimulation of hexosaminidase A-catalyzed degradation of ganglioside GM2 and glycolipid GA2, *Proc. Natl. Acad. Sci. USA* **75**:3979–3983.

Conzelmann, E., Burg, J., Stephan, G., and Sandhoff, K., 1982, Complexing of glycolipids and their transfer between membranes by the activator protein for degradation of lysosomal ganglioside GM2, *Eur. J. Biochem.* **123**:455–465.

Critchley, D. R., Habig, W. H., and Fishman, P. H., 1986, Reevaluation of the role of gangliosides as receptors for tetanus toxin, *J. Neurochem.* **47**:213–222.

Cuello, A. C., Stephens, P. H., Tagari, P. C., Sofroniew, M. V., and Pearson, R. C. A., 1986, Retrograde changes in the nucleus basalis of the rat, caused by cortical damage, are prevented by exogenous ganglioside GM1, *Brain Res.* **376**:373–377.

Cuello, A. C., Garofolo, L., Maysinger, D., and Pioro, E., 1987, Gangliosides and nerve growth factor: Effects on plastic changes after cortical lesions, *J. Neurochem.* **48**(Suppl.):S156.

Dal Toso, R., Presti, D., Benvegnu, D., Tettamanti, G., Toffano, G., and Leon, A., 1986, Primary neural cell cultures and GM1 monosialoganglioside: A model for comprehension of the mechanisms underlying GM1 effects in CNS repair process in vivo, in: *Gangliosides and Neuronal Plasticity*, Vol. 6 (G.

Tettamanti, R. W. Ledeen, K. Sandhoff, Y. Nagai, and G. Toffano, eds.), pp. 245–255, Liviana Press, Padova.

Davis, C. W., and Daly, J. W., 1980, Activation of rat cerebral cortical 3',5'-cyclic nucleotide phosphodiesterase activity by gangliosides, *Mol. Pharmacol.* **17**:206–211.

Dawson, G., 1979, Complex carbohydrates of cultured neuronal and glial cell lines, in: *Complex Carbohydrates of Nervous Tissue* (R. U. Margolis and R. K. Margolis, eds.), pp. 291–326, Plenum Press, New York.

Dawson, G., and Vartanian, T., 1988, Glycolipids as the source and modulator of receptor-mediated second messengers, in: *New Trends in Ganglioside Research: Neurochemical and Neuroregenerative Aspects* (R. W. Ledeen, E. L. Hogan, G. Tettamanti, A. J. Yates, and R. K. Yu, eds.), Liviana Press, Padova, pp. 219–228.

Dawson, G., McLawhon, R., and Miller, R. J., 1980, Inhibition of sialoglycosphingolipid (ganglioside) biosynthesis in mouse clonal lines N4T61 and NG108-15 by β-endorphin, enkephalins, and opiates, *J. Biol. Chem.* **255**:129–137.

Dawson, G., Hancock, L. W., and Vartanian, T., 1986, Regulation of GM2 ganglioside metabolism in cultured cells, *Chem. Phys. Lipids* **42**:105–116.

d'Azzo, A., Proia, R. L., Kolodny, E. H., Kaback, M. M., and Neufeld, E. F., 1984, Faulty association of α- and β-subunits in some forms of β-hexosaminidase A deficiency, *J. Biol. Chem.* **259**:11070–11074.

DeGasperi, R., Li, S.-C., and Li, Y.-T., 1988, A GM2-specific beta-hexosaminidase from the roe of striped · mullet (Mugil cephalus), *J. Biol. Chem.* **263**:1325–1328.

Dimpfel, W., and Habermann, E., 1977, Binding characteristics of ^{125}I-labelled tetanus toxin to primary tissue cultures from mouse embryonic CNS, *J. Neurochem.* **29**:1111–1120.

Dimpfel, W., Moller, W., and Mengs, U., 1981, Ganglioside-induced neurite formation in cultured neuroblastoma cells, in: *Gangliosides in Neurological and Neuromuscular Function, Development, and Repair* (M. M. Rapport and A. Gorio, eds.), pp. 119–134, Raven Press, New York.

Doherty, P., and Walsh, F. S., 1987, Ganglioside GM1 antibodies and B-cholera toxin bind specifically to embryonic chick dorsal root ganglion neurons but do not modulate neurite regeneration, *J. Neurochem.* **48**:1237–1244.

Doherty, P., Dickson, J. G., Flanigan, T. P., Leon, A., Toffano, G., and Walsh, F. S., 1985a, Molecular specificity of ganglioside effects on neurite regeneration of sensory neurons *in vitro, Neurosci. Lett.* **62**:193–198.

Doherty, P., Dickson, J. G., Flanigan, T. P., and Walsh, F. S., 1985b, Ganglioside GM$_1$ does not initiate, but enhances neurite regeneration of nerve growth factor-dependent sensory neurones, *J. Neurochem.* **44**:1259–1265.

Dreyfus, H., Harth, S., Yusufi, A. N. K., Urban, P. F., and Mandel, P., 1980a, Sialyltransferase activities in two neuronal models: Retina and cultures of isolated neurons, in: *Structure and Function of Gangliosides* (L. Svennerholm, P. Mandel, H. Dreyfus, and P.-F. Urban, eds.), pp. 227–237, Plenum Press, New York.

Dreyfus, H., Louis, J. C., Harth, S., and Mandel, P., 1980b, Gangliosides in cultured neurons, *Neuroscience* **5**:1647–1655.

Dreyfus, H., Harth, S., Massarelli, R., and Louis, J. C., 1981, Mechanisms of differentiation in cultured neurons: Involvement of gangliosides, in: *Gangliosides in Neurological and Neuromuscular Function, Development and Repair* (M. M. Rapport and A. Gorio, eds.), pp. 151–170, Raven Press, New York.

Dreyfus, H., Ferret, B., Harth, S., Gorio, A., Freysz, L., and Massarelli, R., 1984, Effect of exogenous gangliosides on the morphology and biochemistry of cultured neurons, in: *Ganglioside Structure, Function, and Biomedical Potential* (R. W. Ledeen R. K. Yu, M. M. Rapport, and K. Suzuki, eds.), pp. 513–524, Plenum Press, New York.

Ehrlich, Y. H., Davis, T. B., Bock, E., Kornecki, E., and Lenox, R. H., 1986, Ectoprotein kinase activity on the external surface of neural cells, *Nature* **320**:67–70.

Esmann, M., Marsh, D., Schwarzmann, G., and Sandhoff, K., 1988, Ganglioside–protein interactions: Spin-label electron spin resonance studies with (Na$^+$,K$^+$)-ATPase membranes, *Biochemistry* **27**:2398–2403.

Evans, W. H., and McIlwain, H., 1967, Excitability and ion content of cerebral tissues treated with alkylating agents, tetanus toxin, or neuraminidase, *J. Neurochem.* **14**:35–44.

Facci, L., Leon, A., Toffano, G., Sonnino, S., Ghidoni, R., and Tettamanti, G., 1984, Promotion of neuritogenesis in mouse neuroblastoma cells by exogenous gangliosides. Relationship between the effect and the cell association of ganglioside GM1, *J. Neurochem.* **42**:299–305.

Ferrari, G., Fabris, M., and Gorio, A., 1983, Gangliosides enhance neurite outgrowth in PC12 cells, *Dev. Brain Res.* **8**:215–222.

Ferret, B., Massarelli, R., Freysz, L., and Dreyfus, H., 1987, Effect of exogenous gangliosides on the metabolism of inositol compounds in chick neurons in culture, *C. R. Acad. Sci.* **304**:97–99.

Ferwerda, W., Blok, C. M., and Heijlman, J., 1981, Turnover of free sialic acid, CMP-sialic acid, and bound sialic acid in rat brain, *J. Neurochem.* **36**:1492–1499.

Fishman, P. H., 1982, Role of membrane gangliosides in the binding action of bacterial toxins, *J. Membr. Biol.* **69**:85–97.

Fishman, P. H., 1986, Recent advances in identifying the functions of gangliosides, *Chem. Phys. Lipids* **42**:137–151.

Fishman, P. H., Simmons, J. L., Brady, R. O., and Freese, E., 1974, Induction of glycolipid biosynthesis by sodium butyrate in HeLa cells, *Biochem. Biophys. Res. Commun.* **59**:292–299.

Fleischer, B., 1977, Localization of some glycolipid glycosylating enzymes in the Golgi apparatus of rat kidney, *J. Supramol. Struct.* **7**:79–89.

Forman, D. S., and Ledeen, R. W., 1972, Axonal transport of gangliosides in the goldfish optic nerve, *Science* **177**:630–633.

Freysz, L., Farooqui, A. A., Adamczewska-Goncerzewicz, Z., and Mandel, P., 1979, Lysosomal hydrolases in neuronal, astroglial, and oligodenroglial enriched fractions of rabbit and beef brain, *J. Lipid Res.* **20**:503–508.

Fujibayashi, S., and Wenger, D. A., 1986, Biosynthesis of the sulfatide/GM1 activator protein (SAP-1) in control and mutant cultured skin fibroblasts, *Biochim. Biophys. Acta* **875**:554–562.

Furst, W., Vogel, A., Lee-Vaupel, M., Conzelmann, E., and Sandhoff, K., 1986, Glycosphingolipid activator proteins, in: *Enzymes of Lipid Metabolism II* (L. Freysz, H. Dreyfus, R. Massarelli, and S. Gatt, eds.), pp. 314–338, Plenum Press, New York.

Furst, W., Machleidt, W., and Sandhoff, K., 1988, The precursor of sulfatide activator protein is processed to three different proteins, *Biol. Chem. Hoppe-Seyler* **369**:317–328.

Furukawa, K., Higgins, T., and Roth, S., 1985, An affinity-purified major histocompatibility complex (MCH) antigen with high N-acetylgalactosaminyltransferase activity, *J. Cell Biol.* **101**:309a.

Gammon, C. M., Goodrum, J. F., Toews, A. D., Okabe, A., and Morell, P., 1985, Axonal transport of glycoconjugates in the rat visual system, *J. Neurochem.* **44**:376–387.

Gammon, C. M., Vaswani, K. K., and Ledeen, R. W., 1987, Isolation of two glycolipid transfer proteins from bovine brain: Reactivity toward gangliosides and neutral glycosphingolipids, *Biochemistry* **26**:6239–6243.

Gershoni, J. M., Lapidot, M., Zakai, N., and Loyter, A., 1986, Protein blot analysis of virus receptors: Identification and characterization of the Sendai virus receptor, *Biochim. Biophys. Acta* **856**:19–26.

Ghidoni, R., Trinchera, M., Venerando, B., Fiorilli, A., and Tettamanti, G., 1986a, Metabolism of exogenous GM1 and related glycolipids in the rat, in: *Gangliosides and Neuronal Plasticity* (G. Tettamanti, R. W. Ledeen, K. Sandhoff, Y. Nagai, and G. Toffano, eds.), pp. 183–200, Liviana Press, Padova.

Ghidoni, R., Trinchera, M., Venerando, B., Fiorilli, A., Sonnino, S., and Tettamanti, G., 1986b, Incorporation and metabolism of exogenous GM1 ganglioside in rat liver, *Biochem. J.* **237**:147–155.

Goldenring, J. R., Otis, L. C., Yu, R. K., and DeLorenzo, R. J., 1985, Calcium/ganglioside-dependent protein kinase activity in rat brain membrane, *J. Neurochem.* **44**:1129–1134.

Gorio, A., Carmignoto, G., Facci, L., and Finesso, M., 1980, Motor nerve sprouting induced by ganglioside treatment. Possible implications for gangliosides on neuronal growth, *Brain Res.* **197**:236–241.

Gorio, A., Marini, P., and Zanoni,R., 1983, Muscle reinnervation. III. Motoneuron sprouting capacity, enhancement by exogenous gangliosides, *Neuroscience* **8**:417–429.

Gradkowska, M., Skup, M., Kiedrowski, L., Calzolari, S., and Oderfeld-Nowak, B., 1986, The effect of GM1 ganglioside on cholinergic and serotoninergic systems in the rat hippocampus following partial denervation is dependent on the degree of fiber degeneration, *Brain Res.* **375**:417–422.

Grafstein, B., Yip, H. K., and Meiri, H., 1982, Techniques for improving axonal regeneration: Assay in goldfish optic nerve, in: *Nervous System Regeneration* (A. M. Giuffrida-Stella, B. Haber, G. Hashim, and J. R. Perez-Polo, eds.), pp. 105–118, Liss, New York.

Griffiths, S. L., Perkins, R. M., Strueli, C. H., and Critchley, D. R., 1986, Variants of BALB/c 3T3 cells lacking complex gangliosides retain a fibronectin matrix and spread normally on fibronectin-coated substrates, *J. Cell Biol.* **102**:469–476.

Hadjiconstantinou, M., and Neff, N. H., 1988, Treatment with GM1 ganglioside restores striatal dopamine in the 1-methyl-4-phenyl-1,2,3,6-tetrahydropyridine-treated mouse, *J. Neurochem.*, **51**:1190–1196.

Hadjiconstantinou, M., Cavalla, D., Anthopoulou, E., Laird, H. E., II, and Neff, N. H., 1985, N-methyl-4-phenyl-1,2,3,6-tetrahydropyridine increases acetylcholine and decreases dopamine in mouse striatum, both responses are blocked by anticholinergic drugs, *J. Neurochem.* **45**:1957–1959.

Hakomori, S., 1970, Cell density-dependent changes of glycolipid concentrations in fibroblasts, and loss of this response in virus-transformed cells, *Proc. Natl. Acad. Sci. USA* **67**:1741–1747.

Hakomori, S.-I., 1981, Glycosphingolipids in cellular interaction, differentiation, and oncogenesis, *Annu. Rev. Biochem.* **50**:733–764.

Hakomori, S., and Kannagi, R., 1983, Glycosphingolipids as tumor-associated and differentiation markers, *J. Natl. Cancer Inst.* **71**:231–251.

Halder, S., and Majumder, C. C., 1986, Phosphorylation of external cell surface protein by an endogenous ecto-protein kinase of goat epididymal intact spermatozoa, *Biochim. Biophys. Acta* **887**:291–303.

Haley, J. E., Wisniewski, H. M., and Ledeen, R. W., 1979, Extra-axonal diffusion in the rabbit optic system: A caution in axonal transport studies, *Brain Res.* **179**:69–76.

Hanai, N., Nores, G., Torres-Mendez, C.-R., and Hakomori, S.-I., 1987, Modified ganglioside as a possible modulator of transmembrane signaling mechanism through growth factor receptors: A preliminary note, *Biochem. Biophys. Res. Commun.* **147**:127–134.

Handa, S., and Burton, R. M., 1969, Biosynthesis of glycolipids: Incorporation of N-acetyl galactosamine by rat brain particulate preparation, *Lipids* **4**:589–598.

Hannun, Y. A., and Bell, R. M., 1987, Lysosphingolipids inhibit protein kinase C: Implications for the sphingolipidoses, *Science* **235**:670–674.

Harris, P. L., and Thornton, E. R., 1978, Carbon-13 and proton nuclear magnetic resonance studies of gangliosides, *J. Am. Chem. Soc.* **100**:6738–6745.

Harry, G. J., Goodrum, J. F., Toews, A. D., and Morell, P., 1987, Axonal transport characteristics of gangliosides in sensory axons of rat sciatic nerve, *J. Neurochem.* **48**:1529–1536.

Hashimoto, Y., Suzuki, A., Yamakawa, T., Miyashita, N., and Moriwaki, K., 1983, Expression of GM_1 and GD_{1a} in mouse liver is linked to the H-2 complex on chromosome 17, *J. Biochem.* **94**:2043–2048.

Hauw, J. J., Fenelon, S., Boutry, J.-M., Nagai, Y., and Escourolle, R., 1981, Effects of brain gangliosides on neurite growth in guinea pig spinal ganglia tissue cultures and on fibroblast cell cultures, in: *Gangliosides in Neurological and Neuromuscular Function, Development, and Repair* (M. M. Rapport and A. Gorio, eds.), pp. 171–176, Raven Press, New York.

Higashi, H., Basu, M., and Basu, S., 1985, Biosynthesis *in vitro* of disialosylneolactotetraosyl-ceramide by a solubilized sialyltransferase from embryonic chicken brain, *J. Biol. Chem.* **260**:824–828.

Holm, M., and Svennerholm, L., 1972, Biosynthesis and biodegradation of rat brain gangliosides studied in vivo, *J. Neurochem.* **19**:609–622.

Igarashi, M., Komiya, Y., and Kurokawa, M., 1985, CMP-sialic acid, the sole sialosyl donor, is intra-axonally transported, *FEBS Lett.* **192**:239–242.

Jeserich, G., Breer, H., and Duvel, M., 1981, Effect of exogenous gangliosides on synaptosomal membrane ATPase activity, *Neurochem. Res.* **6**:465–474.

Jones, J. P., Ramsey, R. B., Aexel, R. T., and Nicholas, H. J., 1972, Lipid biosynthesis in neuron-enriched and glial-enriched fractions of rat brain: Ganglioside biosynthesis, *Life Sci.* **11**:309–315.

Jonsson, G., Gorio, A., Hallman, H., Janigro, D., Kojima, H., Luthman, J., and Zanoni, R., 1984, Effects of GM1 ganglioside on developing and mature serotonin and noradrenaline neurons lesioned by selective neurotoxins, *J. Neurosci. Res.* **12**:459–475.

Kalia, M., and DiPalma, J. R., 1982, Ganglioside-induced acceleration of axonal transport following nerve crush injury in the rat, *Neurosci. Lett.* **34**:1–5.

Kanda, S., Inoue, K., Nojima, S., Utsumi, H., and Weigandt, H., 1982, Incorporation of a ganglioside and a spin-labeled ganglioside analogue into cell and liposomal membranes, *J. Biochem.* **91**:2095–2098.

Karpiak, S. E., 1983, Ganglioside treatment improves recovery of alternation behavior after unilateral entorhinal cortex lesion, *Exp. Neurol.* **81**:330–339.

Karpiak, S. E., Li, Y. S., and Mahadik, S. P., 1987, Gangliosides (GM1 and AGF2) reduce mortality due to ischemia: Protection of membrane function, *Stroke* **18**:184–187.

Kato, I., and Naiki, M., 1976, Ganglioside and rabbit erythrocyte membrane receptor for staphylococcal alpha-toxin, *Infect. Immun.* **13**:289–291.

Katoh-Semba, R., Skaper, S. D., and Varon, S., 1984, Interaction of GM1 ganglioside with PC12 phe-ochromocytoma cells: Serum and NGF-dependent effects on neuritic growth (and proliferation), *J. Neurosci. Res.* **12**:299–310.

Katoh-Semba, R., Facci, L., Skaper, S. D., and Varon, S., 1986, Gangliosides stimulate astroglial cell proliferation in the absence of serum, *J. Cell. Physiol.* **126**:147–153.

Kaufman, B., Basu, S., and Roseman, S., 1967, Studies on the biosynthesis of gangliosides, in: *Inborn Disorders of Sphingolipid Metabolism* (A. M. Aronson and B. W. Volk, eds.), pp. 193–214, Pergamon Press, New York.

Kaufman, B., Basu, S., and Roseman, S., 1968, Enzymatic synthesis of disialogangliosides from monosial-ogangliosides by sialyltransferases from embryonic chicken brain, *J. Biol. Chem.* **243**:5804–5807.

Keenan, T. W., and Morre, D. J., 1975, Glycosyltransferases: Do they exist on the surface membrane of mammalian cells? *FEBS Lett.* **55**:8–13.

Keenan, T. W., Morre, D. J., and Basu, S., 1974, Ganglioside biosynthesis. Concentration of glycosphingolipid glycosyltransferases in Golgi apparatus from rat liver, *J. Biol. Chem.* **249**:310–315.

Kim, J. Y. H., Goldenring, J. R., DeLorenzo, R. J., and Yu, R. K., 1986, Gangliosides inhibit phos-pholipid-sensitive Ca^{2+}-dependent kinase phosphorylation of rat myelin basic protein, *J. Neurosci. Res.* **15**:159–166.

Kitamura, M., Iwamori, M., and Nagai, Y., 1980, Interaction between clostridium botulinum neurotoxin and gangliosides, *Biochim. Biophys. Acta* **628**:328–335.

Klein, D., Leinekugel, P., Pohlentz, G., Schwarzmann, G., and Sandhoff, K., 1988, Metabolism and intracellular transport of gangliosides in cultured fibroblasts, in: *New Trends in Ganglioside Research: Neurochemical and Neuroregenerative Aspects* (R. W. Ledeen, E. L. Hogan, G. Tettamanti, A. J. Yates, and R. K. Yu, eds.), Liviana Press, Padova, pp. 247–258.

Kleinbeckel, D., 1982, Acceleration of muscle re-innervation in rats by ganglioside treatment: An electromyographic study, *Eur. J. Pharmacol.* **80**:243–245.

Kleinman, H. K., Martin, G. R., and Fishman, P. H., 1979, Ganglioside inhibition of fibronectin-mediated cell adhesion to collagen, *Proc. Natl. Acad. Sci. USA* **76**:3367–3371.

Kojima, H., Gorio, A., Janigro, D., and Jonsson, G., 1984, GM1 ganglioside enhances regrowth of noradrenaline nerve terminals in rat cerebral cortex lesioned by the neurotoxin 6-hydroxydopamine, *Neuroscience* **13**:1011–1022.

Koulakoff, A., Bizzini, B., and Berwald-Netter, Y., 1983, Neuronal acquisition of tetanus toxin binding sites: Relationship with the last mitotic cycle, *Dev. Biol.* **100**:350–357.

Kreutter, D., Kim, J. Y. H., Goldenring, J. R., Rasmussen, H., Ukomadu, C., DeLorenzo, R. J., and Yu, R. K., 1987, Regulation of protein kinase C activity by gangliosides, *J. Biol. Chem.* **262**:1633–1637.

Kytzia, H.-J., and Sandhoff, K., 1985, Evidence for two different active sites on human beta-hexosaminidase A, *J. Biol. Chem.* **260**:7568–7572.

Kytzia, H.-J., Hinrichs, U., Maire, I., Suzuki, K., and Sandhoff, K., 1983, Variant of GM2-gangliosidosis with hexosaminidase A having a severely changed substrate specificity, *EMBO J.* **2**:1201–1205.

Laitinen, J., Lopponen, R., Merenmies, J., and Rauvala, H., 1987, Binding of laminin to brain gangliosides and inhibition of laminin–neuron interaction by the gangliosides, *FEBS Lett.* **217**:94–100.

Landa, C. A., Maccioni, H. J. F., and Caputto, R., 1979, The site of synthesis of gangliosides in the chick optic system, *J. Neurochem.* **33**:825–838.

Landa, C. A., Defilpo, S. S., Maccioni, H. J. F., and Caputto, R., 1981, Disposition of gangliosides and sialosylglycoproteins in neuronal membranes, *J. Neurochem.* **37**:813–823.

Lang, W., 1981, Pharmacokinetic studies of ³H-labeled exogenous gangliosides injected intramuscularly into rats, in: *Gangliosides in Neurological and Neuromuscular Function, Development, and Repair* (M. M. Rapport and A. Gorio, eds.), pp. 241–251, Raven Press, New York.

Langenbach, R., and Kennedy, S., 1978, Gangliosides and their cell density-dependent changes in control and chemically transformed C3H/10T1/2 cells, *Exp. Cell Res.* **112**:361–372.

Ledeen, R. W., 1983, Gangliosides, in: *Handbook of Neurochemistry*, Vol. 3 (A. Lajtha, ed.), pp. 41–90, Plenum Press, New York.

Ledeen, R. W., 1984, Biology of gangliosides: Neuritogenic and neuronotrophic properties, *J. Neurosci. Res.* **12**:147–159.

Ledeen, R. W., and Cannella, M. S., 1987, The neuritogenic effect of gangliosides in cell cultures, in: *Gangliosides and Modulation of Neuronal Functions* (H. Rahmann, ed.), pp. 491–500, Springer-Verlag, Berlin.

Ledeen, R. W., Skrivanek, J. A., Tirri, L. J., Margolis, R. K., and Margolis, R. U., 1976, Gangliosides of the neuron: Localization and origin, *Adv. Exp. Med. Biol.* **71**:83–104.

Ledeen, R. W., Skrivanek, J. A., Nunez, J., Sclafani, J. R., Norton, W. T., and Farooq, M., 1981, Implications of the distribution and transport of gangliosides in the nervous system, in: *Gangliosides in Neurological and Neuromuscular Function, Development, and Repair* (M. M. Rapport and A. Gorio, eds.), pp. 211–223, Raven Press, New York.

Ledeen, R. W., Aquino, D. A., Sbaschnig-Agler, M., Gammon, C. M., and Vaswani, K. K., 1987, Fundamentals of neuronal transport of gangliosides. Functional implications, in: *Gangliosides and Modulation of Neuronal Functions*, Vol. H7 (H. Rahmann, ed.), pp. 259–274, Springer-Verlag, Berlin.

Ledeen, R. W., Hogan, E. L., Tettamanti, G., Yates, A. J., and Yu, R. K. (eds.), 1988, *New Trends in Ganglioside Research: Neurochemical and Neuroregenerative Aspects*, Vol. 14, Liviana Press, Padova.

Lee, A. G., 1977, Annular events: Lipid–protein interactions, *Trends Biochem. Sci.* **2**:231–233.

Leon, A., Facci, L., Toffano, G., Sonnino, S., and Tettamanti, G., 1981, Activation of (Na$^+$,K$^+$)ATPase by nanomolar concentrations of GM1 ganglioside, *J. Neurochem.* **37**:350–357.

Leon, A., Facci, L., Benvegnu, D., and Toffano, G., 1982, Morphological and biochemical effects of gangliosides in neuroblastoma cells, *Dev. Neurosci.* **5**:108–114.

Leon, A. D., Benvegnu, D., Dal Toso, R., Presti, D., Facci, L., Giorgi, O., and Toffano, G., 1984, Dorsal root ganglia and nerve growth factor: A model for understanding the mechanism of GM1 effects on neuronal repair, *J. Neurosci. Res.* **12**:277–287.

Leon, A., Dal Toso, R., Presti, D., Benvegnu, D., Faci, L., Kirschner, G., Tettamanti, G., and Toffano, G., 1988, Development and survival of neurons in dissociated fetal mesencephalic serum-free cell cultures. II. Modulatory effects of gangliosides, *J. Neurosci.* **8**:746–753.

Leskawa, K. C., and Hogan, E. L., 1985, Quantitation of the in vitro neuroblastoma response to exogenous, purified gangliosides, *J. Neurosci. Res.* **13**:539–550.

Li, S.-C., Nakamura, T., Ogamo, A., and Li, Y.-T., 1979, Evidence for the presence of two separate protein activators for the enzymic hydrolysis of GM_1 and GM_2 gangliosides, *J. Biol. Chem.* **254**:10592–10595.

Li, S.-C., Sonnino, S., Tettamanti, G., and Li, Y.-T., 1988, Characterization of a nonspecific activator protein for the enzymatic hydrolysis of glycolipids, *J. Biol. Chem.* **263**:6588–6591.

Li, Y. S., Mahadik, S. P., Rapport, M. M., and Karpiak, S. E., 1986, Acute effects of GM1 ganglioside: Reduction in both behavioral asymmetry and loss of Na$^+$-K$^+$ ATPase after nigrostriatal transection, *Brain Res.* **377**:292–297.

Li, Y.-T., Mazzotta, M. Y., Wan, C. C., Orth, R., and Li, S.-C., 1973, Hydrolysis of Tay–Sachs ganglioside by β-hexosaminidase A of human liver and urine, *J. Biol. Chem.* **248**:7512–7515.

Maccioni, H. J., Arce, A., and Caputto, R., 1971, The biosynthesis of gangliosides, labelling of rat brain gangliosides in vivo, *Biochem. J.* **125**:1131–1137.

Maccioni, H. J. F., Defilpo, S. S., Landa, C. A., and Caputto, R., 1978, The biosynthesis of brain gangliosides. Ganglioside-glycosylating activity in rat brain neuronal perikarya fraction, *Biochem. J.* **174**:673–680.

Maccioni, H. J. F., Panzetta, P., Arrieta, D., and Caputto, R., 1984a, Ganglioside glycosyltransferase activities in the cerebral hemispheres from developing rat embryos, *Int. J. Dev. Neurosci.* **2**:13–19.

Maccioni, H. J. F., Panzetta, P., and Arrieta, D., 1984b, Some properties of uridine-5′-diphospho-N-acetylgalactosamine:hematoside N-acetylgalactosaminyltransferase at early and late stages of embryonic development of chicken retina, *Int. J. Dev. Neurosci.* 2:259–266.

Macher, B. A., Lockney, M., Moskal, J. R., Fung. Y. K., and Sweeley, C. C., 1978, Studies on the mechanism of butyrate-induced morphological changes in KB cells, *Exp. Cell Res.* 117:95–102.

Mahuran, D. A., Novak, A., and Lowden, J. A., 1985, The lysosomal hexosaminidase isozymes, in *Isozymes* (M. O. Rattazzi, J. G. Scandalios, and G. S. Whitt, eds.), pp. 229–288, Liss, New York.

Maier, C. E., and Singer, M., 1984, Gangliosides stimulate protein synthesis, growth, and axon number of regenerating limb buds, *J. Comp. Neurol.* 230:459–464.

Marchase, R. B., 1977, Biochemical investigations of retinotectal adhesive specificity, *J. Cell Biol.* 75: 237–257.

Margolis, R. K., Salton, S. R. J., and Margolis, R. U., 1983, Complex carbohydrates of cultured PC12 pheochromocytoma cells. Effects of nerve growth factor and comparison with neonatal and mature rat brain, *J. Biol. Chem.* 258:4110–4117.

Marini, P., Vitadello, M., Biachi, R., Triban, C., and Gorio, A., 1986, Impaired axonal transport of acetylcholinesterase in the sciatic nerve of alloxan-diabetic rats: Effect of ganglioside treatment, *Diabetologia* 29:254–258.

Markwell, M. A. K., Svennerholm, L., and Paulson, J. C., 1981, Specific gangliosides function as host cell receptors for Sendai virus, *Proc. Natl. Acad. Sci. USA* 78:5406–5410.

Massarelli, R., Ferret, B., Gorio, A., Durand, M., and Dreyfus, H., 1985, The effect of exogenous gangliosides on neurons in culture: A morphometric analysis, *Int. J. Dev. Neurosci.* 3:341–348.

Matsui, Y., Lombard, D., Hoflack, B., Harth, S., Massarelli, R., Mandel, P., and Dreyfus, H., 1983, Ectoglycosyltransferase activities at the surface of cultured neurons, *Biochem. Biophys. Res. Commun.* 113:446–453.

Matta, S. G., Yorke, G., and Roisen, F. J., 1986, Neuritogenic and metabolic effects of individual gangliosides and their interaction with nerve growth factor in cultures of neuroblastoma and pheochromocytoma, *Dev. Brain Res.* 27:243–252.

McIlwain, H., 1960, *Chemical Exploration of the Brain: A Study of Excitability and Ion Movement*, Elsevier, Amsterdam.

McIlwain, H., 1961, Characterization of naturally occurring materials which restore excitability to isolated cerebral tissues, *Biochem. J.* 78:24–32.

McLawhon, R. W., Schoon, G. S., and Dawson, G., 1981, Possible role of cyclic AMP in the receptor-mediated regulation of glycosyltransferase activities in neurotumor cell lines, *J. Neurochem.* 37:132–139.

Mengs, U., and Stotzem, C. D., 1987, Ganglioside treatment and nerve regeneration: A morphological study after nerve crush in rats, *Eur. J. Pharmacol.* 142:419–424.

Miller-Podraza, H., and Fishman, P. H., 1982, Translocation of newly synthesized gangliosides to the cell surface, *Biochemistry* 21:3265–3270.

Miller-Podraza, H., Bradley, R. M., and Fishman, P. H., 1982, Biosynthesis and localization of gangliosides in cultured cells, *Biochemistry* 21:3260–3265.

Mirsky, R., Wendon, L. M. B., Black, P., Stolkin, C., and Bray, D., 1978, Tetanus toxin: A cell surface marker for neurones in culture, *Brain Res.* 148:251–259.

Morgan, J. I., and Seifert, W., 1979, Growth factors and gangliosides: A possible new perspective in neuronal growth control, *J. Supramol. Struct.* 10:111–124.

Morre, D. J., Kartenbeck, J., and Franke, W. W., 1979, Membrane flow and interconversions among endomembranes, *Biochim. Biophys. Acta* 559:71–152.

Moskal, J. R., Gardner, D. A., and Basu, S., 1974, Changes in glycolipid glycosyltransferases and glutamate decarboxylase and their relationship to differentiation in neuroblastoma cells, *Biochem. Biophys. Res. Commun.* 61:751–758.

Moss, J., Fishman, P. H., Manganiello, V. C., Vaughan, M., and Brady, R. O., 1976, Functional incorporation of ganglioside into intact cells: Induction of choleragen responsiveness, *Proc. Natl. Acad. Sci. USA* 73:1034–1037.

Moss, J., Osborne, J. C., Jr., Fishman, P. H., Nakaya, S., and Robertson, D. C., 1981, Escherichia coli heat-labile enterotoxin. Ganglioside specificity and ADP-ribosyltransferase activity, *J. Biol. Chem.* 256:12861–12865.

Mugnai, G., Tombaccini, D., and Ruggieri, S., 1984, Ganglioside composition of substrate-adhesion sites of normal and virally-transformed BALB/C 3T3 cells, *Biochem. Biophys. Res. Commun.* **125:**142–148.

Murakami-Murofushi, K., Tadano, K., Koyama, I., and Ishizuka, I., 1981, A trisialoganglioside GT$_3$ of hog kidney. Structure and biosynthesis in vitro, *J. Biochem.* **90:**1817–1820.

Murray, N., and Steck, A. J., 1986, Activation of myelin protein kinase by diacylglycerol and 4β-phorbol 12-myristate 13-acetate, *J. Neurochem.* **46:**1655–1657.

Myerowitz, R., and Proia, R. L., 1984, cDNA clone for the α chain of human β-hexosaminidase: Deficiency of α chain mRNA in Ashkenazi Tay–Sachs fibroblasts, *Proc. Natl. Acad. Sci. USA* **81:**5394–5398.

Myerowitz, R., Piekarz, R., Neufeld, E. F., Shows, T. B., and Suzuki, K., 1985, Human β-hexosaminidase α-chain: Coding sequence and homology with the α-chain, *Proc. Natl. Acad. Sci. USA* **82:** 7830–7834.

Nagai, Y., and Tsuji, S., 1988, Cell biological significance of gangliosides in neural differentiation and development: Critique and proposals, in: *New Trends in Ganglioside Research: Neurochemical and Neuroregenerative Aspects* (R. W. Ledeen, E. L. Hogan, G. Tettamanti, A. J. Yates, and R. K. Yu, eds.), Liviana Press, Padova, pp. 329–350.

Nagai, Y., Nakaishi, H., and Sanai, Y., 1986, Gene transfer as a novel approach to the gene-controlled mechanism of the cellular expression of glycosphingolipids, *Chem. Phys. Lipids* **42:**91–103.

Nakaishi, H., Sanai, Y., Shiroki, K., and Nagai, Y., 1988a, Analysis of cellular expression of gangliosides by gene transfection. I. GD3 expression in *myc*-transfected and transformed 3Y1 correlates with anchorage-independent growth activity, *Biochem. Biophys. Res. Commun.* **150:**760–765.

Nakaishi, H., Sanai, Y., Shibuya, M., and Nagai, Y., 1988b, Analysis of cellular expression of gangliosides by gene transfection. II. Rat 3Y1 cells transformed with several DNAs containing oncogenes (*fes, fps, ras & src*) invariably express sialosylparagloboside, *Biochem. Biophys. Res. Commun.* **150:** 766–774.

Nakajima, J., Tsuji, S., and Nagai, Y., 1986, Bioactive gangliosides: Analysis of functional structures of the tetrasialoganglioside GQ1b which promotes neurite outgrowth, *Biochim. Biophys. Acta* **876:**65–71.

Navon, R., Argov, Z., and Frisch, A., 1986, Hexosaminidase A deficiency in adults, *Am. J. Med. Genet.* **24:**179–196.

Neuenhofer, S., Conzelmann, E., Schwarzmann, G., Egge, H., and Sandhoff, K., 1986, Occurrence of lysoganglioside lyso-GM$_2$ (II^3-Neu-5-Ac-gangliotriaosylsphingosine) in GM$_2$ gangliosidosis brain, *Biol. Chem. Hoppe-Seyler* **367:**241–244.

Ng, S.-S., and Dain, J. A., 1977, Sialyltransferases in rat brain: Reaction kinetics, product analyses, and multiplicities of enzyme species, *J. Neurochem.* **29:**1075–1083.

Norden, A. G., and O'Brien, J. S., 1975, An electrophoretic variant of β-galactosidase with altered catalytic properties in a patient with GM1 gangliosidosis, *Proc. Natl. Acad. Sci. USA* **72:**240–244.

Nores, G. A., and Caputto, R., 1984, Inhibition of the UDP-N-acetylgalactosamine: GM$_3$ N-acetylgalactosaminyl transferase by gangliosides, *J. Neurochem.* **42:**1205–1211.

Norido, F., Canella, R., and Aporti, F., 1981, Acceleration of nerve regeneration by gangliosides estimated by the somatosensory evoked potentials (SEP), *Experientia* **37:**301–302.

Ohman, R., Rosenberg, A., and Svennerholm, L., 1970, Human brain sialidase, *Biochemistry* **9:**3774–3782.

Ohno, K., and Suzuki, K., 1988a, Mutation in GM2-gangliosidosis B1 variant, *J. Neurochem.* **50:**316–318.

Ohno, K., and Suzuki, K., 1988b, A splicing defect due to an exon–intron junctional mutation results in abnormal β-hexosaminidase chain mRNAs in Ashkenazi Jewish patients with Tay–Sachs disease, *Biochem. Biophys. Res. Commun.* **153:**463–469.

Okada, S., and O'Brien, J. S., 1968, Generalized gangliosidosis: Beta-galactosidase deficiency, *Science* **160:**1002–1004.

Pacuszka, T., Duffard, R. O., Nishimura, R. N., Brady, R. O., and Fishman, P. H., 1978, Biosynthesis of bovine thyroid gangliosides, *J. Biol. Chem.* **253:**5839–5846.

Panzetta, P., Maccioni, H. J. F., and Caputto, R., 1980, Synthesis of retinal gangliosides during chick embryonic development, *J. Neurochem.* **35:**100–108.

Partington, C. R., and Daly, J. W., 1979, Effect of gangliosides on adenylate cyclase activity in rat cerebral cortical membranes, *Mol. Pharmacol.* **15:**484–491.

Patt, L. M., Itaya, K., and Hakomori, S.-I., 1978, Retinol induces density-dependent growth inhibition and changes in glycolipids and LETs, *Nature* **273**:379–381.

Perkins, R. M., Kellie, S., Patel, B., and Critchley, D. R., 1982, Gangliosides as receptors for fibronectin? *Exp. Cell Res.* **141**:231–243.

Pohlentz, G., Klein, D., Schwarzmann, G., Schmitz, D., and Sandhoff, K., 1988, Both GA2-, GM2-, and GD2 synthetases and GM1b-, GD1a-, and GT1b synthases are single enzymes in Golgi vesicles from rat liver, *Proc. Natl. Acad. Sci. USA*, **85**:7044–7048.

Preti, A., Fiorilli, A., Lombardo, A., Caimi, L., and Tettamanti, G., 1980, Occurrence of sialyltransferase activity in the synaptosomal membranes prepared from calf brain cortex, *J. Neurochem.* **35**:281–296.

Proia, R. L., and Neufeld, E. F., 1982, Synthesis of β-hexosaminidase in cell-free translation and in intact fibroblasts: An insoluble procedure α-chain in a rare form of Tay–Sachs disease, *Proc. Natl. Acad. Sci. USA* **79**:6360–6364.

Purpura, D. P., and Suzuki, K., 1976, Distortion of neuronal geometry and formation of aberrant synapses in neuronal storage disease, *Brain Res.* **116**:1–21.

Radin, N. S., Brenkert, A., Arora, R., Sellinger, O. Z., and Flangas, A. I., 1972, Glial and neuronal localization of cerebroside-metabolizing enzymes, *Brain Res.* **39**:163–169.

Ragahavan, S. S., Rhoads, D. B., and Kanfer, J. N., 1972, Acid hydrolases in neuronal and glial enriched fractions of rat brain, *Biochim. Biophys. Acta* **268**:755–760.

Rahmann, H., 1983, Functional implications of gangliosides in synaptic transmission (critique), *Neurochem. Int.* **5**:539–547.

Rahmann, H., 1987, Brain gangliosides, bio-electrical activity and post-stimulation effects, in: *Gangliosides and Modulation of Neuronal Functions* (H. Rahmann, ed.), pp. 501–521, Springer, Berlin.

Ramirez, J. J., Fass, B., Karpiak, S. E., and Steward, O., 1987, Ganglioside treatments reduce locomotor hyperactivity after bilateral lesions of the entorhinal cortex, *Neurosci. Lett.* **75**:283–287.

Reith, M., Morgan, I. G., Gombos, G., Breckenridge, W. C., and Vincendon, G., 1972, Synthesis of synaptic glycoproteins. I. The distribution of UDP-galactose:N-acetyl glucosamine galactosyl transferase and thiamine diphosphatase in adult rat brain subcellular fractions, *Neurobiology* **2**:169–175.

Richardson, C. L., Keenan, T. W., and Morre, D. J., 1977, Ganglioside biosynthesis. Characterization of CMP-N-acetylneuraminic acid:lactosylceramide sialyltransferase in Golgi apparatus from rat liver, *Biochim. Biophys. Acta* **488**:88–96.

Robb, G. A., and Keynes, R. J., 1984, Stimulation of nodal and terminal sprouting of mouse motor nerves by gangliosides, *Brain Res.* **295**:368–371.

Rogers, T. B., and Snyder, S. H., 1981, High affinity binding of tetanus toxin to mammalian brain membranes, *J. Biol. Chem.* **255**:2402–2407.

Roisen, F. J., Bartfeld, H., Nagele, R., and Yorke, G., 1981, Ganglioside stimulation of axonal sprouting in vitro, *Science* **214**:577–578.

Roisen, F. J., Matta, S. G., Yorke, G., and Rapport, M. M., 1986, The role of gangliosides in neurotrophic interaction in vitro, in: *Gangliosides and Neuronal Plasticity* (G. Tettamanti, R. W. Ledeen, K. Sandhoff, Y. Nagai, and G. Toffano, eds.), Vol. 6, pp. 281–293, Liviana Press, Padova.

Roseman, S., 1970, The synthesis of complex carbohydrates by multiglycosyltransferase systems and their potential function in intercellular adhesion, *Chem. Phys. Lipids* **5**:270–297.

Rosengren, B., Mansson, J.-E., and Svennerholm, L., 1987, Composition of gangliosides and neutral glycosphingolipids of brain in classical Tay–Sachs and Sandhoff disease: More lyso-G_{M2} in Sandhoff disease? *J. Neurochem.* **49**:834–840.

Rösner, H., 1975, Incorporation of sialic acid into gangliosides and glycoproteins of the optic pathway following an intraocular injection of [N^3H]acetylmannosamine in the chicken, *Brain Res.* **97**:107–116.

Rösner, H., and Merz, G., 1982, Uniform distribution and similar turnover rates of individual gangliosides along axons of retinal ganglion cells in the chicken, *Brain Res.* **236**:63–75.

Rösner, H., Wiegandt, H., and Rahmann, H., 1973, Sialic acid incorporation into gangliosides and glycoproteins of the fish brain, *J. Neurochem.* **21**:655–665.

Roth, S., 1985, Are glycosyltransferases the evolutionary antecedents of the immunoglobulins? *Q. Rev. Biol.* **60**:145–153.

Roth, S., and Marchase, R. B., 1976, An in vitro assay for retinotectal specificity, in: *Neuronal Recognition* (S. H. Barondes, ed.), pp. 227–248, Plenum Press, New York.

Rybak, S., Ginzburg, I., and Yavin, E., 1983, Gangliosides stimulate neurite outgrowth and induce tubulin mRNA accumulation in neural cells, *Biochem. Biophys. Res. Commun.* **116**:974–980.

Sabel, B. A., Slavin, M. D., and Stein, D. G., 1984, GM1 ganglioside treatment facilitates behavioral recovery from bilateral brain damage, *Science* **225**:340–342.

Sabel, B. A., Dunbar, G. L., Butler, W. W., and Stein, D. G., 1985, GM1 ganglioside stimulates neuronal reorganization and reduces rotational asymmetry after hemitransections of the nigro-striatal pathway, *Exp. Brain Res.* **60**:27–37.

Saito, M., Saito, M., and Rosenberg, A., 1984, Action of monensin, a monovalent cationophore, on cultured human fibroblast: Evidence that it induces high cellular accumulation of glucosyl- and lactosylceramide (gluco- and lactocerebroside), *Biochemistry* **23**:1043–1046.

Sandhoff, K., and Christomanou, H., 1979, Biochemistry and genetics of gangliosidoses, *Hum. Genet.* **50**:107–143.

Sandhoff, K., and Conzelmann, E., 1984, The biochemical basis of gangliosidoses, *Neuropediatrics* **15**:85–92.

Sandhoff, K., Conzelmann, E., Neufeld, E. F., Kaback, M. M., and Suzuki, L., 1989, The G_{M2} gangliosidoses, in: *The Metabolic Basis of Inherited Disease*, 6th ed. (C. R. Scriver, A. L. Beaudet, W. S. Sly, and D. Valle, eds.), McGraw-Hill, New York, in press.

Sbaschnig-Agler, M., Dreyfus, H., Norton, W. T., Sensenbrenner, M., Farooq, M., Byrne, M. C., and Ledeen, R. W., 1988, Gangliosides of cultured astroglia, *Brain Res.*, **461**:98–106.

Schachter, H., and Roseman, S., 1980, Mammalian glycosyltransferases. Their role in the synthesis and function of complex carbohydrates and glycolipids, in: *The Biochemistry of Glycoproteins and Proteoglycans* (W. J. Lennarz, ed.), pp. 85–160, Plenum Press, New York.

Schauer, R., Veh, R. W., Sander, M., Corfield, A. P., and Wiegandt, H., 1980, "Neuraminidase-resistant" sialic acid residues of gangliosides, *Adv. Exp. Med. Biol.* **125**:283–294.

Scheel, G., Schwarzmann, G., Hoffmann-Bleihauer, P., and Sandhoff, K., 1985, The influence of ganglioside insertion into brain membranes on the rate of ganglioside degradation by membrane-bound sialidase, *Eur. J. Biochem.* **153**:29–35.

Scheideler, M. A., and Dawson, G., 1986, Direct demonstration of the activation of UDP-N-acetylgalactosamine:[G_{M3}]N-acetylgalactosaminyltransferase by cyclic AMP, *J. Neurochem.* **46**:1639–1643.

Schengrund, C. L., and Rosenberg, A., 1970, Intracellular location and properties of bovine sialidase, *J. Biol. Chem.* **245**:6196–6200.

Schengrund, C. L., Rosenberg, A., and Repman, M. A., 1976, Ecto-ganglioside-sialidase activity of herpes simplex virus transformed hamster embryo fibroblast, *J. Cell Biol.* **70**:555–561.

Schengrund, C. L., Repman, M. A., and Nelson, J. T., 1979, Distribution in spleen subcellular organelles of sialidase active towards natural sialoglycolipid and sialoglycoprotein substrates, *Biochim. Biophys. Acta* **568**:377–385.

Schwartz, M., and Spirman, N., 1982, Sprouting from chicken embryo dorsal root ganglia induced by nerve growth factor is specifically inhibited by affinity-purified antiganglioside antibodies, *Proc. Natl. Acad. Sci. USA* **79**:6080–6083.

Seifert, W., and Fink, H.-J., 1984, *In-vitro* and *in-vivo* studies on gangliosides in the developing and regenerating hippocampus of the rat, in: *Ganglioside Structure, Function, and Biomedical Potential* (R. W. Ledeen, R. K. Yu, M. M. Rapport, and K. Suzuki, eds.), pp. 535–545, Plenum Press, New York.

Sharom, F. J., and Grant, C. W. M., 1978, A model for ganglioside behavior in cell membranes, *Biochim. Biophys. Acta* **507**:280–293.

Shur, B. D., and Bennette, D., 1978, A specific defect in galactosyltransferase regulation on sperm bearing mutant alleles of the T/t locus, *Dev. Biol.* **71**:243–259.

Sinha, A. K., and Rose, S. P. R., 1973, β-N-acetyl D-galactosaminidase in bulk separated neurons and neurapil from rat cerebral cortex, *J. Neurochem.* **20**:39–44.

Skaper, S. D., and Varon, S., 1985, Ganglioside GM1 overcomes serum inhibition of neuritic outgrowth, *Int. J. Dev. Neurosci.* **3**:187–198.

Skaper, S. D., Katoh-Semba, R., and Varon, S., 1985, GM1 ganglioside accelerates neurite outgrowth from primary peripheral and central neurons under selective culture conditions, *Dev. Brain Res.* **23**:19–26.

Skaper, S. D., Facci, L., Rudge, J., Katoh-Semba, R., Manthorpe, M., and Varon, S., 1986, Morphological modulation of cultured rat brain astroglial cells: Antagonism by ganglioside GM1, *Dev. Brain Res.* **25**:21-31.

Skaper, S. D., Favaron, M., Facci, L., and Leon, A., 1987, Gangliosides stimulate the breakdown of polyphosphoinositides in CNS neurons in vitro, *Soc. Neurosci. Abstr.* **13**:1119.

Skaper, S. D., Favaron, M., Facci, L., Vantini, G., Fusco, M., Ferrari, G., and Leon, A., 1988, Ganglioside involvement in neuronotrophic interactions, in: *New Trends in Ganglioside Research: Neurochemical and Neuroregenerative Aspects* (R. W. Ledeen, E. L. Hogan, G. Tettamanti, A. J. Yates, and R. K. Yu, eds.), Liviana Press, Padova, pp. 351-360.

Sommers, L. W., and Hirschberg, C. B., 1982, Transport of sugar nucleotides into rat liver Golgi. A new Golgi marker activity, *J. Biol. Chem.* **257**:10811-10817.

Sonderfeld, S., Conzelmann, E., Schwarzmann, G., Burg, J., Hinrichs, U., and Sandhoff, K., 1985, Incorporation and metabolism of ganglioside GM2 in skin fibroblasts from normal and GM2 gangliosidosis subjects, *Eur. J. Biochem.* **149**:247-255.

Sonnino, S., Ghidoni, R., Masserini, M., Aporti, F., and Tettamanti, G., 1981, Changes in rabbit brain cytosolic and membrane-bound gangliosides during prenatal life, *J. Neurochem.* **36**:227-232.

Sparrow, J. R., and Grafstein, B., 1982, Sciatic nerve regeneration in ganglioside-treated rats, *Exp. Neurol.* **77**:230-235.

Spero, D. A., and Roisen, F. J., 1984a, Ganglioside-induced neuronal surface activity, *Anat. Rec.* **208**:172A-173A.

Spero, D. A., and Roisen, F. J., 1984b, Ganglioside-mediated enhancement of the cytoskeletal organization and activity in neuro-2a neuroblastoma cells, *Dev. Brain Res.* **13**:37-48.

Spiegel, S., and Fishman, P. H., 1987, Gangliosides as bimodal regulators of cell growth, *Proc., Natl. Acad. Sci. USA* **84**:141-145.

Spiegel, S., Schlesinger, J., and Fishman, P. H., 1984, Incorporation of fluorescent gangliosides into human fibroblats: Mobility, fate, and interaction with fibronectin, *J. Cell Biol.* **99**:699-704.

Spiegel, S., Yamada, K. M., Hom, B. E., Moss, J., and Fishman, P. H., 1985, Fluorescent gangliosides as probes for the retention and organization of fibronectin by ganglioside-deficient mouse cells, *J. Cell Biol.* **100**:721-726.

Spiegel, S., Yamada, K., Hom, B. E., Moss, J., and Fishman, P. H., 1986, Fibrillar organization of fibronectin is expressed coordinately with cell surface gangliosides in a variant murine fibroblast, *J. Cell Biol.* **102**:1898-1906.

Spirman, N., Sela, B.-A., and Schwartz, M., 1982, Antiganglioside antibodies inhibit neuritic outgrowth from regenerating goldfish retinal explants, *J. Neurochem.* **39**:874-877.

Spirman, N., Sela, B.-A., Gilter, C., Calef, E., and Schwartz, M., 1984, Regenerative capacity of the goldfish visual system is affected by antibodies specific to gangliosides injected intraocularly, *J. Neuroimmunol.* **6**:197-207.

Spoerri, P. E., 1983, Effects of gangliosides on the in vitro development of neuroblastoma cells: An ultrastructural study, *Int. J. Dev. Neurosci.* **1**:383-391.

Spoerri, P. E., Rapport, M. M., Mahadik, S. P., and Roisen, F. J., 1988, Inhibition of conditioned media-mediated neuritogenesis of sensory ganglia by monoclonal antibodies to GM1 ganglioside, *Dev. Brain Res.* **41**:71-77.

Stoffyn, A., Stoffyn, P., Farooq, M., Snyder, D. S., and Norton, W. T., 1981, Sialosyltransferase activity and specificity in the biosynthesis in vitro of sialosylgalacto-sylceramide (GM$_4$) and sialosyllactosylceramide (GM$_3$) by rat astrocytes, neuronal perikarya, and oligodendroglia, *Neurochem. Res.* **6**:1149-1157.

Suzuki, K., 1967, Formation and turnover of the major brain gangliosides during development, *J. Neurochem.* **14**:917-925.

Suzuki, K., 1970, Formation and turnover of myelin gangliosides, *J. Neurochem.* **17**:209-213.

Suzuki, K., 1987, Gangliosides and neuropathy, in: *Gangliosides and Modulation of Neuronal Functions* (H. Rahmann, ed.), pp. 531-546, Springer-Verlag, Berlin.

Suzuki, K., Tanaka, H., Yamanaka, T., and Van Damme, O., 1980, The specificity of β-galactosidase in the degradation of gangliosides, *Adv. Exp. Med. Biol.* **125**:307-318.

Suzuki, T., Harada, M., Suzuki, Y., and Matsumoto, M., 1984, Incorporation of sialoglycoprotein containing lacto-series oligosaccharides into chicken asialoerythrocyte membranes and restoration of receptor activity toward hemagglutinating virus of Japan (Sendai virus), *J. Biochem.* **95**:1193–1200.

Takeda, Y., Takeda, T., Honda, T., Sakurai, J., Ohtomo, N., and Miwatani, T., 1975, Inhibition of hemolytic activity of the thermostable direct hemolysin of Vibrio parahaemolyticus by ganglioside, *Infect. Immun.* **12**:931–933.

Tanaka, K., Dora, E., Urbanics, R., Greenberg, J. H., Toffano, G., and Reivich, M., 1986, Effect of the ganglioside GM1 on cerebral metabolism, microcirculation, recovery kinetics of ECoG and histology, during the recovery period following focal ischemia in cats, *Stroke* **17**:1170–1178.

Tettamanti, G., Morgan, I. G., Gombos, G., Vincendon, G., and Mandel, P., 1972, Sub-synaptosomal localization of brain particulate neuraminidase, *Brain Res.* **47**:515–518.

Tettamanti, G., Venerando, B., Roberti, S., Chigorno, V., Sonnino, S., Ghidoni, R., Orlando, P., and Massari, P., 1981, The fate of exogenously administered brain gangliosides, in: *Gangliosides in Neurological and Neuromuscular Function, Development, and Repair* (M. M. Rapport and A. Gorio, eds.), pp. 225–240, Raven Press, New York.

Tettamanti, G., Ledeen, R. W., Sandhoff, K., Nagai, Y., and Toffano, G. (eds.), 1986, *Gangliosides and Neuronal Plasticity*, Vol. 6, Liviana Press, Padova.

Toffano, G., Benvengnu, D., Bonetti, A. C., Facci, L., Leon A., Orlando, P., Ghidoni, R., and Tettamanti, G., 1980, Interactions of GM1 ganglioside with crude rat brain neuronal membranes, *J. Neurochem.* **35**:861–866.

Toffano, G., Savoini, G., Morono, F., Lombardi, G., Calza, L., and Agnati, L. F., 1983, GM1 ganglioside stimulates the regeneration of dopaminergic neurons in the central nervous system, *Brain Res.* **261**:163–166.

Tsuji, S., Arita, M., and Nagai, Y., 1983, GQ1b, a bioactive ganglioside that exhibits novel nerve growth factor (NGF)-like activities in the two neuroblastoma cell lines, *J. Biochem.* **94**:303–306.

Tsuji, S., Nakajima, J., Sasaki, T., and Nagai, Y., 1985, Bioactive gangliosides. IV. Ganglioside GQ1b/ca^{2+} dependent protein kinase activity exists in the plasma membrane fraction of neuroblastoma cell line, GOTO, *J. Biochem.* **97**:969–972.

Tsuji, S., Yamashita, T., Tanaka, M., and Nagai, Y., 1988, Synthesis of sialyl compounds as well as natural gangliosides induce neuritogenesis in a mouse neuroblastoma cell line (neuro 2a), *J. Neurochem.* **50**:414–423.

Usuki, S., Lyu, S.-C., and Sweeley, C. C., 1988a, Sialidase activities of cultured human fibroblasts and the metabolism of GM3 ganglioside, *J. Biol. Chem.*, **263**:6847–6857.

Usuki, S., Hoops, P., and Sweeley, C. C., 1988b, Growth control of human foreskin fibroblasts and inhibition of extracellular sialidase activity by 2-deoxy-2,3-dehydro-N-acetyl-neuraminic acid, *J. Biol. Chem.*, **263**:10595–10599.

Van den Eijnden, D. H., and van Dijk, W., 1974, Properties and regional distribution of cerebral CMP-N-acetylneuraminic acid: Glycoprotein sialyltransferase, *Biochim. Biophys. Acta* **362**:136–149.

van Heyningen, W. E., 1984, Gangliosides as membrane receptors for tetanus toxin, cholera toxin and serotonin, *Nature* **249**:415–417.

Varon, S., Manthorpe, M., and Williams, L. R., 1984, Neuronotrophic and neurite-promoting factors and their clinical potentials, *Dev. Neurosci.* **6**:73–100.

Varon, S., Skaper, S. D., and Katoh-Semba, R., 1986, Neuritic responses to GM1 ganglioside in several in vitro systems, in: *Gangliosides and Neuronal Plasticity* (G. Tettamanti, R. Ledeen, K. Sandhoff, Y. Nagai, and G. Toffano, eds.), pp. 215–230, Liviana Press, Padova.

Varon, S., Pettmann, B., and Manthorpe, M., 1988, Extrinsic regulations of neuronal maintenance and repair, in: *New Trends in Ganglioside Research: Neurochemical and Neuroregenerative Aspects* (R. W. Ledeen, E. L. Hogan, C. Tettamanti, A. J. Yates, and R. K. Yu, eds.), Liviana Press, Padova, pp. 607–623.

Vaswani, K. K., and Ledeen, R. W., 1988, Gangliosides stimulate phosphoinositide breakdown in neuro-2A cells, *Trans. Am. Soc. Neurochem.* **19**:180.

Vaswani, K. K., Wu, G., and Ledeen, R. W., 1989, Gangliosides stimulate calcium influx and hydrolysis of phosphoinositides in neuro-2A cells, submitted for publication.

Verghese, J. P., Bradley, W. G., Mitsumoto, H., and Chad, D., 1982, A blind controlled trial of adrenocorticotropin and cerebral gangliosides in nerve regeneration in the rat, *Exp. Neurol.* **77**:455–458.

Vyskocill, F., Di Gregorio, F., and Gorio, A., 1985, The facilitating effect of gangliosides on the electrogenic (Na$^+$/ pump and on the resistance of the membrane potenetial to hypoxia in neuromuscular preparation, *Pfluegers Arch.* **403**:1–6.

Walkley, S. U., Wurzelmann, S., and Purpura, D. P., 1981, Ultrastructure of neurites and meganeurites of cortical pyramidal neurons in feline gangliosidosis as revealed by the combined Golgi–EM technique, *Brain Res.* **211**:393–398.

Wiegandt, H., 1985, Gangliosides, in: *Glycolipids,* Vol. 10 (A. Neuberger and L. L. M. van Deenen, eds.), pp. 199–260, Elsevier, Amsterdam.

Willinger, M., and Schachner, M., 1980, GM1 ganglioside as a marker for neuronal differentiation in mouse cerebellum, *Dev. Biol.* **74**:101–117.

Wojcik, M., Ulas, J., and Oderfeld-Nowak, B., 1982, The stimulating effect of ganglioside injections on the recovery of choline acetyltransferase and acetylcholinesterase activities in the hippocampus of the rat after septal lesions, *Neuroscience* **7**:495–499.

Yamada, K. H., Critchley, D. R., Fishman, P. H., and Moss, J., 1983, Exogenous gangliosides enhance the interaction of fibronectin with ganglioside-deficient cells, *Exp. Cell Res.* **143**:295–302.

Yamada, K., Abe, A., and Sasaki, T., 1985, Specificity of the glycolipid transfer protein from pig brain, *J. Biol. Chem.* **260**:4615–4621.

Yamakawa, T., and Nagai, Y., 1978, Glycolipids at the cell surface and their biological functions, *Trends Biochem. Sci.* **3**:128–131.

Yanagisawa, K., Taniguchi, N., and Makita, A., 1987, Purification and properties of GM2 synthase, UDP-N-acetylgalactosamine: GM3 β-N-acetylgalactosaminyltransferase from rat liver, *Biochim. Biophys. Acta* **919**:213–220.

Yasuda, Y., Tiemeyer, M., Blackburn, C. C., and Schnaar, R. L., 1988, Neuronal recognition of gangliosides: Evidence for a brain ganglioside receptor, in: *New Trends in Ganglioside Research: Neurochemical and Neuroregenerative Aspects* (R. W. Ledeen, E. L. Hogan, G. Tettamanti, A. J. Yates, and R. K. Yu, eds.), Liviana Press, Padova, pp. 229–243.

Yates, A. J., 1986, Gangliosides in the nervous system during development and regeneration, *Neurochem. Pathol.* **5**:309–329.

Yates, A. J., Tipnis, U. R., Hofteig, J. H., and Warner, J. K., 1984, Biosynthesis and transport of gangliosides in peripheral nerve, in: *Ganglioside Structure, Function and Biomedical Potential* (R. Ledeen, R. K. Yu, M. M. Rapport, and K. Suzuki, eds.), pp. 155–168, Plenum Press, New York.

Yavin, E., and Yavin, Z., 1979, Ganglioside profiles during neural tissue development. Acquisition in the prenatal rat brain and cerebral cell cultures, *Dev. Neurosci.* **2**:25–37.

Yavin, E., Gil, S., Consolazione, A., dal Toso, R., and Leon, A., 1987, Selective enhancement of tubulin gene expression and increase in oligo(dT)-bound RNA in the rat brain after nigrostriatal pathway unilateral lesion and treatment with ganglioside, *J. Neurosci. Res.* **18**:615–620.

Yip, M. C. M., and Dain, J. A., 1969, The enzymic synthesis of ganglioside. 1. Brain uridine diphosphate D-galactose:N-acetyl-galactosaminyl-galactosyl-glucosyl-ceramide galactosyl transferase, *Lipids* **4**:270–277.

Yogeeswaran, G., Murray, R. K., Pearson, M. L., Sanwal, B. D., McMorris, F. A., and Ruddle, F. H., 1973, Glycosphingolipids of clonal lines of mouse neuroblastoma and neuroblastoma × L cell hybrids, *J. Biol. Chem.* **248**:1231–1239.

Yohe, H. C., Jacobson, R. I., and Yu, R. K., 1983, Ganglioside–basic protein interaction: Protection of gangliosides against neuraminidase action, *J. Neurosci. Res.* **9**:401–412.

Yu, R. K., and Ando, S., 1980, Structures of some new complex gangliosides of fish brain, in: *Structure and Function of Gangliosides* (L. Svennerholm, P. Mandel, H. Dreyfus, and P.-F. Urban, eds.), pp. 33–45, Plenum Press, New York.

Yusuf, H. K. M., Pohlentz, G., Schwarzmann, G., and Sandhoff, K., 1983a, Ganglioside biosynthesis in Golgi apparatus of rat liver, *Eur. J. Biochem.* **134**:47–54.

Yusuf, H. K. M., Pohlentz, G., and Sandhoff, K., 1983b, Tunicamycin inhibits ganglioside biosynthesis by blocking sugar nucleotide transport across the membrane vesicles, *Proc. Natl. Acad. Sci. USA* **80**:7075–7079.

Yusuf, H. K. M., Schwarzmann, G., Pohlentz, G., and Sandhoff, K., 1987, Oligosialogangliosides inhibit GM2- and GD3-synthesis in isolated Golgi vesicles from rat liver, *Biol. Chem. Hoppe-Seyler* **368**:455–462.

Structure and Localization of Glycoproteins and Proteoglycans

Renée K. Margolis and Richard U. Margolis

1. INTRODUCTION

Glycosaminoglycans are high-molecular-weight linear carbohydrate polymers that are generally composed of disaccharide repeating units of a uronic acid (D-glucuronic acid or L-iduronic acid) and a hexosamine (GlcNAc or GalNAc). Chondroitin sulfate and heparan sulfate occur as proteoglycans in which the polysaccharide chains are covalently linked at their reducing ends to the hydroxyl groups of serine residues in a protein moiety.

The glycoproteins differ from the glycosaminoglycans in the following major respects: (1) they do not usually contain uronic acid; (2) they lack a serially repeating unit; (3) they contain a relatively low number of sugar residues in the heterosaccharide, which is often branched; and (4) they contain several sugars that are not characteristic components of the glycosaminoglycans (e.g., fucose, sialic acid, galactose, mannose). However, there are exceptions to the general description given above, and in certain cases the distinction between glycosaminoglycans and glycoproteins is not well defined.

Several recent monographs and reviews have been devoted to the biochemistry, cell biology, and structural analysis of glycoproteins and proteoglycans, and the reader is referred to these for more general background information (Kennedy, 1979;

Abbreviations used in this chapter: Con A, concanavalin A; NGF, nerve growth factor; NILE, NGF-inducible large external glycoprotein; AcNeu, *N*-acetylneuraminic acid; Fuc, L-fucose; Gal, D-galactose; GalNAc, N-acetyl-D-galactosamine; GlcNAcol, N-acetyl-D-glucosaminitol; GalNAcol, N-acetyl-D-galactosaminitol; GcNeu, N-glycolylneuraminic acid; Glc, D-glucose; GlcNAc, N-acetyl-D-glucosamine; Man, D-mannose; Manol, mannitol; DBH, dopamine β-hydroxylase.

Renée K. Margolis • Department of Pharmacology, State University of New York, Health Science Center at Brooklyn, Brooklyn, New York 11203.   *Richard U. Margolis* • Department of Pharmacology, New York University Medical Center, New York, New York 10016.

Ginsburg, 1982, 1987; Schauer, 1982; Höök *et al.*, 1984; Sweeley and Nunez, 1985; Hassell *et al.*, 1986; Osawa and Tsuji, 1987; Wight and Mecham, 1987).

2. STRUCTURE

2.1. Glycoproteins

2.1.1. *N*-Glycosidic Oligosaccharides

Approximately 85–90% of the carbohydrate in brain glycoproteins is linked via GlcNAc to the amide nitrogen of asparagine residues in the protein moiety. Several different types of such *N*-glycosidically linked carbohydrate units are known to occur, and can be partially fractionated by affinity chromatography on Con A–Sepharose under standardized conditions (Finne and Krusius, 1982; Merkle and Cummings, 1987). The "high-mannose" glycans, which contain only GlcNAc and Man as their sugar components, are strongly bound to Con A. A typical structure for this type of oligosaccharide from brain is shown in Figure 1. However, most of the *N*-glycosidically linked carbohydrate units in nervous tissue glycoproteins are of the "complex" type, which contain other sugars (e.g., Gal, Fuc, and sialic acid) in addition to Man and GlcNAc, and are generally more highly branched than the high-mannose glycans. Those having two peripheral branches (biantennary) are bound only weakly to Con A, whereas the tri- and tetraantennary complex oligosaccharides (Figure 2) are not bound at all.

"Hybrid" oligosaccharides, which contain both complex and high-mannose peripheral branches, account for less than 1% of the oligosaccharides in brain and PC12 pheochromocytoma cell glycoproteins (R. K. Margolis *et al.*, 1983a). In brain, a significant portion of these oligosaccharides may be present at Asn-98 of the Thy-1 glycoprotein, where half of the oligosaccharides are reported to be of the hybrid type (Parekh *et al.*, 1987).*

Substitution pattern of brain glycoprotein oligosaccharides: The branch point in the core pentasaccharide consists of a 3,6-di-*O*-substituted Man residue together with a small proportion (12% of the total Man in rat brain glycoproteins) which is 3,4,6-tri-*O*-substituted, while the more peripheral Man residues are mono- or disubstituted at the 2, 4, and 6 positions (Krusius and Finne, 1977). The tri-*O*-substituted Man presumably reflects the presence in brain of biantennary complex oligosaccharides having a "bisecting" β1-4-linked GlcNAc residue, such as the biantennary oligosaccharides at Asn-74 of the brain Thy-1 glycoprotein (almost 90% of which are reported to be of this type; Parekh *et al.*, 1987). Although glycopeptides from brain and PC12 cell glycoproteins generally do not bind to the erythroagglutinating lectin (E-PHA) of *Phaseolus vulgaris* (R. K. Margolis *et al.*, 1983a,b), which is known to bind certain bisected biantennary oligosaccharides, the other structural determinants required for binding by this lectin (Yamashita *et al.*, 1983a) are probably not present in most bisected oligosaccharides of brain and PC12 cell glycoproteins.

Over half of the GlcNAc residues are 4-*O*-substituted, while approximately one quarter are terminal and a small proportion are 3,4- and 4,6-di-*O*-substituted (Krusius and Finne, 1977). Gal apparently accounts for one of the substituents found at C-3 of GlcNAc residues, since significant amounts of Gal(β1-3)GlcNAcol can be detected after partial acid hydrolysis and reduction of a brain glycopeptide fraction (Krusius and Finne, 1978).

With respect to Gal, 53% is 3-*O*-substituted (largely by sialic acid) and most of the remainder (38%) is

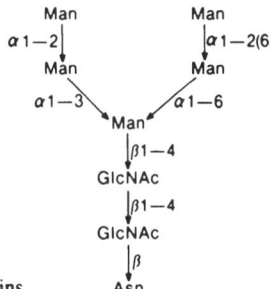

Figure 1. Typical high-mannose-type oligosaccharide of brain glycoproteins.

2.1.2. O-Glycosidic Oligosaccharides

Ten to fifteen percent of the carbohydrate in brain glycoproteins is O-glyco-sidically linked via GalNAc to the hydroxyl groups of Ser and Thr residues in the protein moiety. These O-glycosidically linked oligosaccharides consist of the core disaccharide Gal(β1-3)GalNAc, which may occur either as such, or substituted with sialic acid residues at C-3 of Gal and/or C-6 of GalNAc. The tetrasaccharide accounts for half of the total O-glycosidically linked carbohydrate in rat brain glycoproteins (for references and further information, see Margolis and Margolis, 1979). The α anomer of the core disaccharide (which has been detected only in a nonsialylated form) accounts for approximately 13% of the O-glycosidically linked oligosaccharides of rat brain and appears to be present only in nervous tissue (Finne and Krusius, 1976). The chondroitin sulfate proteoglycans of brain also contain a series of novel oligosac-charides which are O-glycosidically linked via Man to Ser and Thr residues in the protein core (Finne et al., 1979; Krusius et al., 1986, 1987). These are described in the subsection of Section 2.2.

---

terminal (Krusius and Finne, 1977). The 8% of 6-O-substituted Gal is largely due to the presence of sialic acid residues at this position. The 6-O-substituted Gal and the 5% of 6-O-substituted GlcNAc residues are probably also partially accounted for by the fact that brain glycoproteins contain sulfate which is linked at C-6 of these two sugars, and that a sulfated trisaccharide containing Gal, glucosamine, and Man can be obtained by partial acid hydrolysis of brain glycopeptides (Margolis and Margolis, 1970).

In the complex oligosaccharides of brain glycoproteins, much of the Fuc is linked to peripheral GlcNAc residues in an oligosaccharide with the structure AcNeu(α2-3)Gal(β1-4)[Fuc(α1-3)]GlcNAc(β1-(Krusius and Finne, 1978). Only 29% of the Fuc in rat brain glycoproteins was found to be linked to the proximal GlcNAc residue in the di-N-acetylchitobiose unit and less than 2% to penultimate Gal residues (Krusius and Finne, 1977). Fuc is commonly found in these latter two locations in soluble (secreted) glycoproteins such as those of plasma, as compared to the largely membrane-bound glycoproteins present in nervous tissue.

While most of the sialic acid in brain glycoproteins is in the expected nonreducing terminal position, in adult rat brain approximately 9% has been shown to be 8-O-substituted by another sialic acid residue. Disialosyl groups are present in particularly large amounts in glycoproteins of certain microsomal and plasma membrane preparations from brain (Finne et al., 1977; Krusius et al., 1978).

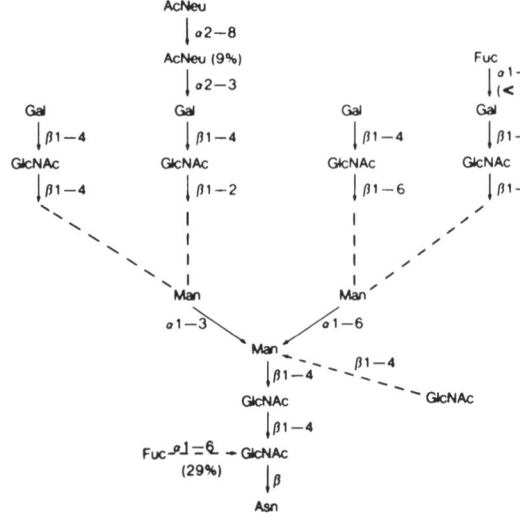

```
AcNeu
 |α2—8
AcNeu (9%)                                    Fuc
 |α2—3                                         |α1—2
Gal          Gal          Gal          Gal    |(< 2%)
 |β1—4        |β1—4         |β1—4        |β1—4
GlcNAc       GlcNAc       GlcNAc       GlcNAc
 |β1—4        |β1—2         |β1—6        |β1—2
```

Figure 2. Structural features of complex oligosaccharides. This composite figure illustrates the most common sugar sequences and linkages found in the complex oligosaccharides of brain glycoproteins. However, it should be noted that the structural features summarized here have not been demonstrated to occur on a *single* identified oligosaccharide. Mono- and oligosaccharides that occur as substituents on the "core" structure are connected by dashed lines to the sugar residues to which they may be linked. The percentages given in parentheses refer to the proportion of the total fucose in rat brain glycoproteins which is present at two specific linkage sites, and indicate the small proportion of sialic acid present in disialosyl linkages, rather than its more common occurrence as single residues at the nonreducing termini of Gal(β1-4)GlcNAc (β1-)Man chains. For more detailed information on the substitution patterns of the different monosaccharides, see the footnote on p. 86.

2.1.3. Polysialosyl Oligosaccharides

In embryonic and early postnatal brain, a small proportion of Asn-linked tri- and tetraantennary oligosaccharides contain up to 12 or more AcNeu residues joined by α2-8 linkages in polysialic acid chains (Finne, 1985; Finne and Mäkelä, 1985). These oligosaccharides, which in brain are apparently confined to the neural cell adhesion molecule, also contain ester sulfate residues (Margolis and Margolis, 1983). In addition to brain, polysialosyl oligosaccharides are present in neural tumor cell lines such as PC12 and neuroblastoma cells (Margolis and Margolis, 1983; Livingston *et al.*, 1988), in the voltage-sensitive sodium channel from *Electrophorus electricus* (James and Agnew, 1987), as well as in developing postnatal rat kidney (Roth *et al.*, 1987) and probably other nonnervous tissues. Their possible biological roles in cell–cell interactions and other processes are discussed in Chapter 12.

2.1.4. Poly(*N*-acetyllactosaminyl) Oligosaccharides

Oligosaccharides of the poly(*N*-acetyllactosamine) series have been identified in glycoproteins and glycolipids which express a number of developmentally regulated and tumor-associated antigens. They are characterized by the presence of up to six or more disaccharide repeating units having the structure $(Gal\beta1\text{-}4GlcNAc\beta1\text{-}3)_n$, which in the case of glycoproteins are linked to a conventional $(Man)_3(GlcNAc)_2$ complex oligosaccharide core. These carbohydrate chains were originally detected in erythrocytes and their precursor cells, and have more recently been found in a number of other cell types, including sympathetic neurons, PC12 cells, and neuroblastoma cells (R. K.

Margolis *et al.*, 1986; Spillmann and Finne, 1987). Susceptibility to several bacterial endo-β-galactosidases is considered diagnostic of the presence of such oligosaccharide chains, although sulfation and branching at internal β-galactosidic linkages is known to confer resistance to enzymatic hydrolysis. In neuroblastoma and PC12 cells, the poly(N-acetyllactosaminyl) oligosaccharides have a predominantly linear structure (Spillmann and Finne, 1987).

Endo-β-galactosidase treatment of glycopeptides derived from the trypsinate and membranes of PC12 pheochromocytoma cells and cultured sympathetic neurons demonstrated the presence of poly(N-acetyllactosaminyl) units on tri- and tetraantennary oligosaccharides, some of which have a core Fuc and a 2,6-substituted α-linked Man (R. K. Margolis *et al.*, 1986). NGF-induced differentiation of the PC12 cells led to a small but significant decrease in the proportion of these oligosaccharides. After selective labeling of poly(N-acetyllactosaminyl) glycans on the PC12 cell surface, SDS-PAGE and fluorography revealed that they are present in a number of distinct glycoproteins having apparent molecular sizes in the range of 35,000 to 250,000 Da. One of these is a major 230,000-Da cell-surface glycoprotein [the NGF-inducible large external (NILE) glycoprotein] of PC12 cells, where they appear to account for much or all of its larger size in PC12 cells as compared to the immunochemically cross-reactive 205,000-Da species present in postnatal brain (see Section 3.1). Poly(N-acetyllactosaminyl) oligosaccharides are generally not present in the glycoproteins of postnatal brain, although they do occur in the form of keratan sulfate chains in the chondroitin sulfate proteoglycans (see the subsection of Section 2.2).

In other studies, poly(N-acetyllactosaminyl) oligosaccharides were detected in a subpopulation of large tri- and tetraantennary glycopeptides derived from chromaffin granule membrane glycoproteins. After specific labeling of these oligosaccharides in intact membranes followed by two-dimensional SDS-PAGE, it was found that most of the poly(N-acetyllactosaminyl) units are present on Glycoprotein IV, a quantitatively minor glycoprotein whose function has not been established, together with lesser labeling of Glycoproteins II and III, whereas these oligosaccharides are not present in the major glycoproteins (dopamine β-hydroxylase and carboxypeptidase H) of chromaffin granule membranes (R. U. Margolis *et al.*, 1988).

2.1.5. Sulfation of Glycoprotein Oligosaccharides

In the first report of sulfated tissue (as distinguished from mucin) glycoproteins, it was concluded that the N-glycosidic oligosaccharides of rat brain glycoproteins contained sulfated residues which were tentatively identified as galactose 6-sulfate and N-acetylglucosamine 6-sulfate (Margolis and Margolis, 1970). Although, for a period of some years, brain appeared to be a relatively special case in this regard, by the early 1980s a large number of reports began to appear which clearly demonstrated that sulfation of glycoprotein oligosaccharides is a rather general phenomenon. The list of examples is by now quite long and diverse, both structurally and phylogenetically, including membrane glycoproteins (from rat ascites mammary adenocarcinoma, the basement membrane glycoprotein entactin, corneal epithelium, rat liver plasma membranes, murine zona pellucida, endothelial cells, calf thyroid plasma membranes); as

well as pituitary hormones (lutropin and proopiomelanocortin, the common precursor
to β-endorphin and α-melanocyte-stimulating hormone); various other nonmucin secreted glycoproteins (such as hen egg albumin and a chick chorioallantoic fluid
glycoprotein); sea urchin glycoproteins; lysosomal hydrolases of the cellular slime
mold, *Dictyostelium discoideum; Halobacteria* cell surface and flagellar glycoproteins;
and viral coat proteins. In those cases where more detailed structural studies have been
performed, the sulfated residues have most commonly been identified as *N*-acetyl-
glucosamine 6-sulfate and galactose 6-sulfate.*

Although most of the sulfated oligosaccharides in brain and other tissues have
been found to be *N*-glycosidically linked, it should be noted that the chondroitin sulfate
proteoglycans of brain contain a series of novel Man-linked *O*-glycosidic tri- to pentasaccharides, some of which are sulfated, in addition to sulfated tri- and tetraantennary Asn-linked oligosaccharides (see the subsection of Section 2.2).

The possibility that as yet uncharacterized sulfated sugar residues are present in
nervous tissue glycoproteins assumes added importance from the recent identification
of glucuronic acid 3-sulfate as the epitope recognized by the HNK-1 monoclonal
antibody (see also Chapters 1, 8, and 11), and the widespread occurrence of this novel
sulfated sugar in nervous tissue and other glycoproteins and proteoglycans. These
include the myelin associated glycoprotein, the ependymins (a family of soluble
glycoproteins of goldfish and mammalian brain), and several neural cell adhesion
molecules such as N-CAM, the NILE glycoprotein (also referred to as the Ll antigen or
as Ng-CAM), and the Jl antigen. The HNK-1 glucuronic acid 3-sulfate epitope is also
present in several glycoproteins of PC12 pheochromocytoma cells and chromaffin
granule membranes, as well as on poly(*N*-acetyllactosaminyl) and other sulfated
oligosaccharides in the chondroitin sulfate proteoglycans of brain, cartilage, and
chondrosarcoma (but not in the heparan sulfate proteoglycan of brain, or in either of
two chondroitin sulfate/dermatan sulfate proteoglycans in the chromaffin granule matrix). (For further details and references, see R. K. Margolis *et al.*, 1987b; Gowda *et
al.*, 1989a.)

2.2. Glycosaminoglycans and Proteoglycans

The structures of the glycosaminoglycans are summarized in Table 1. The concentration of total glycosaminoglycans in adult rat brain (in terms of hexosamine) is
0.4 μmole/100 mg lipid-free dry weight, comprising 63% chondroitin sulfate, 25%
hyaluronic acid, and 12% heparan sulfate. (For the concentration and composition of
brain glycosaminoglycans in other species, see Margolis and Margolis, 1979.)

**N*-acetylglucosamine 4- and 6-sulfate in calf thyroid plasma membrane (Edge and Spiro, 1984); *N*-acetyl-
glucosamine sulfate in endothelial cell membranes (Merkle and Heifetz, 1984); *N*-acetylglucosamine 6-
sulfate (91% of sulfate residues) and galactose 6-sulfate in chick chorioallantoic fluid (Choi and Meyer,
1975); *N*-acetylglucosamine 6-sulfate in hog gastric mucosa (Slomiany and Meyer, 1972); *N*-acetylhex-
osamine mono- and disulfate and galactose monosulfate in rat gastric mucosa (Liau and Horowitz, 1982);
and galactose 6-sulfate in human bronchial mucin (Roussel *et al*, 1975). However, *N*-acetylgalactosamine
4-sulfate (bovine pituitary lutropin, follitropin, and thytropin; Green and Baenziger, 1988); mannose 4-
sulfate (hen egg albumin; Yamashita *et al.*, 1983b); and glucuronic acid 2- or 3-sulfate (halobacterial cell
surface and flagellar glycoproteins; Lechner *et al.*, 1985; Wieland *et al.*, 1985) have also been reported.

Table 1. Structures of the Glycosaminoglycans

Glycosaminoglycan[a]	Sugars present in disaccharide repeating unit and linkage	Minor or atypical monosaccharide components[b]
Hyaluronic acid	D-Glucuronic acid (β1→3) N-Acetyl-D-glucosamine (β1→4)	
Chondroitin 4-sulfate (chondroitin sulfate A)	D-Glucuronic acid (β1→3) N-Acetyl-D-galactosamine 4-O-sulfate (β1→4)	
Chondroitin 6-sulfate (chondroitin sulfate C)	D-Glucuronic acid (β1→3) N-Acetyl-D-galactosamine 6-O-sulfate (β1→4)	
Dermatan sulfate (chondroitin sulfate B)	L-Iduronic acid (α1→3)[c] N-Acetyl-D-galactosamine 4-O-sulfate (β1→4)	D-Glucuronic acid (β1→3) N-Acetyl-D-glucosamine 6-O-sulfate
Keratan sulfate (keratosulfate)	D-Galactose and D-galactose 6-O-sulfate (β1→4) N-Acetyl-D-glucosamine 6-O-sulfate ((β1→3)	L-Fucose D-Mannose Sialic acid
Heparan sulfate (heparitin sulfate)	D-Glucuronic acid (β1→4)[c,d] N-Acetyl-D-glucosamine (α1→4) N-Sulfo-D-glucosamine and N-sulfo-D-glucosamine 6-O-sulfate (α1→4)	L-Iduronic acid (α1→4)[c] L-Iduronic acid 2-O-sulfate N-Acetyl-D-glucosamine 6-O-sulfate
Heparin	L-Iduronic acid (α1→4) and L-Iduronic acid 2-O-sulfate N-Sulfo-D-glucosamine and N-sulfo-D-glucosamine 6-O-sulfate (α1→4)	D-Glucuronic acid N-Acetyl-D-glucosamine 6-O-sulfate

[a] Older nomenclature in parentheses.
[b] Chondroitin 4-sulfate, chondroitin 6-sulfate, dermatan sulfate, heparin, and heparan sulfate also contain D-galactose and D-xylose in the carbohydrate–peptide linkage (-glucuronosyl-galactosyl-galactosyl-xylosyl-O-serine.
[c] L-Iduronic acid is the 5-epimer of D-glucuronic acid, and therefore the iduronosyl group has α-configuration according to the convention of carbohydrate chemistry. Linkage is equatorial, carboxyl group axial.
[d] In heparin and heparan sulfate, the two glucuronic acid residues at the reducing end are linked β1→3 and β1→4.

Chondroitin sulfate is the predominant glycosaminoglycan of adult brain, and in rat brain is present almost exclusively as the 4-sulfate isomer. Dermatan sulfate, which is similar to chondroitin 4-sulfate but contains L-iduronic acid rather than D-glucuronic acid, is present in only very small amounts, if at all, in normal brain (Kiang et al., 1981). However, dermatan sulfate constitutes half or more of the glycosaminoglycans present in the matrix and membranes of adrenal chromaffin granules (Kiang et al., 1982). Although keratan sulfate proteoglycans do not appear to occur as such in nervous tissue, Man-linked keratan sulfate chains are present as minor components (< 3% by weight) of the chondroitin sulfate proteoglycans of brain (see below).

Heparan sulfate has a more complex composition than the other glycosamino- glycans insofar as it contains GlcNAc, glucosamine-N-sulfate, glucosamine-N,O-dis- ulfate, D-glucuronic acid, L-iduronic acid, L-iduronic acid 2-sulfate, and probably small amounts of GlcNAc 6-sulfate residues. Moreover, there is known to be a consid- erable variation in the structures of heparan sulfates isolated from different sources. Heparan sulfate is distinguished from the closely related glycosaminoglycan heparin (a product of mast cells) mainly by its lower degree of sulfation, larger proportion of N- acetylated as compared to N-sulfated glucosamine residues, and by its containing predominantly glucuronic acid as its uronic acid, rather than the 70–80% iduronic acid found in most preparations of heparin. While heparan sulfate is a well-documented constituent of nervous tissue (Klinger et al., 1985; R. K. Margolis et al., 1987a; Ripellino and Margolis, 1989), there is no evidence that heparin is present in brain parenchyma. (However, traces of heparin probably do occur in brain as a component of mast cells, which are known to be associated with blood vessels and meninges.)

In all tissues studied including brain, chondroitin sulfate and heparan sulfate are O-glycosidically linked to a protein core through a sequence consisting of -glucurono- syl-galactosyl-galactosyl-xylosyl-serine. Therefore, these sulfated glycosaminoglycans occur in tissues as *proteoglycans*. In contrast, hyaluronic acid does not occur in a covalent complex with protein, and chain initiation takes place by an apparently unique mechanism which is described in Chapter 5.

Chondroitin Sulfate Proteoglycans of Brain

The chondroitin sulfate proteoglycans of brain range in molecular size from approximately 260 to 325 kDa, based on their gel filtration behavior under dissociative conditions (Krusius et al., 1987). They are mostly soluble in a phosphate-buffered saline extract, and account for less than 1% of the soluble brain protein (Kiang et al., 1981). Only a relatively small portion of the native chondroitin sulfate proteoglycans of brain enter a 6–12% SDS-polyacrylamide gel. However, after treatment of the proteoglycans with chondroitinase ABC (or chondroitinase and endo-β-galactosidase) in the presence of protease inhibitors, seven bands with molecular sizes ranging from 67 to > 350 kDa appear in Coomassie blue-stained gels and in fluorographs of sodium [^{35}S]sulfate- labeled proteoglycans. Most of these components probably represent individual pro- teoglycan species rather than different degrees of non-chondroitin sulfate/keratan sul- fate glycosylation of a single protein core, since [^{35}S]methionine-labeled proteins of comparable molecular size were synthesized by an *in vitro* translation system (Gowda et *al.*, 1989a). These findings suggest that chondroitin sulfate proteoglycans which differ in molecular size and composition (but are immunochemically cross-reactive) may be specific to particular cell types in brain. Although biochemical assays have demonstrated only a limited degree of aggregation with hyaluronic acid, comparative studies on the localization of hyaluronic acid, hyaluronic acid-binding region, link protein, and chondroitin sulfate proteoglycans in developing rat cerebellum suggest that much of the chondroitin sulfate proteoglycan of brain may occur in the form of aggregates with hyaluronic acid (and link protein) *in situ* (Ripellino *et al.*, 1988, 1989b).

The chondroitin sulfate proteoglycans contain an average of 56% protein, 24% glycosaminoglycans (predominantly chondroitin 4-sulfate), and 20% *N*- and *O*-glycosidic glycoprotein oligosaccharides (Kiang *et al.*, 1981; Klinger *et al.*, 1985; Krusius *et al.*, 1986, 1987). The Asn-linked oligosaccharides are almost exclusively of the tri- and tetraantennary types (Klinger *et al.*, 1985), whereas the *O*-glycosidic oligosaccharides consist both of conventional Gal(β1-3)GalNAc units and their mono- and disialyl derivatives, as well as a series of novel mannosyl-*O*-serine/threonine-linked oligosaccharides which can be isolated by mild alkaline-borohydride treatment of the proteoglycan glycopeptides, and have the sequence GlcNAc(β1-3)Manol at their proximal ends (Finne *et al.*, 1979; Krusius *et al.*, 1986, 1987). These latter include GlcNAc(β1-3)Manol itself, Gal(β1-4)[Fuc(α1-3)]GlcNAc(β1-3)Manol, AcNeu(α2-3) Gal(β1-4)GlcNAc(β1-3)Manol, and both short (3000–4500 Da) and long (10,000 Da) Man-linked keratan sulfate chains composed of disaccharide repeating units consisting of Gal(β1-4)GlcNAc-6-SO$_4$(β1-3).

The structural properties of the less abundant proteoglycan oligosaccharides were studied by methylation analysis and fast-atom-bombardment mass spectrometry (Krusius *et al.*, 1987). Five fractions containing [^{35}S]sulfate-labeled oligosaccharides were obtained by ion exchange chromatography, each of which eluted from Sephadex G-50 as two well-separated peaks. The larger-molecular-size fractions contained short Man-linked keratan sulfate chains, together with some Asn-linked oligosaccharides. The smaller oligosaccharides obtained by gel filtration represent tri- to pentasaccharides of 800–1400 Da that contain one to two residues of sialic acid and/or sulfate.

Alkaline borohydride treatment of the neutral and monosialyl glycopeptide fractions followed by amino acid and carbohydrate analyses demonstrated that over half of the carbohydrate–peptide linkages in the chondroitin sulfate proteoglycans are of the Man-*O*-Ser/Thr type. Although the latter linkages are found in certain microbial and invertebrate glycoproteins, they have not been reported in other mammalian tissues, and mannitol has not been detected following alkaline borohydride treatment of the chondroitin sulfate proteoglycan of cartilage (which is, however, known to contain other types of both *N*- and *O*-glycosidically linked oligosaccharides). This distinctive oligosaccharide composition may be related to specialized functional roles for chondroitin sulfate proteoglycans in nervous tissue, since immunoelectron microscopy has demonstrated that they are predominantly extracellular in early postnatal rat brain, but progressively assume an intracellular (cytoplasmic) localization in neurons and astrocytes during the first month of brain development (see Section 3.4.1).

Some of the Asn-linked oligosaccharides are sulfated and presumably carry glucuronic acid 3-sulfate residues, since the proteoglycan (both before and after removal of the chondroitin sulfate polysaccharide chains) reacts strongly with the HNK-1 monoclonal antibody, which recognizes this carbohydrate epitope (R. K. Margolis *et al.*, 1987b). Endo-β-galactosidase treatment of the chondroitin sulfate proteoglycans resulted in the almost complete disappearance of staining of proteoglycan immunoblots by the HNK-1 monoclonal antibody, indicating that a significant portion of the glucuronic acid 3-sulfate residues recognized by this antibody are present on poly(*N*-acetyllactosaminyl) oligosaccharides. However, after treatment with chondroitinase ABC followed by endo-β-galactosidase, several proteoglycan species retained HNK-1 reactivity, presumably due to the presence of glucuronic acid 3-sulfate on other oligosaccharides which are both resistant to endo-β-galactosidase, and inaccessible to the antibody in the native proteoglycan. Immunostaining of the endo-β-galactosidase degradation products after separation by thin-layer chromatography demonstrated that HNK-1 reactivity was confined to a minor population of large oligosaccharides having a low chromatographic mobility (Gowda *et al.*, 1989a).

3. LOCALIZATION AND PROPERTIES OF SPECIFIC NERVOUS TISSUE GLYCOPROTEINS AND PROTEOGLYCANS

Approximately 90% of the brain glycoproteins are integral components of cell surface and internal membranes, whereas much of the glycosaminoglycan content of nervous tissue (especially chondroitin sulfate and hyaluronic acid) is either initially soluble or easily releasable from the particulate fraction. Their distribution is also considerably more limited than that of the glycoproteins, insofar as little or no glycosaminoglycans are found in highly purified nerve endings and synaptosomal subfractions (R. K. Margolis *et al.*, 1975; Simpson *et al.*, 1976), or in mitochondria (R. K. Margolis *et al.*, 1975a), myelin (Margolis, 1967), or oligodendroglia (Margolis and Margolis, 1974). (For more detailed information concerning the levels of glycoconjugates in cellular and subcellular fractions of nervous tissue, see Margolis and Margolis, 1979.) Although a number of specific glycoproteins have recently been identified by electrophoretic and immunochemical techniques, in most cases these have not been characterized nor has their purity been well established. This section will therefore be limited to a few specific glycoproteins and proteoglycans which have been more intensively studied with respect to their structure, localization, and biological functions.

3.1. The NILE Glycoprotein (L1 Antigen, Ng-CAM)

In nervous tissue and several types of cultured neural tumor cells, plasma membrane glycoproteins of approximately 200,000 Da have been detected and partially characterized. A clonal line of rat adrenal medullary PC12 pheochromocytoma cells contains a major 230,000-Da cell surface glycoprotein whose concentration and labeling by radioactive sugar or amino acid precursors is increased in the presence of NGF, which induces the PC12 cells to extend long processes and acquire other properties similar to those of sympathetic neurons (Greene and Tischler, 1982). This surface glycoprotein is the NILE glycoprotein (McGuire *et al.*, 1978), and using indirect immunofluorescence it was found that polyclonal antisera raised against NILE glycoprotein stained the surfaces of living NGF-treated and -untreated PC12 cells in a uniformly distributed granular pattern. Immunofluorescent staining of primary cell cultures and tissue whole mounts revealed that immunochemically cross-reactive NILE glycoprotein is expressed on the cell surfaces (somas and neurites) of all peripheral and central neurons examined. NILE-cross-reactive material was also found to a small extent on Schwann cell surfaces but not at all on a wide variety of other cell types, suggesting that immunoreactive NILE glycoprotein is widely and selectively distributed on neuronal cell surfaces in all four mammalian species tested (Salton *et al.*, 1983b). It appears that the NILE glycoprotein is identical to the high-molecular-weight component of the Ll antigen (Bock *et al.*, 1985; Sajovic *et al.*, 1986), and to the Ng-CAM cell adhesion molecule (Friedlander *et al.*, 1986). Recent reports indicate that the NILE glycoprotein is involved in fasciculation of neurites in culture (Stallcup and Beasley, 1985), and the time course of its appearance during brain development suggests that it plays a specific role in the formation of fiber bundles (Stallcup *et al.*, 1985; Beasley and Stallcup, 1987). (For a review of the NILE glycoprotein, see Greene and Shelanski, 1989.)

Tri- and tetraantennary complex oligosaccharides are the predominant carbohydrate units present in the NILE glycoprotein isolated by immunoprecipitation from detergent extracts of both brain and PC12 cells, where they represent 77 to 90% of the biosynthetically labeled oligosaccharides (R. K. Margolis *et al.*, 1983b). Most of these are not substituted by Fuc on the core GlcNAc which is linked to Asn, and are accompanied by smaller proportions of biantennary and high-mannose oligosaccharides. Sequential lectin-agarose affinity chromatography, together with neuraminidase treatment of the fractionated glycopeptides, demonstrated a moderate degree of microheterogeneity among the predominant tri- and tetraantennary oligosaccharide units with respect to the presence of core Fuc, outer Gal and sialic acid residues, and the substitution positions on the α-linked Man residues. The greater molecular size of the PC12 cell NILE glycoprotein as compared to the immunochemically cross-reactive 205,000-Da species present in brain is due to the greater size of the PC12 cell tri- and tetraantennary complex oligosaccharides. Much of this size difference can be attributed to the presence of poly(*N*-acetyllactosaminyl) oligosaccharides in the PC12 cell but not the brain form of the NILE glycoprotein (R. K. Margolis *et al.*, 1986).

The recent cloning of the NILE/Ll glycoprotein has demonstrated that (like N-CAM, Thy-1, and MAG) it is also a member of the immunoglobulin superfamily (see Section 3.2), and that it shares three type III domains with fibronectin (Moos *et al.*, 1988).

3.2. Brain Thy-1 Glycoprotein

The Thy-1 antigen was first identified in the mouse as a cell-surface alloantigen of thymus and brain, and was later shown to be present in rat, canine, and human brain. In all species studied, the Thy-1 antigen is a major constituent of neuronal membranes and is also present in some astrocytes (Hooghe-Peters and Hooghe, 1982), while in lymphoid tissues it is probably the most abundant cell-surface molecule of mouse and rat thymocytes, but is present in lesser amounts in dog thymocytes and not at all in human thymocytes. Its tissue distribution follows unusual patterns within a species and shows considerable differences between species (for references, see Campbell et al., 1981).

The concentration of Thy-1 in brain increases almost 100-fold during early postnatal development (Morris, 1985), and in cultured PC12 rat pheochromocytoma cells its synthesis is significantly stimulated by NGF, but with a time course which precedes that of neurite outgrowth (Richter-Landsberg et al., 1985). During development, previously Thy-1-negative cells can acquire high levels of Thy-1, and conversely, detectable antigen can be lost from cells which at earlier stages stained intensely (Morris, 1985; Bolin and Rouse, 1986; Seiger et al., 1986). For example, at birth Thy-1 is faintly detected in Purkinje cells and in the molecular layer of rat or mouse cerebellum. The intensity of staining at these two sites increases to a maximum at day 9; this subsequently decreases in the Purkinje cell cytoplasm until most are negative by day 21, but persists in the molecular layer into adulthood (Bolin and Rouse, 1986). Recent radioimmunohistochemical studies have confirmed the primarily neuronal localization of Thy-1, and suggested that it is expressed differentially on specific subsets of neurons which are abundant and widespread throughout the brain (Weber et al., 1988).

All the Thy-1 antigenic determinants from rat or mouse brain and thymus are expressed in glycoproteins having molecular weights (determined by sedimentation equilibrium measurements) of 17,500 and 18,700 for brain and thymocyte Thy-1, respectively. In each case the molecular weight of the polypeptide is 12,500 and the large percentage of carbohydrate results in an erroneously high apparent molecular weight of 25,000 on SDS-PAGE. In the rat the mature protein contains 111 amino acids and is attached to the membrane by a glycosyl-phosphatidylinositol tail (see below). The peptide sequence is identical in the brain and thymocyte forms (Williams and Gagnon, 1982), and three oligosaccharides are linked at Asn residues 23, 74, and 98.

The oligosaccharide structures have been examined at each of the three glycosylation sites in the rat brain and thymus glycoproteins (Parekh et al., 1987). The results show that there is tissue-specificity of glycosylation and that superimposed on this is a significant degree of site-specificity. Based on the site distribution of oligosaccharides, no Thy-1 molecules were found in common between the two tissues despite the amino acid sequences being identical. These results suggest that by controlling N-glycosylation, a tissue creates a unique set of "glycoforms" (i.e., identical polypeptides having oligosaccharides that differ either in structure or in disposition).

In brain, Asn-23 contains only $(Man)_{5,6}$ high-mannose oligosaccharides, but the Thy-1 molecules differ in the combination of oligosaccharide structures present at sites 74 and 98. Asn-74 contains predominantly (88%) core-fucosylated, bisected biantennary oligosaccharides without sialic acid, while the remaining 12% are sialylated but not bisected. Sialic acid is apparently the only anionic residue present on Thy-1 oligosaccharides, since all oligosaccharides are converted to neutral species after digestion with neuraminidase. A portion of the biantennary oligosaccharides also contain an incompletely characterized outer-arm $Gal\alpha(\beta)1-3(4)[Fuc\alpha1-3(4)]GlcNAc$ sequence which is resistant to exoglycosidase digestion. Approximately equal proportions of

(Man)$_5$ high-mannose and hybrid oligosaccharides are present at Asn-98. All of the biantennary and hybrid oligosaccharides have peripheral 3,4-disubstituted GlcNAc residues. In agreement with the previous finding that poly(N-acetyllactosaminyl) oligosaccharides (which occur in PC12 pheochromocytoma cells and sympathetic neurons, see Section 2.1.4.) are generally not present in brain glycoproteins (R. K. Margolis et al., 1986), a significant proportion of these oligosaccharides are found at Asn-98 of thymocyte Thy-1 but are absent in its counterpart from brain.

Analysis of purified human and canine brain Thy-1 showed that these glycoproteins contain 36 to 38% carbohydrate (McKenzie et al., 1981), and have amino acid and carbohydrate compositions which are similar to those reported for rat and mouse brain Thy-1 glycoprotein. The small amounts of GalNAc detected in all of these samples are probably attributable to the presence of this hexosamine in some of the glycosyl-phosphatidylinositol membrane protein anchors.

Thy-1 is the smallest known member of the immunoglobulin superfamily (Williams and Barclay, 1988). It has been suggested that both Thy-1 and immunoglobulin evolved from primitive molecules, with an immunoglobulin fold, which mediated cell–cell interactions, and that the present role of Thy-1 may be similar to that of the primitive immunoglobulin domain. According to this hypothesis, the cell interactions in which Thy-1 is involved are not uniquely concerned with the immune system, but rather are the basic interactions which lead to tissue formation by cells, and the molecules mediating tissue interactions are not necessarily tissue-specific. Thus, the involvement of immunoglobulin-related structures in tissue interactions is more primitive than their involvement in the immune system, and the immune functions may have evolved from those sets of molecules mediating tissue interactions.

Glycosyl-phosphatidylinositol-Anchored Glycoproteins

Most integral membrane proteins are anchored by a balance of noncovalent interactions between relatively hydrophobic and polar polypeptide domains with the hydrophobic core of the lipid bilayer and the surrounding medium, respectively. However, it has recently been recognized that an extremely diverse group of proteins is anchored to the plasma membrane through a covalently attached glycosyl-phosphatidylinositol moiety. These include various enzymes (e.g., alkaline phosphatase, 5′-nucleotidase, acetylcholinesterase, alkaline phosphodiesterase I), a number of lymphocyte antigens including the Thy-1 glycoprotein (see above), decay accelerating factor (a complement regulatory protein), the trypanosomal variant surface glycoprotein, the smallest (120 kDa) component of the neural cell adhesion molecule, and a rat liver heparan sulfate proteoglycan (for references and a review, see Low and Saltiel, 1988).

Based on those proteins for which detailed information is available, the anchoring structure generally consists of a phosphatidylinositol molecule whose 1,2-diacylglycerol moiety is embedded in the membrane bilayer and is responsible for anchoring, and an oligosaccharide of variable structure and composition. The oligosaccharide is linked to the membrane phosphatidylinositol via a glycosidic linkage with a glucosamine that has a free amino group, and via a nonreducing terminal mannose 6-

phosphate to the hydroxyl of an ethanolamine which is amide-linked via its amino group to the α-carboxyl of the C-terminal amino acid (Figure 3).

In a study of phosphatidylinositol-anchored proteins of nervous tissue, PC12 pheochromocytoma cells and cultures of early postnatal rat cerebellum were treated with a phosphatidylinositol-specific phospholipase C (R. K. Margolis *et al.*, 1988). Enzyme treatment of [³H]glucosamine- or [³H]fucose-labeled PC12 cells led to a 15-fold increase in released glycoproteins. The major component was identified as Thy-1, and was accompanied by a second glycoprotein having an apparent molecular size of 158,000 Da, and which was demonstrated to contain fucosylated poly(*N*-acetyllactosaminyl) oligosaccharides. The phospholipase also released labeled Thy-1 and the 158-kDa glycoprotein from PC12 cells cultured in the presence of [³H]ethanolamine, which specifically labels this component of the phosphatidylinositol membrane anchoring sequence, while in the lipid-free protein residue of cells not treated with phospholipase, Thy-1 and a doublet at 46/48 kDa were the only labeled proteins. Several early postnatal rat brain glycoproteins also appear to be anchored to the membrane by phosphatidylinositol. Sulfated glycoproteins of 155, 132/134, 61, and 21 kDa are the predominant species released by phospholipase, which does not affect a major 44- or 46-kDa protein seen in [³H]ethanolamine-labeled brain cultures and PC12 cells, respectively. The 155/158-kDa glycoproteins may be common to both PC12 cells and brain, whereas the 44- to 46-kDa proteins (which incorporated [³H]ethanolamine but not [³H]glucosamine and were not released by phospholipase) are probably not phosphatidylinositol-anchored, but rather represent a cytoplasmic 46-kDa ethanolamine-containing protein recently described in a number of cell lines (Tisdale and Tartakoff, 1988). It has also been reported that a 140-kDa phosphatidylinositol-anchored protein is present in astrocytes and C6 glioma cells (Amblard *et al.*, 1988).

The physiological functions of this novel anchoring mechanism are still unclear, and are likely to be different among many of the highly diverse proteins which employ this membrane anchor. However, it is reasonable to assume that a glycosyl-phosphatidylinositol anchor would confer a greater degree of lateral mobility within the outer leaflet of the membrane bilayer, and would also facilitate the selective release

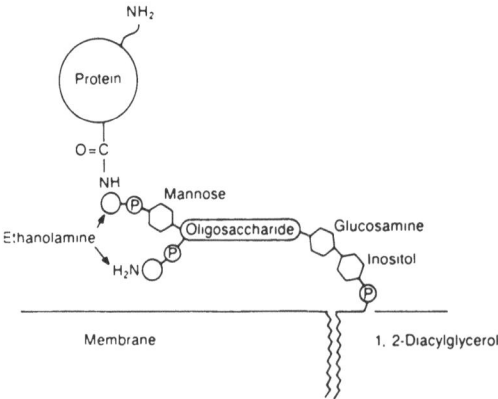

Figure 3. Schematic representation of the phosphatidylinositol membrane protein anchor (Low and Saltiel, 1988). In certain cases there is only a single ethanolamine, and other sugars may be present in the oligosaccharide (e.g., mannose, 1–2 mole/mole; galactose, 1–8 mole/mole; and galactosamine, 1 mole/mole).

and/or uptake of certain enzymes, cell adhesion proteins, antigens, or other proteins having a role in cell surface events. The generation at the cell surface by specific phospholipases C of 1,2-diacylglycerol (an activator of protein kinase C) and of glycosyl-inositol phosphates also suggests that these products may serve as second messengers, which could be delivered to their intracellular target sites by diffusion across the lipid bilayer or by receptor-mediated endocytosis, respectively.

3.3. The F3-87-8 Glycoprotein

Lakin and Fabre (1981) have described a monoclonal antibody (F3-87-8) which recognizes a developmentally regulated brain-specific glycoprotein which is present in mouse, rat, dog, and human brain, but is absent from frog and chicken brain. After purification by immunoaffinity chromatography from rat and human brain, the glycoproteins from both species were found to contain approximately 47% carbohydrate, to have virtually identical amino acid compositions (containing 11–13% each of Asp/Asn, Ser, Thr, and Leu), and to migrate as a doublet on SDS-PAGE (Lakin et al., 1983). In human brain the two components have molecular sizes of 130,000 and 100,000 Da, whereas they are slightly smaller in rat brain and the larger-molecular-size component is more prominent. There is complete identity of the first 20 amino acids at the termini of the rat and human glycoproteins, and an unusually high and phylogenetically conserved number of cysteine residues in this region, which, however, showed no homology to other known sequences (Flanagan et al., 1986). In recent studies from our laboratories, glycopeptides (biosynthetically labeled with [^{3}H]glucosamine) were prepared from immunoaffinity-purified F3-87-8 glycoprotein and fractionated by lectin–agarose affinity chromatography (Ripellino et al., 1989a). These studies indicate that the F3-87-8 glycoprotein from rat brain contains approximately 65% tri- and tetraantennary oligosaccharides, 8% biantennary oligosaccharides, 10% high-mannose oligosaccharides, and 18% O-glycosidic oligosaccharides (based on alkali treatment of the glycopeptides not bound by Con A–agarose).

The initial report on the F3-87-8 glycoprotein stated that 16-week human fetal brain and neonatal dog brain had little or no antigen activity, which, in immunofluorescence studies of adult human brain, was found to be particularly localized in what appeared to be fiber tracts in the thalamus, basal ganglia, and dentate nucleus, although all regions including gray matter were stained (Lakin and Fabre, 1981). In a more systematic immunocytochemical study of the localization of this glycoprotein in developing rat brain (Ripellino et al., 1989a), we have found that in one-week postnatal cerebellum there is light staining of the internal granule cell layer and surrounding the Purkinje cells. By two weeks, an intense staining of myelinating fiber tracts appears, accompanied by much lighter staining in the granule cell layer and at the base of the molecular layer. Staining of the white matter remains strong at three weeks postnatal, together with significant staining throughout the molecular layer, and then decreases in both areas by one month. In adult brain there is relatively uniform staining of approximately equal intensity in the white matter, granule cell layer and molecular layer, while the Purkinje cell bodies remain unstained throughout development. In agreement with the previous immunochemical analysis, no staining was seen in other tissues, confirming the brain-specific localization of this glycoprotein. These findings suggest a role for the F3-87-8 glycoprotein in one or more developmental processes. However, attempts to further clarify this question by its localization at the electron microscopic level have been hindered by the fact that the epitope recognized by this monoclonal antibody appears to be denatured by even relatively mild aldehyde fixation, although antigenicity was preserved in the cold acetone-fixed cryostat sections used in our studies.

3.4. Hyaluronic Acid and the Chondroitin Sulfate Proteoglycans of Brain

3.4.1. Chondroitin Sulfate Proteoglycans

Most of the chondroitin sulfate and hyaluronic acid in brain is either present in an initially soluble form or loosely associated with smooth, low-density microsomal membrane subfractions (Kiang et al., 1978, 1981), whereas heparan sulfate proteoglycans appear to be largely integral components of plasma membranes. Indirect types of evidence suggested that the chondroitin sulfate proteoglycans were primarily cytoplasmic constituents of neurons and astrocytes in adult brain. For example, when neurons are isolated in bulk from rat brain and lysed by a change in tonicity or pH, 82% of the chondroitin sulfate (but only 20–25% of the total cell protein and glycoprotein hexosamine) is released, together with over 90% of the lactate dehydrogenase (R. K. Margolis et al., 1979).

The structural features of the chondroitin sulfate proteoglycans of brain were described in Section 2.2. Immunocytochemical studies at the light and electron microscopic levels, employing affinity-purified $F(ab')_2$ prepared from polyclonal antibodies to the chondroitin sulfate proteoglycans, have provided more detailed information concerning their localization in nervous tissue. The proteoglycans were found to be exclusively intracellular in adult cerebellum, cerebrum, brain stem, and spinal cord (Aquino et al., 1984a). Some neurons and astrocytes (including Golgi epithelial cells and Bergmann fibers) show strong cytoplasmic staining. Although in the central nervous system there is heavy axoplasmic staining of many myelinated and unmyelinated fibers, not all axons stain. Staining was also seen in retinal neurons and glia (ganglion cells, horizontal cells, and Müller cells), but several central nervous tissue elements are consistently unstained, including Purkinje cells, oligodendrocytes, myelin, optic nerve axons, nerve endings, and synaptic vesicles. These findings are in good agreement with earlier biochemical studies on the localization of glycosaminoglycans in cellular and subcellular fractions of brain (Margolis and Margolis, 1974; R. K. Margolis et al., 1975a, 1979).

In sympathetic ganglion and peripheral nerve there is no staining of neuronal cell bodies, axons, myelin, or Schwann cells, but in sciatic nerve the Schwann cell basal lamina is stained, as is the extracellular matrix surrounding collagen fibrils. Staining was also observed in connective tissue, which is consistent with immunochemical studies demonstrating that antibodies to the chondroitin sulfate proteoglycans of brain cross-react to varying degrees with certain connective tissue proteoglycans.

In contrast to the intracellular (cytoplasmic) localization of chondroitin sulfate proteoglycans in adult brain, immunoelectron microscopic studies in immature (7 day postnatal) rat cerebellum (Aquino *et al.*, 1984b) demonstrated predominantly extracellular staining in the granule cell and molecular layers. Staining was also extracellular and/or associated with plasma membranes in the region of the presumptive white matter. Axons, which are unmyelinated at this age, generally did not stain, although faint intracellular staining was present in some astrocytes. At 10 and 14 days postnatal there was a significant decrease in extracellular space and staining, and by 21 days distinct cytoplasmic staining of neurons and astrocytes appeared. This intracellular staining further increased by 33 days so as to closely resemble the pattern seen in adult brain. Analyses of the proteoglycans isolated from 7-day-old and adult brain demonstrated that they have essentially identical biochemical compositions, immunochemical reactivity, size, charge, and density, indicating that the antibodies used in these studies recognize the same macromolecules in both early postnatal and adult brain, and that the localization of the chondroitin sulfate proteoglycans changes progressively from a predominantly extracellular to an intracellular location during brain development.

In other studies, it has been reported that a cell surface antigen of brain termed NG2 is a high-molecular-weight chondroitin sulfate proteoglycan (Stallcup *et al.*, 1983). Later immunocytochemical localization of this antigen in rat cerebellum led to the conclusion that it is associated with smooth protoplasmic astrocytes (Levine and Card, 1987), and is expressed on bipotential glial precursor cells in cultures of postnatal rat optic nerve (Stallcup and Beasley, 1987). However, the NG2 antigen has not been well characterized biochemically, and the only evidence for its proteoglycan nature is that in NG2 immunoprecipitates, labeled material above the major 300-kDa band is removed by chondroitinase treatment. Since we have found that immunoprecipitates (especially of [^{35}S]sulfate-labeled glycoproteins) are frequently contaminated with coprecipitated chondroitin sulfate proteoglycans, definitive identification of this antigen as a chondroitin sulfate proteoglycan must await more detailed biochemical evidence, such as that the 300-kDa NG2 band *itself* decreases in size after chondroitinase treatment, or β-xyloside inhibition of proteoglycan biosynthesis. This question is especially interesting in view of the recent report that an antiserum to the unsaturated disaccharide "stubs" remaining after chondroitinase digestion of chondroitin sulfate proteoglycans stains a subpopulation of cultured astrocytes (Gallo *et al.*, 1987). The antiserum also stains the bipotential precursors of these cells, which are capable of differentiating into GFAP-positive astrocytes or into galactocerebroside-positive oligodendrocytes, depending on the culture conditions. However, the oligodendrocytes lost their reactivity with the antiserum to chondroitin sulfate disaccharides.

It has also recently been reported that cytotactin, a glycoprotein involved in neuron–glia adhesion, binds to a chondroitin sulfate proteoglycan with a 280-kDa core protein and HNK-1 epitopes (Hoffman and Edelman, 1987). Immunochemical studies indicate that cytotactin contains the same polypeptide chain(s) as the myotendinous antigen and tenascin, two previously identified extracellular matrix glycoproteins, and like molecules isolated from fibroblasts and cartilage, cytotactin appears in electron micrographs as six-armed structures called hexabrachions (Hoffman *et al.*, 1988).

During early development of the chicken embryo, cytotactin and the proteoglycan become differentially distributed in the rostral and caudal halves of the developing sclerotome (Tan *et al.*, 1987). Their differential expression is crest cell-independent, although they inhibit crest cell migration, and it is proposed that cytotactin and the chondroitin sulfate proteoglycan may contribute to the localization of neural crest cells in the rostral half of the sclerotome.

In later studies using antibodies to cytotactin and the cytotactin-binding proteoglycan, it was found that during neural development both the levels and molecular forms of each molecule varied, following different time courses (Hoffman *et al.*, 1988). A novel 250-kDa form of cytotactin that contained chondroitin sulfate was also detected. Unlike the molecules from neural tissue, cytotactin and the cytotactin-binding proteoglycan from nonneural tissues such as fibroblasts lacked the HNK-1 epitope, which, like the chondroitin sulfate glycosaminoglycan chains, does not appear to be directly involved in binding. Cell culture experiments indicate that cytotactin is specifically synthesized by glia whereas the cytotactin-binding proteoglycan appears to be a product of neurons. Both molecules were found either associated with the cell surface or in an intracellular, perinuclear pattern.

3.4.2. Hyaluronic Acid

It has recently become clear that hyaluronic acid, which contains a disaccharide repeating unit composed of glucuronic acid(β1-3)N-acetylglucosamine(β1-4), does not occur as a covalent complex with protein. In cartilage and certain other tissues, a specialized region of the core protein in the chondroitin sulfate proteoglycan monomers (called the hyaluronic acid binding region) is responsible for their alignment along a central filament of hyaluronic acid to form very large proteoglycan aggregates. The active binding site has a high affinity, with a K_d of approximately 10^{-6} M, for decasaccharide or larger oligomers of hyaluronic acid but a much lower affinity for shorter oligomers, and this interaction is highly specific, insofar as no other anionic biopolymer has been found which will compete with hyaluronic acid for binding. When a third constituent, called link protein, is present, the proteoglycan molecules are essentially locked into place on the hyaluronic acid, with a dissociation constant which is too low to measure conveniently (Hascall and Hascall, 1982). We have made use of the properties of cartilage proteoglycan aggregates to develop a sensitive and specific biological probe for the localization of hyaluronic acid in tissue sections at the light and electron microscopic levels (Ripellino *et al.*, 1985, 1988), since because of its highly conserved nature it has not yet been possible to raise high-affinity antibodies to undegraded hyaluronic acid by conventional techniques.

The hyaluronic acid binding region was prepared by trypsin digestion of chondroitin sulfate proteoglycan aggregates isolated by CsCl density gradient centrifugation from the Swarm rat chondrosarcoma, and biotinylated in the presence of hyaluronic acid and link protein. After the removal of these two components by gel filtration followed by high-performance liquid chromatography (both in 4 M guanidine HCl), the biotinylated hyaluronic acid binding region was used, in conjunction with avidin–peroxidase, as a specific probe for the light and electron microscopic localization of hyaluronic acid in developing and mature rat cerebellum. Control experiments demonstrated that both hyaluronic acid oligosaccharides containing greater than ten monosaccharide residues (which are known to inhibit its binding to macromolecular hyaluronic acid), and pretreatment of tissue sections with *Streptomyces* hyaluronidase (which is specific for hyaluronic acid and shows no protease activity under our conditions) completely block staining by the biotinylated hyaluronic acid binding region.

At postnatal week 1 there was strong staining of extracellular hyaluronic acid in the presumptive white matter, in the internal granule cell layer, and as a dense band of staining at the base of the molecular layer, surrounding the parallel fibers. This band progressively moved toward the pial surface during postnatal week 2, and extracellular staining remained predominant through postnatal week 3. In adult brain there was no significant extracellular staining of hyaluronic acid, which was most apparent in the granule cell cytoplasm, and intraaxonally in parallel fibers and some myelinated axons. The white matter was also unstained in adult brain, and staining was not seen in Purkinje cell bodies or dendrites at any age. The localization of hyaluronic acid and its developmental changes are very similar to those previously found in immunocytochemical studies of the chondroitin sulfate proteoglycans in nervous tissue (Aquino *et al.*, 1984a,b), and to recent results from studies employing monoclonal antibodies to the hyaluronic acid binding region and link protein. The presence of brain hyaluronic acid in the form of aggregates with chondroitin sulfate proteoglycans would provide a reasonable explanation for their similar localizations and coordinate developmental changes. Although the intracellular occurrence of hyaluronic acid (and chondroitin sulfate proteoglycans) is unusual, our results are consistent with preliminary evidence for the axonal transport of hyaluronic acid (Gammon *et al.*, 1985), and with earlier biochemical analyses demonstrating the presence of hyaluronic acid in myelin-free axons isolated in bulk from rat and bovine brain (R. K. Margolis *et al.*, 1975).

3.4.3. Cytoplasmic Glycoconjugates

A number of other studies have recently appeared which indicate the presence of cytoplasmic glycoproteins in a variety of cell types. Several laboratories have reported that nuclear pore complex glycoproteins contain cytoplasmically disposed *O*-glycosidically linked GlcNAc residues (Holt *et al.*, 1987a; Hanover *et al.*, 1987; Davis and Blobel, 1987). Biosynthetic studies of p62 (a 62-kDa nuclear pore complex protein containing *O*-linked GlcNAc residues) in cultured rat liver cells demonstrated that most of the GlcNAc residues are added within 5 min from the start of translation and perhaps even contraslationally. The fact that at early time points all of the newly synthesized p62 was found in the postmicrosomal supernatant suggests that it is synthesized as a soluble cytosolic protein that was never associated with microsomal or Golgi membranes, and that it receives most of its sugar residues while it is in this soluble form (Davis and Blobel, 1987).

In addition to being found in abundance on proteins associated with the cytoplasmic and nucleoplasmic faces of the nuclear pore complex, *O*-linked GlcNAc residues have also been demonstrated in an as yet unidentified 65-kDa protein present in the cytoplasm of erythrocytes, and in Band 4.1, a protein which serves as a bridge

joining the cytoskeleton to the inner surface of the erythrocyte plasma membrane (Holt
et al., 1987b). Moreover, Abeijon and Hirschberg (1988) have reported the presence
of cytoplasmically oriented *O*-linked GlcNAc residues in several smooth and rough
endoplasmic reticulum glycoproteins. A recent report demonstrating different substrate
specificities of rat kidney lysosomal and cytosolic α-D-mannosidases (Tulsiani and
Touster, 1987) also strongly suggests the existence of a cytoplasmic pathway for
glycoprotein biosynthesis and catabolism.

Considered together, these reports indicate the existence of proteins which are
probably synthesized by free (rather than membrane-bound) ribosomes, and are then
released into the cytoplasm where they are glycosylated by soluble glycosyltransferases. This novel pathway would therefore be clearly distinct from the usual mechanism whereby proteins are glycosylated by membrane-bound glycosyltransferases after
irreversible segregation into the lumen of the endoplasmic reticulum and Golgi compartments. It is also possible that chondroitin sulfate proteoglycans are synthesized by
conventional pathways and taken up into the cytoplasm after release at the cell surface.

3.4.4. Roles of Hyaluronic Acid and Chondroitin Sulfate Proteoglycans in Nervous Tissue Development

Developmental studies are particularly pertinent to the localization and possible
functions of glycosaminoglycans in brain. It was found that during the development of
rat brain, the levels of all three glycosaminoglycans increased postnatally to reach a
peak at 7 days, after which they declined steadily, attaining by 30 days concentrations
within 10% of those present in adult brain (R. U. Margolis *et al.*, 1975). The greatest
change occurred in hyaluronic acid, which decreased by 50% between 7 and 10 days,
and declined to adult levels (28% of the peak concentration) by 18 days of age (see
Figure 1 of the Appendix to this chapter). In 7-day-old rats almost 90% of the
hyaluronic acid in brain is water-extractable, as compared to only 15% in adult animals, and this large amount of soluble hyaluronic acid in young brain is relatively
inactive metabolically.

During early stages of brain development, it is likely that extracellular hyaluronic
acid provides a highly hydrated and easily penetrable matrix through which neuronal
migration may take place, before the intercellular space decreases to its much smaller
dimensions in mature brain. It is also possible that extracellular hyaluronic acid and
chondroitin sulfate proteoglycans must in some manner be removed or relocalized
before various adhesive processes may begin. Aside from this type of passive or
"facilitative" role, histochemical studies on the spatial and temporal distribution of
extracellular matrix in developing cerebral cortex of normal and reeler mutant mice
(Nakanishi, 1983) suggest that glycosaminoglycans and proteoglycans may also have a
more direct involvement in relation to afferent axon targeting.

Other evidence that chondroitin sulfate proteoglycans play a role in developmental processes and cell behavior comes from studies in which proteoglycan biosynthesis
is inhibited by supplying an exogenous β-xyloside acceptor (e.g., 4-methylum-
belliferyl-β-D-xyloside or *p*-nitrophenyl-β-D-xyloside), which competes with the proteoglycan core protein for glycosaminoglycan chain initiation. In the presence of 1 mM

β-xyloside (but not β-galactoside), NGF-treated PC12 cells show a significant increase in process length and a decrease in the number, branching, diameter, and adhesion of processes to the collagen substratum, suggesting that sulfated proteoglycans are involved in determining the surface properties and morphology of PC12 cells. Analogous effects are seen in monolayer and aggregate primary cultures of early postnatal rat cerebellum, and biochemical analyses confirm that the β-xyloside almost completely inhibits chondroitin sulfate (and, to a lesser extent, heparan sulfate) proteoglycan biosynthesis in these cell culture systems (R. K. Margolis *et al.*, unpublished results). Moreover, addition of F(ab′)$_2$ fragments prepared from specific antisera to the chondroitin sulfate proteoglycans of brain (Aquino *et al.*, 1984a) also results in a decrease in process outgrowth, cell migration along processes, and adhesion to either plastic or polylysine substrata in primary cerebellar cultures, whereas the same concentration of preimmune F(ab′)$_2$ has little or no effect (Ripellino *et al.*, unpublished results).

3.5. Heparan Sulfate Proteoglycans of Nervous Tissue

3.5.1. Brain

There has been an increasing interest in heparan sulfate proteoglycans, which are ubiquitous components of the plasma membrane and basement membranes in many tissues, where they are thought to play important roles in the regulation of membrane permeability and in cell–cell interactions (for reviews, see Höök *et al.*, 1984; Gallagher *et al.*, 1986). Heparan sulfate proteoglycans, extracted from rat brain microsomal membranes or whole forebrain with deoxycholate, were purified from accompanying chondroitin sulfate proteoglycans and membrane glycoproteins by ion-exchange chromatography, affinity chromatography on lipoprotein lipase–Sepharose, and gel filtration (Klinger *et al.*, 1985). Their structural properties are summarized in Table 2. In brain, most of the heparan sulfate glucosamine residues are *N*-sulfated (yielding predominantly di- and tetrasaccharide nitrous acid degradation products), and there is very little change in the *N*-sulfation of brain heparan sulfate during the first 30 days after birth (Klinger *et al.*, 1985; R. K. Margolis *et al.*, 1987a). Comparison of the effects of heparinase and heparitinase treatment revealed that the heparan sulfate proteoglycans of brain contain a significant proportion of relatively short *N*-sulfoglucosaminyl 6-*O*-sulfate(α1-4)iduronosyl 2-*O*-sulfate(α1-4) repeating units, and that the portions of the heparan sulfate chains in the vicinity of the carbohydrate–protein linkage region are characterized by the presence of D-glucuronic acid rather than L-iduronic acid (Ripellino and Margolis, 1989).

In [³H]glucosamine-labeled heparan sulfate proteoglycans, approximately 22% of the radioactivity is present in glycoprotein oligosaccharides, consisting predominantly of *N*-glycosidically linked tri- and tetraantennary complex oligosaccharides (60%, some of which are sulfated) and *O*-glycosidic oligosaccharides (33%). Small amounts of chondroitin sulfate (approximately 6–10% of the total glycosaminoglycan present in heparan sulfate proteoglycan preparations) copurified with the heparan sulfate proteoglycans through a variety of fractionation procedures. This chondroitin sulfate can be at least partially accounted for by the existence of a small proportion of hybrid

Table 2. Properties of Nervous Tissue Proteoglycans

	Adult rat brain[a]		PC12 pheochromocytoma cells[b]		Chick embryo retina[c]	
	HSPG[d]	CSPG	HSPG	CSPG	HSPG	CSPG[e]
Proteoglycan (kDa)	220	260–325	100–170	50–100, 120–190	250–300	>400
Core proteins (kDa)	55	67–>350	65	14, 105	60–65	220–230
Glycosaminoglycan chains (kDa)	14–15	19	16	33–34	15–25	60–70
Glycoprotein oligosaccharides[f]						
N-Glycosidic						
Tri- and tetraantennary	60	66	71 (58)[g]	73 (70)		
Biantennary	5.4	5.5	3 (17)	3 (2)		
High mannose	0.8	0.6	—	—		
O-Glycosidic	34	28	26 (25)	24 (28)		

[a]Data for rat brain are from Klinger et al., 1985; Krusius et al., 1987; Ripellino et al., 1989; and Gowda et al., 1989a.
[b]Data for PC12 cells are from Gowda et al., 1989c.
[c]Data for retina are from Morris et al., 1987.
[d]HSPG, heparan sulfate proteoglycan; CSPG, chondroitin sulfate proteoglycan.
[e]Also contains dermatan sulfate regions.
[f]Expressed as percent of total glycoprotein oligosaccharides.
[g]PC12 cell proteoglycan oligosaccharide composition is given for the cell-associated proteoglycans. Values for proteoglycans released into the medium are shown in parentheses.

proteoglycan molecules containing predominantly heparan sulfate as well as a few percent of chondroitin sulfate chains on the same protein core (Ripellino and Margolis, 1989).

Incubation of [^{35}S]sulfate-labeled microsomes with heparin or 2 M NaCl released approximately 21 and 13%, respectively, of the total heparan sulfate, as compared to the 8–9% released by buffered saline or chondroitin sulfate, and the 83% which is extracted by 0.2% deoxycholate. It therefore appears that there are at least two distinct types of association of heparan sulfate proteoglycans with brain membranes. Some structural features and binding properties of a significantly smaller (approximately 55 kDa) heparan sulfate proteoglycan enzymatically labeled by calf brain microsomes have also been described (Miller and Waechter, 1984). However, the characteristics of this proteoglycan suggest that it represents a subpopulation or biosynthetic precursor of the major heparan sulfate proteoglycans of brain (for further discussion of these differences, see Klinger et al., 1985). It should also be noted that plasma membrane heparan sulfate proteoglycans, which presumably account for most of those present in brain, differ considerably in structure and immunochemical reactivity from basement membrane heparan sulfate proteoglycans, such as those found in peripheral nerve and other tissues.

3.5.2. PC12 Cells

Heparan sulfate comprises 3–5% of the total glucosamine radioactivity incorporated into glycoproteins and proteoglycans by cultured PC12 pheochromocytoma cells, and three-quarters of this heparan sulfate is present in the membranes (R. K. Margolis et al., 1983a). Data on the proteoglycan, core protein, and glycosaminoglycan chain sizes, and the glycoprotein oligosaccharide composition, of PC12 cell proteoglycans are summarized in Table 2.

In the presence of nanogram levels of NGF, PC12 cells cease to divide, extend long microtubule-containing processes, and acquire electrical excitability and increased sensitivity to acetylcholine (Greene and Tischler, 1982). NGF treatment of PC12 cells led to a 70% decrease in the cellular concentration of heparan sulfate (affecting the soluble and membrane pools to an equal extent), and produced a corresponding increase in heparan sulfate released from the cells into the medium (R. K. Margolis et al., 1983a). No change was observed in the cellular chondroitin sulfate following NGF treatment, although there was a small (18%) increase in glycoproteins, which probably reflected the increased membrane area resulting from NGF-induced neurite extension.

In view of these findings, the size, charge, and sulfation pattern of heparan sulfate were examined in the cell soluble fraction, membranes, and culture medium of PC12 cells cultured in the presence and absence of NGF, and the structural features of PC12 cell heparan sulfate were compared with those of rat brain at several stages of early postnatal development (R. K. Margolis et al., 1987a). Nitrous acid degradation studies revealed significant differences in the distribution of N-sulfate and N-acetyl groups in heparan sulfate present in the PC12 cell soluble fraction, membranes, and medium, and demonstrated that NGF treatment led to an increased proportion of N-sulfated segments in the cell-associated heparan sulfate, although no such change was seen in that released into the culture medium. The overall charge and size (approximately 15,000 Da) of heparan sulfate chains were similar in the different PC12 cell fractions (and in brain), although NGF treatment led to a decrease in the proportion of less charged chains in the PC12 cell membranes, and a small increase in molecular size. The finding that neuronally differentiated NGF-treated PC12 cells contain a more highly sulfated heparan sulfate than untreated cells is consistent with other reports that cell-surface heparan sulfate synthesized by virally transformed cells and by transformed cell lines is undersulfated compared to their untransformed (i.e., more differentiated) counterparts (for a review, see Höök *et al.*, 1984).

3.5.3. Neuroblastoma Cells

It is also instructive to compare the properties of heparan sulfate in PC12 cells with that synthesized by cultured human CHP100 neuroblastoma cells (Hampson *et al.*, 1983, 1984; Fransson *et al.*, 1985), which also exhibit many properties of sympathetic neurons. Mild trypsinization of either neuroblastoma cells (Hampson *et al.*, 1983) or [^{35}S]sulfate-labeled PC12 cells (Gowda *et al.*, 1989c) released over half of the cell-associated heparan sulfate, which was the predominant glycosaminoglycan in the neuroblastoma cells. In both cell types, heparan sulfate represented only 22% of the glycosaminoglycans released into the culture medium (27% in the case of NGF-treated PC12 cells), the remainder being accounted for by chondroitin sulfate and dermatan sulfate (or chondroitin sulfate alone in PC12 cells, which do not produce detectable amounts of dermatan sulfate; R. K. Margolis *et al.*, 1983a). There is relatively little or no heterogeneity in the size or overall charge of the heparan sulfate chains produced by the two cell types, although PC12 cells synthesized a small proportion of less highly sulfated chains, whose concentration decreased in the membranes of NGF-treated cells. On the other hand, nitrous acid treatment demonstrated that the PC12 cell soluble and membrane heparan sulfate contain large blocks of N-acetylated disaccharide units whose relative proportion is significantly decreased by NGF treatment, whereas neuroblastoma cells more closely resemble brain in yielding predominantly di- and tetrasaccharide nitrous acid degradation products. In both neuroblastoma and PC12 cell heparan sulfate, there are essentially no O-sulfate residues in the N-acetylated segments, although a small proportion of O-sulfate groups are found outside of N-sulfated di- or tetrasaccharides in the heparan sulfate of brain.

Heparan sulfate proteoglycans of Platt human neuroblastoma cells have also been fractionated by affinity chromatography on agarose columns derivatized with platelet factor-4 (Maresh *et al.*, 1984). It was found that in addition to free heparan sulfate chains, these cells contained several relatively small molecular size heparan sulfate proteoglycans which could be partially resolved by affinity chromatography on platelet factor-4.

3.5.4. Functional Roles of Heparan Sulfate Proteoglycans

A number of studies have indicated the importance of cell-surface and basement membrane heparan sulfate proteoglycans in nervous tissue. A heparan sulfate proteoglycan present in the basal lamina of skeletal muscle fibers is highly concentrated at the neuromuscular junction, where it may be involved in regulating the spatial organization and localization of acetylcholine receptors (Anderson and Fambrough, 1983; Anderson et al., 1984). There is evidence that a heparan sulfate proteoglycan mediates the anchorage of the collagen-tailed (asymmetric) form of acetylcholinesterase to the synaptic basal lamina in the electric organ of the electric ray (Brandan et al., 1985), and in mutant PC12 cells lacking a cell-surface heparan sulfate proteoglycan, there are large changes in the amount and cellular localization of 16 S asymmetric acetylcholinesterase (Inestrosa et al., 1985).

Other studies have indicated that a heparan sulfate proteoglycan is involved in cell–substratum adhesion of embryonic chick neural retina cells (Cole et al., 1985), and neuronal cell–cell adhesion (Cole et al., 1986) through its interactions with a neuronal cell adhesion glycoprotein (N-CAM). Since it has also been shown that the heparin-binding domains of laminin and fibronectin can promote neurite outgrowth (Edgar et al., 1984; Rogers et al., 1985), and since a heparan sulfate proteoglycan has been identified as an externally exposed integral plasma membrane protein that is anchored to the Schwann cell cytoskeleton (Carey and Todd, 1986), it would appear that heparan sulfate proteoglycans may play a critical regulatory role in a wide variety of cell–cell interactions during nervous tissue development.

It was originally reported that a heparan sulfate proteoglycan secreted into the medium of PC12 cells and nonneuronal dorsal root ganglion cells induces rapid neurite outgrowth by primary sympathetic neurons after adsorption onto polylysine-coated surfaces (Matthew et al., 1985). The proteoglycan found in the PC12 cell medium appears to be derived from a cell-surface heparan sulfate proteoglycan with which it is immunochemically cross-reactive, and from which it can be obtained by mild proteolysis. However, more recent studies indicate that laminin associated with the heparan sulfate proteoglycan may be largely or entirely responsible for the neurite outgrowth-promoting activity, since it was later found that laminin was present in conditioned media from each of six cell types (including PC12 cells) known to secrete a neurite outgrowth-promoting factor, and that when laminin was selectively removed, there was a corresponding loss of neurite outgrowth-promoting activity (Lander et al., 1985). Similarly, it is probable that in dorsal root ganglion neurons and brain membranes, cell-surface heparan sulfate proteoglycans serve as an anchor for growth factors having Schwann cell mitogenic activity (Ratner et al., 1988).

Although it has been suggested (based on studies of PC12 cells) that the amyloid β protein precursor is a 65-kDa heparan sulfate proteoglycan core protein (Schubert et al., 1988), the evidence for this conclusion is not particularly strong (Gowda et al., 1989b), and data from other laboratories indicate that the precursor protein is in the 110- to 135-kDa range (Gandy et al., 1988; Selkoe et al., 1988).

Finally, with regard to the possible biological functions of heparan sulfate proteoglycans in nervous tissue, some preliminary data suggest that the period clock gene of D. melanogaster, which is involved in the generation of biological rhythms, codes for a heparan sulfate proteoglycan (Reddy et al., 1986). The most striking feature of the predicted coding sequence is an extensive run of Gly-Thr residues, which are homologous to the series of Gly-Ser repeats found in certain chondroitin sulfate proteoglycans (see Chapter 5). However, any conclusions concerning the possible role of proteoglycans in the generation or maintenance of biological rhythms must await a more thorough biochemical characterization of this putative heparan sulfate proteoglycan.

3.6. Retina Proteoglycans

Extracellular matrix molecules of the retina have recently been reviewed by Hewitt (1986). Early studies by Bach and Berman (1971a,b) of the acid-soluble glycoconjugates extracted from cattle retina by overnight agitation in unbuffered saline indicated the presence of chondroitin sulfate (about 75% of which was undersulfated) and sialoglycopeptides, both of which may represent partial proteolytic products of native proteoglycans and glycoproteins. Many subsequent studies have employed embryonic chicken retinas, labeled with radioactive precursors due to their relatively low concentration of proteoglycans (which can be estimated at 2–4 μg per 14-day-old chick embryo retina, based on the efficiency of extraction, the amount of total glycosaminoglycan per retina, and the carbohydrate/protein ratio of the proteoglycans; Morris et al., 1977, 1984, 1987).

After incubation of 14-day chicken embryo retinas in medium containing sodium [^{35}S]sulfate, approximately equal proportions (28%) of [^{35}S]sulfate-labeled glycosaminoglycans (predominantly chondroitin sulfate and dermatan sulfate) were found in the culture medium and in a subsequent saline extract, whereas almost all of the residual material (enriched in heparan sulfate) was extracted with 4 M guanidine HCl (Morris and Ting, 1981). It has not been determined whether the dermatan sulfate-like regions are distinct molecules or are dispersed within the chondroitin sulfate chains. Most of the proteoglycans were of low buoyant density (although 30–50% of the heparan sulfate proteoglycans had a density > 1.6 g/ml), and there was no significant aggregation with hyaluronic acid (Morris et al., 1984). Further characterization demonstrated that the chondroitin sulfate/dermatan sulfate proteoglycans had a molecular size > 400 kDa and contained two or three glycosaminoglycan chains of 15–25 kDa, while the heparan sulfate proteoglycans had a size of 250–300 kDa and contained 9–12 chains of 15–25 kDa (Morris et al., 1987). Free heparan sulfate chains were also present.

The retina proteoglycans are located predominantly within the interphotoreceptor matrix and at the inner limiting membrane, with minor amounts in the synaptic regions (Porrello et al., 1986; Johnson and Hageman, 1987). Chondroitin 6-sulfate has been identified in the cone sheath of primates (Hageman and Johnson, 1987) and in the region of the outer and inner photoreceptor segments of rats (Porrello and LaVail, 1986). Differential extraction of 14-day chicken embryo retinas revealed that whereas the chondroitin sulfate/dermatan sulfate proteoglycans are present in the culture medium and saline extract, release of the heparan sulfate proteoglycans required extraction with 4 M guanidine HCl/2% Triton X-100, or trypsinization (Morris et al., 1987). These data suggest that the chondroitin sulfate/dermatan sulfate proteoglycans are present in the extracellular matrix, and that the heparan sulfate proteoglycans are inserted into the plasma membrane. Recent studies have shown that both heparan sulfate and chondroitin sulfate proteoglycans are synthesized by cultures of chick embryo neural retinal cells and photoreceptors, free of flat, glial-like cells (Needham *et al.*, 1988).

Between 5 and 14 days of embryonic development, the proportion of heparan sulfate in chicken retina decreased from 74% to 53% of the total [^{35}S]sulfate-labeled glycosaminoglycans, and chondroitin sulfate increased from 17% to 40%, with a sixfold increase in the ratio of chondroitin 4-sulfate to chondroitin 6-sulfate (Morris *et al.*, 1977). This differentiation-dependent transition could be blocked by the thymidine analogue 5-bromo-2'-deoxyuridine, which also interferes with normal cytodifferentiation. It has been suggested that proteoglycans may act together with other components of the extracellular matrix in the interphotoreceptor space to influence photoreceptor outgrowth (Hewitt, 1986), and there is also evidence for interactions between heparan sulfate proteoglycans and a cell–substratum adhesion protein in cultured chick neural retina (Cole *et al.*, 1985).

3.7. Glycoproteins and Proteoglycans of Chromaffin Granules and Large Dense-Cored Vesicles

Complex carbohydrates of secretory organelles were previously reviewed by Giannattasio *et al.* (1979). Since that time, detailed reports have appeared concerning the characterization and intracellular transport of glycosaminoglycans and glycoproteins in prolactin granules (Zanini *et al.*, 1980; Giannattasio *et al.*, 1980), and on the glycoproteins and proteoglycans of chromaffin granules. Synaptic vesicle glycoproteins and proteoglycans are discussed in Chapter 10.

3.7.1. Chromaffin Granules

Forty to sixty distinct proteins have been recognized by electrophoretic analysis of chromaffin granule membranes (Abbs and Phillips, 1980), and later investigations utilizing radioiodinated lectins have revealed that over 20 of these are glycoproteins with molecular sizes ranging from 15,000 to 200,000 Da (Gavine *et al.*, 1984). Three of the major glycoproteins (dopamine β-hydroxylase, glycoprotein II, and glycoprotein III) have been isolated and their carbohydrate content determined (Fischer-Colbrie *et al.*, 1982, 1984). (For a review of the composition and molecular function of chromaffin granules, see Winkler *et al.*, 1986.)

It appears that the carbohydrate (Fischer-Colbrie *et al.*, 1982) and protein (Lamouroux *et al.*, 1987) moieties of the soluble and membrane-bound forms of dopamine β-hydroxylase (DBH) are fundamentally identical in their structures (or sugar composition). DBH isolated from the soluble contents of bovine chromaffin granules contains biantennary complex oligosaccharides and high-mannose oligosaccharides (the latter having an average of six Man residues) in a molar ratio of 2 : 1 (R. K. Margolis *et al.*, 1984). The recent isolation of a cDNA clone containing the complete coding sequence of human DBH demonstrates that the polypeptide chains comprise 578 amino acids corresponding to an unmodified protein of 64,862 Da, and is preceded by a cleaved signal peptide of 25 residues (Lamouroux *et al.*, 1987). Although DBH exists in both soluble and membrane-bound forms, the hydropathy plot reveals no obvious hydrophobic segment (except the signal peptide), suggesting that the membrane attachment of DBH probably results from a posttranslational modification, such as the addition of phosphatidylinositol anchoring sequence (see Section 3.2). The oligosaccharide composition of DBH, which contains an average of six Asn-linked oligosaccharides per tetrameric molecule (R. K. Margolis *et al.*, 1984), requires at least two putative glycosylation sites per monomer. In fact, four potential sites were found (Lamouroux *et al.*, 1987), indicating that glycosylation is limited by the availability of some of these sites to the necessary glycosyltransferases.

Although DBH is a major membrane protein in chromaffin granules, over 70% of the total carbohydrate in the membrane glycoproteins is present in the form of large tri- and tetraantennary complex oligosaccharides (R. K. Margolis *et al.*, 1984). After isolation of DBH and two other chromaffin granule membrane glycoproteins (GP-II and GP-III) by sequential lectin affinity chromatography, it was found that GP-II and GP-III contain 25–32% carbohydrate, as compared to only 5% present in DBH (Fischer-Colbrie *et al.*, 1982). Glycoprotein IV (and to a lesser extent, glycoproteins II and III) contain poly(*N*-acetyllactosaminyl) oligosaccharides, which are not present on the major chromaffin granule membrane glycoproteins (R. U. Margolis *et al.*, 1988; see also Section 2.1.4). Glycoprotein IV, which is present in chromaffin granules in a concentration too low to be stained by Coomassie blue (Pryde and Phillips, 1986), appears to account for most of the poly(*N*-acetyllactosaminyl) oligosaccharides. Although the function of these glycoproteins is still unknown, it is noteworthy that they are not confined to chromaffin granules, but are also present in the pituitary (glycoproteins II and III; Fischer-Colbrie *et al.*, 1984; Obendorf *et al.*, 1988), and in exocrine tissues (glycoprotein II; Obendorf *et al.*, 1988).

A 53/56-kDa glycoprotein doublet present in the soluble lysate and membranes of chromaffin granules has been identified as carboxypeptidase H, an enzyme involved in the processing of neuropeptides such as the enkephalins (Laslop *et al.*, 1986). The identification as enzymes of two chromaffin granule glycoproteins (dopamine β-hydroxylase and carboxypeptidase H) which occur in both membrane-bound and soluble forms suggests that the other glycoprotein known to have these properties (i.e., glycoprotein III; Fischer-Colbrie *et al.*, 1984) may also have a still unknown enzymatic activity.

Most studies indicate that the oligosaccharides of chromaffin granule membrane glycoproteins are exposed exclusively on the inner surface of the membrane (Huber *et al.*, 1979; Abbs and Phillips, 1980), in agreement with the general rule that carbohydrate is present on the inside of intracellular membranes and organelles (e.g., endoplasmic reticulum) but on the outer leaflet of plasma (cell surface) membranes. (However, see Section 3.4.3 on cytoplasmic glycoconjugates.)

The chromogranins (a class of related acidic glycoproteins), two chondroitin sulfate/dermatan sulfate proteoglycans, and a fraction highly enriched in DBH were isolated from the chromaffin granule matrix (Kiang *et al.*, 1982). The chromogranins contain 5.4% carbohydrate consisting of *O*-glycosidically linked tri- and tetrasaccharides composed of GalNAc, Gal, and sialic acid (AcNeu and/or GcNeu).* After fractionation of the chromogranins by gel filtration, it was found that these oligosaccharides are not uniformly distributed as a function of chromogranin size, insofar as the larger chromogranins (i.e., chromogranin A) contain predominantly trisaccharides, whereas the tetrasaccharides are concentrated in the chromogranins of intermediate molecular size (Kiang et al., 1982).

The major chromogranin is chromogranin A, whose primary structure has been sequenced from its complementary DNA (Benedum et al., 1986; Iacangelo et al., 1986). Bovine chromogranin A has an apparent molecular weight of 75,000 on SDS-PAGE, but an actual M_r of 48,000 for the unmodified polypeptide chain. In agreement with the oligosaccharide composition described above, it was found not to contain an Asn-X-Ser/Thr consensus sequence for N-glycosylation. Chromogranin B (also known as secretogranin I) is a tyrosine-sulfated glycoprotein, with a carbohydrate composition similar to that of chromogranin A, and which is also present in secretory granules of a wide variety of peptidergic endocrine cells and neurons (Winkler et al., 1986; Benedum et al., 1987). However, its concentration in the soluble chromaffin granule matrix is only 1–2% that of chromogranin A. It has an apparent M_r of 100,000 to 120,000 on SDS-PAGE (100 kDa in bovine chromaffin granules; a 105/113 kDa doublet in rat PC12 pheochromocytoma cells; and 120 kDa in human pheochromocytoma), but the primary structure of the protein from human pheochromocytoma (derived from the sequence of its cDNA) reveals that it is a 76-kDa polypeptide containing 657 amino acids (Benedum et al., 1987). Comparison of the predicted amino acid sequence of human secretogranin I (chromogranin B) with that of bovine chromogranin A revealed that these two proteins have homologous terminal domains and a large intervening variable region. The biological function of these highly acidic proteins having a widespread distribution in endocrine and nervous tissues is still unclear. However, it is thought that they may serve as precursors for regulatory peptides, or, in the case of chromogranin A, in the sequestration and mobilization of calcium from secretory vesicles during stimulus–secretion coupling (see references above).

The chromaffin granule matrix also contains two proteoglycans (I and II) in which the glycosaminoglycan component consists of 48% dermatan sulfate, 23–24% each of chondroitin 4- and 6-sulfate, and 5% heparan sulfate (Kiang et al., 1982). N- and O-glycosidically linked glycoprotein oligosaccharides are also present. Although peptide mapping studies (Banerjee and Margolis, 1982) indicate that the two proteoglycans have very similar or identical protein moieties, proteoglycan II has twice the concentration of glycosaminoglycans and one half the concentration of O-glycosidically linked oligosaccharides [primarily disialyl derivatives of Gal(β1-3)GalNAc] as compared to proteoglycan I. Eighty percent of the glycosaminoglycans (proteoglycans) of chro-

---

*One half of the O-glycosidically linked carbohydrate is released by alkaline borohydride in the form of the trisaccharides AcNeu(α2-3)Gal(β1-3)GalNAcol (40%) and GcNeu(α2-3)Gal(β1-3)GalNAcol (10%). The tetrasaccharide components consist of AcNeu(α2-3)Gal(β1-3)[AcNeu(α2-6)]GalNAcol (20%), GcNeu (α2-8)GcNeu(α2-6)[Gal(β1-3)]GalNAcol (7%), as well as 16 and 7%, respectively, of the tetrasaccharide structures given above but in which the two types of sialic acid cannot be individually assigned to specific linkage positions.

maffin granules are present in the soluble contents (Geissler *et al.*, 1977), whose glycosaminoglycan composition (see above) is considerably different from that of the membranes (11% heparan sulfate, 12% chondroitin 4-sulfate, 10% chondroitin 6-sulfate, and 67% dermatan sulfate).

PC12 pheochromocytoma cells display many of the neurotransmitter properties associated with normal adrenal chromaffin cells and contain chromaffin granules which have been isolated and partially characterized. Potassium-induced depolarization of PC12 cells labeled with [^{3}H]glucosamine leads to a six-fold increase in the release of several labeled glycoproteins and proteoglycans, which together account for approximately 7% of the soluble cell radioactivity (Salton *et al.*, 1983a), and have properties very similar to those previously characterized in the bovine chromaffin granule matrix (Kiang *et al.*, 1982). The released complex carbohydrates include chromogranins, DBH, and two chondroitin sulfate/heparan sulfate proteoglycan fractions.

3.7.2. Large Dense-Cored Vesicles

In sympathetic nerves, norepinephrine is stored in small (45–55 nm diameter) and large (75–90 nm) dense-cored vesicles. In biochemical composition and properties, the large vesicles (LDCV) closely resemble adrenal chromaffin granules, insofar as both contain common specific proteins such as chromogranins, DBH, cytochrome b_{561}, and opioid peptides. Like chromaffin granules, LDCV also accumulate catecholamines by a Mg^{2+}-adenosine 5'-triphosphate-dependent process. Highly purified noradrenergic LDCV were isolated from bovine sympathetic nerve endings by sucrose–D_2O density gradient centrifugation, and their glycoprotein, glycosaminoglycan, and ganglioside composition was examined (R. U. Margolis *et al.*, 1987). Their concentration of glycoprotein hexosamine and sialic acid was 6.6 and 3.9 μmole/100 mg lipid-free dry weight, respectively, values which are similar to those previously found in bovine chromaffin granules. However, whereas chromaffin granule glycoproteins are characterized by their high proportion of GalNAc-containing *O*-glycosidically linked oligosaccharides (present in the chromogranins), such oligosaccharides accounted for only 17% of those in noradrenergic synaptic vesicle glycoproteins. Fractionation of *N*-^{3}H-acetylated glycopeptides by sequential lectin affinity chromatography demonstrated that approximately two-thirds of the oligosaccharides were of the tri- and tetraantennary complex type, accompanied by 14% biantennary oligosaccharides and 3% high-mannose oligosaccharides. The vesicles had a relatively low concentration of chondroitin sulfate (less than 5% of that in chromaffin granules) but significant amounts of heparan sulfate (0.4 μmole GlcNAc/100 mg lipid-free dry weight). No hyaluronic acid was detected. Considered together with other information concerning the complex carbohydrates of cholinergic vesicles of marine electric organ (see Chapter 10) and of prolactin granules of the anterior pituitary (Zanini *et al.*, 1980; Giannattasio *et al.*, 1980), our data suggest that these glycoconjugates have at least several functional roles, which do not require a unique composition or concentration of complex carbohydrates common even to such closely related storage organelles as chromaffin granules and LDCV.

4. BIOLOGICAL FUNCTION OF THE CARBOHYDRATE UNITS IN GLYCOPROTEINS AND PROTEOGLYCANS

The oligosaccharide moieties of glycoproteins are thought to have a wide diversity of functional roles, whose more precise identification has recently attracted considerable attention from cell biologists (for reviews, see Olden *et al.*, 1985; West, 1986). There is evidence that posttranslational modifications such as glycosylation represent a significant alteration in the size and structure of the polypeptide, thereby modifying such physicochemical properties as tertiary conformation, solubility, viscosity, and charge. The glycan units are known to protect the polypeptide component of glycoproteins against uncontrolled proteolysis, and we have shown that the presence of sialic acid residues in dopamine β-hydroxylase contributes to its resistance to proteolytic degradation (Aquino *et al.*, 1980), even though the oligosaccharides are apparently not required for catalytic activity (Hamos *et al.*, 1987). Since many proteins and glycoproteins are proteolytic products of larger-molecular-weight precursors, oligosaccharides on exposed domains can restrict the sites of proteolytic cleavage and in this way contribute to the specificity of the cleavage pattern during the conversion of a precursor protein to the active polypeptide forms.

While it has been demonstrated that the glycan moiety is not necessary for the biological activity of many glycoproteins, other studies have shown that the carbohydrate moiety is required for the biological activity of human chorionic gonadotropin and other glycoproteins (Chen *et al.*, 1982; Manjunath and Sairam, 1982; Kalyan and Bahl, 1983; Gralnick *et al.*, 1983), that glycoprotein oligosaccharides present in the acetylcholine receptor play a role in determining the assembly and folding of the newly synthesized polypeptides to assume a conformation compatible with its characteristic metabolic properties and ligand interactions (Merlie *et al.*, 1982; Prives and Bar-Sagi, 1983), and that glycosylation is required for the maintenance of functional sodium channels in neuroblastoma cells (Waechter *et al.*, 1983). Glycosylation is not required for the export or secretion of most glycoproteins, but in a number of cases it does affect their intracellular sorting and externalization. Several studies also indicate that inhibitors of glycosylation block embryonic development and differentiation, but these effects may be indirect, such as by promoting the unfolding or proteolytic degradation of essential nonglycosylated proteins.

Certain glycoproteins and glycosaminoglycans (such as hyaluronic acid) are important constituents of the extracellular matrix, and in addition to their obvious involvement in modulating cell–cell interactions in developmental processes (see Chapters 11–14), they may also play a more general role in the regulation of the brain cell microenvironment. The finding of proteoglycans intracellularly in the nucleus (R. K. Margolis *et al.*, 1976a; Aquino *et al.*, 1984a; Ripellino *et al.*, 1988) and cytoplasm was unexpected, and their functional roles at these sites still remain to be determined.

ACKNOWLEDGMENTS

Research cited in this review which was performed in the authors' laboratories was supported by grants from the National Institutes of Health (NS-09348 and

NS-13876) and the National Institute of Mental Health (MH-00129). We thank Dr. John Morris for providing a summary of research on retinal proteoglycans, and Julia Cohen and Sabine Pasemann for assistance in preparing the manuscript.

5. APPENDIX

5.1. Developmental Changes of Brain Glycosaminoglycans and Glycoproteins

5.1.1. Glycosaminoglycans

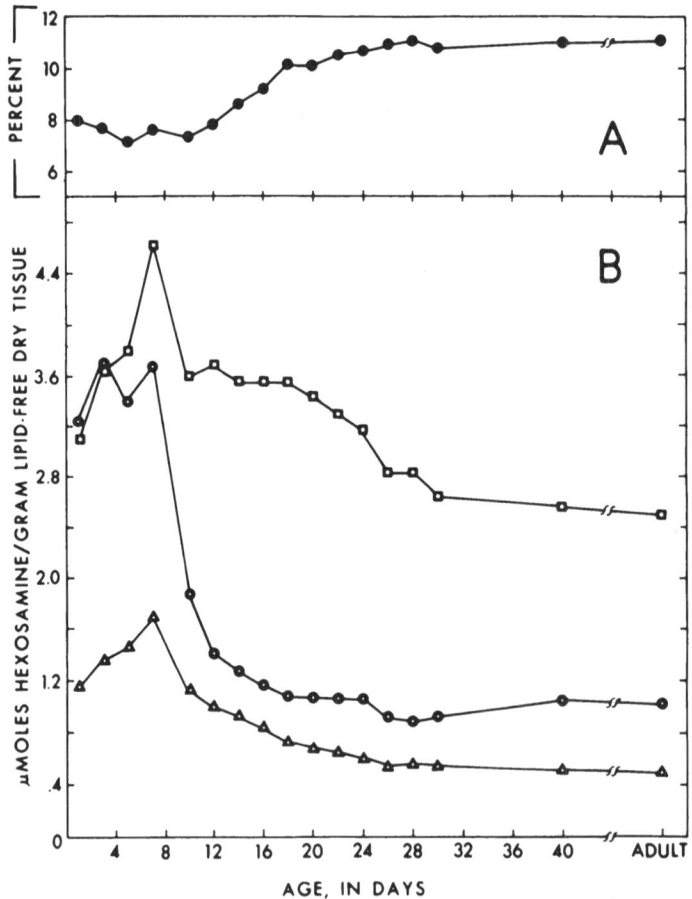

(A) Lipid-free weight as percentage of fresh weight of rat brain during postnatal development. (B) Concentration of glycosaminoglycans, expressed as micromoles of the constituent hexosamine/gram lipid-free dry weight, as a function of increasing postnatal age. (□) Chondroitin sulfate; (○) hyaluronic acid; (△)heparan sulfate. From R. U. Margolis *et al.* (1975).

5.1.2. Glycoproteins

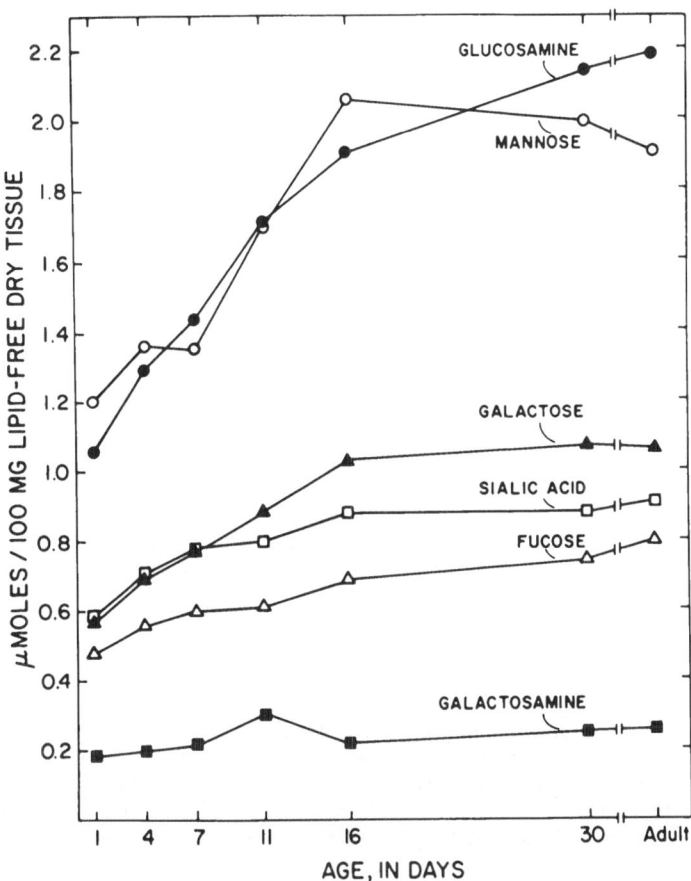

Concentration of glycoprotein carbohydrate in rat brain as a function of increasing age from neonatal to adult (R. K. Margolis *et al.*, 1976b). For comparable values for human brain, see Korpeinen *et al.*, 1982.

5.2. *Turnover of Glycoproteins and Glycosaminoglycans in Rat Brain*

Glycoproteins[a]

Glycoprotein constituent	$t_{\frac{1}{2}}$ (days)[b]
Hexosamine	4.6, 15
Sialic acid	6, 30
Fucose	1, 30
Sulfate	2.5, 14
Threonine[c]	13, 38

[a]For references, see R. K. Margolis *et al.*, 1975b.
[b]The longer turnover half-time listed for each glycoprotein constituent may partially reflect reutilization of labeled precursor (Ferwerda *et al.*, 1981).
[c]Representative of the protein moiety.

Glycosaminoglycans[a]

Glycosaminoglycan	$t_{\frac{1}{2}}$ (days)	
	Sulfate	Hexosamine
Hyaluronic acid	—	9, 45
Chondroitin sulfate	7	21
Heparan sulfate	3	3.7, 10

[a]From Margolis and Margolis (1972, 1973).

6. REFERENCES

Abbs, M. T., and Phillips, J. H., 1980, Organization of the proteins of the chromaffin granule membrane, *Biochim. Biophys. Acta* **595**:200–221.

Abeijon, C., and Hirschberg, C. B., 1988, Intrinsic membrane glycoproteins with cytosol-oriented sugars in the endoplasmic reticulum, *Proc. Natl. Acad. Sci. USA* **85**:1010–1014.

Amblard, F., He, J.-T., Barbet, J., Goridis, C., and Prochiantz, A., 1988, A 140-kilodalton protein is released from cultured astrocytes by phosphatidylinositol phospholipase C, *J. Neurochem.* **50**:486–489.

Anderson, M. J., and Fambrough, D. M., 1983, Aggregates of acetylcholine receptors are associated with plaques of a basal lamina heparan sulfate proteoglycan on the surface of skeletal muscle fibers, *J. Cell Biol.* **97**:1396–1411.

Anderson, M. J., Klier, F. G., and Tanguay, K. E., 1984, Acetylcholine receptor aggregation parallels the deposition of a basal lamina proteoglycan during development of the neuromuscular junction, *J. Cell Biol.* **99**:1769–1784.

Aquino, D., Wong, R., Margolis, R. U., and Margolis, R. K., 1980, Sialic acid residues inhibit proteolytic degradation of dopamine β-hydroxylase, *FEBS Lett.* **112**:195–198.

Aquino, D. A., Margolis, R. U., and Margolis, R. K., 1984a, Immunocytochemical localization of a chondroitin sulfate proteoglycan in nervous tissue. I. Adult brain, retina and peripheral nerve, *J. Cell Biol.* **99**:1117–1129.

Aquino, D. A., Margolis, R. U., and Margolis, R. K., 1984b, Immunocytochemical localization of a chondroitin sulfate proteoglycan in nervous tissue. II. Studies in developing brain, *J. Cell Biol.* **99**:1130–1139.

Bach, G., and Berman, E. R., 1971a, Amino sugar-containing compounds of the retina. I. Isolation and identification, *Biochim. Biophys. Acta* **252**:453–461.

Bach, G., and Berman, E. R., 1971b, Amino sugar-containing compounds of the retina. II. Structural studies, *Biochim. Biophys. Acta* **252**:461–471.

Banerjee, S., and Margolis, R. U., 1982, Glycoproteins and proteoglycans of the chromaffin granule matrix, *J. Neurochem.* **39**:1700–1703.

Beasley, L., and Stallcup, W. B., 1987, The nerve growth factor-inducible large external (NILE) glycoprotein and neural cell adhesion molecule (N-CAM) have distinct patterns of expression in the developing rat central nervous system, *J. Neurosci.* **7**:708–715.

Benedum, U. M., Baeuerle, P. A., Konecki, D. S., Frank, R., Powell, J., Mallet, J., and Huttner, W. B., 1986, The primary structure of bovine chromogranin A: A representative of a class of acidic secretory proteins common to a variety of peptidergic cells, *EMBO J.* **5**:1495–1502.

Benedum, U. M., Lamouroux, A., Konecki, D. S., Rosa, P., Hille, A., Baeuerle, P. A., Frank, R., Lottspeich, F., Mallet, J., and Huttner, W. B., 1987, The primary structure of human secretogranin I (chromogranin B): Comparison with chromogranin A reveals homologous terminal domains and a large intervening variable region, *EMBO J.* **6**:1203–1211.

Bock, E., Richter-Landsberg, C., Faissner, A., and Schachner, M., 1985, Demonstration of immunochemical identity between the nerve growth factor-inducible large external (NILE) glycoprotein and the cell adhesion molecule Ll, *EMBO J.* **4**:2765–2768.

Bolin, L. M., and Rouse, R. V., 1986, Localization of Thy-1 expression during postnatal development of the mouse cerebellar cortex, *J. Neurocytol.* **15**:29–36.

Brandan, E., Maldonado, M., Garrido, J., and Inestrosa, N. C., 1985, Anchorage of collagen-tailed acetylcholinesterase to the extracellular matrix is mediated by heparan sulfate proteoglycans, *J. Cell Biol.* **101**:985–992.

Campbell, D. G. Gagnon, J., Reid, K. B. M., and Williams, A. F., 1981, Rat brain Thy-1 glycoprotein. The amino acid sequence, disulphide bonds and an unusual hydrophobic region, *Biochem. J.* **195**:15–30.

Carey, D. J., and Todd, M. S., 1986, A cytoskeleton-associated plasma membrane heparan sulfate proteoglycan in Schwann cells, *J. Biol. Chem.* **261**:7518–7525.

Chen, H.-C., Shimohigashi, Y., Dufau, M. L., and Catt, K. J., 1982, Characterization and biological properties of chemically deglycosylated human chorionic gonadotropin, *J. Biol. Chem.* **257**:14446–14452.

Choi, H. U., and Meyer, K., 1975, The structure of a sulfated glycoprotein of chick allantoic fluid: Methylation and periodate oxidation, *Carbohyd. Res.* **40**:77–88.

Cole, G. J., Schubert, D., and Glaser, L., 1985, Cell–substratum adhesion in chick neural retina depends upon protein–heparan sulfate interactions, *J. Cell Biol.* **100**:1192–1199.

Cole, G. J., Loewy, A., and Glaser, L., 1986, Neuronal cell–cell adhesion depends on interactions of N-CAM with heparin-like molecules, *Nature* **320**:445–447.

Davis, L. I., and Blobel, G., 1987, Nuclear pore complex contains a family of glycoproteins that includes p62: Glycosylation through a previously unidentified cellular pathway, *Proc. Natl. Acad. Sci. USA* **84**:7552–7556.

Edgar, D., Timpl, R., and Thoenen, H., 1984, The heparin-binding domain of laminin is responsible for its effects on neurite outgrowth and neuronal survival, *EMBO J.* **3**:1463–1468.

Edge, A. S. B., and Spiro, R. G., 1984, Presence of sulfate in N-glycosidically linked carbohydrate units of calf thyroid plasma membrane glycoproteins, *J. Biol. Chem.* **259**:4710–4713.

Ferwerda, W., Blok, C. M., and Heijlman, J., 1981, Turnover of free sialic acid, CMP-sialic acid, and bound sialic acid in rat brain, *J. Neurochem.* **36**:1492–1499.

Finne, J., 1985, Polysialic acid—a glycoprotein carbohydrate involved in neural adhesion and bacterial meningitis, *Trends Biochem. Sci.* **10**:129–132.

Finne, J., and Krusius, T., 1976, O-glycosidic carbohydrate units from glycoproteins of different tissues: Demonstration of a brain-specific disaccharide, α-galactosyl(1-3)N-acetylgalactosamine, *FEBS Lett.* **66**:94–97.

Finne, J., and Krusius, T., 1982, Preparation and fractionation of glycopeptides, *Methods Enzymol.* **83**:269–277.

Finne, J., and Mäkelä, P. H., 1985, Cleavage of the polysialosyl units of brain glycoproteins by a bacteriophage endosialidase, *J. Biol. Chem.* **260**:1265–1270.

Finne, J., Krusius, T., Rauvala, H., and Hemminki, K., 1977, The disialosyl group of glycoproteins: Occurrence in different tissues and cellular membranes, *Eur. J. Biochem.* **77**:319–323.

Finne, J., Krusius, T., Margolis, R. K., and Margolis, R. U., 1979, Novel mannitol-containing oligosaccharides obtained by mild alkaline borohydride treatment of a chondroitin sulfate proteoglycan from brain, *J. Biol. Chem.* **254**:10295–10300.

Fischer-Colbrie, R., Schachinger, M., Zangerle, R., and Winkler, H., 1982, Dopamine β-hydroxylase and other glycoproteins from the soluble content and the membranes of adrenal chromaffin granules: Isolation and carbohydrate analysis, *J. Neurochem.* **38**:725–732.

Fischer-Colbrie, R., Zangerle, R., Frischenschlager, I., Weber, A., and Winkler, H., 1984, Isolation and immunological characterization of a glycoprotein from adrenal chromaffin granules, *J. Neurochem.* **42**:1008–1016.

Flanagan, B. F., Teplow, D. B., Dreyer, W. J., and Fabre, J. W., 1986, Unusual phylogenetic conservation of the N-terminal amino acid sequence of the central nervous system-specific membrane glycoprotein F3-87-8 (CNSgp130), *J. Neurochem.* **46**:542–544.

Fransson, L.-Å., Hampson, I., Kumar, S., and Gallagher, J., 1985, Chemical heterogeneity of heparan sulfate from a human neuroblastoma cell line, *Acta Chem. Scand. B* **39:**305–313.

Friedlander, D., Grumet, M., and Edelman, G., 1986, Nerve growth factor enhances expression of neuron-glia cell adhesion molecule in PC12 cells, *J. Cell Biol.* **102:**413–419.

Gallagher, J. T., Lyon, M., and Steward, W. P., 1986, Structure and function of heparan sulfate proteoglycans, *Biochem. J.* **236:**313–325.

Gallo, V., Bertolotto, A., and Levi, G., 1987, The proteoglycan chondroitin sulfate is present in a subpopulation of cultured astrocytes and in their precursors, *Dev. Biol.* **123:**282–285.

Gammon, C. M., Goodrum, J. F., Toews, A. D., Okabe, A., and Morell, P., 1985, Axonal transport of glycoconjugates in the rat visual system, *J. Neurochem.* **44:**376–387.

Gandy, S., Czernick, A. J., and Greengard, P., 1988, Phosphorylation of Alzheimer disease amyloid precursor peptide by protein kinase C and Ca^{2+}/calmodulin-dependent protein kinase II, *Proc. Natl. Acad. Sci. USA* **85:**6218–6221.

Gavine, F. S., Pryde, J. G., Deane, D. L., and Apps, D. K., 1984, Glycoproteins of the chromaffin granule membrane: Separation by two-dimensional electrophoresis and identification by lectin binding, *J. Neurochem.* **43:**1243–1252.

Geissler, D., Martinek, A., Margolis, R. U., Margolis, R. K., Skrivanek, J. A., Ledeen, R. W., König, P., and Winkler, H., 1977, Composition and biogenesis of complex carbohydrates of adrenal chromaffin granules, *Neuroscience* **2:**685–693.

Giannattasio, G., Zanini, A., and Meldolesi, J., 1979, Complex carbohydrates of secretory organelles, in: *Complex Carbohydrates of Nervous Tissue* (R. U. Margolis and R. K. Margolis, eds.), pp. 327–345, Plenum Press, New York.

Giannattasio, G., Zanini, A., Rosa, P., Meldolesi, J., Margolis, R. K., and Margolis, R. U., 1980, Molecular organization of prolactin granules–III. Intracellular transport of glycosaminoglycans and glycoproteins of the bovine prolactin granule matrix, *J. Cell Biol.* **86:**273–279.

Ginsburg, V. (ed.), 1982, *Complex Carbohydrates*, Part D, (*Meth. Enzymol.*, Vol. 83), Academic Press, New York.

Ginsburg, V. (ed.), 1987, *Complex Carbohydrates*, Part E, (*Meth. Enzymol.*, Vol. 138), Academic Press, New York.

Gowda, D. C., Margolis, R. U., and Margolis, R. K., 1989a, Presence of the HNK-1 epitope on poly(*N*-acetyllactosaminyl) oligosaccharides and identification of multiple core proteins in the chondroitin sulfate proteoglycans of brain, *Biochemistry*, in press.

Gowda, D. C., Margolis, R. K., Frangione, B., Ghiso, J., Larrondo-Lillo, M., and Margolis, R. U., 1989b, Relation of the amyloid β protein precursor to heparan sulfate proteoglycans, *Science*, in press.

Gowda, D. C., Goossen, B., Margolis, R. K., and Margolis, R. U., 1989c, Chondroitin sulfate and heparan sulfate proteoglycans of PC12 pheochromocytoma cells, *J. Biol. Chem.*, in press.

Gralnick, H. R., Williams, S. B., and Rick, M. E., 1983, Role of carbohydrate in multimeric structure of factor VIII/von Willebrand factor protein, *Proc. Natl. Acad. Sci. USA* **80:**2771–2774.

Green, E. D., and Baenziger, J. U., 1988, Asparagine-linked oligosaccharides on lutropin, follitropin, and thyrotropin. I. Structural elucidation of the sulfated and sialylated oligosaccharides on bovine, ovine, and human pituitary glycoprotein hormones, *J. Biol. Chem.* **263:**25–35.

Greene, L. A., and Shelanski, M. L., 1989, The nerve growth factor-inducible large external (NILE) glycoprotein: Biochemistry and regulation of synthesis, in: *Morphoregulatory Molecules* (G. M. Edelman, B. A. Cunningham, and J.-P. Thiery, eds.), Wiley, New York.

Greene, L. A., and Tischler, A. S., 1982, PC12 pheochromocytoma cultures in neurobiological research, in: *Advances in Cellular Neurobiology* (S. Fedoroff and L. Hertz, eds.), Vol. 3, pp. 373–414, Academic Press, New York.

Hageman, G. S., and Johnson, L. V., 1987, Chondroitin 6-sulfate glycosaminoglycan is a major constituent of primate cone photoreceptor matrix sheaths, *Curr. Eye Res.* **6:**639–646.

Hamos, J., Desai, P. R., and Villafranca, J. J., 1987, Characterization and kinetic studies of deglycosylated dopamine β-hydroxylase, *FASEB J.* **1:**143–148.

Hampson, I. N., Kumar, S., and Gallagher, J. T., 1983, Differences in the distribution of *O*-sulphate groups of cell-surface and secreted heparan sulphate produced by human neuroblastoma cells in culture, *Biochim. Biophys. Acta* **763:**183–190.

Hampson, I. N., Kumar, S., and Gallagher, J. T., 1984, Heterogeneity of cell-associated and secretory heparan sulphate proteoglycans produced by cultured human neuroblastoma cells, *Biochim. Biophys. Acta* **801**:306–313.

Hanover, J. A., Cohen, C. K., Willingham, M. C., and Park, M. K., 1987, *O*-Linked *N*-acetylglucosamine is attached to proteins of the nuclear pore. Evidence for cytoplasmic and nucleoplasmic glycoproteins, *J. Biol. Chem.* **262**:9887–9894.

Hascall, V. C., and Hascall, G. K., 1982, Proteoglycans, in: *Cell Biology of Extracellular Matrix* (E. D. Hay, ed.), pp. 39–63, Plenum Press, New York.

Hassell, J. R., Kimura, J. H., and Hascall, V. C., 1986, Proteoglycan core protein families, *Annu. Rev. Biochem.* **55**:539–567.

Hewitt, A. T., 1986, Extracellular matrix molecules: Their importance in the structure and function of the retina, in: *The Retina: A Model for Cell Biology Studies* (R. Adler and D. Farber, eds.), Part II, pp. 169–214, Academic Press, New York.

Hoffman, S., and Edelman, G. M., 1987, A proteoglycan with HNK-1 antigenic determinants is a neuron-associated ligand for cytotactin, *Proc. Natl. Acad. Sci. USA* **84**:2523–2527.

Hoffman, S., Crossin, K. L., and Edelman, G. M., 1988, Molecular forms, binding functions, and developmental expression patterns of cytotactin and cytotactin-binding proteoglycan, an interactive pair of extracellular matrix molecules, *J. Cell Biol.* **106**:519–532.

Holt, G. D., Snow, C. M., Senior, A., Haltiwanger, R. S., Gerace, L., and Hart, G. W., 1987a, Nuclear pore complex glycoproteins contain cytoplasmically disposed *O*-linked *N*-acetylglucosamine, *J. Cell Biol.* **104**:1157–1164.

Holt, G. D., Haltiwanger, R. S., Torres, C.-R., and Hart, G. W., 1987b, Erythrocytes contain cytoplasmic glycoproteins, *J. Biol. Chem.* **262**:14847–14850.

Hooghe-Peters, E. L., and Hooghe, R. J., 1982, The Thy-1 glycoprotein on nerve cells in culture, *J. Neuroimmunol.* **2**:191–200.

Höök, M., Kjellén, L., Johansson, S., and Robinson, J., 1984, Cell surface glycosaminoglycans, *Annu. Rev. Biochem.* **53**:847–869.

Huber, E., König, P., Schuler, G., Aberer, W., Plattner, H., and Winkler, H., 1979, Characterization and topography of the glycoproteins of adrenal chromaffin granules, *J. Neurochem.* **32**:35–47.

Iacangelo, A., Affolter, H.-U., Eiden, L. E., Herbert, E., and Grimes, M., 1986, Bovine chromogranin A sequence and distribution of its messenger RNA in endocrine tissues, *Nature* **323**:82–86.

Inestrosa, N. C., Matthew, W. D., Reiness, C. G., Hall, Z. W., and Reichardt, L. F., 1985, Atypical distribution of asymmetric acetylcholinesterase in mutant PC12 pheochromocytoma cells lacking a cell surface heparan sulfate proteoglycan, *J. Neurochem.* **45**:86–94.

James, W. M., and Agnew, W. S., 1987, Multiple oligosaccharide chains in the voltage-sensitive Na channel from *Electrophorus electricus:* Evidence for α-2,8 linked polysialic acid, *Biochem. Biophys. Res. Commun.* **148**:817–826.

Johnson, L. V., and Hageman, G. S., 1987, Enzymatic characterization of peanut agglutinin-binding components in the retinal interphotoreceptor matrix, *Exp. Eye Res.* **44**:553–566.

Kalyan, N. K., and Bahl, O. P., 1983, Role of carbohydrate in human chorionic gonadotropin, *J. Biol. Chem.* **258**:67–74.

Kennedy, J. F., 1979, *Proteoglycans—Biological and Chemical Aspects in Human Life,* Elsevier, Amsterdam.

Kiang, W.-L., Crockett, C. P., Margolis, R. K., and Margolis, R. U., 1978, Glycosaminoglycans and glycoproteins associated with microsomal subfractions of brain and liver, *Biochemistry* **17**:3841–3848.

Kiang, W.-L., Margolis, R. U., and Margolis, R. K., 1981, Fractionation and properties of a chondroitin sulfate proteoglycan and the soluble glycoproteins of brain, *J. Biol. Chem.* **256**:10529–10537.

Kiang, W.-L., Krusius, T., Finne, J., Margolis, R. U., and Margolis, R. K., 1982, Glycoproteins and proteoglycans of the chromaffin granule matrix, *J. Biol. Chem.* **257**:1651–1659.

Klinger, M. M., Margolis, R. U., and Margolis, R. K., 1985, Isolation and characterization of the heparan sulfate proteoglycans of brain. Use of affinity chromatography on lipoprotein lipase-agarose, *J. Biol. Chem.* **260**:4082–4090.

Korpeinen, T., Mononen, I., Krusius, T., and Järnefelt, J., 1982, Glycosylation of proteins in developing human brain, *J. Neurochem.* **39**:1737–1739.

Krusius, T., and Finne, J., 1977, Structural features of tissue glycoproteins. Fractionation and methylation analysis of glycopeptides derived from rat brain, kidney and liver, *Eur. J. Biochem.* **78**:369–379.

Krusius, T., and Finne, J., 1978, Characterization of a novel sugar sequence from rat brain glycoproteins containing fucose and sialic acid, *Eur. J. Biochem.* **84**:395–403.

Krusius, T., Finne, J., Margolis, R. U., and Margolis, R. K., 1978, Structural features of microsomal, synaptosomal, mitochondrial, and soluble glycoproteins of brain, *Biochemistry* **17**:3849–3854.

Krusius, T., Finne, J., Margolis, R. K., and Margolis, R. U., 1986, Identification of an *O*-glycosidic mannose-linked sialylated tetrasaccharide and keratan sulfate oligosaccharides in the chondroitin sulfate proteoglycan of brain, *J. Biol. Chem.* **261**:8237–8242.

Krusius, T., Reinhold, V. N., Margolis, R. K., and Margolis, R. U., 1987, Structural studies on sialylated and sulfated *O*-glycosidic mannose-linked oligosaccharides in the chondroitin sulfate proteoglycan of brain, *Biochem. J.* **245**:229–234.

Lakin, K. H., and Fabre, J. W., 1981, Identification with a monoclonal antibody of a phylogenetically conserved brain-specific determinant on a 130,000 molecular weight glycoprotein of human brain, *J. Neurochem.* **37**:1170–1178.

Lakin, K. H., Allen, A. K., and Fabre, J. W., 1983, Purification and preliminary biochemical characterization of the human and rat forms of the central nervous system-specific molecule, F3-87-8, *J. Neurochem.* **41**:385–394.

Lamouroux, A., Vigny, A., Biguet, N. F., Darmon, M. C., Franck, R., Henry, J.-P., and Mallet, J., 1987, The primary structure of human dopamine-β-hydroxylase: Insights into the relationship between the soluble and the membrane-bound forms of the enzyme, *EMBO J.* **6**:3931–3937.

Lander, A. D., Fujii, D. K., and Reichardt, L. F., 1985, Laminin is associated with the "neurite outgrowth-promoting factors" found in conditioned medium, *Proc. Natl. Acad. Sci. USA* **82**:2183–2187.

Laslop, A., Fischer-Colbrie, R., Hook, V., Obendorf, D., and Winkler, H., 1986, Identification of two glycoproteins of chromaffin granules as the carboxypeptidase H, *Neurosci. Lett.* **72**:300–304.

Lechner, J., Wieland, F., and Sumper, M., 1985, Biosynthesis of sulfated saccharides *N*-glycosidically linked to the protein via glucose, *J. Biol. Chem.* **260**:860–866.

Levine, J. M., and Card, J. P., 1987, Light and electron microscopic localization of a cell surface antigen (NG2) in the rat cerebellum: Association with smooth protoplasmic astrocytes, *J. Neurosci.* **7**:2711–2720.

Liau, Y. H., and Horowitz, M. I., 1982, Incorporation *in vitro* of [³H]glucosamine or [³H]glucose and [³⁵S]SO₄²⁻ into rat gastric mucosa, *J. Biol. Chem.* **257**:4709–4718.

Livingston, B. D., Jacobs, J. L., Glick, M. C., and Troy, F. A., 1988, Extended polysialic acid chains ($n >$ 55) in glycoproteins from human neuroblastoma cells, *J. Biol. Chem.* **263**:9443–9448.

Low, M. G., and Saltiel, A. R., 1988, Structural and functional roles of glycosyl-phosphatidylinositol in membranes, *Science* **239**:268–275.

Manjunath, P., and Sairam, M. R., 1982, Biochemical, biological, and immunological properties of chemically deglycosylated human choriogonadotropin, *J. Biol. Chem.* **257**:7109–7115.

Maresh, G. A., Chernoff, E. A. G., and Culp, L. A., 1984, Heparan sulfate proteoglycans of human neuroblastoma cells: Affinity fractionation on columns of platelet factor-4, *Arch. Biochem. Biophys.* **233**:428–437.

Margolis, R. U., 1967, Acid mucopolysaccharides and proteins of bovine whole brain, white matter and myelin, *Biochim. Biophys. Acta* **141**:91–102.

Margolis, R. K., and Margolis, R. U., 1970, Sulfated glycopeptides from rat brain glycoproteins, *Biochemistry* **9**:4389–4396.

Margolis, R. U., and Margolis, R. K., 1972, Sulfate turnover in mucopolysaccharides and glycoproteins of brain, *Biochim. Biophys. Acta* **264**:426–431.

Margolis, R. K., and Margolis, R. U., 1973, The turnover of hexosamine and sialic acid in glycoproteins and mucopolysaccharides of brain, *Biochim. Biophys. Acta* **304**:413–420.

Margolis, R. U., and Margolis, R. K., 1974, Distribution and metabolism of mucopolysaccharides and glycoproteins in neuronal perikaria, astrocytes, and oligodendroglia, *Biochemistry* **13**:2849–2852.

Margolis, R. K., and Margolis, R. U., 1979, Structure and distribution of glycoproteins and glycosaminoglycans, in: *Complex Carbohydrates of Nervous Tissue* (R. U. Margolis and R. K. Margolis, eds.), pp. 45–73, Plenum Press, New York.

Margolis, R. K., and Margolis, R. U., 1983, Distribution and characteristics of polysialosyl oligosaccharides in nervous tissue glycoproteins, *Biochem. Biophys. Res. Commun.* **116**:889–894.

Margolis, R. K., Margolis, R. U., Preti, C., and Lai, D., 1975a, Distribution and metabolism of glycoproteins and glycosaminoglycans in subcellular fractions of brain, *Biochemistry* **14**:4797–4804.

Margolis, R. K., Preti, C., Chang, L., and Margolis, R. U., 1975b, Metabolism of the protein moiety of brain glycoproteins, *J. Neurochem.* **25**:707–709.

Margolis, R. K., Crockett, C. P., Kiang, W.-L., and Margolis, R. U., 1976a, Glycosaminoglycans and glycoproteins associated with rat brain nuclei, *Biochim. Biophys. Acta* **451**:465–469.

Margolis, R. K., Preti, C., Lai, D., and Margolis, R. U., 1976b, Developmental changes in brain glycoproteins, *Brain Res.* **112**:363–369.

Margolis, R. K., Thomas, M. D., Crockett, C. P., and Margolis, R. U., 1979, Presence of chondroitin sulfate in the neuronal cytoplasm, *Proc. Natl. Acad. Sci. USA* **76**:1711–1715.

Margolis, R. K., Salton, S. R. J., and Margolis, R. U., 1983a, Complex carbohydrates of cultured PC12 pheochromocytoma cells. Effects of nerve growth factor and comparison with neonatal and mature rat brain, *J. Biol. Chem.* **258**:4110–4117.

Margolis, R. K., Salton, S. R. J., and Margolis, R. U., 1983b, Structural features of the nerve growth factor inducible large external glycoprotein of PC12 pheochromocytoma cells and brain, *J. Neurochem.* **41**:1635–1640.

Margolis, R. K., Finne, J., Krusius, T., and Margolis, R. U., 1984, Structural studies on glycoprotein oligosaccharides of chromaffin granule membranes and dopamine β-hydroxylase, *Arch. Biochem. Biophys.* **228**:443–449.

Margolis, R. K., Greene, L. A., and Margolis, R. U., 1986, Poly(N-acetyllactosaminyl) oligosaccharides in glycoproteins of PC12 pheochromocytoma cells and sympathetic neurons, *Biochemistry* **25**:3463–3468.

Margolis, R. K., Salton, S. R. J., and Margolis, R. U., 1987a, Effects of nerve growth factor-induced differentiation on the heparan sulfate of PC12 pheochromocytoma cells and comparison with developing brain. *Arch. Biochem. Biophys.* **257**:107–114.

Margolis, R. K., Ripellino, J. A., Goossen B., Steinbrich, R., and Margolis, R. U., 1987b, Occurrence of the HNK-1 epitope (3-sulfoglucuronic acid) in PC12 pheochromocytoma cells, chromaffin granule membranes, and chondroitin sulfate proteoglycans, *Biochem. Biophys. Res. Commun.* **145**:1142–1148.

Margolis, R. K., Goossen, B., and Margolis, R. U., 1988, Phosphatidylinositol-anchored glycoproteins of PC12 pheochromocytoma cells and brain, *Biochemistry* **27**:3454–3458.

Margolis, R. U., Margolis, R. K., Chang, L., and Preti, C., 1975, Glycosaminoglycans of brain during development, *Biochemistry* **14**:85–88.

Margolis, R. U., Ledeen, R. W., Sbaschnig-Agler, M., Byrne, M. C., Klein, R. L., Douglas, B. H., II, and Margolis, R. K., 1987, Complex carbohydrate composition of large dense-cored vesicles from sympathetic nerve, *J. Neurochem.* **49**:1839–1844.

Margolis, R. U., Fischer-Colbrie, R., and Margolis, R. K., 1988, Poly(N-acetyllactosaminyl) oligosaccharides of chromaffin granule membrane glycoproteins, *J. Neurochem.* **51**:1819–1824.

Matthew, W. D., Greenspan, R. J., Lander, A. D., and Reichardt, L. F., 1985, Immunopurification and characterization of a neuronal heparan sulfate proteoglycan, *J. Neurosci.* **5**:1842–1850.

McGuire, J. C., Greene, L. A., and Furano, A. V., 1978, NGF stimulates incorporation of fucose or glucosamine into an external glycoprotein in cultured rat PC12 pheochromocytoma cells, *Cell* **15**:357–365.

McKenzie, J. L., Allen, A. K., and Fabre, J. W., 1981, Biochemical characterization including amino acid and carbohydrate compositions of canine and human brain Thy-1 antigen, *Biochem. J.* **197**:629–636.

Merkle, R. K., and Cummings, R. D., 1987, Lectin affinity chromatography of glycopeptides, *Methods Enzymol.* **138**:232–259.

Merkle, R. K., and Heifetz, A., 1984, Enzymatic sulfation of N-glycosidically linked oligosaccharides by endothelial cell membranes, *Arch. Biochem. Biophys.* **234**:460–467.

Merlie, J. P., Sebbane, R., Tzartos, S., and Lindstrom, J., 1982, Inhibition of glycosylation with tunicamycin blocks assembly of newly synthesized acetylcholine receptor subunits in muscle cells, *J. Biol. Chem.* **257**:2694–2701.

Miller, R. R., and Waechter, C. J., 1984, Structural features and some binding properties of proteoheparan sulfate enzymatically labeled by calf brain microsomes, *Arch. Biochem. Biophys.* **228**:247–257.

Moos, M., Tacke, R., Scherer, H., Teplow, D., Früh, K., and Schachner, M., 1988, Neural adhesion molecule L1 as a member of the immunoglobulin superfamily with binding domains similar to fibronectin, *Nature* **334**:701–703.

Morris, J. E., and Ting, Y.-P., 1981, Comparison of proteoglycans extracted by saline and guanidinium chloride from cultured chick retinas, *J. Neurochem.* **37**:1594–1602.

Morris, J. E., Hopwood, J. J., and Dorfman, A., 1977, Biosynthesis of glycosaminoglycans in the developing retina, *Dev. Biol.* **58**:313–327.

Morris, J. E., Ting, Y.-P., and Birkholz-Lambrecht, A., 1984, Low buoyant density proteoglycans from saline and dissociative extracts of embryonic chicken retinas, *J. Neurochem.* **42**:798–809.

Morris, J. E., Yanagishita, M., and Hascall, V., 1987, Proteoglycans synthesized by embryonic chicken retina in culture: Composition and compartmentalization, *Arch. Biochem. Biophys.* **258**:206–218.

Morris, R., 1985, Thy-1 in developing nervous tissue, *Dev. Neurosci.* **7**:133–160.

Nakanishi, S., 1983, Extracellular matrix during laminar pattern formation of neocortex in normal and reeler mutant mice, *Dev. Biol.* **95**:305–316.

Needham, L. K., Adler, R., and Hewitt, A. T., 1988, Proteoglycan synthesis in flat cell-free cultures of chick embryo retinal neurons and photoreceptors, *Dev. Biol.* **126**:304–314.

Obendorf, D., Schwarzenbrunner, U., Fischer-Colbrie, R., Laslop, A., and Winkler, H., 1988, Immunological characterization of a membrane glycoprotein of chromaffin granules: Its presence in endocrine and exocrine tissues, *Neuroscience* **25**:343–351.

Olden, K., Bernard, B. A., Humphries, M. J., Yeo, K.-T., White, S. L., Newton, S. A., Bauer, H. C., and Parent, J. B., 1985, Function of glycoprotein glycans, *Trends Biochem. Sci.* **10**:78–82.

Osawa, T., and Tsuji, T., 1987, Fractionation and structural assessment of oligosaccharides and glycopeptides by use of immobilized lectins, *Annu. Rev. Biochem.* **56**:21–42.

Parekh, R. B., Tse, A. G. D., Dwek, R. A., Williams, A. F., and Rademacher, T. W., 1987, Tissue-specific *N*-glycosylation, site-specific oligosaccharide patterns and lentil lectin recognition of rat Thy-1, *EMBO J.* **6**:1233–1244.

Porrello, K., and LaVail, M. M. 1986, Immunocytochemical localization of chondroitin sulfates in the interphotoreceptor matrix of the normal and dystrophic rat retina, *Curr. Eye Res.* **5**:981–994.

Porrello, K., Yasumura, D., and LaVail, M. M., 1986, The interphotoreceptor matrix in RCS rats: Histochemical analysis and correlation with the rate of retinal degeneration, *Exp. Eye Res.* **43**:413–430.

Prives, J., and Bar-Sagi, D., 1983, Effect of tunicamycin, an inhibitor of protein glycosylation, on the biological properties of acetylcholine receptor in cultured muscle cells, *J. Biol. Chem.* **258**:1775–1780.

Pryde, J. G., and Phillips, J. H., 1986, Fractionation of membrane proteins by temperature-induced phase separation in Triton X-114, *Biochem. J.* **233**:525–533.

Ratner, N., Hong, D., Lieberman, M. A., Bunge, R. P., and Glaser, L., 1988, The neuronal cell-surface molecule mitogenic for Schwann cells is a heparin-binding protein, *Proc. Natl. Acad. Sci. USA* **85**:6992–6996.

Reddy, P., Jacquier, A. C., Abovich, N., Petersen, G., and Rosbash, M., 1986, The *period* clock locus of *D. melanogaster* codes for a proteoglycan, *Cell* **46**:53–61.

Richter-Landsberg, C., Greene, L. A., and Shelanski, M. L., 1985, Cell surface Thy-1-cross-reactive glycoprotein in cultured PC12 cells: Modulation of nerve growth factor and association with the cytoskeleton, *J. Neurosci.* **5**:468–476.

Ripellino, J. A., and Margolis, R. U., 1989, Structural properties of the heparan sulfate proteoglycans of brain, *J. Neurochem.* **52**:807–812.

Ripellino, J. A., Klinger, M. M., Margolis, R. U., and Margolis, R. K., 1985, The hyaluronic acid binding region as a specific probe for the localization of hyaluronic acid in tissue sections. Application to chick embryo and rat brain, *J. Histochem. Cytochem.* **33**:1060–1066.

Ripellino, J. A., Bailo, M., Margolis, R. U., and Margolis, R. K., 1988, Light and electron microscopic studies on the localization of hyaluronic acid in developing rat cerebellum, *J. Cell Biol.* **106**:845–855.

Ripellino, J. A., Margolis, R. U., and Margolis, R. K., 1989a, Oligosaccharide composition, localization, and developmental changes of a brain-specific (F3-87-8) glycoprotein, *J. Neurochem.*, in press.

Ripellino, J. A., Margolis, R. U., and Margolis, R. K., 1989b, Immunoelectron microscopic localization of hyaluronic acid binding region and link protein epitopes in brain, *J. Cell Biol.*, in press.

Rogers, S. L., McCarthy, J. B., Palm, S. L., Furcht, L. T., and Letourneau, P. C., 1985, Neuron-specific interactions with two neurite-promoting fragments of fibronectin, *J. Neurosci.* **5:**369–378.

Roth, J., Taatjes, D. J., Bitter-Suermann, D., and Finne, J., 1987, Polysialic acid units are spatially and temporally expressed in developing postnatal rat kidney, *Proc. Natl. Acad. Sci. USA* **84:**1969–1973.

Roussel, P., Lamblin, G., Degand, P., Walker-Nasir, E., and Jeanloz, R. W., 1975, Heterogeneity of the carbohydrate chains of sulfated bronchial glycoproteins isolated from a patient suffering from cystic fibrosis, *J. Biol. Chem.* **250:**2114–2122.

Sajovic, P., Kouvelas, E., and Trenkner, E., 1986, Probable identity of NILE glycoprotein and the high-molecular-weight component of Ll antigen, *J. Neurochem.* **47:**541–546.

Salton, S. R. J., Margolis, R. U., and Margolis, R. K., 1983a, Release of chromaffin granule glycoproteins and proteoglycans from potassium-stimulated PC12 pheochromocytoma cells, *J. Neurochem.* **41:**1165–1170.

Salton, S. R. J., Richter-Landsberg, C., Greene, L. A., and Shelanski, M. L., 1983b, Nerve growth factor-inducible large external (NILE) glycoprotein: Studies of a central and peripheral neuronal marker, *J. Neurosci.* **3:**441–454.

Schauer, R. (ed.), 1982, *Sialic Acids—Chemistry, Metabolism and Function,* Springer-Verlag, Berlin.

Schubert, D., Schroeder, R., LaCorbiere, M., Saitoh, T., and Cole, G., 1988, Amyloid β protein precursor is possibly a heparan sulfate proteoglycan core protein, *Science* **241:**223–226.

Seiger, A., Almqvist, P., Granholm, A.-C., and Olson, L., 1986, On the localization of Thy-1-like immunoreactivity in the rodent and human nervous system, *Med. Biol.* **64:**109–117.

Selkoe, D. J., Podlisny, M. B., Joachim, C. L., Vickers, E. A., Lee, G., Fritz, L. C., and Oltersdorf, T., 1988, β-Amyloid precursor protein of Alzheimer disease occurs as 110- to 135-kilodalton membrane-associated proteins in neural and nonneural tissues, *Proc. Natl. Acad. Sci. USA* **85:**7341–7345.

Simpson, D. L., Thorne, D. R., and Loh, H. H., 1976, Sulfated glycoproteins, glycolipids and glycosaminoglycans from synaptic plasma and myelin membranes: Isolation and characterization of sulfated glycopepetides, *Biochemistry* **15:**5449–5457.

Slomiany, B. L., and Meyer, K., 1972, Isolation and structural studies of sulfated glycoproteins of hog gastric mucosa, *J. Biol. Chem.* **247:**5062–5070.

Spillmann, D., and Finne, J., 1987, Poly-*N*-acetyllactosamine glycans of cellular glycoproteins: Predominance of linear chains in mouse neuroblastoma and rat pheochromocytoma cell lines, *J. Neurochem.* **49:**874–883.

Stallcup, W. B., and Beasley, L., 1985, Involvement of the nerve growth factor-inducible large external glycoprotein (NILE) in neurite fasciculation in primary cultures of rat brain, *Proc. Natl. Acad. Sci. USA* **82:**1276–1280.

Stallcup, W. B., and Beasley, L., 1987, Bipotential glial precursor cells of the optic nerve express the NG2 proteoglycan, *J. Neurosci.* **7:**2737–2744.

Stallcup, W. B., Beasley, L., and Levine, J., 1983, Cell-surface molecules that characterize different stages in the development of cerebellar interneurons, *Cold Spring Harbor Symp. Quant. Biol.* **48:**761–774.

Stallcup, W. B., Beasley, L. L., and Levine, J. M., 1985, Antibody against nerve growth factor-inducible large external (NILE) glycoprotein labels nerve fiber tracts in the developing rat nervous system, *J. Neurosci.* **5:**1090–1101.

Sweeley, C. C., and Nunez, H. A., 1985, Structural analysis of glycoconjugates by mass spectrometry and nuclear magnetic resonance spectroscopy, *Annu. Rev. Biochem.* **54:**765–801.

Tan, S.-S., Crossin, K. L., Hoffman, S., and Edelman, G. M., 1987, Asymmetric expression in somites of cytotactin and its proteoglycan ligand is correlated with neural crest cell distribution, *Proc. Natl. Acad. Sci. USA* **84:**7977–7981.

Tisdale, E. J., and Tartakoff, A. M., 1988, Extensive labeling with [^{3}H]ethanolamine of a hydrophilic protein of animal cells, *J. Biol. Chem.* **263:**8244–8252.

Tulsiani, D. R. P., and Touster, O., 1987, Substrate specificities of rat kidney lysosomal and cytosolic α-D-mannosidases and effects of swainsonine suggest a role of the cytosolic enzyme in glycoprotein catabolism, *J. Biol. Chem.* **262:**6506–6514.

Waechter, C. J., Schmidt, J. W., and Catterall, W. A., 1983, Glycosylation is required for maintenance of functional sodium channels in neuroblastoma cells, *J. Biol. Chem.* **258:**5117–5123.

Weber, R. J., Hill, J. M., and Pert, C. B., 1988, Regional distribution and density of Thy 1.1 in rat brain and its relation to subpopulations of neurons, *J. Neuroimmunol.* **17:**137–145.

West, C. M., 1986, Current ideas on the significance of protein glycosylation, *Mol. Cell. Biochem.* **72:**3–20.

Wieland, F., Paul, G., and Sumper, M., 1985, Halobacterial flagellins are sulfated glycoproteins, *J. Biol. Chem.* **260:**15180–15185.

Wight, T. N., and Mecham, R. P. (eds.), 1987. *Biology of Proteoglycans,* Academic Press, New York.

Williams, A. F., and Barclay, A. N., 1988, The immunoglobulin superfamily—Domains for cell surface recognition, *Annu. Rev. Immunol.* **6:**381–406.

Williams, A. F., and Gagnon, J., 1982, Neuronal cell Thy-1 glycoprotein: Homology with immunoglobulin, *Science* **216:**696–703.

Winkler, H., Apps, D. K., and Fischer-Colbrie, R., 1986, The molecular function of adrenal chromaffin granules: Established facts and unresolved topics, *Neuroscience* **18:**261–290.

Yamashita, K., Hitoi, A., and Kobata, A., 1983a, Structural determinants of *Phaseolus vulgaris* erythroagglutinating lectin for oligosaccharides, *J. Biol. Chem.* **258:**14753–14755.

Yamashita, K., Ueda, I., and Kobata, A., 1983b, Sulfated asparagine-linked sugar chains of hen egg albumin, *J. Biol. Chem.* **258:**14144–14147.

Zanini, A., Giannattasio, G., Nussdorfer, G., Margolis, R. K., Margolis, R. U., and Meldolesi, J., 1980, Molecular organization of prolactin granules. II. Characterization of glycosaminoglycans and glycoproteins of the bovine prolactin granule matrix, *J. Cell Biol.* **86:**260–272.

4

Biosynthesis of Glycoproteins

Charles J. Waechter

1. INTRODUCTION AND BACKGROUND

Since membrane glycoproteins have been implicated in many neurobiological processes and functions (see Chapters 7, 8, and 11), the mechanisms and regulation of biosynthesis of the oligosaccharide units have been of great interest to cell and neurobiologists during the past 20 years. Much of this research activity has focused on the structure, biosynthesis, and function of *N*-linked oligosaccharides because the majority of the carbohydrate chains attached to membrane glycoproteins in nervous tissue are attached via *N*-glycosidic bonds (Krusius and Finne, 1977). For further discussion of the structural details of carbohydrate units of brain glycoproteins, see Chapter 3.

Since the discovery of dolichol-bound oligosaccharide intermediates in liver by Leloir and his co-workers (Parodi *et al.*, 1972), the structure and synthesis of the dolichol-bound precursor oligosaccharide, $Glc_3Man_9GlcNAc_2$, its transfer to nascent polypeptide acceptors, and its subsequent processing to polymannose- or complex-type oligosaccharide units have been elucidated in detail in many mammalian cells (Presper and Heath, 1983; Kornfeld and Kornfeld, 1985).

The purpose of this chapter is to discuss the progress made on glycoprotein biosynthesis in nervous tissue since this subject was last reviewed (Waechter and Scher, 1979). An attempt will be made to summarize the new information available on the synthesis of dolichol-linked saccharide intermediates, the regulation of the lipid intermediate pathway, the processing of *N*-linked oligosaccharides, and the understanding of the role of *N*-glycosylation in the assembly of membrane glycoproteins performing important functions in nervous tissue. Now that the structures of the oligosaccharide chains and the mechanisms for their biosynthesis are understood in considerable detail, much of the research conducted in this area during the past 10 years has been directed at understanding the precise role of glycosylation in the assem-

Charles J. Waechter • Department of Biochemistry, University of Kentucky College of Medicine, A. B. Chandler Medical Center, Lexington, Kentucky 40536.

bly, translocation, and various functions of glycoproteins in nervous tissue. This current research activity could be referred to as "neuroglycobiology," and it is hoped that this chapter will provide a reasonably comprehensive review of recent progress in glycoprotein biosynthesis in nervous tissue.

2. DOLICHOL PATHWAY FOR PROTEIN N-GLYCOSYLATION

The early work on the synthesis and function of dolichol-linked saccharide intermediates in brain was reviewed by Waechter and Scher (1979). The enzymatic studies published since the last review, describing the reactions by which the lipid intermediates are synthesized in brain tissue, are summarized in this section.

2.1. Enzymatic Synthesis of a Dolichol-Linked Oligosaccharide Intermediate in the Rough Endoplasmic Reticulum

Evidence for the synthesis of a dolichol-linked oligosaccharide intermediate in brain was first obtained from *in vitro* studies with microsomal fractions (Waechter and Scher, 1979; Scher and Waechter, 1979). The dolichol-bound oligosaccharide precursor of N-linked carbohydrate chains appears to be $Glc_3Man_9GlcNAc_2$ in brain (Bhat and Waechter, 1988) and bovine retina (Plantner and Kean, 1988), as established in other mammalian tissues (Presper and Heath, 1983; Kornfeld and Kornfeld, 1985). The individual glycosylation reactions leading to the synthesis of $Glc_3Man_9GlcNAc_2$-P-P-Dol are illustrated in Figure 1. Virtually all of the enzymes synthesizing the glycolipid intermediates have been found in the rough ER in brain (Scher *et al.*, 1984). However, the topological orientations of the individual reactions depicted in Figure 1 are based on studies performed with microsomal preparations from other tissues. Although Reactions 2 and 5 (broken arrows, Figure 1) are shown to occur on the cytoplasmic face, the orientation of the active sites of the enzymes catalyzing these reactions in the rough ER is still somewhat controversial. For excellent reviews on the details and interpretations of the topological studies on these enzymes, see Hirschberg and Snider (1987) and Lennarz (1987).

The synthesis of the oligosaccharide–lipid intermediate is initiated by the transfer of GlcNAc 1-P from UDP-GlcNAc to Dol-P (Reaction 1, Figure 1). This enzymatic step is blocked by tunicamycin (see Section 7), thereby preventing the formation of the lipophilic oligosaccharide donor. The second GlcNAc residue is donated directly from UDP-GlcNAc (Waechter and Harford, 1979), forming the N,N'-diacetylchitobiosyl intermediate (Reaction 2). The first mannosyl residue is then transferred from GDP-Man and joined to the chitobiosyl-lipid in a β configuration (Waechter and Harford, 1979). In this stage of biosynthesis, four α-mannosyl units are also acquired directly from GDP-Man (Banerjee *et al.*, 1981) (Reaction set 3).

The remaining mannosyl and glucosyl transfer reactions are mediated by Man-P-Dol and Glc-P-Dol, which are synthesized reversibly from the respective nucleotide derivatives (Reactions 4 and 5). The last four α-mannosyl residues are donated by Man-P-Dol (Reaction set 6), and the oligosaccharide–lipid intermediate is completed

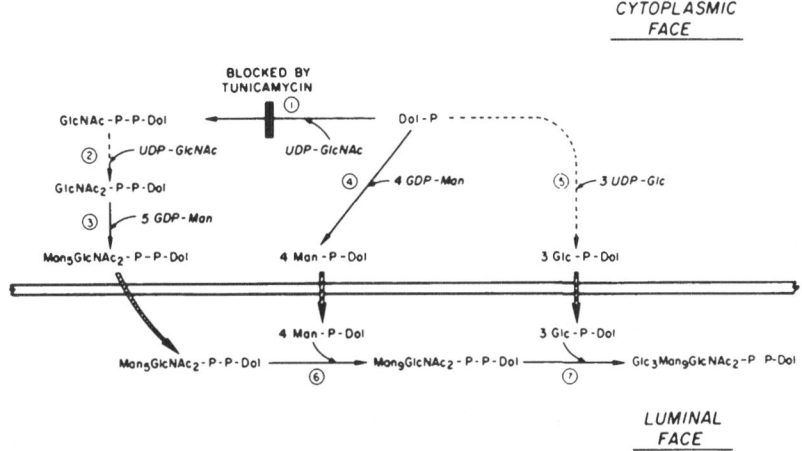

Figure 1. Reactions leading to the biosynthesis of the oligosaccharide–lipid intermediate, Glc$_3$Man$_9$Glc-NAc$_2$-P-P-Dol, in brain tissue. The enzymatic reactions proposed to occur on the cytoplasmic face of the rough ER are illustrated in the upper half of the figure, and the reactions believed to occur on the luminal face are shown on the lower half. The cross-hatched arrows denote the translocation of a glycolipid intermediate from the cytoplasmic face to the luminal face of the rough ER. The dotted arrows (Reactions 2 and 5) indicate that the topological orientation of the enzymes catalyzing these reaction is uncertain. Dol-P, dolichyl monophosphate; Dol-P-P, dolichyl pyrophosphate.

by the direct transfer of three α-glucosyl residues from Glc-P-Dol (Banerjee *et al.*, 1981) (Reaction set 7). *In vitro* experiments, in which Glc-P-Dol labeling by UDP-[^{14}C]-Glc was blocked by amphomycin, a lipopeptide antibiotic, have suggested that glucosyl residues might also be added to *N*-linked oligosaccharides directly from the sugar nucleotide (Banerjee *et al.*, 1981). The presence of a UDP-Glc transporter in the rough ER (Hirschberg and Snider, 1987) would provide a mode of entry for the sugar nucleotide into the luminal compartment. As indicated in Figure 1, the final seven lipid-mediated glycosylation reactions are believed to occur on the luminal surface of the rough ER. The properties of the enzymes and methods for the assay of the lipid-mediated glycosyltransferases in brain have been compiled previously (Waechter and Scher, 1981).

2.2. Enzymatic Transfer of Glucosylated Precursor Oligosaccharide from Dolichyl Pyrophosphate to Asparagine Residues of Nascent Polypeptide Acceptors

The completed dolichol-bound precursor oligosaccharide, Glc$_3$Man$_9$GlcNAc$_2$, is transferred *en bloc* to accessible -Asn-X-Ser/Thr- recognition sites of appropriate nascent polypeptides (Kaplan *et al.*, 1987) (Figure 2). This reaction is believed to occur cotranslationally on the luminal face of the rough ER (Hirschberg and Snider, 1987; Lennarz, 1987). A 57-kDa glycosylation site recognition component of the oligosaccharyltransferase has been labeled with a tripeptide photoprobe in microsomal

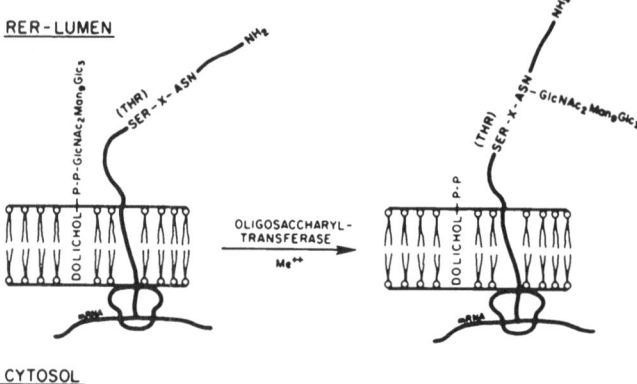

Figure 2. *En bloc* transfer of $Glc_3Man_9GlcNAc_2$ from Dol-P-P to nascent polypeptide acceptors catalyzed by membrane-bound oligosaccharyltransferase on the luminal face of the rough ER.

fractions from brain and other tissues (Kaplan *et al.*, 1988), but at this writing the catalytic component has not been isolated.

Brain microsomes catalyze the transfer of the precursor oligosaccharide from exogenous, partially purified $Glc_3Man_9GlcNAc_2$-P-P-Dol to endogenous polypeptide acceptors in the presence of Ca^{2+} (Scher and Waechter, 1978, 1979), and the oligosaccharyltransferase activity has been shown to be localized in the rough ER (Scher *et al.*, 1984). The *in vitro* transfer of N,N'-diacetylchitobiose from Dol-P-P to glycoprotein was observed with microsomes from gray matter, but not white matter (Harford and Waechter, 1979). It is not known if the disaccharide transfer occurs *in vivo*.

3. PROCESSING OF N-LINKED OLIGOSACCHARIDE CHAINS

Metabolic labeling experiments and extensive enzymological studies, performed in many laboratories, have elucidated the excision–revision scheme by which $Glc_3Man_9GlcNAc_2$ is processed to form a complex-type N-linked oligosaccharide chain (Figure 3). The reactions illustrated in Figure 2 have been characterized primarily by enzymatic studies with microsomal preparations from liver and other extraneural tissues, but there is evidence for similar, if not identical, processing glucosidase and mannosidase activities in brain and peripheral nerve. For a more thorough discussion of the studies that have defined these reactions, see the review by Kornfeld and Kornfeld (1985).

Relatively soon after the precursor oligosaccharide is transferred to the nascent polypeptide chain (Figure 2), the outermost α-glucosyl residue is removed by α1-2 glucosidase I (Reaction 1, Figure 3). The two internal glucosyl residues are cleaved separately by a single α1-3 glucosidase II (Reaction 2). The first evidence for the presence of these two processing glucosidase activities in nervous tissue was obtained with calf brain microsomal preparations (Scher and Waechter, 1978, 1979). In these

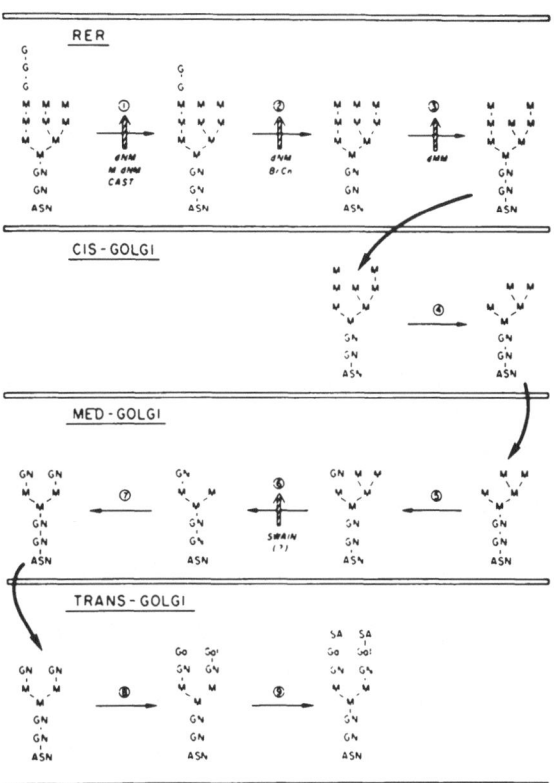

Figure 3. Reactions involved in the processing of glucosylated precursor oligosaccharide to complex-type or polymannose-type N-linked oligosaccharides, and sites of action of various processing inhibitors. Reaction 1 is catalyzed by: α-glucosidase I; (2) α-glucosidase II; (3) ER α1-2 mannosidase I; (4) Golgi α-mannosidase I; (5) N-acetylglucosaminyltransferase I; (6) Golgi α-mannosidase II; (7) N-acetylglucosaminyltransferase II; (8) UDP-Gal:glycoprotein galactosyltransferase; (9) CMP-SA:glycoprotein sialyltransferase. Abbreviations of processing inhibitors: dNM, 1-deoxynojirimycin; M-dNM, N-methyl-1-deoxynojirimycin; CAST, castanospermine; BrCn, bromoconduritol; dMM, 1-deoxymannojirimycin; SWAIN, swainsonine.

earlier studies, competition experiments with α- and β-glucosides suggested that the outermost glucosyl unit might be β-linked and the two inner glucosyl units were α-linked. More recent work (Tulsiani and Touster, 1987) indicates that the anomeric configuration of the three glucosyl residues and the specificity of the corresponding glucosidases are the same in brain as in other mammalian tissues. Touster's group also reported the presence of a novel disaccharidase in brain that cleaves Glcα1-3→mannose from GlcMan$_7$GlcNAc.

Following the removal of the glucosyl residues, and prior to departure from the rough ER, one mannosyl residue is cleaved by an α-mannosidase (Reaction 3). Additional mannosyl units may be trimmed from membrane glycoproteins that remain in the rough ER, forming a variety of polymannose-type oligosaccharide units (Kornfeld and Kornfeld, 1985).

Membrane glycoproteins destined for other components are then translocated to the *cis*-Golgi and the N-linked chains are converted to Man$_5$GlcNAc$_2$ by Golgi mannosidase I (Reaction 4). Movement to the Golgi stacks and through the different Golgi compartments appears to occur by a vesicular transport system (Bergmann and Singer,

1983; Rothman *et al.*, 1984a,b). Pulse-labeling experiments with [³H]mannose have detected mannosylated *N,N'*-diacetylchitobiosyl intermediate oligosaccharides that would be formed at this stage of processing in sciatic nerve (Poduslo, 1985). Clearly, an array of "mature" polymannose-type oligosaccharide units can be assembled by this series of processing mannosidases. However, the factors controlling the extent of mannose removal from individual glycoproteins in the rough ER and Golgi are incompletely understood at this time.

At this stage, polymannose oligosaccharide units of lysosomal enzymes acquire a mannose 6-phosphate recognition marker. The phosphate group is derived from the β position of UDP-GlcNAc via a GlcNAc 1-phosphate transfer, and then the GlcNAc cap is removed by a specific α-*N*-acetylglucosaminidase. Based on the observation that deglycosylated lysosomal enzymes interfere with the transfer of GlcNAc 1-phosphate to polymannose chains of lysosomal enzymes by a rat liver Golgi *N*-acetylglucosaminylphosphotransferase, it has been proposed that the phosphotransferase recognizes a protein domain in lysosomal enzymes (Kornfeld, 1987). These reactions are currently thought to occur in the *cis*-Golgi (Kornfeld and Kornfeld, 1985), although these enzymes have not yet been characterized in nervous tissue.

The first GlcNAc residue of *N*-linked oligosaccharide units is added by the action of *N*-acetylglucosaminyltransferase I in the *medial*-Golgi (Reaction 5). These oligosaccharides could become hybrid structures or be further processed to complex-type oligosaccharides after two more α-mannosyl residues are removed by Golgi mannosidase II (Reaction 6). In the assembly of complex-type chains, the second GlcNAc residue is added by *N*-acetylglucosaminyltransferase II (Reaction 7). At this stage, fucose residues can be added to one of the GlcNAc units of the *N,N'*-diacetylchitobiosyl inner core (Kornfeld and Kornfeld, 1985).

The trisaccharide caps of complex-type *N*-linked chains are completed by the addition of the appropriate number of galactosyl and sialyl (SA) residues by specific galactosyl- and sialyltransferases in the *trans*-Golgi compartment. For excellent discussions on the control of the branching and microheterogeneity of *N*-linked oligosaccharides, see Schachter *et al.* (1983) and Schachter (1986). Once the assembly of the oligosaccharide chains is completed in the *trans*-Golgi, the mature glycoproteins would then egress to the cell body plasma membrane, synaptic plasma membrane, axolemma, or other sites in neuronal or glial cells.

3.1. Effect of Inhibitors of Processing Glycosidases on N-Linked Oligosaccharide Assembly in Nervous Tissue

A number of compounds have been found that effectively inhibit specific reactions catalyzed by the processing glycosidases (see reviews by Fuhrmann *et al.*, 1985, and Elbein, 1987). As denoted in Figure 3, glucosidase I (Reaction 1) is inhibited by 1-deoxynojirimycin (dNM), *N*-methyl-1-deoxynojirimycin (M-dNM), and castanospermine (CAST). Glucosidase II (Reaction 2) is blocked by the presence of dNM and bromoconduritol (BrCn). Rough ER mannosidase I (Reaction 3) is inhibited by 1-deoxymannojirimycin (dMM), and Golgi mannosidase II (Reaction 6) is blocked by swainsonine (SWAIN). The presence of a swainsonine-sensitive mannosidase II in brain is still open to question (Tulsiani *et al.*, 1988). However, these processing inhibitors have been utilized by several groups to investigate the role of oligosaccharide processing on the function and translocation of membrane glycoproteins in brain and muscle cells.

When glucosidase I and II activity was blocked in BC3H-1 cells by dNM, the α subunit of the acetylcholine receptor was degraded more rapidly and only 25% acquired the ability to bind bungarotoxin (McHardy *et al.*, 1986). The cell surface receptor made in the presence of the inhibitor expressed the normal affinity for cholinergic ligands. The authors concluded that trimming of the glucosyl residues may be required for correct conformational changes and possibly other posttranslational modifications. Failure to remove the glucose units may also delay departure from the rough ER, resulting in increased susceptibility to proteolytic degradation. Indeed, Lodish and King (1984) have shown that the translocation of some, but not all, secretory glycoproteins was blocked by dNM in human hepatoma (HepG2) cells.

The glucosidase inhibitors, dNM and CAST, but not the mannosidase inhibitors, dMM and SWAIN, blocked the induction of sulfolipid synthesis and 2',3'-cyclic nucleotide 3'-phosphohydrolase activity in primary embryonic rat brain cultures (Bhat, 1988). Myoblast fusion has also been shown to be blocked by dNM, but not dMM, implicating glucose trimming in this neurobiological process (Holland and Herscovics, 1986).

Less than 20% of the sialic acid was incorporated into the α subunit of the sodium channel in rat brain neurons when deglucosylation was blocked by CAST (Schmidt and Catterall, 1987). CAST also inhibited sulfation, but not palmitoylation of the α subunit. Neither CAST nor SWAIN prevented the covalent association of the α and β2 subunits or the transport of the αβ2 complex to the cell surface.

Fliesler *et al.* (1986) have also inhibited glucose trimming in *Xenopus* retinas with CAST and they concluded that posttranslational processing of *N*-linked oligosaccharides in this system is not essential for normal routing of membrane glycoproteins and disc morphogenesis.

3.2. Sugar Nucleotide:Glycoprotein Glycosyltransferases Involved in the Biosynthesis of N-Linked Oligosaccharide Chains

The properties of the sugar nucleotide:glycosyltransferases responsible for the addition of the trisaccharide caps in complex *N*-linked oligosaccharides of brain glycoproteins were reviewed previously (Waechter and Scher, 1979). Some new information on this class of glycosyltransferases is presented below.

The developmental pattern for UDP-Gal:glycoprotein galactosyltransferase was studied in three regions of embryonic rat brain, and the specific activities were highest in the order: visual cortex > superior colliculus > lateral geniculate nucleus (Braulke and Biesold, 1981). The specific activities in all three regions declined postnatally until day 14. A sheep brain fucosyltransferase has been solubilized with Triton X-100, and its substrate specificity examined by chemical and enzymatic analysis of the fucosylated product (Broquet *et al.*, 1982). The results indicate that the brain enzyme fucosylates the second GlcNAc residue of the oligomannochitobiosyl core. This activity appears to be enriched in the light microsomal and mitochondrial fraction (Broquet *et al.*, 1984). A stimulation in fucokinase activity may explain how dopamine increases the rate of incorporation of fucose into glycoproteins of the hippocampus (Jork *et al.*, 1984).

CMP-NeuNAc:poly-α2-8-sialosyl sialyltransferase activity catalyzing elongation of the novel polysialosyl chains of N-CAM (see Chapter 12) has been demonstrated with a Golgi-enriched fraction from fetal rat brain (McCoy *et al.*, 1985). A similar membrane fraction from adult brain had considerably less polysialosyl sialyltransferase activity. Thus, the sialyltransferase activity was higher at a developmental period when the N-CAM-bound polysialosyl chains are longer (Jorgensen and Moller, 1980; Finne, 1982; Rothbard *et al.*, 1982).

Ecto-galactosyl-, fucosyl-, and sialyltransferases have been found on the external surface of neuronal cultures (Matsui *et al.*, 1983). A possible role in cell–cell interactions was proposed. In a related study, a synaptosomal sialyltransferase was reported that added sialyl residues to surface glycoproteins, and the sialyl residues were found to be inaccessible to membrane-bound sialidases (Breen and Regan, 1986). Synaptic junction fractions isolated from adult chicken brain also contained sialyl-, galactosyl-, and fucosyltransferase activities (Rostas *et al.*, 1981).

3.3. Sugar Nucleotide:Glycoprotein Glycosyltransferases Involved in the Biosynthesis of O-Linked Oligosaccharide Chains

Although *N*-linked oligosaccharides have been the major focus of glycoprotein research in brain during the past 10 years, some significant new developments on the biosynthesis of O-linked chains have also appeared. An *N*-acetylgalactosaminyltransferase activity and its endogenous acceptor from embryonic chick neural retina cells have been separated into three forms by centrifugation on sucrose density gradients (Balsamo and Lilien, 1982). Two of these isoglycosyltransferases are present in plasma membrane preparations, and the authors have proposed a role for the glycosyltransferase/acceptor complex in the formation or maintenance of stable cell surface adhesions.

A brain sialyltransferase has also been reported which apparently adds an α-(2→6) sialyl residue to the terminal GalNAc unit of the O-linked trisaccharide, α-NeuNAc-(2→3)-Gal→GalNAc (Baubichon-Cortay *et al.*, 1986).

4. ENZYMES INVOLVED IN DOLICHYL PHOSPHATE METABOLISM

Relative to the enzymes catalyzing the formation of the glycolipid intermediates, the enzymes involved in the *de novo* biosynthesis of Dol-P are poorly characterized in brain and other mammalian tissues. From the outline of the postulated pathway illustrated in Figure 4, it can be seen that farnesyl pyrophosphate represents a branch point for the biosynthesis of dolichol and cholesterol. There is good evidence for the role of isopentenyl pyrophosphate as the isoprenyl donor (Daleo *et al.*, 1977; Grange and Adair, 1977; Wellner and Lucas, 1979; Wong and Lennarz, 1982) in the biosynthesis

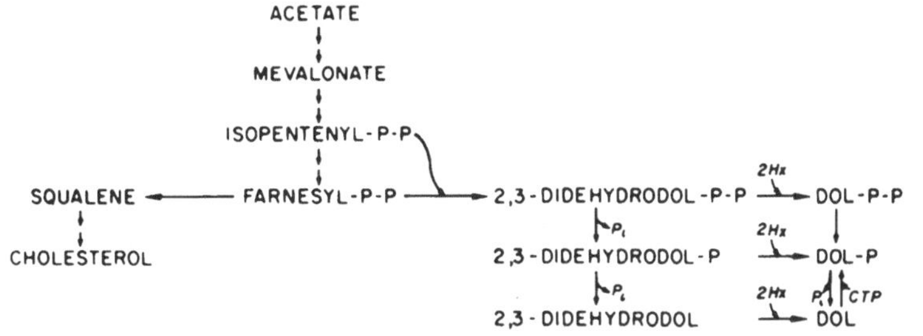

ACETATE
↓
MEVALONATE
↓
ISOPENTENYL-P-P
↓

SQUALENE ← FARNESYL-P-P → 2,3-DIDEHYDRODOL-P-P $\xrightarrow{2Hx}$ DOL-P-P
↓ $\downarrow P_i$ $2Hx \downarrow$
CHOLESTEROL   2,3-DIDEHYDRODOL-P $\xrightarrow{2Hx}$ DOL-P
                          $\downarrow P_i$ $2Hx \quad P_i \uparrow CTP$
                     2,3-DIDEHYDRODOL $\xrightarrow{2Hx}$ DOL

$$DOL = \begin{matrix} CH_3 \\ CH_3 \end{matrix}\!\!>\!C=CHCH_2-\left(CH_2\overset{\overset{CH_3}{|}}{C}=CHCH_2\right)_{\!17}\!-CH_2\overset{\overset{CH_3}{|}}{C}H-CH_2CH_2-OH$$

Figure 4. Outline of postulated pathway for dolichol and dolichyl phosphate biosynthesis and its relationship to cholesterol biosynthesis. Hx, unidentified reductant involved in saturation of α-isoprene unit; DOL, dolichol.

of Dol-P in animal tissues. However, it is still not certain how many *cis*-isoprenyltransferases participate in the elongation of farnesyl pyrophosphate to form C_{95}-dolichol, the most common isoprenologue, or how the elongation process is terminated. Based on studies with hepatocytes, Dallner and his colleagues (Ekström *et al.*, 1987) have proposed that chain elongation may be terminated by the addition of isopentenol. If this is correct, that pool of dolichol could be converted to Dol-P by dolichol kinase, following the reduction of the α-isoprene unit. As yet, the mechanism by which the terminal isoprene unit is reduced has not been elucidated in detail. Reduction of the long-chain polyisoprenyl pyrophosphate could also terminate chain elongation since the allylic pyrophosphate is probably the "activated" intermediate in the isoprenylation reactions.

While there is very little information on the *de novo* pathway for Dol-P synthesis in brain, enzymes catalyzing dolichol phosphorylation, the dephosphorylation of Dol-P, the conversion of Dol-P-P to Dol-P, and the deacylation of dolichyl oleate have been described in membrane preparations from nervous tissue. The metabolic relationship of dolichol kinase, polyisoprenyl phosphate phosphatase, and Dol-P-P phosphatase activity to the *N*-glycosylation pathway is shown in Figure 5.

4.1. Dolichol Kinase

One mechanism for the synthesis of Dol-P involves the direct phosphorylation of dolichol with CTP serving as the phosphoryl donor. It is not clear if this reaction simply represents a means of "activating" reserve pools of the free polyisoprenol or if the kinase catalyzes the last step in the *de novo* pathway for Dol-P biosynthesis. Plainly, it will be difficult to answer this question until the identity of the natural substrate for the reduction of the α-isoprene unit is established conclusively.

The enzymatic phosphorylation of endogenous and exogenous dolichols by brain microsomes was first reported by Burton *et al.* (1979). These studies showed that CTP

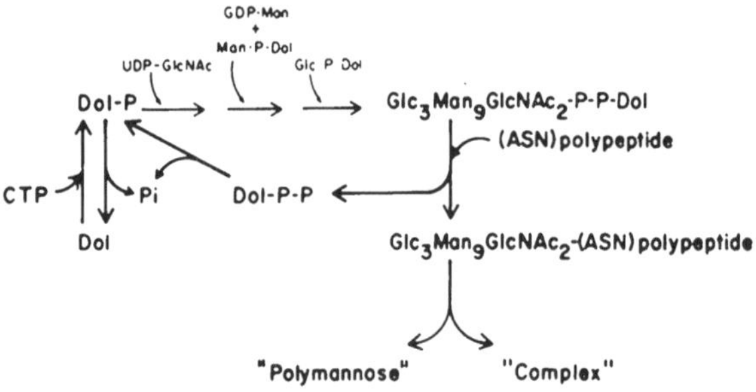

Figure 5. Relationship of dolichol kinase, polyisoprenyl phosphate phosphatase, and Dol-P-P phosphatase to dolichol pathway for protein N-glycosylation.

was the preferred phosphoryl donor for the brain kinase, as reported for other mammalian cells (Presper and Heath, 1983). *In vitro* experiments demonstrated that endogenous dolichyl phosphate, synthesized by dolichol kinase, was available for the synthesis of Man-P-Dol, Glc-P-Dol, and GlcNAc-P-P-Dol (Burton *et al.*, 1979; Scher *et al.*, 1980). Later studies indicated that at least some molecules of Dol-P were present in a common pool that was accessible to the enzymes synthesizing the three dolichol-bound monosaccharide intermediates (Scher *et al.*, 1984). A procedure for solubilizing the brain enzyme has been described (Sumbilla and Waechter, 1985b), and the properties of the particulate and CHAPS-solubilized forms of the brain enzyme have been presented in detail (Sumbilla and Waechter, 1985a).

Dolichol phosphorylation occurs most effectively in brain microsomal fractions in the presence of either 10 mM Ca^{2+} (Sumbilla and Waechter, 1985b) or Zn^{2+} (Sakakihara and Volpe, 1985b). In a study related to the Ca^{2+} requirement for the brain kinase, Gandhi and Keenan (1983) reported that brain dolichol kinase might be regulated by calmodulin. However, this suggestion has been disputed by Sakakihara and Volpe (1985b). More recently, it has been shown that pig brain dolichol kinase exhibits a cofactor requirement for either phosphatidylethanolamine or phosphatidylcholine (Genain and Waechter, 1987).

4.2. Polyisoprenyl Phosphate Phosphatase

Membrane preparations from rat brain neuronal perikarya (Idoyaga-Vargas *et al.*, 1980) and calf brain (Burton *et al.*, 1981) have been shown to catalyze the enzymatic dephosphorylation of exogenous Dol-P. Endogenous Dol-^{32}P, prelabeled by dolichol kinase, was also dephosphorylated by a fluoride-sensitive phosphatase activity in calf brain membranes (Burton *et al.*, 1981). Substrate competition experiments in the latter study indicated that Dol-P is hydrolyzed by a polyisoprenyl phosphate phosphatase(s) in brain, and probably not nonspecifically by phosphatidate phosphatase or the general phosphatase activity assayed by following the hydrolysis of *p*-nitrophenyl phosphate. The properties of calf brain polyisoprenyl phosphatase and a procedure for the solubilization of the calf brain activity have been described (Sumbilla and Waechter,

1985a), but further work will be required to establish firmly the specificity and number of enzymes catalyzing the dephosphorylation reaction in brain membranes.

4.3. Dolichyl Pyrophosphate Phosphatase

The action of a Dol-P-P phosphatase is required to regenerate Dol-P, the "active" form of the glycosyl carrier lipid after $Glc_3Man_9GlcNAc_2$ is transferred to nascent polypeptide acceptors (Figure 4). A pyrophosphate phosphatase activity capable of catalyzing this reaction has been detected in brain microsomal preparations by follow-ing the release of $^{32}P_i$ from $[\beta\text{-}^{32}P]Dol\text{-}P\text{-}P$ (Scher and Waechter, 1984). The exact number and specificity of the enzyme(s) catalyzing this reaction in the brain micro-somal fractions have not been determined. The enzymatic properties of this activity and a procedure for obtaining soluble extracts of the activity with Triton X-100 have been described (Scher and Waechter, 1985). Compatible with this activity being a Dol-P-P phosphatase, undecaprenyl pyrophosphate was found to be a competitive inhibitor.

4.4. Dolichyl Ester Hydrolase

Since significant quantities of fatty acyl esters of dolichol are present in brain membranes (Sakakihara and Volpe, 1984), it is plausible that these pools of dolichyl esters can be recruited for the synthesis of lipid intermediates by sequential deacylation and phosphorylation reactions. Calf brain membranes have been shown to catalyze the lipolytic cleavage of dolichyl oleate (Scher and Waechter, 1981). Although the precise specificity of the lipase(s) has not been established, it appears to be distinct from neutral glyceride lipases and cholesterol ester hydrolase (Sumbilla and Waechter, 1985b).

4.5. Subcellular Locations of Enzymes Involved in Dolichyl Phosphate Metabolism

Subcellular fractionation studies (Scher *et al.*, 1984) indicate that dolichol kinase is located predominantly in the rough ER. In the same study, the highest specific activity for the polyisoprenyl phosphate phosphatase was recovered in the light micro-somal fraction (Figure 6) and in axolemma-rich preparations. Lower levels of phos-phatase activity were also found in the rough ER and mitochondrial–lysosomal frac-tion. More work will be required to establish the exact number and the precise specificities of the enzyme(s) expressing this polyisoprenyl phosphate phosphatase activity. Dolichyl ester hydrolase is present in crude microsomal preparations (Scher and Waechter, 1981), but a detailed subcellular study on this activity has not been reported. Dolichyl pyrophosphate phosphatase activity has been found in the heavy and light microsomes with the highest specific activity recovered in the light microsomal fraction, containing the Golgi apparatus (Wolf *et al.*, 1988). Lower levels were de-tected in the synaptic plasma membrane fractions.

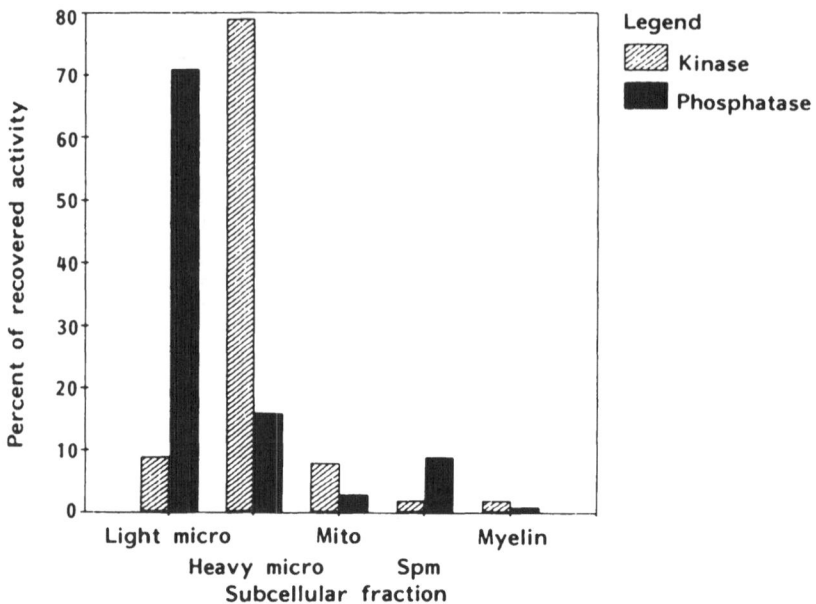

Figure 6. Distribution of dolichol kinase and polyisoprenyl phosphate phosphatase activity in various subcellular fractions from calf brain. Light micro, light microsomes; heavy micro, heavy microsomes; mito, mitochondrial/lysosomal fraction; spm, synaptic plasma membrane fraction. (Data from Scher *et al.*, 1984.)

5. DEVELOPMENTAL CHANGES IN DOLICHOL PATHWAY

Since neuronal and glial cells undergo striking proliferative and differentiative changes that, presumably, require new synthesis of specific membrane glycoproteins, it is quite likely that there are corresponding changes in the rates of biosynthesis of glycolipid intermediates and of protein N-glycosylation during brain development. Recent studies on cellular N-glycosylation activity and the levels of enzymes involved in the synthesis of $Glc_3Man_9GlcNAc_2$-P-P-Dol and Dol-P are reviewed below.

5.1. Metabolic Labeling Studies on Protein N-Glycosylation Activity

Alperin *et al.* (1986) have compared the rates of protein N-glycosylation in cerebral cortex particles between 5- and 30-day-old rat brain by [2-^{3}H]mannose-labeling experiments. These studies indicated that the rate of N-linked membrane glycoprotein biosynthesis was sevenfold greater in the younger animals. A rate of 2.1 ng mannose/mg protein per hr was estimated.

Similar [2-^{3}H]mannose-incorporation studies with cultured embryonic rat brain cells documented a sharp developmental increase in oligosaccharide-lipid synthesis and protein N-glycosylation between 12 and 16 days in culture (Bhat and Waechter, 1988). Maximal labeling of glycoprotein corresponded to the time of induction of the oligodendroglial enzyme marker, 2′,3′-cyclic nucleotide 3′-phosphohydrolase, and the astroglial enzyme marker, glutamine synthetase. The latter results suggest a possible relationship between the appearance of new *N*-linked membrane glycoproteins and the growth and differentiation of glial cells.

Related studies with primary cultures of newborn rat cerebrum have provided evidence that protein *N*-glycosylation is required for DNA synthesis, and that the requirement is expressed late in G_1 phase (Ishii and Volpe, 1987).

5.2. Developmental Changes in Enzyme Activities Related to the Lipid Intermediate Pathway

The level of activity of several enzymes responsible for the synthesis of glycolipid intermediates has been shown to change developmentally in brain. The initial rate of synthesis of Man-P-Dol by white matter microsomes was highest in 3-week-old animals, and declined to lower adult rates by 8 weeks of age (Harford and Waechter, 1980). Enzymological analyses indicated that the developmental increase was due to a higher amount of endogenous Dol-P available to the mannosyltransferase in the rough ER. These results suggested the possibility that Dol-P levels in the rough ER could be one rate-controlling factor in the biosynthesis of the dolichol-linked oligosaccharide intermediate. There is also evidence that Dol-P levels could be rate-limiting for GlcNAc-P-P-Dol synthesis during development in rat sciatic nerve (Smith *et al.*, 1985).

The levels of the enzymes synthesizing Man-P-Dol, GlcNAc-P-P-Dol, and Glc-P-Dol, as well as the oligosaccharyltransferase, were higher in membrane preparations from rat cerebral cortex in 5-day-old animals as compared to 15-day-old animals (Idoyaga-Vargas and Carminatti, 1982). These increases in enzymes participating in the dolichol pathway corresponded to synapse formation in the cerebral cortex. In a related study, GlcNAc-P-P-Dol synthase activity in rat brain microsomes increased approximately fourfold between 15 and 24 days of age (Volpe *et al.*, 1987). These results are compatible with the earlier hypothesis of Harford and Waechter (1980), suggesting that the rate of synthesis of lipid intermediates in adult brian could be limited by the amount of endogenous Dol-P accessible to the glycosyltransferases.

5.3. Developmental Changes in Dolichyl Phosphate Metabolism

The regulatory interrelationships between dolichol and cholesterol biosynthesis have been investigated by comparing the rates of synthesis of the two lipids in [1-^{14}C]acetate-labeling experiments with tissue slices from mouse brain at different ages (James and Kandutsch, 1980). When the data were expressed on the basis of brain weight, the rate of cholesterol synthesis declined sharply beginning 3 days before birth. In contrast to this result, a peak in dolichol synthesis was seen at approximately postnatal day 5, and then the biosynthetic rate declined gradually. These disparate developmental patterns strongly suggest that the two biosynthetic pathways are regulated independently in brain. Although the mechanisms by which these pathways are regulated are not known, separate controls of these pathways would allow brain cells to meet their requirements for maintaining glycoprotein biosynthesis at developmental periods when active *de novo* cholesterol biosynthesis would not be necessary.

Polyisoprenyl phosphate phosphatase activity has been shown to increase substantially with age in white and gray matter from pig brain (Scher *et al.*, 1985). The apparent K_m values for Dol-P were virtually identical for microsomal fractions from young (120 μM) and adult (130 μM) animals, suggesting the presence of a higher level of unmodified phosphatase in brains of older animals. This increase may be one factor contributing to the accumulation of the free polyprenol during aging (see Section 6).

While polyisoprenyl phosphatase activity was higher in gray matter microsomes from 90-day-old animals, compared to 4 days prenatal, dolichol kinase activity was higher in the prenatal brains. Reciprocal changes in the kinase and phosphatase (Figure 6) could be an effective means of modulating the amount of Dol-P available for the synthesis of lipid intermediates.

Corresponding increases in dolichol kinase and Dol-P content have been reported to occur in whole rat brain between 1 and 30 days of age (Volpe *et al.*, 1987). Although it cannot be determined if these increases reflect developmental changes in both white and gray matter, these results also suggest that dolichol kinase activity can play a role in regulating the amount of glycosyl carrier lipid available in the rough ER. It is of interest that dolichol kinase is located predominantly in the rough ER while the phosphatase activity is apparently highest in the light microsomal and plasma membrane fractions (Scher *et al.*, 1984). The distribution of these potentially related metabolic activities raises the possibility that dolichol-carrier proteins may exist. These proteins could facilitate the return of free dolichol formed, perhaps at the Golgi or plasma membrane, to the rough ER where it could be rephosphorylated.

6. CONTENT OF DOLICHOL, DOLICHYL ESTERS, AND DOLICHYL PHOSPHATE IN NERVOUS TISSUE

Dolichol was first isolated from brain tissue by Wolfe and colleagues (Breckenridge *et al.*, 1973). The initial estimates indicated that brain contained approximately 10–15 μg/g wet wt tissue, and analysis by reversed-phase thin-layer chromatography revealed a range of isoprenologues from C_{90} to C_{105} in rat and calf brain. More recent studies utilizing HPLC have indicated that C_{95} is the predominant chain length of brain dolichols (Pullarkat and Reha, 1982) with age-dependent values of 15–300 μg/g tissue reported. The presence of fatty acyl esters of dolichol in brain tissue has also been documented (Sakakihara and Volpe, 1984). Regional studies have shown that dolichol (Sakakihara and Volpe, 1985a; Andersson *et al.*, 1987) and Dol-P (Andersson *et al.*, 1987) are fairly evenly distributed in brain with the content of the free polyprenol increasing substantially with age (see below).

Developmental Changes in Dolichol and Dolichyl Phosphate Levels

Analytical studies by Pullarkat and Reha (1982) indicated that there were striking age-related increases in dolichol levels in gray matter from human brain. Less dramatic increases were observed for white matter. Similar results have been obtained with human brain (Ng Ying Kin *et al.*, 1983) and rat brain (Ng Ying Kin *et al.*, 1983;

Sakakihira and Volpe, 1984). Only modest age-related increases in Dol-P were found in the various regions of human brain (Andersson *et al.*, 1987).

There is also evidence that dolichol levels are elevated in patients with neuronal ceroid-lipofuscinosis and Alzheimer's disease (Ng Ying Kin *et al.*, 1983). Hall and Patrick (1987) reported that dolichol and Dol-P levels are higher in brain tissue from patients with ceroid-lipofuscinosis compared to normal tissue. Dol-P levels are also significantly elevated in a canine model for ceroid-lipofuscinosis (Keller *et al.*, 1984). However, labeling experiments with fibroblasts from normal individuals and patients with Batten's disease provided no evidence for a defect in dolichol metabolism in this disorder (Paton and Poulos, 1984).

7. CONSEQUENCES OF BLOCKING N-GLYCOSYLATION OF MEMBRANE GLYCOPROTEINS PERFORMING SPECIFIC NEUROBIOLOGICAL FUNCTIONS

Tunicamycin, an antibiotic inhibitor of protein N-glycosylation, has proven to be an effective tool for investigating the function of N-linked oligosaccharide units on a variety of secretory and membrane glycoproteins. The general conclusions of these studies are that core N-glycosylation influences the folding of the newly translated polypeptide chains, and that the proper conformation is essential for subsequent post-translational modifications and accurate intracellular routing of some, but not all, glycoproteins. It has been shown clearly that proper folding and subunit assembly are required for the hemagglutinin glycoprotein of influenza virus to exit the rough ER (Gething *et al.*, 1986). There is also evidence that some unglycosylated polypeptides may be relatively susceptible to proteolytic degradation, and therefore metabolically less stable than their glycosylated counterparts. For a more extensive review of this area of research, see Elbein (1987).

Tkacz and Lampen (1975) first demonstrated that tunicamycin inhibits the N-glycosylation process by blocking the transfer of GlcNAc 1-P from UDP-GlcNAc to Dol-P in liver microsomes, thus preventing the formation of GlcNAc-P-P-Dol and the initiation of synthesis of $Glc_3Man_9GlcNAc_2$-P-P-Dol (Reaction 1, Figure 1). *In vitro* studies with microsomal preparations (Waechter and Harford, 1977; Waechter and Harford, 1979) showed that tunicamycin potently and selectively inhibited the transfer of GlcNAc 1-P from UDP-GlcNAc to Dol-P, but not the addition of the second GlcNAc residue to $GlcNAc_2$-P-P-Dol or the direct transfer of GlcNAc residues from UDP-GlcNAc to peripheral sites on N-linked oligosaccharide chains in nervous tissue. Several studies in which the N-glycosylation inhibitor has been used in neurobiological systems are summarized below.

7.1. Acetylcholine Receptors

The nicotinic acetylcholine receptor is a pentameric transmembrane glycoprotein composed of four N-glycosylated subunits ($\alpha_2\beta\gamma\delta$) (Merlie and Smith, 1986). Merlie and co-workers (Merlie *et al.*, 1982) have found that the ability of the α subunit to bind α-bungarotoxin is dramatically impaired when N-glycosylation is inhibited by tunicamycin in the muscle cell line, BCK-1. Subsequent studies have indicated that the structural modifications required for bungarotoxin binding and subunit assembly occur in the ER, with the receptor then rapidly transported to the Golgi (McHardy *et al.*, 1987). Prives and Bar-Sagi (1983) reported that treatment of embryonic skeletal muscle cells with tunicamycin resulted in a reduction in cell surface acetylcholine receptors and speculated that the underglycosylated form of the receptor was susceptible to proteolysis.

Related studies (Liles and Nathanson, 1986) on the neuronal muscarinic acetylcholine receptor have established that N-glycosylation is required for the maintenance of functional receptors on the cell surface of the neuroblastoma cell line, N1E-115.

7.2. Voltage-Sensitive Sodium Channels

The voltage-sensitive sodium channels of nerve and muscle are oligomeric membrane glycoproteins that also perform a vital neurobiological function (Barchi, 1984; Catterall, 1986). Rat brain sodium channels are comprised of three N-glycosylated subunits (αβ1β2) with α and β2 joined by a disulfide bridge, and β1 is associated noncovalently (Hartshorne *et al.*, 1982; Messner and Catterall, 1985). The role of N-glycosylation in the assembly and function of the channel subunits has been investigated by studying the effect of tunicamycin on neuronal and muscle model systems.

The effect of blocking N-glycosylation on the assembly and function of the voltage-sensitive sodium channels in cultured neuroblastoma cells (clone N18) was studied by monitoring the high-affinity saxitoxin (STX) receptor sites. While control cells contained approximately 50,000 receptors/cell, there was a substantial reduction in cell surface STX receptors in cells treated with tunicamycin (Waechter *et al.*, 1983). Consistent with a reduction in cell surface channel number, neurotoxin-activated Na$^+$ influx was also diminished in tunicamycin-treated cells.

Cellular studies indicated that the reduction in channel number at the surface was not due to an accumulation of STX receptors at an intracellular compartment, or to enhanced proteolysis. It is quite likely that N-glycosylation is essential for the folding, subsequent posttranslational modifications, and translocation of the channel subunits to occur properly. The ability of the glycosylation inhibitor to prevent the assembly of new sodium channels provided a means of making initial estimates for the rate of appearance at the cell surface (1900/cell per hr) and a half-life (26 hr) of functional channels in the neuroblastoma system. Later studies (Schmidt and Catterall, 1986) showed that when N-glycosylation was blocked by tunicamycin in primary cultures of rat brain neurons, the unglycosylated α subunit was not disulfide-linked to β2 and was degraded at a faster rate.

Tunicamycin also produced a reduction in the expression of surface Na$^+$ channels in cultured chick skeletal muscle cells (Bar-Sagi and Prives, 1983). This effect was partially reversed by leupeptin, suggesting that the unglycosylated channel was degraded at an increased rate.

7.3. Opiate Receptors

Exposure of the mouse neuroblastoma cell line, N4TG1, and the neuroblastoma hybrid cell line, NCB-20, to tunicamycin caused a 50% loss of opiate receptor sites, as assessed by enkephalin binding assays (McLawhon et al., 1983). A Scatchard analysis indicated that there was a lower number of receptors with unaltered ligand affinity. Similar results were reported later for the effect of tunicamycin treatment on [³H]diprenorphine binding sites in the neuroblastoma × glioma hybrid, NG108-15 (Law et al., 1985).

7.4. Rhodopsin

Two conflicting reports have been published on the requirement of N-glycosylation for the insertion of opsin into rod outer segment membranes. Plantner et al. (1980) found that when N-glycosylation of opsin was effectively blocked by tunicamycin, the unglycosylated protein was still incorporated into the disk membranes of bovine retinas. In contrast, Fliesler and Basinger (1985) found that incubation of frog retinas with tunicamycin inhibited N-glycosylation and also prevented the incorporation of the unglycosylated protein into rod outer segment membranes.

7.5. P0 of PNS Myelin

N-Glycosylation of P0, the major membrane glycoprotein of PNS myelin, has been shown to be blocked by tunicamycin in tissue slices from sciatic nerve or spinal roots (Smith and Sternberger, 1982; Rapaport et al., 1982). The unglycosylated P0 is also apparently incorporated as efficiently as the glycoprotein into a myelin-containing subcellular fraction. It is not known, however, if the unglycosylated protein is inserted in the proper conformation. The incorporation of P0 into myelin was impaired by monensin (Rapaport et al., 1982).

7.6. Other Neurobiological Processes

Several other interesting studies have been reported on the effect of the N-glycosylation inhibitor on a variety of neurobiological processes, although the roles of specific membrane glycoproteins are not implicated. For example, Olden et al. (1981) observed that tunicamycin blocks fusion of embryonic muscle cells in culture, suggesting that membrane glycoproteins are involved in this differentiation process. Since the inhibitory effect was partially reversed by leupeptin or pepstatin, the authors speculated that N-glycosylation stabilized the proteins in a conformation that was relatively resistant to proteolytic degradation.

Intracranial injection of the glycosylation inhibitor in rat brain was found to impair neuronal development, indicated by a reduction in glutamate decarboxylase activity (Kohsaka et al., 1985). When protein N-glycosylation was blocked in goldfish retinal explants by tunicamycin, neurite outgrowth was reduced (Heacock, 1982).

Consistent with an inhibition of glycosylation, the binding capacity of the neurites for several lectins was considerably lower. Although the neurites produced in the presence of tunicamycin were shorter and less abundant, they appeared to be normal morphologically. The effect of preventing protein N-glycosylation on specific neuronal functions was not examined in this study.

There are also cases in which the addition of N-linked oligosaccharide units does not appear to be essential. For example, blocking N-glycosylation of the β subunit had no apparent effect on subunit assembly, intracellular transport, or rate of degradation of the Na^+/K^+-ATPase in cultured chick sensory neurons (Tamkun and Fambrough, 1986).

8. SUMMARY AND PROSPECTUS

There is now convincing evidence that N-glycosylation is essential for the proper folding, posttranslational modification, accurate translocation, and function of many membrane glycoproteins involved in vital neurobiological processes (see examples cited in Section 7). It is, therefore, critically important to understand the control of oligosaccharide–lipid intermediate synthesis, the core glycosylation event, and the processing of N-linked carbohydrate chains. Although substantially more work has been done on the mechanism for protein N-glycosylation in extraneural tissues, the enzymatic studies reported for nervous tissue indicate that the pathway for oligosaccharide–lipid synthesis and N-glycosylation is virtually identical in brain and peripheral nerve.

A number of the objectives for future studies outlined in a previous review (Waechter and Scher, 1979) have been accomplished during the past decade, but several aspects of the regulation of lipid intermediate synthesis and protein N-glycosylation, the posttranslational processing of the N-linked oligosaccharide chains, as well as the translocation and regulation of function of the membrane glycoproteins mediating important neurobiological processes will require further clarification. Some specific areas that need to be studied in more detail in brain are: (1) the *de novo* pathway for dolichol and Dol-P biosynthesis, including determining the number and properties of the *cis*-isoprenyltransferases and elucidating the mechanism for the reduction of the α-isoprene unit; (2) the metabolic controls governing the levels of Dol-P during development in brain; (3) the characterization of the enzymes involved in the excision–revision of $Glc_3Man_9GlcNAc_2$ to either a variety of polymannose-type or complex-type oligosaccharide chains; (4) the regulation of the processing enzymes; and (5) the structural features and mechanisms by which membrane glycoproteins are directed to the diverse membrane compartments and processes in nerve and glial cells.

The application of recombinant DNA technology and other recently developed analytical methodology should provide new technical impetus to these studies. The discovery of an inhibitor for selectively blocking the assembly of O-linked oligosaccharides, as effectively as tunicamycin prevents N-glycosylation, would also provide a powerful tool for investigating the role of this somewhat neglected class of protein-bound oligosaccharide units.

Although an effort has been made to provide a comprehensive account of the contributions of the numerous laboratories working on glycoprotein biosynthesis in nervous tissue, some significant information may regrettably have been omitted due to limitations in space, time of preparation, or an unintentional oversight.

Acknowledgments

The work performed by the author, cited in this review, was supported by National Institutes of Health Grant GM-36065. The author expresses his appreciation to Dr. J. Rush, Dr. N. Bhat, David Frank, Martha J. Wolf, Jean H. Overmeyer, and Mary Fern Waechter for their careful editing of the manuscript and helpful suggestions.

9. REFERENCES

Alperin, D. M., Idoyaga-Vargas, V. P., and Carminatti, H., 1986, Rate of protein glycosylation in rat cerebral cortex, *J. Neurochem.* **47**:355–362.

Andersson, M., Appelkvist, E.-L., Kristensson, K., and Dallner, G., 1987, Distribution of dolichol and dolichyl phosphate in human brain, *J. Neurochem.* **49**:685–691.

Balsamo, J., and Lilien, J., 1982, An N-acetylgalactosaminyltransferase and its acceptor in embryonic chick neural retina exist in interconvertible particulate forms depending on their cellular location, *J. Biol. Chem.* **257**:349–354.

Banerjee, D. K., Scher, M. G., and Waechter, C. J., 1981, Amphomycin: Effect of the lipopeptide antibiotic on the glycosylation and extraction of dolichyl monophosphate in calf brain membranes, *Biochemistry* **20**:1561–1568.

Barchi, R. L., 1984, Voltage-sensitive Na$^+$ ion channels: Molecular properties and functional reconstitution, *Trends Biochem. Sci.* **9**:358–361.

Bar-Sagi, D., and Prives, J., 1983, Tunicamycin inhibits the expression of surface Na$^+$ channels in cultured muscle cells, *J. Cell. Physiol.* **11**:77–81.

Baubichon-Cortay, H., Serres-Guillaumond, M., Louisot, P., and Broquet, P., 1986, A brain sialyltransferase having a narrow specificity for O-glycosyl-linked oligosaccharide chains, *Carbohydr. Res.* **149**:209–223.

Bergmann, J. E., and Singer, S. J., 1983, Immunoelectron microscopic studies of the intracellular transport of the membrane glycoprotein (G) of vesicular stomatitis virus in infected Chinese hamster ovary cells, *J. Cell Biol.* **97**:1777–1787.

Bhat, N. R., 1988, Effects of inhibitors of glycoprotein processing on oligodendroglial differentiation in primary cultures of embryonic rat brain cells, *J. Neurosci. Res.*, **20**:158–164.

Bhat, N. R., and Waechter, C. J., 1988, Induction of N-glycosylation activity in cultured embryonic rat brain cells, *J. Neurochem.* **50**:375–381.

Braulke, T., and Biesold, D., 1981, Developmental patterns of galactosyltransferase activity in various regions of rat brain, *J. Neurochem.* **36**:1289–1291.

Breckenridge, W. C., Wolfe, L. S., and Ng Ying Kin, N. M. K., 1973, The structure of brain polyisoprenols, *J. Neurochem.* **21**:1311–1318.

Breen, K. C., and Regan, C. M., 1986, Synaptosomal sialyltransferase glycosylates surface proteins that are inaccessible to the action of membrane-bound sialidase, *J. Neurochem.* **47**:1176–1180.

Broquet, P., Leon, M., and Louisot, P., 1982, Substrate specificity of cerebral GDP-fucose:glycoprotein fucosyltransferase, *Eur. J. Biochem.* **123**:9–13.

Broquet, P., Serres-Guillaumond, M., Baubichon-Cortay, H., Peschard, M.-J., and Louisot, P., 1984, Subcellular localisation of cerebral fucosyltransferase, *FEBS Lett.* **174**:43–46.

Burton, W. A., Scher, M. G., and Waechter, C. J., 1979, Enzymatic phosphorylation of dolichol in central nervous tissue, *J. Biol. Chem.* **254:**7129–7136.

Burton, W. A., Scher, M. G., and Waechter, C. J., 1981, Enzymatic dephosphorylation of endogenous and exogenous dolichyl monophosphate by calf brain membranes, *Arch. Biochem. Biophys.* **208:**409–417.

Catterall, W. A., 1986, Molecular properties of voltage-sensitive sodium channels, *Annu. Rev. Biochem.* **55:**953–985.

Daleo, G. R., Hopp, H. E., Romero, P. A., and Pont Lezica, R., 1977, Biosynthesis of dolichol phosphate by subcellular fractions from liver, *FEBS Lett.* **81:**411–414.

Ekström, T. J., Ericsson, J., and Dallner, G., 1987, Localization and terminal reactions of dolichol biosynthesis, *Chem. Scr.* **27:**39–47.

Elbein, A. D., 1987, Inhibitors of the biosynthesis and processing of N-linked oligosaccharide chains, *Annu. Rev. Biochem.* **56:**497–534.

Finne, J., 1982, Occurrence of unique polysialosyl carbohydrate units in glycoproteins of developing brain, *J. Biol. Chem.* **257:**11966–11970.

Fliesler, S. J., and Basinger, S. F., 1985, Tunicamycin blocks the incorporation of opsin into retinal rod outer segment membranes, *Proc. Natl. Acad. Sci. USA* **82:**1116–1120.

Fliesler, S. J., Rayborn, M. E., and Hollyfield, J. G., 1986, Inhibition of oligosaccharide processing and membrane morphogenesis in retinal rod photoreceptor cells, *Proc. Natl. Acad. Sci. USA* **83:**6435–6439.

Fuhrmann, U., Bause, E., and Ploegh, H., 1985, Inhibitors of oligosaccharide processing, *Biochim. Biophys. Acta* **825:**95–110.

Gandhi, C. R., and Keenan, R. W., 1983, The role of calmodulin in the regulation of dolichol kinase, *J. Biol. Chem.* **258:**7639–7643.

Genain, C., and Waechter, C. J., 1987, Activation of pig brain dolichol kinase by phospholipids, *Fed. Proc.* **46:**980 (Abstr.).

Gething, M.-J., McCammon, K., and Sambrook, J., 1986, Expression of wild-type and mutant forms of influenza hemagglutinin: The role of folding in intracellular transport, *Cell* **46:**939–950.

Grange, D. K., and Adair, W. L., 1977, Studies on the biosynthesis of dolichyl phosphate: Evidence for the *in vitro* formation of 2,3-dehydrodolichyl phosphate, *Biochem. Biophys. Res. Commun.* **79:**734–740.

Hall, N. A., and Patrick, A. D., 1987, Accumulation of phosphorylated dolichol in several tissues in ceroid-lipofuscinosis (Batten disease), *Clin. Chim. Acta* **170:**323–330.

Harford, J. B., and Waechter, C. J., 1979, Transfer of N,N'-diacetylchitobiose from dolichyl diphosphate into a gray matter membrane glycoprotein, *Arch. Biochem. Biophys.* **197:**424–435.

Harford, J. B., and Waechter, C. J., 1980, A developmental change in dolichyl phosphate mannose synthase activity in pig brain, *Biochem J.* **188:**481–490.

Hartshorne, R. P., Messner, D. J., Coppersmith, J. C., and Catterall, W. C., 1982, The saxitoxin receptor of the sodium channel from rat brain: Evidence for two nonidentical β subunits, *J. Biol. Chem.* **257:**13888–13891.

Heacock, A. M., 1982, Glycoprotein requirement for neurite outgrowth in goldfish retina explants: Effects of tunicamycin, *Brain Res.* **241:**307–315.

Hirschberg, C. B., and Snider, M. D., 1987, Topography of glycosylation in the rough endoplasmic reticulum and Golgi apparatus, *Annu. Rev. Biochem.* **56:**63–87.

Holland, P. C., and Herscovics, A., 1986, Inhibition of myoblast fusion by the glucosidase inhibitor N-methyl-1-deoxynojirimycin, but not by the mannosidase inhibitor 1-deoxymannojirimycin, *Biochem. J.* **238:**335–340.

Idoyaga-Vargas, V., and Carminatti, H., 1982, Postnatal changes in dolichol-pathway enzyme activities in cerebral cortex neurons, *Biochem. J.* **202:**87–95.

Idoyaga-Vargas, V., Belocopitow, E., Mentaberry, A., and Carminatti, H., 1980, A phosphatase acting on dolichyl phosphate in membranes from neuronal perikarya, *FEBS Lett.* **112:**63–66.

Ishii, S., and Volpe, J. J., 1987, Dolichol-linked glycoprotein synthesis in G_1 is necessary for DNA synthesis in synchronized primary cultures of cerebral glia, *J. Neurochem.* **49:**1606–1612.

James, M. J., and Kandutsch, A. A., 1980, Evidence for independent regulation of dolichol and cholesterol synthesis in developing mouse brain, *Biochim. Biophys. Acta* **619:**432–435.

Jorgensen, O. S., and Moller, M., 1980, Immunocytochemical demonstration of the D2 protein in the presynaptic complex, *Brain Res.* **194**:419–429.

Jork, R., Schmitt, M., Lossner, B., and Matthies, H., 1984, Dopamine stimulated L-fucose incorporation into brain proteins is related to an increase in fucokinase activity, *Biomed. Biochim. Acta* **43**:261–270.

Kaplan, H. A., Welply, J. K., and Lennarz, W. J., 1987, Oligosaccharyl transferase: The central enzyme in the pathway of glycoprotein assembly, *Biochim. Biophys. Acta* **906**:161–173.

Kaplan, H. A., Naider, F., and Lennarz, W. J., 1988, Partial characterization and purification of the glycosylation site recognition component of oligosaccharyltransferase, *J. Biol. Chem.* **263**:7814–7820.

Keller, R. K., Armstrong, D., Crum, F. C., and Koppang, N., 1984, Dolichol and dolichyl phosphate levels in brain tissue from English setters with ceroid lipofuscinosis, *J. Neurochem.* **42**:1040–1047.

Kohsaka, S., Mita, K., Matsuyama, M., Mizuno, M., and Tsukada, Y., 1985, Impaired development of rat cerebellum induced by neonatal injection of the glycoprotein synthesis inhibitor, tunicamycin, *J. Neurochem.* **44**:406–410.

Kornfeld, R., and Kornfeld, S., 1985, Assembly of asparagine-linked oligosaccharides, *Annu. Rev. Biochem.* **54**:631–664.

Kornfeld, S., 1987, Trafficking of lysosomal enzymes, *FASEB J.* **1**:462–468.

Krusius, T., and Finne, J., 1977, Structural features of tissue glycoproteins: Fractionation and methylation analyses of glycopeptides derived from rat brain, kidney and liver, *Eur. J. Biochem.* **78**:369–380.

Law, P. Y., Ungar, H. G., Hom, D. S., and Loh, H. H., 1985, Effects of cycloheximide and tunicamycin on opiate receptor activities in neuroblastoma × glioma NG108-15 hybrid cells, *Biochem. Pharmacol.* **34**:9–17.

Lennarz, W. J., 1987, Protein glycosylation in the endoplasmic reticulum: Current topological issues, *Biochemistry* **26**:7205–7210.

Liles, W. C., and Nathanson, N. M., 1986, Regulation of neuronal muscarinic acetylcholine receptor number by protein glycosylation, *J. Neurochem.* **46**:89–95.

Lodish, H. F., and King, N., 1984, Glucose removal from N-linked oligosaccharides is required for efficient maturation of certain secretory glycoproteins from the rough endoplasmic reticulum to the Golgi complex, *J. Cell Biol.* **98**:1720–1729.

Matsui, Y., Lombard, D., Hoflack, B., Harth, S., Massarelli, R., Mandel, P., and Dreyfus, H., 1983, Ectoglycosyltransferase activities at the surface of cultured neurons, *Biochem. Biophys. Res. Commun.* **113**:446–453.

McCoy, R. D., Vimr, E. R., and Troy, F. A., 1985, CMP-NeuNAc:poly-α-2,8-sialosyl sialyltransferase and the biosynthesis of polysialosyl units in neural cell adhesion molecules, *J. Biol. Chem.* **260**:12695–12699.

McHardy, M. S., Schlesinger, S., Lindstrom, J., and Merlie, J. P., 1986, The effects of inhibiting oligosaccharide trimming by 1-deoxynojirimycin on the nicotinic acetylcholine receptor, *J. Biol. Chem.* **261**:14825–14832.

McHardy, S. M., Lindstrom, J., and Merlie, J. P., 1987, Formation of the α-bungarotoxin binding site and assembly of the nicotinic acetylcholine receptor subunits occur in the endoplasmic reticulum, *J. Biol. Chem.* **262**:4367–4376.

McLawhon, R. W., Cermak, D., Ellory, J. C., and Dawson, G., 1983, Glycosylation-dependent regulation of opiate (enkephalin) receptors in neurotumor cells, *J. Neurochem.* **41**:1286–1296.

Merlie, J. P., and Smith, M. M., 1986, Synthesis and assembly of acetylcholine receptor, a multisubunit membrane glycoprotein, *J. Membr. Biol.* **91**:1–10.

Merlie, J. P., Sebbane, R., Tzartos, S., and Lindstrom, J., 1982, Inhibition of glycosylation with tunicamycin blocks assembly of newly synthesized acetylcholine receptor subunits in muscle cells, *J. Biol. Chem.* **257**:2694–2701.

Messner, D. J., and Catterall, W. A., 1985, The sodium channel from rat brain: Separation and characterization of subunits, *J. Biol. Chem.* **260**:10597–10604.

Ng Ying Kin, N. M. K., Palo, J., Haltia, M., and Wolfe, L. S., 1983, High levels of brain dolichols in neuronal ceroid-lipofuscinosis and senescence, *J. Neurochem.* **40**:1465–1473.

Olden, K., Law, J., Hunter, V., Romain, R., and Parent, J. B., 1981, Inhibition of fusion of embryonic muscle cells in culture by tunicamycin is prevented by leupeptin, *J. Cell Biol.* **88**:199–204.

Parodi, A. J., Behrens, N. H., Leloir, L. F., and Carminatti, H., 1972, The role of polyprenol-bound saccharides as intermediates in glycoprotein synthesis in liver, *Proc. Natl. Acad. Sci. USA* **69:** 3268-3272.

Paton, B. C., and Poulos, A., 1984, Dolichol metabolism in cultured skin fibroblasts from patients with "neuronal" ceroid lipofuscinosis (Batten's disease), *J. Inher. Metab. Dis.* **7:**112-116.

Plantner, J. J., and Kean, E. L., 1988, The dolichol pathway in the retina: Oligosaccharide-lipid biosynthesis, *Exp. Eye Res.* **46:**785-800.

Plantner, J. J., Poncz, L., and Kean, E. L., 1980, Effect of tunicamycin on the glycosylation of rhodopsin, *Arch. Biochem. Biophys.* **201:**527-532.

Poduslo, J. F., 1985, Posttranslational protein modification: Biosynthetic control mechanisms in the glycosylation of the major myelin glycoprotein by Schwann cells, *J. Neurochem.* **44:**1194-1206.

Presper, K. A., and Heath, E. C., 1983, Glycosylated lipid intermediates involved in glycoprotein biosynthesis, in: *The Enzymes,* Vol. XVI (P. D. Boyer, ed.), pp. 449-488, Academic Press, New York.

Prives, J., and Bar-Sagi, D., 1983, Effect of tunicamycin, an inhibitor of protein glycosylation, on the biological properties of acetylcholine receptor in cultured muscle cells, *J. Biol. Chem.* **258:**1775-1780.

Pullarkat, R. K., and Reha, H., 1982, Accumulation of dolichols in brains of elderly, *J. Biol. Chem.* **257:** 5991-5993.

Rapaport, R. N., Benjamins, J. A., and Skoff, R. P., 1982, Effects of monensin on assembly of P_0 protein into peripheral nerve myelin, *J. Neurochem.* **39:**1101-1110.

Rostas, J. A. P., Leung, W. N., and Jeffrey, P. L., 1981, Glycosyltransferase activities in chicken brain synaptic junctions, *Neurosci. Lett.* **24:**155-160.

Rothbard, J. B., Brackenbury, R., Cunningham, B. A., and Edelman, G. M., 1982, Differences in the carbohydrate structures of neural cell-adhesion molecules from adult and embryonic chicken brains, *J. Biol. Chem.* **257:**11064-11069.

Rothman, J. E., Miller, R. L., and Urbani, L. J., 1984a, Intercompartmental transport in the Golgi complex is a dissociative process: Facile transfer of membrane protein between two Golgi populations, *J. Cell Biol.* **99:**260-271.

Rothman, J. E., Urbani, L. J., and Brands, R., 1984b, Transport of protein between cytoplasmic membranes of fused cells: Correspondence to processes reconstituted in a cell-free system, *J. Cell Biol.* **99:** 248-259.

Sakakihara, Y., and Volpe, J. J., 1984, Dolichol deposition in developing mammalian brain: Content of free and fatty acylated dolichol and proportion of specific isoprenologues, *Dev. Brain Res.* **14:**255-262.

Sakakihara, Y., and Volpe, J. J., 1985a, Dolichol in human brain: Regional and developmental aspects, *J. Neurochem.* **44:**1535-1540.

Sakakihara, Y., and Volpe, J. J., 1985b, Zn^{2+}, not Ca^{2+}, is the most effective cation for activation of dolichol kinase of mammalian brain, *J. Biol. Chem.* **260:**15413-15419.

Schachter, H., 1986, Biosynthetic controls that determine the branching and microheterogeneity of protein-bound oligosaccharides, *Can. J. Biochem. Cell Biol.* **64:**163-181.

Schachter, H., Narasimhan, S., Gleeson, P. A., and Vella, G., 1983, Control of branching during the biosynthesis of asparagine-linked oligosaccharides, *Can. J. Biochem. Cell Biol.* **61:**1049-1066.

Scher, M. G., and Waechter, C. J., 1978, Possible role of membrane-bound glucosidase in the processing of calf brain glycoproteins, in: *Proc. Eur. Soc. Neurochem.*, Vol. 1 (V. Neuhoff, ed.), p. 559, Verlag Chemie, Weinheim.

Scher, M. G., and Waechter, C. J., 1979, A glucosylated oligosaccharide lipid intermediate in calf brain: Evidence for the transfer of oligosaccharide into membrane glycoprotein and subsequent removal of glucosyl residues, *J. Biol. Chem.* **254:**2630-2637.

Scher, M. G., and Waechter, C. J., 1981, Lipolytic cleavage of dolichyl oleate catalyzed by calf brain membranes, *Biochem. Biophys. Res. Commun.* **99:**675-681.

Scher, M. G., and Waechter, C. J., 1984, Brain dolichyl pyrophosphate phosphatase: Solubilization, characterization and differentiation from dolichyl monophosphate phosphatase activity, *J. Biol. Chem.* **259:**14580-14585.

Scher, M. G., and Waechter, C. J., 1985, Dolichyl pyrophosphate phosphatase in brain, *Methods Enzymol.* **111:**547-553.

Scher, M. G., Burton, W. A., and Waechter, C. J., 1980, Enzymatic glucosylation of dolichyl monophosphate formed *via* cytidine triphosphate in calf brain membranes, *J. Neurochem.* **35**:844–849.

Scher, M. G., DeVries, G. H., and Waechter, C. J., 1984, Subcellular sites of enzymes catalyzing the phosphorylation–dephosphorylation of dolichol in the central nervous system, *Arch. Biochem. Biophys.* **231**:293–302.

Scher, M. G., Sumbilla, C. M., and Waechter, C. J., 1985, Dolichyl phosphate metabolism in brain: Developmental increase in polyisoprenyl phosphate phosphatase activity, *J. Biol. Chem.* **260**:13742–13746.

Schmidt, J. W., and Catterall, W. A., 1986, Biosynthesis and processing of the α-subunit of the voltage-sensitive sodium channel in rat brain neurons, *Cell* **46**:437–445.

Schmidt, J. W., and Catterall, W. A., 1987, Palmitylation, sulfation and glycosylation of the α subunit of the sodium channel, *J. Biol. Chem.* **262**:13713–13723.

Smith, M. E., and Sternberger, N. H., 1982, Glycoprotein biosynthesis in peripheral nervous system myelin: Effect of tunicamycin, *J. Neurochem.* **38**:1044–1049.

Smith, M. E., Somera, F. P., and Sims, T. J., 1985, Enzymatic regulation of glycoprotein synthesis in peripheral nervous system myelin, *J. Neurochem.* **45**:1205–1212.

Sumbilla, C., and Waechter, C. J., 1985a, Dolichol kinase, phosphatase and esterase activity in calf brain, *Methods Enzymol.* **111**:471–482.

Sumbilla, C., and Waechter, C. J., 1985b, Properties of brain dolichol kinase activity solubilized with a zwitterionic detergent, *Arch. Biochem. Biophys.* **238**:75–82.

Tamkun, M. M., and Fambrough, D. M., 1986, The $(Na^+ + K^+)$-ATPase of chick sensory neurons: Studies on biosynthesis and intracellular transport, *J. Biol. Chem.* **261**:1009–1019.

Tkacz, J. S., and Lampen, J. O., 1975, Tunicamycin inhibition of polyisoprenyl N-acetylglucosaminyl pyrophosphate formation in calf-liver microsomes, *Biochem. Biophys. Res. Commun.* **65**:248–257.

Tulsiani, D. R. P., and Touster, O., 1987, Processing of N-linked glycoproteins in rat brain by glucosidases I and II and a glucosyl mannosidase, *Fed. Proc.* **46**:2151.

Tulsiani, D. R. P., Broquist, H. P., James, L. F., and Touster, O., 1988, Production of hybrid glycoproteins and accumulation of oligosaccharides in the brain of sheep and pigs administered swainsonine or locoweed, *Arch. Biochem. Biophys.* **264**:607–617.

Volpe, J. J., Sakakihara, Y., and Ishii, S., 1987, Dolichol-linked glycoprotein synthesis in developing mammalian brain: Maturational changes of the N-acetylglucosaminylphosphotransferase, *Dev. Brain Res.* **33**:277–284.

Waechter, C. J., and Harford, J. B., 1977, Evidence for the enzymatic transfer of N-acetylglucosamine into dolichol derivatives and glycoproteins by calf brain membranes, *Arch. Biochem. Biophys.* **181**:185–198.

Waechter, C. J., and Harford, J. B., 1979, A dolichol-linked trisaccharide from central nervous tissue: Structure and biosynthesis, *Arch. Biochem. Biophys.* **192**:380–390.

Waechter, C. J., and Scher, M. G., 1979, Biosynthesis of glycoproteins, in: *Complex Carbohydrates of Nervous Tissue* (R. U. Margolis and R. K. Margolis, eds.), pp. 75–102, Plenum Press, New York.

Waechter, C. J., and Scher, M. G., 1981, Methods for studying lipid-mediated glycosyltransferases involved in the assembly of glycoproteins in nervous tissue, in: *Research Methods in Neurochemistry*, Vol. 5 (N. Marks and R. Rodnight, eds.), pp. 201–231, Plenum Press, New York.

Waechter, C. J., Schmidt, J. W., and Catterall, W. A., 1983, Glycosylation is required for maintenance of functional sodium channels in neuroblastoma cells, *J. Biol. Chem.* **258**:5117–5123.

Wellner, R. B., and Lucas, J. J., 1979, Evidence for a compound with the properties of 2,3-dehydrodolichyl pyrophosphate, *FEBS Lett.* **104**:379–383.

Wolf, M. J., Scher, M. G., and Waechter, C. J., 1988, Subcellular location of dolichyl pyrophosphate phosphatase activity in brain, *FASEB J.* **2**:1528 (Abstr.).

Wong, T. K., and Lennarz, W. J., 1982, The site of biosynthesis and intracellular deposition of dolichol in rat liver, *J. Biol. Chem.* **257**:6619–6624.

Biosynthesis of Glycosaminoglycans and Proteoglycans

Nancy B. Schwartz and Neil R. Smalheiser

1. INTRODUCTION

The class of polysaccharides known as glycosaminoglycans consists of repeated disaccharides, usually containing a sulfated hexosamine and uronic acid. As abundant constituents of extracellular matrices, most work on glycosaminoglycans has focused on their roles in cartilage, bone, and synovial fluid, where they are thought to be important for tissue turgor and hydration, ion binding and buffering, and tensile strength and resiliency. Early interest in the biosynthesis of glycosaminoglycans was generated by the discovery of lysosomal storage diseases and disorders of connective tissue metabolism such as arthritis.

Glycosaminoglycans are all synthesized as components of proteoglycans, with the possible exception of hyaluronate. In proteoglycans, the glycosaminoglycans as well as *N*- and *O*-linked oligosaccharides are covalently attached to a polypeptide backbone or protein core, which may vary greatly in size and composition. Some proteoglycans contain only one or more chains of a single type of glycosaminoglycan, while others are hybrids consisting of more than one glycosaminoglycan chain, e.g., heparan sulfate and chondroitin sulfate from mouse mammary epithelial cells (Rapraeger *et al.*, 1985), or keratan sulfate and chondroitin sulfate from cartilage (Rodén, 1980). The proteoglycans and glycosaminoglycans are present in all tissues and are found in several subcellular compartments (membranes, cytoplasm, nucleus) as well as in the extracellular matrix. The major glycosaminoglycans found in nervous tissue are chondroitin sulfate, heparan sulfate, and hyaluronic acid (Margolis and Margolis, 1979).

Nancy B. Schwartz • Departments of Pediatrics and Biochemistry and Molecular Biology, and the Kennedy Mental Retardation Research Center, University of Chicago, Chicago, Illinois 60637.   *Neil R. Smalheiser* • Department of Pediatrics and the Kennedy Mental Retardation Research Center, University of Chicago, Chicago, Illinois 60637.

From the outset, it is essential to emphasize the tremendous diversity of the proteoglycans which results both from the number and variety of core proteins, and from the number and variety of carbohydrate substituents on the core protein. The latter arise from a large number of posttranslational modification reactions, estimated to be greater than 25,000 individual reactions, to produce a single chondroitin sulfate proteoglycan. Biosynthesis starts with translation of the core protein, followed by initiation and elongation of the carbohydrate chains. The monosaccharide units are added individually by stepwise transfer from a donor nucleotide sugar to an appropriate acceptor of the growing molecule, in reactions catalyzed by a host of glycosyltransferases. Additional modification of the sugars (i.e., sulfation, epimerization, or deacetylation) often occurs after polymerization. Presumably the general features of this biosynthetic process of the diverse families of proteoglycans are common to all tissues, including the nervous system. Unfortunately, neither dynamic patterns of synthesis of these complex macromolecules, nor the specific details of synthesis of a single proteoglycan have been explored in nervous tissue. Most likely, this information gap is due to the (perceived) low levels of proteoglycans, the heterogeneity of cell type, and/or difficulties in handling brain or nervous tissue. Therefore, in order to provide a thorough treatment of proteoglycan biosynthesis, information derived mostly from systems other than neural tissue must be considered. A comprehensive treatment of early metabolic studies of glycosaminoglycans in nervous tissue was included in a previous review (Jourdian, 1979).

2. PROTEOGLYCAN ASSEMBLY

2.1. Core Proteins

Since all the proteoglycans (with the exception of hyaluronate) are initiated on a protein core, genesis of the core is the critical factor in determining the nature and amount of proteoglycan produced. However, in contrast to the relatively large amount of information available on the carbohydrate components, much less is known about the structure or synthesis of proteoglycan core proteins. It is becoming increasingly well established that there are several different core proteins for each major type of proteoglycan and that different cell types express different proteoglycans (Hassell *et al.*, 1986). The general paucity of information on core protein sequence and synthesis is due mostly to the heterogeneity and complexity of proteoglycan molecules *per se*, and the inherent technical difficulties in obtaining intact core protein free of carbohydrate. Increasingly, the primary amino acid sequences of proteoglycan core proteins are being deduced from cDNA clones.

2.1.1. Core Protein Cloning

Since the protein cores represent the primary gene product, valuable information on the relatedness of different proteoglycans can be derived from molecular cloning studies. This has recently become an active area for investigation of proteoglycan biosynthesis. For instance, the cloning and sequencing of cDNA encoding the core protein of several low-molecular-weight proteoglycans has been reported, e.g., the major chondroitin/dermatan sulfate proteoglycan (PG40) produced by human fibroblasts (Krusius and Ruoslahti, 1986), a chondroitin/dermatan sulfate proteoglycan (PG19) that is an abundant product of a rat yolk sac carcinoma (Bourdon *et al.*, 1985, 1986), and the chondroitin sulfate proteoglycan from bovine bone (Day *et al.*, 1986, 1987). Some of these low-molecular-weight proteoglycans have been shown by Northern blot analysis to have a rather wide tissue distribution, e.g., skin, cornea, smooth muscle, articular cartilage, and tendon (Day *et al.*, 1986). Selection of similar mRNAs in brain has not been examined. More recently, the structure and sequence of the core protein for the high-molecular-weight aggregating species of proteoglycan commonly found in cartilage has been deduced from cDNA clones isolated from chick embryo sternal cartilage (Sai *et al.*, 1986), bovine articular cartilage (Oldberg *et al.*, 1987), and the Swarm rat chondrosarcoma (Doege *et al.*, 1986, 1987). In at least one study, using cytoplasmic extracts obtained from chick embryos, several tissues including liver, heart, skeletal muscle, and brain demonstrated low levels of hybridizable RNA, whereas prechondrogenic mesenchyme did not contain any detectable RNA species (Sai *et al.*, 1986). In recent studies from our laboratory, a cDNA clone for chick cartilage core protein has been shown to hybridize to mRNA from 14-day embryonic chick brain (unpublished).

2.1.2. Cell-Free Translation

Another approach to obtaining information about the synthesis and structure of proteoglycan core proteins involves isolation and characterization of core protein precursors. A single component of $M_r \sim 340,000$ has been translated from RNA prepared from embryonic chick chondrocytes in a cell-free protein-synthesizing system, and immunoprecipitated with a monoclonal antibody to chondroitin sulfate proteoglycan core protein (Upholt *et al.*, 1979; Vertel *et al.*, 1984). The 340kDa protein was present in cell-free translation products directed by RNA prepared from chick limb buds and sterna but not from liver or calvaria. A slightly smaller product has been found using RNA isolated from rat chondrosarcoma (Vertel *et al.*, 1984), and a product of M_r greater than 300,000 was obtained with RNA from bovine cartilage (Treadwell *et al.*, 1980).

2.1.3. Core Protein Precursors and Processing

Since the molecular size of the cell-free translated product is substantially larger than that reported for the core protein of a completed proteoglycan (Hascall and Riolo, 1972; Schwartz *et al.*, 1985), it is possible that processing of the core protein to a smaller size may occur concomitant with the glycosylation that results in a completed, secreted proteoglycan. Thus, several studies have examined the initially synthesized core protein precursor and its intracellular fate on the way to becoming a proteoglycan. A single high-molecular-weight (~ 376kDa) precursor that already contains N-linked oligosaccharides and probably xylose, has been localized to the rough ER of embryonic chick chondrocytes (Schwartz *et al.*, 1985; Geetha-Habib *et al.*, 1984). This precursor form then moves to the Golgi where extensive glycosylation occurs leading to addition of chondroitin sulfate, keratan sulfate, and *O*-linked oligosaccharides. Similar studies have been carried out in the rat chondrosarcoma system, suggesting a similar topological sequence of events (Kimura *et al.*, 1981, 1984). The temporal characteristics of proteoglycan assembly have also been established in rat chondrosarcoma. In these cells, the core protein appears to reside in the rough ER for a rather long time ($t_{1/2}$ $\sim$ 60–90 min) with very little modification, and then extensive processing occurs very rapidly in the Golgi ($t_{1/2}$ $\sim$ 10 min) (Fellini *et al.*, 1984; Mitchell and Hardingham, 1981). In recent studies in embryonic chick chondrocytes, the kinetics of core protein processing were found to be three to four times faster than in rat chondrosarcoma, exhibiting a $t_{1/2}$ in both the rough ER and Golgi of less than 10 min (Campbell and Schwartz, 1988). At least one other proteoglycan, proteodermatan sulfate, is synthesized, processed, and secreted by human skin fibroblasts fairly rapidly (Glössl *et al.*, 1984), with kinetics similar to that of the chick chondrocytes, rather than the rat chondrosarcoma. Most impressive, however, is the ability of all of the cells examined to carry out the multitude of processing and modification reactions in an efficient and rapid manner in the Golgi. Similar studies, detailing the kinetic parameters and processing steps of the synthesis of proteoglycans in brain or nervous system tissues or cells, have not been reported. However, we have detected proteoglycan in embryonic chick brain whose core protein is similar in size and antigenicity to the cartilage core protein (Krueger and Schwartz, 1988). Although sharing an immunological determinant (S103L) in the protein core, there are differences in the overall properties of the proteoglycans; most significant is the lack of any keratan sulfate chains on the brain species (Norling *et al.*, 1984). With these differences in posttranslational modifications, it is difficult to speculate on which features of the biosynthetic process may be common to cartilage and brain.

2.1.4. Consensus Sequences

It is well established that there are several different core proteins all of which bear the same class of glycosaminoglycan chains and that many core proteins contain more than one glycosaminoglycan chain per molecule, which may be of different classes. This leads to a major question in synthesis of these molecules: What is the primary recognition signal in the protein backbone for initiation of glycosylation? In order to answer this question for any one glycosaminoglycan in any particular proteoglycan, it is essential to know the primary amino acid sequence around glycosylation sites. It has long been known that most attachment sites (for chondroitin sulfate, dermatan sulfate, heparan sulfate, and heparin) involve a xylose-serine linkage and that the substituted serines are followed by a glycine residue (Rodén *et al.*, 1985; Robinson *et al.*, 1978; Bourdon *et al.*, 1985; Pearson *et al.*, 1983). In some proteoglycans, such as the yolk sac tumor chondroitin/dermatan sulfate molecule (PG19), and a heparin proteoglycan, the substituted serines are part of a series of repeated serine-glycine dipeptides (Robinson *et al.*, 1978; Bourdon *et al.*, 1985; Oldberg *et al.*, 1981), while in the fibroblast proteodermatan sulfate proteoglycan (PG40), the chondroitin/dermatan chain is at-

Table 1. Xylosyltransferase Activity in 14-Day Chick Brain
or Cartilage

Enzyme source	Specific activity (cpm/hr per mg protein)		
	No acceptor	PGHF	Peptide
Brain 10k supernatant	1,120	2,000	4,230
Brain 10k pellet	1,029	2,120	—
Cartilage 10k supernatant	15,560	54,440	121,100
Cartilage 10k pellet	20,000	21,200	—

tached to the serine of a single serine-glycine dipeptide (Pearson *et al.*, 1983; Brennan *et al.*, 1984; Krusius and Ruoslahti, 1986; Chopra *et al.*, 1985). However, many proteins contain the serine-glycine dipeptide sequence, yet are not proteoglycans; conversely, not all serine-glycine sequences within proteoglycans receive glycosaminoglycan chains. Thus, additional signals must be involved in the recognition of core proteins by the chain-initiating enzymes during synthesis. Using information derived from cloned core protein genes, a conserved 12-amino-acid sequence has been found in several core proteins. Using model peptides containing this sequence, it has been shown that several features of the sequence are essential for xylosyltransferase acceptor activity including: acidic-X-serine-glycine-X-glycine (Bourdon *et al.*, 1987). It was previously shown that neither bovine nasal chondroitin sulfate proteoglycan nor hyaluronidase-treated chondroitin sulfate proteoglycan (both of which still contain substituted serine residues) served as xylose acceptors with chick brain homogenates, while Smith degraded proteoglycan was a good acceptor (Jourdian, 1979). This finding indicates that a specific acceptor sequence is required for initiation of chondroitin sulfate chains in brain. That chain-initiating xylosyltransferases which recognize the consensus sequence are present in embryonic chick cartilage, liver, and brain is demonstrated by the data in Table 1. Whether additional or unique recognition sequences pertain to nervous tissue proteoglycans is presently unknown.

2.2. Nucleotide Sugar Precursors

Once translation of the protein core of a proteoglycan has been completed (or while still continuing), the covalent attachment of carbohydrate substituents may begin. As with all sugar transfer reactions, an appropriate acceptor (discussed above), donor, and specific glycosyltransferase are required. All glycosaminoglycan synthesis proceeds via donor nucleotide sugars. The nucleotide sugars contain two phosphoryl bonds with a large ΔG of hydrolysis, which contribute to the energized character of these compounds in the transfer reaction, and which confer specificity on the enzymes catalyzing the individual reactions. The pathways by which the various sugar nucleotide precursors are synthesized are well established (for reviews, see Jourdian, 1979; Schwartz, 1986). After the initial transfer reaction that covalently links carbohydrate and protein, subsequent reactions use the nonreducing end of an acceptor sugar, thereby linking sugar residues to each other by glycosidic bonds. The nature of the bond formed is determined by the specificity of an individual glycosyltransferase and is unique for the sugar acceptor, the sugar transferred, and the linkage formed. Thus, polysaccharide synthesis is controlled by a nontemplate mechanism in which genes code for specific glycosyltransferases. Furthermore, the large and diverse numbers of molecules that can be generated by such a system suggest that oligosaccharides have potential for great informational content.

2.3. Glycosaminoglycan Chain Assembly

Of the several glycosaminoglycans found in nervous tissue, the synthesis of chondroitin sulfate (as representative of those with the characteristic serine-xylose-galactose-galactose linkage region) and hyaluronic acid have been studied in some detail, and will be considered separately. For the most part, the assembly of chondroitin sulfate chains in brain resembles the process in other tissues, particularly cartilage. As with all glycosaminoglycans, synthesis proceeds by single sugar addition (rather than by synthesis and transfer *en bloc* of a preformed oligosaccharide unit, as in glycoprotein biogenesis). In this process each reaction is catalyzed by a specific glycosyltransferase, and each product serves as an acceptor for the subsequent sugar added (Figure 1).

2.3.1. Chain Initiation

Synthesis of chondroitin sulfate, dermatan sulfate, heparin, or heparan sulfate are each initiated by the addition of xylose to specific serine residues, catalyzed by a xylosyltransferase. This is the only glycosyltransferase of complex carbohydrates that has been purified to homogeneity from embryonic chick cartilage (Schwartz and Rodén, 1974b) and rat chondrosarcoma (Schwartz and Dorfman, 1975). The enzyme from cartilage has a molecular weight of approximately 100kDa, and is composed of two pairs of dissimilar subunits. Although never purified from any other source, xylosyltransferase activity has been reported in embryonic chick brain (Brandt *et al.*, 1975). Total activity increased from day 7 of development to hatching, while specific activity remained constant (Brandt *et al.*, 1975). The activities have been determined in soluble and particulate fractions prepared from homogenates of embryonic chick cartilage and brain (Table 1). While the activity levels are roughly 20-fold lower in brain than in cartilage, a substantial portion of xylosyltransferase activity is soluble in both tissues. In a previous examination of the distribution of glycosyltransferases in subcellular fractions of chick brain, a majority (65%) of the xylosyltransferase was also found in the cytosol (Jourdian, 1979). Similar enzyme activity levels were determined using the deglycosylated intact core protein (PGHF) as acceptor (Krueger *et al.*, 1985) and the artificial peptide containing the serine acceptor consensus sequence (Bourdon *et al.*, 1987), suggesting that the brain form of xylosyltransferase recognizes the same features of the peptide substrate as have been studied in detail in cartilage.

The question of when and where the core protein is xylosylated and hence how glycosaminoglycan synthesis is initiated remains somewhat controversial. In an early

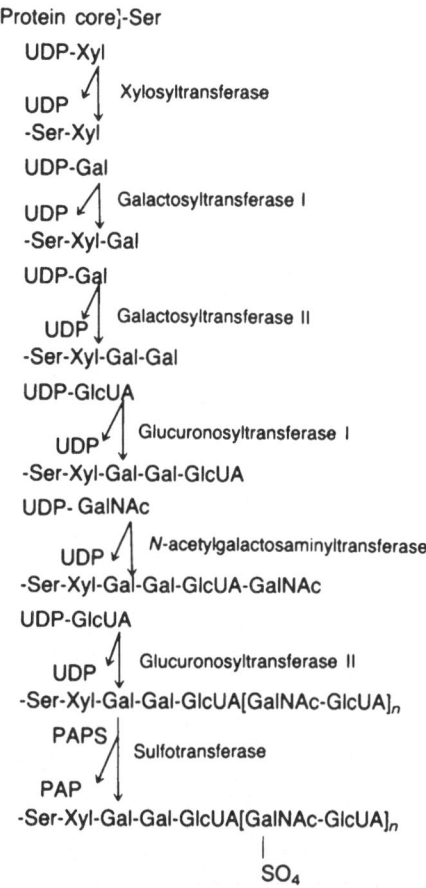

Figure 1. Synthesis of chondroitin sulfate proteoglycan.

study it was suggested that xylosylation may occur while the core protein is nascent (Sugahara *et al.*, 1981). However, further evidence indicates that xylosylation sites are still available after a completed peptide containing the *N*-linked oligosaccharides is synthesized (Geetha-Habib *et al.*, 1984). The rough ER was suggested as the site of xylosylation in embryonic chick cartilage by subcellular fractionation which colocalized large amounts of the chain-initiating xylosyltransferase and endogenous core protein acceptor (Schwartz *et al.*, 1985; Geetha-Habib *et al.*, 1984). The topology of xylosylation was subsequently confirmed by an ultrastructural study using antibodies to localize xylosyltransferase to the cisternae of the rough ER (Hoffman *et al.*, 1984). In contrast to these studies in chick, indirect evidence for xylosylation to be a rather late event (probably occurring in the Golgi, where the remainder of the chain elongation reactions occur) has been obtained in rat chondrosarcoma (Kimura *et al.*, 1984). Because xylosyltransferase is so abundant and entirely soluble in this tumor (Schwartz and Dorfman, 1975), subcellular localization of the enzyme to any membrane compartment, as has been done in chick cartilage, is not possible. No information is available on the site of xylosylation of core proteins in brain or nervous tissues.

2.3.2. Linkage Region

The completion of the carbohydrate protein linkage region (galactose-galactose-glucuronic acid) is accomplished by three glycosyltransferases, each specific with respect to the sugar acceptor, donor, and linkage formed (Figure 1). For instance, the two sequential galactosyltransferases are distinguishable on the basis of the acceptor structures; the first transfers galactose to xylosyl-serine, free xylose, or various xyloside derivatives, and the second requires the disaccharide galactosyl-xylose for transfer. Similarly, the linkage region glucuronosyltransferase, which transfers glucuronic acid to terminal galactose residues, is distinct from the subsequent glucuronosyltransferase involved in synthesis of the repeating disaccharide units. As before, most information on these enzymes comes from studies in cartilage, where each of the activities has been solubilized and partially purified (Schwartz and Rodén, 1975). These enzymes are all tightly membrane-bound and progress in purifying them has thus been slow.

A few interesting aspects about their properties, behavior, and regulation have been elucidated, however. A specific interaction between galactosyltransferase I and the chain-initiating xylosyltransferase has been demonstrated by two independent approaches (Schwartz and Rodén, 1974a; Schwartz, 1975). The specificity and strength of this association has led to speculation that xylosyltransferase may serve to carry and "dock" the core protein to the proper region of the Golgi for further glycosylation (Schwartz, 1982). This association between these first two enzymes that catalyze consecutive reactions in the pathway of glycosaminoglycan synthesis is especially intriguing in light of the significant differences in their properties. Xylosyltransferase is a very soluble protein, found predominantly in the lumen of the ER, and is readily extracted from tissues with low-ionic-strength buffers (Schwartz and Rodén, 1974a; Schwartz and Dorfman, 1975). In contrast, galactosyltransferase I is extremely tightly bound to Golgi membranes, requires detergent and salt to solubilize it, and after solubilization galactosyltransferase remains distinctly hydrophobic and is dependent on lipids or detergent for enzyme activity (Schwartz and Rodén, 1975; Schwartz, 1976). The other glycosyltransferases exhibit properties intermediate between these two with regard to solubility and effect of ionic and hydrophobic forces on enzyme activity (Schwartz and Rodén, 1975). From additional studies (see below) it appears that the galactosyltransferase step may be an important regulatory site in glycosaminoglycan synthesis. For instance, rat chondrosarcoma chondrocytes are severely deficient in galactosyltransferase activity, which may explain the large pool size of core protein and long dwell time in the rough ER in rat chondrosarcoma (Schwartz and Rodén, 1975; Campbell and Schwartz, 1988). In recent studies, galactosyltransferase I, but not the other linkage region or chain elongation glycosyltransferases, has been shown to be sensitive to certain regulatory hormones like cortisol (Bird *et al.*, 1986). Lastly, this appears to be the only site in the pathway of synthesis of a glycosaminoglycan chain that artificial acceptors, β-D-xylosides, can be used by cells instead of the natural xylosylated core protein (Schwartz, 1977, 1979). Although no studies on the properties or purification of these enzymes from brain or nervous tissue have been reported, the activities of both galactosyltransferases and the linkage region glucuronosyltransferase have been demonstrated in 13-day chick embryo brains. The first galactosyltransferase has been solubilized and partially purified (15-fold) and the substrate specificity confirmed (Jourdian, 1979; Brandt et al., 1975); the second galactosyltransferase has been detected although the glycosidic linkage has not been confirmed (Jourdian, 1979). The first glucuronosyltransferase was also solubilized and purified approximately 65-fold, and shown to have slightly broader substrate specificity than its counterpart in cartilage, since the brain form was able to use lactose comparably to galactosylgalactose as acceptor (Jourdian, 1979; Brandt et al., 1969). Interestingly, of several embryonic chick tissues examined, brain preparations exhibited the highest specific activities of glucuronosyltransferase I (Brandt et al., 1969). It was also reported in that early study that the specific activity of glucuronosyltransferase I was 7-fold higher in gray matter than in white matter; the former also has the higher glycosaminoglycan content in brain (Brandt et al., 1969).

2.3.3. Polymerization

The characteristic repeating polymer of chondroitin sulfate is synthesized by the concerted action of an N-acetylgalactosaminyltransferase and a glucuronosyltransferase (Figure 1). In vitro transfer, substrate specificity, and linkages formed have all been demonstrated (Rodén et al., 1972); the polymerizing enzymes have been solubilized and partially purified (30-fold) (Schwartz and Rodén, 1975), and localized to the Golgi apparatus in cartilage (Geetha-Habib et al., 1984). Both activities have also been demonstrated in developing chick embryo brain, although the substrate specificity of these brain preparations has not been studied in detail (Jourdian, 1979; Brandt et al., 1975). Although little is known about these specific reactions of the overall polymerization process in either tissue, it must be a highly regulated, well-organized process since hybrids of chondroitin sulfate and hyaluronic acid (both normal components of brain) have never been observed. Furthermore, certain of the glycosaminoglycans undergo specific modification reactions concomitant with or following polymerization, also requiring a highly coordinated system (see below).

2.4. Polysaccharide Modification

Although the mechanisms responsible for termination of glycosaminoglycan chains are not well understood, certain modification reactions, including sulfation, phosphorylation, epimerization, and deacetylation, may occur concomitant with or shortly after polymer synthesis. Of these modifications, sulfation is the most common since N- or O-linked sulfate groups are found on all of the glycosaminoglycans except hyaluronate, and is the only further modification of chondroitin sulfate chains. The sulfation process, like glycosylation, requires several components, including an acceptor, a donor, and a sulfotransferase.

2.4.1. Sulfate Activation and Utilization

Of two natural donor substrates that have been identified, it is thought that the universal sulfate donor in all higher organisms is phosphoadenosine 5′-phosphosulfate (PAPS). PAPS is formed from ATP and inorganic sulfate in two steps:

$$ATP + SO_4{}^{2-} \xrightleftharpoons[\text{ATP sulfurylase}]{} APS + PPi \qquad (1)$$

$$ATP + APS \xrightleftharpoons[\text{APS kinase}]{} PAPS + ADP \qquad (2)$$

The first reaction, formation of adenosine phosphosulfate (APS), is catalyzed by ATP sulfurylase and is driven in the direction of production of APS by coupling with an inorganic pyrophosphatase. The second stage of activation, formation of PAPS, is catalyzed by APS kinase. In early work, these reactions were shown to occur in rabbit brain (Gregory and Lipmann, 1957) and rat brain (Balasubramanian and Bachhawat, 1961), but most information on the nature and properties of the PAPS-synthesizing enzymes has come from studies in microorganisms, liver (Burnell and Roy, 1971), or cartilage (Geller *et al.*, 1987).

In recent work, a substantial purification (> 2000-fold) of each of the sulfate-activating enzymes has been achieved from rat chondrosarcoma and embryonic chick cartilage (Geller *et al.*, 1987). Unexpectedly, the PAPS-synthesizing enzymes were found to copurify, and in fact were not separable by gel filtration, ion-exchange or affinity chromatography. From preliminary characterization and kinetic analysis, evidence has been accumulated to suggest that ATP sulfurylase and APS kinase function as a kinetically coordinated complex (unpublished), facilitating the rapid and efficient production of PAPS.

Enzymatic formation of sulfate esters is then mediated by transfer of sulfate groups from PAPS to an appropriate acceptor molecule. Naturally occurring chondroitin sulfate has been found sulfated in either the 4 or 6 position on N-acetylgalactosamine residues. The nature of the substrate, i.e., whether the polysaccharide is free or bound to protein, and the size of the oligosaccharide influence the efficiency and type of sulfation (for review, see De *et al.*, 1978). Specific sulfotransferases have been identified that catalyze the transfer of sulfate from PAPS to the 4 and 6 position of N-acetylgalactosamine residues, and some progress on purification and characterization has been made (Kimata *et al.*, 1973; Schwartz, 1982).

Although less information is available on the enzyme system responsible for generation of PAPS, and on the nature of endogenous sulfate acceptors and sulfotransferases in brain (Burkart *et al.*, 1981), sulfate has been shown to fully penetrate the blood–brain barrier, and is available at the intracellular site of the PAPS-generating system. In developing rat brain, PAPS formation was highest at day 1, with a second, but smaller, peak at day 12 (Balasubramanian and Bachhawat, 1965). In rat (Guha *et al.*, 1960) and mouse (Burkart and Weismann, 1987) brain, the sulfation rate of glycosaminoglycans is highest after birth and then decreases, paralleling the pattern of PAPS formation. A second increase in PAPS synthesis from day 7 to 12 reflects the *in vivo* formation of sulfogalactosyl glycerolipids (Burkart *et al.*, 1983) and sulfatides (Burkart *et al.*, 1981). Because of the correlation of PAPS synthesis with functional activity of glycosaminoglycan- and sulfolipid-glycosyltransferases, it was suggested that sulfation may be controlled not by the activity of the synthesizing enzymes but by the limitation of availability of PAPS. We have also suggested such a regulatory role for PAPS in synthesis of proteoglycans following studies of a naturally occurring defect in the sulfation pathway (Sugahara and Schwartz, 1979, 1982a,b). PAPS is also required for N- and O-sulfation reactions involved in synthesis of dermatan sulfate, heparin, and heparan sulfate, and may in part regulate the extent of these other polymer modifications (see below).

2.4.2. Synthesis of Dermatan Sulfate, Heparin, and Heparan Sulfate

Certain additional modification reactions occur at the polymer level leading to production of some of the other glycosaminoglycans. For instance, epimerization at C-5 of internal glucuronic acid residues of the nonsulfated chondroitin precursor to form iduronic acid residues leads to synthesis of dermatan sulfate (Malmström *et al.*, 1975). The epimerization reaction is tightly coupled with sulfation, so that the conversion to iduronic acid residues is favored in the presence of PAPS. Since a chondroitin chain is precursor to both chondroitin and dermatan sulfate, the glycosyltransferase reactions involved in synthesizing the polymeric chain prior to modification are most likely comparable.

Heparin and heparan sulfate are similar to chondroitin sulfate and dermatan sulfate in that they all share the identical four-sugar linkage region sequence and presumably synthesis of the linkage region occurs in a similar fashion for all four. The acceptor sequence in the core proteins to which these types of glycosaminoglycans is attached has not been elucidated, although heparin does have long serine-glycine repeats, the majority of which are bound to polysaccharide chains (Robinson *et al.*, 1978). A distinct polymeric structure consists of N-acetylglucosamine and uronic acid residues, formed by the alternating action of an N-acetylglueosaminyltransferase and glucuronosyltransferase. This nonsulfated polymer then undergoes a series of modification reactions which have largely been elucidated through the elegant work of Lindahl and co-workers, using the mastocytoma system. As outlined in Figure 2, N-deacetylation of certain of the N-acetylglucosamine residues initially occurs catalyzed by a deacetylase (Riesenfeld *et al.*, 1982). This appears to be a commitment step and the deacetylase may be a key regulatory enzyme since the following series of reactions are unable to continue without deacetylation.

Subsequently, the free amino groups of the deacetylated glucosamine residues are rapidly sulfated by a specific N-sulfotransferase. This enzyme has been demonstrated in several tissues including rat brain (Balasubramanian and Bachhawat, 1964; Balasubramanian *et al.*, 1968). Specific glucuronic acid residues linked to the N-sulfated glucosamine residues are then converted to the C-5 epimer L-iduronic acid by an epimerase (Jacobsson *et al.*, 1979; Malmström *et al.*, 1980). This enzyme is largely soluble, and has been extensively purified and the reaction mechanism well characterized. The epimerases involved in dermatan sulfate and heparin/heparan sulfate conversions appear to be distinct (Malmström and Öberg, 1981). Interestingly, this epimerization occurs concomitantly with a second sulfation reaction, at the 2 position of the newly created iduronic acid residues. The final step of polymer modification is another O-sulfation of the glucosamine residues, catalyzed by an O-sulfotransferase which appears to be distinct from the sulfotransferase that transfers sulfate to iduronic acid residues (Jansson *et al.*, 1975). A recent study has shown that N-sulfation and O-

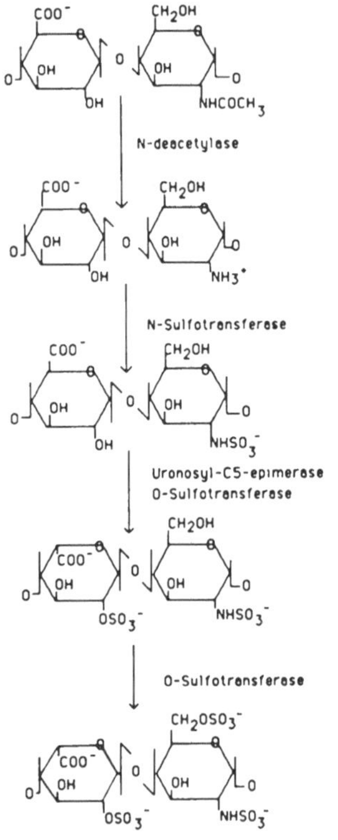
COO⁻
CH2OH
OH
OH
O
O
OH
NHCOCH3
N-deacetylase
NH3+
N-Sulfotransferase
NHSO3⁻
Uronosyl-C5-epimerase
O-Sulfotransferase
OSO3⁻
O-Sulfotransferase
CH2OSO3⁻

Figure 2. Polymer modification reactions involved in synthesis of heparin.

sulfation of proteoglycans are catalyzed by separate enzymes in nervous tissue (Miller and Waechter, 1988). Furthermore, the detection of two chromatographically distinct *N*-sulfotransferases raises the possibility that *N*-sulfation of proteoheparan sulfates is catalyzed by more than one enzyme. Although a proteoglycan appears to undergo all these modifications, biologically active heparin does not remain protein-bound, and is usually found in the molecular weight range of 10–15kDa. Therefore, subsequent enzymatic cleavage at specific glucuronic acid residues must occur to release the heparin fragments (Ögren and Lindahl, 1975).

The overall process of polymer modification appears to occur in a strictly ordered fashion, with a distinct pattern of intermediates that can be induced to accumulate at the individual steps by substrate limitation. This experimental approach has actually been extremely useful in elucidating the steps of polymer modification. Heparan sulfate is similar structurally to heparin, although it contains fewer *N*-sulfate groups, L-iduronic acid residues, and *O*-sulfate groups. If it is assumed that the general order of reactions, substrate specificities, and requisite enzymes are similar for both molecules, then an initial reduction in the level of *N*-deacetylation would account for the observed pattern of higher *N*-acetyl groups, and lower *N*-sulfate, iduronic acid, and *O*-sulfate groups in the heparan sulfates. Notwithstanding the structural similarities, heparin and heparan sulfate are biologically very different molecules, synthesized by different cells, distributed in different cellular compartments, and playing different functional roles. The factors that control the production and activity of the modifying enzymes, and hence determine the fate of the unsulfated polymeric precursor toward becoming either a heparin or a heparan sulfate molecule, remain unknown.

2.5. Biosynthesis of Hyaluronic Acid

Hyaluronc acid, although by definition a glycosaminoglycan, has such significantly different properties and unresolved problems with respect to its mechanism of synthesis, that it deserves special consideration. From all available information, hyaluronate consists only of a repeating disaccharide unit of N-acetylglucosamine and glucuronic acid, suggesting two important differences from the other glycosaminoglycans; i.e., it is not sulfated, and is not bound to protein and therefore is not a constituent of a proteoglycan. Whether protein is covalently bound to hyaluronate has been repeatedly addressed; neither puromycin nor cycloheximide, at concentrations that inhibited protein synthesis by greater than 95%, has any effect on hyaluronate synthesis in streptococci (Stoolmiller and Dorfman, 1967) or human skin fibroblasts (Siewert and Strominger, 1967; Stoolmiller and Dorfman, 1969). Similarly, bacitracin (Siewert and Strominger, 1967; Stoolmiller and Dorfman, 1969), or tunicamycin (Appel et al., 1979; Sugahara et al., 1979) inhibitors used to elucidate the roles of polyisoprenol carrier lipids in polysaccharide and glycoprotein synthesis, have no effect on hyaluronate synthesis in bacteria and eukaryotes, respectively. These findings are critically important for the biosynthesis of hyaluronate, for they leave wanting a ready explanation for initiation of hyaluronate chain synthesis.

The molecular events involved in elongation of hyaluronate chains also remain obscure. Enzymatic synthesis of hyaluronate was first demonstrated with UDP-glucuronic acid and UDP-N-acetylglucosamine in a cell-free preparation from Rous sarcoma (Glaser and Brown, 1955). Subsequently, a model was proposed of synthetase activity for the steps occurring during elongation of hyaluronate at the nonreducing end in a prokaryotic system (Markovitz et al., 1959). Although several mammalian tissue preparations have been used to catalyze the synthesis of high-molecular-weight hyaluronate from the two nucleotide sugars, a model for hyaluronate synthesis in eukaryotes has not been postulated. Evidence has been presented for elongation occurring at the reducing end in the hybrid cell B6 line (Prehm, 1983a,b); however, this interesting finding has not been confirmed in at least two other studies of hyaluronate synthesis (Philipson et al., 1985; Mian, 1986).

Although this paucity of knowledge concerning the mechanism of hyaluronate synthesis precludes a detailed discussion of the process, some interesting information has recently been forthcoming. For most of our recent studies, we have used mouse oligodendroglioma cells. These represent an excellent model system for the study of hyaluronate synthesis since the cells grow rapidly, do not produce a fibroblastic extracellular matrix (and therefore do not adhere tightly to tissue culture plates), and produce large amounts of hyaluronate but small amounts of other glycosaminoglycans (Philipson and Schwartz, 1984). Using this cell line, we showed that hyaluronate synthetase is localized at the inner surface of the plasma membrane in the glioma cells in culture (Philipson and Schwartz, 1984). This subcellular localization was subsequently confirmed (Prehm, 1984). Although the presence of hyaluronate in nervous tissue has long been recognized and the presence of hyaluronate in cells of neural origin was demonstrated almost 20 years ago (Dorfman and Ho, 1970), this represents the first conclusive evidence of hyaluronate synthetase in a neural cell line.

The presence of the synthetic machinery for a glycosaminoglycan in the plasma membrane (rather than in the Golgi apparatus) is novel, and does help explain some of the enigmas concerning the properties and synthesis of hyaluronate. If hyaluronate is not an intracellular product, then lack of involvement of protein synthesis and other postsynthetic modifications common to the other glycosaminoglycans (like sulfation) is immediately explained. Extrusion directly from the plasma membrane also precludes a packaging or secretion mechanism for these extremely large molecules. The findings pertaining to localization of hyaluronate synthetase in the plasma membrane have not, however, been reconciled with recent intriguing results showing a distinctive change in hyaluronate from an extracellular to an intracellular localization during postnatal development (Ripellino *et al.*, 1988). This evidence for the presence of hyaluronate in the cytoplasm raises further unresolved issues concerning the site of synthesis and mechanism of entry into intracellular compartments.

Another aspect of hyaluronate synthesis which we have recently attempted to address involves the number and nature of the components in the synthetase complex. Since hyaluronate is composed of only two sugars, the supposition of at least two glycosyltransferases is obvious. However, the use of oligosaccharide acceptors, which was the most direct and conclusive means of elucidating the biosynthetic pathways of chondroitin sulfate and heparin (Rodén, 1980), has not been a successful approach; therefore, the individual glycosyltransferase reactions have not been demonstrated conclusively. Moreover, any attempts to dissociate the two activities from each other, or from the membrane components to which they are tenaciously bound, have resulted in inactivation. In order to circumvent some of these problems, a steady-state analysis of hyaluronate synthetase kinetic parameters has been carried out (Philipson *et al.*, 1985). This study has led for the first time to a kinetic model of hyaluronate synthetase in a eukaryotic system. The formal mechanism predicts the existence of an oligosaccharide intermediate, with the participation of at least two and possibly three activities in the complex; presents an explanation for internal labeling and therefore direction of chain synthesis; and suggests a mechanism for *de novo* synthesis (Philipson *et al.*, 1985). Confirmation of this model and resolution of additional important problems in hyaluronate synthesis require the solubilization and separation of the individual activities. Toward this end, solubilization of the synthetase from *Streptococcus* has recently been reported (Prehm and Mausolf, 1986; Triscott and van der Rijn, 1986). We have also been successful in solubilizing and partially purifying a functional hyaluronate synthetase from oligodendroglioma cells (Ng and Schwartz, 1987).

2.6. Regulation of Synthesis

From a regulatory viewpoint, it might be expected that the synthesis of proteoglycans would be most efficiently controlled at the level of synthesis of core protein. Unfortunately, no information is available on the transcriptional or translational control of core protein synthesis. Only recently have cDNA clones become available for sequencing, allowing the construction of a model for the domain structure of the large cartilage core protein. This has given some insight into the function and evolution of these structural domains, and the sequence relationship between cartilage core protein and some other proteoglycan core proteins (Doege *et al.*, 1987). Interestingly, certain structural features have already emerged that show relationships to other protein families; the basement membrane heparan sulfate proteoglycan (BM-1) shares regions of homology with laminin and link proteins, including EGF-like repeat domains found commonly in extracellular proteins (Noonan *et al.*, 1987). The globular domains at the N- and C-terminus of the large chondroitin sulfate core protein of cartilage are related to link protein and lectin gene families, respectively (Doege *et al.*, 1986, 1987; Sai *et al.*, 1986). Simple Ser-Gly repeats of the type found in heparin core protein have also been described within a novel class of proteoglycans, exemplified by the *per* locus of *Drosophila* (Shin *et al.*, 1985). With the introduction of molecular techniques to this field, the process of assembly of the proteoglycan core proteins, and the regulation of the synthesis of this large and diverse group of matrix proteins, should be elucidated in a timely fashion.

Little is known about how the different core proteins, with very diverse yet distinct biological functions, are selected and delivered to the polysaccharide synthesizing machinery (which then adds the characteristic carbohydrate chains) that makes a core protein into a proteoglycan. Until the existence of hybrid proteoglycans (carrying chondroitin sulfate and keratan sulfate, or chondroitin sulfate and heparan sulfate) (Rapraeger *et al.*, 1985) was demonstrated, it was tempting to speculate that different intracellular routes via compartmentalization was the predominant mechanism for sorting proteoglycans. Now it is more difficult to explain the successful synthesis of a hybrid proteoglycan with our current concept of the temporal and spatial arrangements of the biosynthetic apparatus. In this context, it is also unknown which specific features of a core protein are recognized for initiation of the various glycosaminoglycan chains, four of which (chondroitin sulfate, dermatan sulfate, heparan sulfate, and heparin) use the same serine-bound tetrasaccharide linkage region. The initial work in this area elucidating a consensus sequence for chondroitin/dermatan sulfate (Bourdon *et al.*, 1987) looked promising. However, it does not appear to be a predominant sequence in rat chondrosarcoma core protein (Doege *et al.*, 1987), while in embryonic chick it is more prevalent. Greater emphasis on protein and nucleic acid sequencing of core proteins and their genes will be required in order to identify specific recognition sequences for initiating each type of glycosaminoglycan chain. Since the core proteins are the primary gene products, resolution of these basic questions should enhance significantly our understanding of the overall control of proteoglycan synthesis.

The temporal and topological relationship between formation of the core protein and carbohydrate addition may encompass critical regulatory elements, yet undefined. By analogy with glycoprotein systems that have been examined (Braell and Lodish, 1981), it is possible that initiation of polysaccharide chains begins while the core protein is being synthesized on ribosomes or shortly thereafter. The finding of xylosyltransferase in the lumen of the rough ER (Hoffman *et al.*, 1984; Geetha-Habib *et al.*, 1984; Schwartz *et al.*, 1985) suggests that the addition of xylose to the core protein may occur early in the biogenesis of proteoglycan in chick chondrocytes. However, a different conclusion was arrived at in studies on rat chondrosarcoma (Kimura *et al.*, 1984). It has also been shown that the translocation rate of UDP-xylose into vesicles of the Golgi apparatus was two- to fivefold higher than into vesicles of the rough ER, leading to the suggestion that biosynthesis of proteoglycan linkage region occurs in the Golgi in rat liver (Nuwayhid *et al.*, 1986). These studies clearly illustrate the difficulties in assigning the intracellular site of synthesis based on localization of substrates, products, or enzymes, especially in different cell types where the nature and quantity of proteoglycans synthesized vary significantly. A further issue that makes the localization of nucleotide sugar precursors problematic involves the long-proposed regulatory role of UDP-xylose in controlling synthesis of proteoglycans. Neufeld and Hall (1965) showed that UDP-xylose is a potent inhibitor of UDP-glucose dehydrogenase, which has the immediate effect of depressing synthesis of UDP-glucuronic acid and UDP-xylose (via decarboxylation of UDP-glucuronic acid). Thus, when a situation occurs such that inadequate levels of core protein are available for xylosylation, the accumulated UDP-xylose inhibits further synthesis of UDP-glucuronic acid and UDP-xylose. It is difficult to reconcile whether this important feedback control mechanism aids in coupling core protein and polysaccharide synthesis, since UDP-xylose is presumably produced and utilized in separate cellular compartments.

Following addition of xylose to the protein core, addition of the linkage region sugars, characteristic disaccharide repeats of individual polymers, and any postpolymerization modification reactions presumably occur predominantly in the Golgi in a rapid and efficient manner. How this overall process, composed of a multitude of enzymatic reactions, is regulated remains unknown. It has long been suggested that the glycosyltransferases may be organized into a multienzyme complex (Rodén, 1980) in which enzymes catalyzing consecutive transfer steps are located in adjacent positions. An organization of this type would require interactions between the individual glycosyltransferases, as well as possible interaction with the Golgi membranes. The specific interaction between the enzymes catalyzing the first two steps of chondroitin sulfate synthesis (Schwartz, 1975; Schwartz and Rodén, 1974a) supported the concept of a multienzyme complex. It has likewise been suggested that heparin and heparan sulfate may be synthesized in different cellular locations by multienzyme systems where the relative proportions of the polymer-modifying enzymes, or other factors yet undefined, regulate the activities of the requisite enzymes (Rodén, 1980). It remains unknown how the glycosaminoglycan glycosyltransferase and modifying enzymes are organized intracellularly and what specific interactions contribute to the amazing efficiency and fidelity of the synthetic process. These questions await further purification (and perhaps cloning) of these enzymes.

The latter approach would also enhance our understanding of the mechanisms which regulate the synthesis of the enzymes themselves. Some early studies on developing chondrocyte cultures showed that the activity levels of all the glycosyltransferases increased coordinately, reaching maxima about 2 days before the maximum level of proteoglycan synthesis was achieved (Schwartz, 1976). Although these enzymes appear to be coordinately modulated during development, in fact their turnover rates varied significantly. The metabolism of xylosyltransferase seems to be controlled in a fashion coupled more with that of core protein than with the other glycosyltransferases. Again, these results suggest that a critical control mechanism may function at this site in the biosynthetic scheme, making the interaction between xylosyltransferase and galactosyltransferase particularly important. It should be mentioned again that this is the site where synthesis of xylosylated core protein can be uncoupled from the rest of the glycosaminoglycan chain by inserting the artificial xylosides (Galligani *et al.*, 1975; Schwartz, 1979). At the least, all of these studies suggest that the glycosaminoglycan biosynthetic enzyme system is not a static entity containing the same set of molecules in a fixed ratio throughout its entire life cycle. But the information available is still too limited to propose a model for regulating proteoglycan synthesis at the level of the glycosyltransferases and modifying enzymes.

Whether any control mechanisms at the latter stages of synthesis influence the overall rates or pattern of proteoglycan production is not well understood, although there are several modification reactions that occur along with or shortly after polymerization. In this context, proteoglycans have been shown to be phosphorylated, both on the protein and on the carbohydrate (specifically the 2 position of the linkage region xylose residues) (Oegema *et al.*, 1984). The significance and function of proteoglycan phosphorylation remain to be established. In contrast, sulfation is the most common modification and the one where some information on how the process may be controlled is available. Of the various components of a sulfation reaction, there is increasing evidence that the rate and extent of sulfation are not controlled by the activity of the sulfotransferases (Schwartz, 1982; Hart, 1978), but rather by the availability of the sulfate donor, PAPS (Burkart *et al.*, 1981, 1983; Sugahara and Schwartz, 1979, 1982a,b). For the latter studies, a naturally occurring mutant defective in the sulfation process has proven invaluable. In this mutant mouse, the PAPS concentration is not adequate for synthesis of large amounts of highly sulfated proteoglycans found in cartilage and therefore is expressed predominantly as a growth disorder of skeletal tissue. However, the tissue distribution of the defect, as well as a normal sulfation pattern of certain molecules, suggest a more complex explanation for the regulation of the sulfation process overall.

3. CELLULAR ASPECTS OF PROTEOGLYCAN BIOSYNTHESIS

While studies have traced the biosynthesis and fate of classes of glycosaminoglycan chains in brain (reviewed in Margolis and Margolis, 1979), only recently have individual, defined proteoglycan molecules been identified in nervous tissue; detailed studies of their biosynthesis have not been carried out. Nevertheless, by comparing the situation in brain to that of other tissues which have been examined in more detail, it is still possible to discuss some of the cellular aspects of proteoglycan biosynthesis.

Brain differs from most other tissues in that extracellular matrix does not play a structural role in maintaining tissue integrity. Indeed, for many years it was unclear whether brain possessed any matrix at all, apart from the basal laminae which line blood vessels and meninges. Krayanek (1980) and Nakanishi (1983) showed persuasively that the enlarged, apparently empty extracellular spaces seen in aldehyde-fixed sections of developing brain were actually filled with proteoglycan-rich matrices when appropriate fixation and staining techniques were employed. In both retina and tectum, the matrices formed oriented channels along which optic axons grow (Krayanek, 1980; Silver and Sidman, 1980; Krayanek and Goldberg, 1981). Such morphological evidence, together with cell culture studies and immunocytochemical data, strongly suggested that the outgrowth of long tract axons in developing brain is regulated by the discrete, transient expression of matrix components (see Chapter 13). Although it is now clear that several types of matrices are present during neural development, their distinct composition, physiological roles, and regulation have not been reviewed previously. This section will describe some of the cellular compartments of brain, and consider how control at the level of proteoglycan biosynthesis may potentially be involved in regulating their biological activities during neural development.

3.1. Basal Laminae versus Oriented Extracellular Channels

In both fetal and adult CNS, basal laminae are found wherever astrocytic endfeet appose blood vessels and the pial surface. As elsewhere in the body, their components include laminin, fibronectin, basement membrane proteoglycans, entactin/nidogen, and nonfibrillary collagens. (Brain tissue is devoid of fibrillary collagen except in the connective tissue stroma of meningeal cells; Sievers *et al.*, 1987.) Both astrocytes and endothelial cells may contribute to the basal laminae. The polarized secretion, assembly, insolubility, and persistence of these basal laminae throughout life clearly differ from the transient extracellular matrix-filled channels which are apparently synthesized by the same type of cell, i.e., astrocytes, and which contain at least some of the same components. The two matrices therefore appear to represent distinct cellular compartments, regulated independently. The organized basal lamina is controlled in part at the level of biosynthesis and secretion of collagen (e.g., Bunge and Bunge, 1983), but little is known of the assembly and turnover of matrix components in brain extracellular spaces.

It is intriguing that vertebrate CNS neurites have rarely been observed to make direct contact with basal laminae (Easter *et al.*, 1984), despite evidence that CNS neurites can grow profusely on isolated basement membranes and basal laminae *in vitro* (Smalheiser *et al.*, 1982, 1984; Halfter *et al.*, 1987). This is all the more striking since peripheral nerves, and neurites in invertebrates, often encounter basal laminae (e.g., Sanes, 1983; Hedgecock *et al.*, 1987). Furthermore, astrocytes in brain almost always contact basal laminae, either forming endfeet or *en passant* contacts; some types of glial cells grow along blood vessel basal laminae for long distances (Ramón y Cajal, 1984). It is as if CNS neurites are uniquely, perhaps actively, prevented from contacting basal laminae—this might allow them to be guided by weaker but more specific cues present within the extracellular channels and along cell surfaces.

3.2. Adherons

Adherons are extracellular complexes which are shed from the surfaces of a variety of neural and nonneural cell types (reviewed in Schubert, 1984). A possible exception are primary neurons, which have not yet unequivocally been shown to synthesize matrix of any kind. Adherons are a particulate microcosm of cell-associated extracellular matrix, for they contain matrix proteins, soluble forms of membrane proteins, and matrix-bound factors (proteases, protease inhibitors, and growth factors) apparently all bound together (Schubert and LaCorbiere, 1980a,b, 1982a,b, 1985, 1986; Schubert et al., 1983a,b, 1986). It has been suggested that adherons are vertebrate analogues of the large proteoglycan-containing particles which act as cell aggregation factors in sponges (Henkart et al., 1973). It has also been proposed that adherons correspond to matrix granules observed within organized basement membranes and other matrices. However, shedding of adherons cannot be a major contributor to matrix assembly, since it is clear in several cell types that heparan sulfate proteoglycans of matrix have core proteins distinct from those found on the cell surface (Dziadek et al., 1985; Keller and Furthmayr, 1986; Jalkanen et al., 1988). Secreted basement membrane proteins, which self-assemble in solution, form large coaggregates called "matrisomes" which might also resemble matrix granules (Kleinman et al., 1986).

As a specialized cell-associated compartment of matrix, adherons play a major role in cell–substratum and cell–cell adhesion (Schubert, 1984); in turn, proteoglycans are critical in the structural integrity of adherons (Schubert and LaCorbiere, 1980b; Schubert et al., 1983a). Proteoglycans also have a bewildering array of potential interactions with other components of cells and adherons. Heparan sulfate chains, present on cells or within adherons, may bind to N-CAM (within adherons or on cells, respectively) (Cole and Glaser, 1986); to other heparan sulfate chains (Fransson et al., 1981); to a variety of heparin-binding growth-, survival-, and neurite-promoting factors, including purpurin (Schubert et al., 1986), FGF (Gospodarowicz and Neufeld, 1987), and nexin I (Guenther et al., 1985; Gloor et al., 1986); and to matrix components such as fibronectin and laminin (see Chapter 13). The heparan sulfate proteoglycan BM-1, found in glial conditioned medium, also appears to have direct neurite-promoting effects (Hantaz-Ambroise et al., 1987). It is evident that control at the levels of biosynthesis or postsynthetic modification of proteoglycans could, in principle, regulate many of the functions attributed to adherons. At present, however, no details are known of the biosynthesis or fate of these proteoglycans, nor of adherons as a whole.

3.3. Cell-Associated Matrix Molecules

Though adherons are associated with, and derived from, cell surfaces, it is not clear whether they are a major means by which cells bear cell-surface molecules. In any event, evidence suggests that neuronal migration and axonal growth are guided by direct contact with the surfaces of radial glia and neuroepithelial endfeet. Even where basal laminae and overt channels are absent, many cell surfaces express matrix components such as laminin in a punctate pattern which correlates spatiotemporally with the active growth of long tract axons in a variety of both developing and regenerating CNS systems (reviewed in Riggott and Moody, 1987; see also Chapter 13).

A punctate pattern of laminin staining also occurs in other tissues during morphogenesis, including peripheral nerve ganglia and neurite outgrowth pathways (Rogers *et al.*, 1986; Riggott and Moody, 1987), early kidney (Ekblom *et al.*, 1980), and preimplantation mouse embryo, where laminin and matrix heparan sulfate proteoglycan BM-1 are known to be expressed in the absence of both entactin and collagen IV at preblastocyst stages (Leivo *et al.*, 1980; Wu *et al.*, 1983; Dziadek and Timpl, 1985). Moreover, this laminin is structurally different from the form found in basement membranes, in that only B chains are expressed initially (Cooper and MacQueen, 1983). While punctate deposits might often represent primitive, early stages in the assembly of organized matrices, it is likely in some cases (including brain) that punctate distributions of cell-surface matrix components such as laminin serve a biological role distinct from that of the same components found in channels and basal laminae. To test this notion critically, it will be important to elucidate how punctate laminin is bound to glial cell surfaces, what its structure is, and what other matrix proteins and proteoglycans are bound to it. Low concentrations of laminin may interact selectively with neurites by a high-affinity mechanism, serving an informational role rather than acting as an adhesive substratum or scaffold for growth *per se*. Such an informational role for laminin has been proposed by several workers based on cell culture experiments (Davis *et al.*, 1985; Adler *et al.*, 1985; Gundersen, 1987; Smalheiser and Schwartz, 1987; Smalheiser, 1989).

Neurobiologists have often cited parallels between the nervous system and the immune system, based on theoretical grounds as well as on direct molecular homologies between neural and immune molecules. However, workers have not yet begun to explore whether the *cellular* mechanisms of presenting molecules on neural cell surfaces might also be related to the antigen-presenting function of immune cells. The analogy should be considered carefully, since the same cells involved in major cell-surface guidance events in the CNS (astrocytes and Müller cells) can act as antigen-presenting cells in immune assays when they are induced to express Ia antigens (Hirsch *et al.*, 1983; Wekerle *et al.*, 1986). As in macrophages (Unanue, 1984), it is reasonable to expect that glial cells may also express the requisite cellular machinery for antigen processing during normal development, even when Ia antigen is not induced. Conversely, follicular dendritic cells in lymphoid tissues shed particulate antigen-containing complexes from their surfaces in a manner that to some extent resembles the shedding of adherons (Szakal *et al.*, 1988).

From the perspective of biosynthesis as well as function, the cellular analogy between neural and immune signaling suggests several new avenues for future work. For example, it raises the possibility that some neural information molecules need to be internalized and processed (in either the transmitting or receiving cell) to exert their effects. For example, laminin is internalized by receptor-mediated endocytosis in at least certain melanoma cell lines (Wewer *et al.*, 1987), and recently it has been claimed that laminin is present intracellularly within certain mature neurons (Yamamoto *et al.*, 1987). If internalization and proteolytic processing do occur within neural cells, this could imply novel intracellular roles not only for matrix proteins, but also for proteases and protease inhibitors in modulating neurite outgrowth (Patterson, 1985; Zurn et al., 1988). The immune system also provides an important precedent in which the addition of chondroitin sulfate chains to a minority of Ii proteins on the cell surface has a central role in the overall ability of Ia-bearing cells to present antigen effectively. The chondroitin sulfate chain is needed not for the initial binding of antigen to the cells, but most likely acts as a routing signal for antigen internalization and processing (Rosamond et al., 1987). In this context, note that the HNK-1 sulfated carbohydrate epitope is present on a minority of N-CAM and certain other neural "cell adhesion molecules" (see Chapter 11). These CAMs do not require the HNK-1 moiety to bind at least some ligands; HNK-1 may act as an additional or accessory ligand-binding site (Kunemund et al., 1987), or possibly as a routing signal.

It is also fascinating to compare the biological role of cell-bound complexes of matrix molecules with that of cell-bound complexes of immune molecules:

1. In the immune system, several different proteins must be co-recognized as a complex in order to present antigen to the recipient cell (Grey and Chesnut, 1985). Similarly, in fibroblasts and neural cells, active adhesive responses to fibronectin also require co-recognition of several interacting matrix domains and several cell-surface proteins, including a heparan sulfate proteoglycan (Lark et al., 1985; Woods et al., 1986; Waite et al., 1987).

2. In the immune system, certain lymphocytic cell types are preferentially stimulated by native conformational epitopes on antigens, while others respond better to epitopes on denatured or processed, degraded antigen (Unanue, 1984). Again, there is a possible parallel in the nervous system, for a high-affinity neurite-promoting domain of laminin is conformationally sensitive, while a second low-affinity neurite-promoting domain becomes functionally exposed only when laminin is cleaved or denatured (Edgar et al., 1983).

3. Complexes of immune molecules involve both polymorphic and nonpolymorphic moieties, and antigen is only recognized in the context of the proper MHC antigen. It is not clear whether any analogous molecular "restriction" event is required for cell–cell recognition in the nervous system. However, at least on a formal level, there is a parallel in that certain trophic agents exert neurite-promoting and survival effects upon neural cells much more effectively when the cells are cultured on certain substrata, e.g., on laminin (Baron-Van Evercooren et al., 1982; Edgar et al., 1983; Pixley and Cotman, 1986). Substratum-bound cues can thus be said to gate or "restrict" the ability of soluble cues to act effectively upon developing neurons (Smalheiser and Schwartz, 1987).

3.4. Mesenchymelike Matrices

Immature cartilage, muscle, and other mesenchymal tissues contain loosely organized matrices rich in hyaluronic acid, a variety of proteoglycans (especially hyaluronic acid-binding chondroitin sulfate proteoglycans), and variable amounts of proteins such as fibronectin, J1/cytotactin/tenascin, thrombospondin, and hyaluronectin. Such matrix components have also been shown conclusively to exist in the developing CNS. A local matrix is associated transiently with cortical subplate neurons (Stewart and Pearlman, 1987; Chun and Shatz, 1988); in adult brain, matrices are also found locally at nodes of Ranvier (Delpech et al., 1982; Rieger et al., 1986) and in subependymal networks (Torack and Grawe, 1980). However, individual components such as hyaluronic acid, cytotactin/tenascin, or specific proteoglycans are distributed rather extensively in developing brain, even in regions that have no morphologically apparent matrices as assessed by routine light or electron microscopy (Aquino et al., 1984; Crossin et al., 1986; Margolis et al., 1986; Ripellino et al., 1988). The patterns observed suggest that extracellular spaces in developing brain contain a family of regional-specific specialized matrices of the mesenchymal type. Based on cell culture work, it is likely that astrocytes and oligodendrocytes both synthesize components within these matrices (e.g., Dorfman and Ho, 1970; Gallo et al., 1987).

Three distinct chondroitin sulfate proteoglycans are currently known to be associated with mesenchymelike matrices of brain: the large cartilage- or muscle-type proteoglycan (Norling et al., 1984), which cross-reacts with the S103L core protein epitope seen in the cartilage and muscle proteoglycans (N. Schwartz and A. Caplan, unpublished); a proteoglycan of similar size and general properties which bears the HNK-1 sulfated carbohydrate epitope but not the S103L epitope (Hoffman and Edelman, 1987; N. Schwartz, unpublished); and a smaller proteoglycan described by Margolis and co-workers (see Chapter 3). All three appear to complex with hyaluronic acid, and it is likely that at least the first two can bind to cytotactin/tenascin (Chiquet and Fambrough, 1984; Hoffman and Edelman, 1987). It is presently unclear whether their core proteins are related, nor is it known whether these proteoglycans exhibit region- or stage-specific modifications. Progress in this area should be rapid in the next few years, since detailed structural knowledge is already in hand concerning the peptides, carbohydrates, and cDNA sequences of the homologous chondroitin sulfate proteoglycan of cartilage (see Section 1).

3.5. Intracellular Proteoglycans

Using histochemical methods, hyaluronic acid and glycosaminoglycans have long been localized to the cytoplasm of neurons and glial cells in the adult brain (e.g., Castejón, 1970). However, this localization has only recently been confirmed with more specific immunocytochemical probes (Aquino et al., 1984), forcing neurobiologists to confront this puzzling finding. It is not clear whether hyaluronic acid and proteoglycans are synthesized by neurons (at either fetal or adult stages), or whether these are internalized from embryonic extracellular spaces and stored. The turnover and fate of these glycosaminoglycans are unknown, as is their biological role, if any (see Chapter 3).

In many types of cells, exogenous heparan sulfate inhibits cell proliferation. A number of possible mechanisms have been invoked to explain this effect; for example, the binding of heparan sulfate to cell-surface thrombospondin or other heparin-binding growth factors (Majack et al., 1986). However, recent work by Conrad and co-workers has revealed that both exogenous and newly synthesized heparan sulfate can be actively taken up by cells and transported to the nucleus of hepatocytes (Ishihara *et al.*, 1986). As understood at present, a novel intracellular pathway appears to exist: Cells synthesize a specific heparan sulfate proteoglycan, which becomes linked to the plasma membrane via a phosphatidylinositol-lipid tail; this proteoglycan is cleaved from the cells, becomes bound to another cell-surface receptor, and is internalized (Ishihara *et al.*, 1987), upon which it loses its protein core (in a nonlysosomal compartment) and enters the nucleus as free glycosaminoglycan chains (Ishihara *et al.*, 1986). Nuclear heparan sulfate chains have a markedly unusual composition, being highly modified and rich in iduronic acid 2-*O*-sulfate (Fedarko and Conrad, 1986). Conrad and coworkers have demonstrated that the amount of nuclear heparan sulfate correlates inversely with the log growth rate of these cells under a variety of physiological conditions (Fedarko and Conrad, 1986; Ishihara *et al.*, 1987). Such an elaborate pathway for growth control would presumably allow a cell to sense conditions occurring at its own extracellular matrix. Moreover, it is exciting to consider that specific glycosaminoglycan sequences within these unusual heparan sulfate chains might be used as recognition or routing signals as part of this pathway; these specific sequences might also function as negative growth factors, interacting with specific protein or DNA sequences within the nucleus. Before these hypotheses can be taken too seriously, however, it will be necessary to have a more complete description of the path and mechanism by which newly synthesized heparan sulfate travels to the nucleus, and its interactions and fate within the nucleus.

4. BIOSYNTHETIC ALTERATIONS AS AN EXPERIMENTAL TOOL

4.1. Biosynthetic Mutants in Whole Organisms

Several mouse mutants have been proven or postulated to affect the biosynthesis of proteoglycans. The cmd mouse is a recessive lethal in which the major large core protein of cartilage is not expressed, leading to short limbs and poor tissue integrity (Kimata *et al.*, 1981; Brennan *et al.*, 1983); death occurs with the first breath after birth, as the tracheal cartilage collapses. The bm mouse exhibits undersulfated proteoglycans due to a defect in synthesis of PAPS which is the rate-limiting step in sulfation (Sugahara and Schwartz, 1979, 1982a). Because of the partial nature of the defect, tissues such as cartilage which have very high sulfation requirements are more severely affected than most other tissues (Sugahara and Schwartz, 1982b). Histology of bm brains appear normal, albeit Alcian blue staining of tissue sections is reduced. The ch mouse is named for its congenital hydrocephalus, exhibiting gross morphogenetic abnormalities in many different tissues including brain. It has been proposed that the underlying defect is one of reduced biosynthesis of matrix proteoglycans (Richardson, 1985; Melvin and Schwartz, 1988). In none of these mutants have the brains been examined carefully for consequences for neural development and function, at the level of behavior, pathology, or biochemistry.

Several mutations in man also affect both brain and connective tissue, and might possibly involve abnormalities in matrix components: In both tuberous sclerosis and certain types of dwarfism such as thanatophoric dysplasia (Ho *et al.*, 1984), neuronal migration occurs aberrantly, with ectopic nodules being frequent. Lysosomal storage diseases comprise a family of mutations in which one or another of the lysosomal catabolic enzymes are defective (see Chapter 15). While it remains a puzzle how the initial lysosomal defect leads to the overall dysfunction of brain cells, there is a good correlation between the nature of the storage product and the presence of CNS deterioration; interestingly, only those syndromes in which heparan sulfate accumulates are associated with primary mental retardation.

The *per* locus of *Drosophila* was first identified through isolating behavioral mutant flies, which had abnormally short or long biological rhythms (reviewed in Hall and Rosbash, 1988). Rather surprisingly, study of the cloned *per* locus revealed that the gene product was predicted to be a proteoglycan, expressed both in brain and in certain other tissues (Jackson *et al.*, 1986; Reddy *et al.*, 1986). This conclusion has been verified by examining the *per* protein directly (Reddy *et al.*, 1986; Bargiello *et al.*, 1987). Using an antibody raised to a synthetic peptide (corresponding to a portion of the predicted C-terminal region of the *per* protein), it appears tentatively to be expressed on the cell surface or extracellular matrix (Bargiello *et al.*, 1987). This proteoglycan has been proposed to regulate the permeability of gap junctions (Bargiello *et al.*, 1987), which in turn appears to modulate the rate of oscillations within pattern-generating networks of cells (Baylies *et al.*, 1987; Liu *et al.*, 1988). Homologues of the *per* gene have also been identified in mammalian species (Shin *et al.*, 1985), but nothing is known of their structure or function. The *per* locus has generated a renaissance of interest in proteoglycans among neurobiologists; this should be one of the most exciting areas to be explored in the next few years, aided by the availability of molecular genetic techniques.

4.2. Biosynthetic Mutants in Cultured Cells

Recently, Esko and co-workers described a general method for generating and selecting mutant CHO cells which are deficient in sulfate incorporation (Esko *et al.*, 1985). Some of these mutants simply cannot take up sulfate (Esko *et al.*, 1986), while more interesting mutants are deficient in incorporating galactose as well as sulfate and may have deficient synthesis of glycosaminoglycan chains. In particular, mutants have been found which are deficient in xylosyltransferase or in galactosyltransferase I (Esko *et al.*, 1985). This genetic approach has provided evidence that both chrondroitin sulfate and heparan sulfate may share the same xylosyltransferase enzyme in their synthesis, and that cells can utilize other glycosyltransferases besides galactosyltransferase I in adding the first galactose to xylosylated core proteins (Esko *et al.*, 1985, 1987).

The genetic approach shows promise for analyzing the roles of proteoglycans in nervous tissue as well. The isolation of mutant muscle cells deficient in both sulfate and galactose incorporation was recently reported, and one of the mutants was shown to have lost clustering of its acetylcholine receptors (Gordon *et al.*, 1987). This finding is congruent with other evidence that heparan sulfate proteoglycans are critical at the neuromuscular junction for local matrix assembly, which in turn directs the organization and clustering of the postsynaptic membrane (reviewed in Anderson and Swenarchuk, 1987). In another study, the distribution of acetylcholinesterase (AChE) was examined within phenotypic variants of PC12 cells, which were selected because they fail to express a normal cell surface heparan sulfate proteoglycan (Inestrosa *et al.*, 1985). Variant cells expressed increased amounts of the high-molecular-weight, collagen-tailed form of AChE but most of this was aberrantly distributed, being within intracellular compartments instead of at the cell surface. This suggests that heparan sulfate proteoglycan might not only bind AChE extracellularly, but may also be involved in the initial assembly or subsequent secretion of the high-molecular-weight form of AChE (Inestrosa *et al.*, 1985).

4.3. β-Xylosides

A popular experimental method for examining the contribution of proteoglycans to a biological process is to treat cells with exogenous β-xylosides. These agents competitively inhibit the addition of glycosaminoglycan chains to xylosylated core proteins, resulting in carbohydrate-deficient proteoglycans. In addition, β-xylosides initiate the elongation of free chondroitin sulfate chains (and apparently no other class of chains) in the absence of core protein (Galligani *et al.*, 1975; Schwartz, 1977, 1979). Moreover, at least in principle, these free chains may compete with nascent proteoglycans for access to modifying enzymes, resulting not only in fewer glycosaminoglycan chains per core protein, but also in undermodified glycosaminoglycan chains upon the core protein (Schwartz, 1979).

β-Xylosides have been employed by neurobiologists in a variety of experimental systems (e.g., Ratner *et al.*, 1985; Inestrosa *et al.*, 1985; Carey *et al.*, 1987). However, it can be difficult to pinpoint the underlying molecular basis of the β-xyloside effect in any given case, since: (1) a single cell may be induced to show abnormalities in several types of proteoglycans; (2) both the number and the degree of modifications of glycosaminoglycan chains are altered; (3) the assembly, routing, secretion, or turnover of core proteins (and other proteins as well) may become altered; (4) a discrete change in a single matrix proteoglycan may cause secondary effects upon other matrix proteins, resulting in gross changes in the integrity of the entire extracellular matrix; (5) secretion of free chondroitin sulfate chains adds another complicating factor, which can alter the pH or osmolarity of the extracellular medium, or affect cell constituents by virtue of its highly negative charge; (6) gross ultrastructural changes in the Golgi apparatus can also be produced in cells by treatment with xylosides (Kanwar *et al.*, 1986). The results of β-xyloside treatment are thus most persuasive when combined with independent evidence (e.g., enzymatic treatments or genetic techniques) implicating a specific proteoglycan in the cell type under study.

5. SPECIFIC GLYCOSAMINOGLYCAN CHAIN MODIFICATIONS

All glycosaminoglycan chains are polymers consisting of simple disaccharide repeats, but this monotonous backbone can be modified in a large number of ways.

Heparan sulfate chains are particularly heterogeneous, varying in sites of N-deacetylation, N-deamidation, N-sulfation, O-sulfation, and epimerization of glucuronic acid to iduronic acid (Figure 2). This might suggest that glycosaminoglycan sequences constitute a specific informational code, in line with the variable/constant structure of other informational molecules such as DNA and immunoglobulins. It is natural to wonder if a glycosaminoglycan-based recognition code could operate in the brain.

For glycosaminoglycan sequences to form an informational code, four criteria must be satisfied: (1) Specific sequences must be shown to exist. (2) These sequences must be created in a manner that is tightly regulated by cells, and not simply generated at random. (3) These sequences must be shown to recognize specific ligands. (4) The different possible patterns of glycosaminoglycan sequences must form a *language;* that is, there must be a significant number of different patterns created, each with a distinct biological meaning. The first three criteria have been satisfied for heparin sequences binding to antithrombin III. This binding event underlies the anticoagulant activity of heparin; by binding antithrombin III, heparin greatly increases the rate with which antithrombin III binds thrombin, in turn inhibiting the clotting of blood (Rosenberg, 1985). As elucidated by Lindahl, Rosenberg and their co-workers (reviewed in Rosenberg, 1985; Lindahl *et al.*, 1986), heparin binds antithrombin III via a specific pentasaccharide sequence. This pentasaccharide structure occurs very infrequently within heparin chains, probably being restricted at or near the nonreducing (free) ends (Radoff and Danishefsky, 1984). The sequence is synthesized in the Golgi apparatus by the partial modification of completed heparin chains, in a series of steps occurring in a fixed order (Figure 2). Each step is a prerequisite for the following modification, suggesting that the partially modified sugar sequences are themselves recognition signals for sequential passage along the multienzyme pathway. A final 3-O-sulfation step is unique to the pentasaccharide, and is not known to occur elsewhere within heparin (with the exception of one recent report, which also found 3-O-sulfate within heparin sequences that possess growth-inhibitory effects on cells (Castellot *et al.*, 1986). It is unclear what role, if any, is played by the protein core in determining which glycosaminoglycan chains or sites become modified; certain (but not all) heparan sulfate proteoglycans also synthesize chains which contain the pentasaccharide sequence, as inferred by their ability to bind antithrombin III (Pejler *et al.*, 1987).

When heparan sulfate chains are compared among different cell types or among different cellular compartments, they often vary greatly in their degree of modification, and according to whether the modifications occur in clusters versus uniformly along the chains. These differences may correlate with the presence of different core proteins, routing of certain chains to different cellular compartments, or different states of cellular growth (e.g., Keller *et al.*, 1980; Hampson *et al.*, 1983; Fedarko and Conrad, 1986; Margolis *et al.*, 1987). In many cases, the variable extent of modifications observed could simply reflect different levels of activity in N-deacetylase, which is the first and obligatory step preceding all further modifications (see Section 1). Therefore, despite these intriguing correlative data, there is no compelling evidence that the heterogeneity of heparan sulfate chains reflects the presence of unique or specific glycosaminoglycan sequences.

The daunting technical difficulties encountered presently in the direct sequencing of glycosaminoglycan chains have prevented workers from obtaining the critical data—the sequences themselves. Other approaches which may be fruitful for detecting the possible presence and bioactivity of specific glycosaminoglycan sequences include the chemical synthesis of specific modified oligosaccharides (Choay et al., 1983); the raising of monoclonal antibodies against glycosaminoglycan sequences; and the study of heparan sulfate/heparan binding sequences which bind to ligands other than anti-thrombin III. For example, the neurite-promoting protease nexin I is closely homologous to antithrombin III (Gloor et al., 1986), and it will be interesting to learn if heparin binds nexin I in a way that is subtly different from that of antithrombin III. Other neural heparin binding ligands include laminin, fibronectin, purpurin and other growth factors, N-CAM, and MAG (Fahrig et al., 1987). In no case is it known whether any specific heparan sulfate modifications are required for these binding events, beyond the observation that undersulfated variant heparan sulfate proteoglycans bind fibronectin poorly compared to chains of wild-type cells (Stamatoglou and Keller, 1983; Robinson et al., 1984).

6. CONCLUSION

In the past decade, the field of proteoglycan biosynthesis has undergone a marked shift in emphasis, from studies predominantly of the glycosaminoglycan moieties of proteoglycans, to the current focus on their core proteins. Investigators can now identify and follow the expression of individual, defined proteoglycan molecules in cartilage and basement membranes. However, the nervous system (particularly the CNS) remains a fertile, unexplored area for future inquiry. It is anticipated that the study of proteoglycans will become an exciting topic for molecular neurobiologists over the coming decade.

7. REFERENCES

Adler, R., Jerdan, J., and Hewitt, A. T., 1985, Responses of cultured neural retinal cells to substratum-bound laminin and other extracellular matrix molecules, Dev. Biol. 112:110–114.

Anderson, M. J., and Swenarchuk, L. E., 1987, Nerve induced remodeling of basal lamina during formation of the neuromuscular junction in cell culture, Prog. Brain Res. 71:409–421.

Appel, A., Horwitz, A. L., and Dorfman, A., 1979, Synthesis of hyaluronic acid in Marfan syndrome, J. Biol. Chem. 254:12199–12203.

Aquino, D., Margolis, R. U., and Margolis, R. K., 1984, Immunocytochemical localization of a chondroitin sulfate proteoglycan in nervous tissue. II. Studies in developing brain, J. Cell Biol. 99: 1130–1139.

Balasubramanian, A. S., and Bachhawat, B. K., 1964, Enzyme transfer of sulfate from 3'-phosphoadenosine 5'-phosphosulfate to mucopolysaccharides in rat brain, J. Neurochem. 11:877–881.

Balasubramanian, A. S., and Bacchawat, B. K., 1965, Formation of cerebroside sulfate from 3' phosphoadenosine 5'-phosphosulfate in sheep brain, Biochim. Biophys. Acta 106:218–220.

Balasubramanian, A. S., and Bacchawat, B. K., 1961, Formation of active sulfate in rat brain, *J. Sci. Ind. Res.* **20C:**202–204.

Balasubramanian, A. S., Joun, N. S., and Marx, W., 1968, Sulfation of *N*-desulfoheparin and heparan sulfate by a purified enzyme from mastocytoma, *Arch. Biochem. Biophys.* **128:**623–628.

Bargiello, T. A., Saez, L., Baylies, M. K., Gasic, G., Young, M. W., and Spray, D. C., 1987, The *Drosophila* clock gene *per* affects intercellular junctional communication, *Nature* **328:**686–691.

Baron-Van Evercooren, A., Kleinman, H., Ohno, S., Marangos, P., Schwartz, J., and Dubois-Dalcq, M., 1982, Nerve growth factor, laminin, and fibronectin promote neurite growth in human fetal sensory ganglia cultures, *J. Neurosci. Res.* **8:**179–193.

Baylies, M., Bargiello, T., Jackson, F., and Young, M., 1987, Changes in abundance or structure of the *per* gene product can alter periodicity of the *Drosophila* clock, *Nature* **326:**390–392.

Bird, T. A., Schwartz, N. B., and Peterkovsky, B., 1986, Mechanism for the decreased biosynthesis of cartilage proteoglycans in the scorbutic guinea pig, *J. Biol. Chem.* **261:**11166–11172.

Bourdon, M. A., Oldberg, A., Pierschbacher, M., and Ruoslahti, E., 1985, Molecular cloning and sequence analysis of a chondroitin sulfate proteoglycan cDNA, *Proc. Natl. Acad. Sci. USA* **82:**1321–1325.

Bourdon, M. A., Shiga, M., and Ruoslahti, E., 1986, Identification from cDNA of the precursor form of a chondroitin sulfate proteoglycan core protein, *J. Biol. Chem.* **261:**12534–12537.

Bourdon, M. A., Krusius, T., Campbell, S., Schwartz, N. B., and Ruoslahti, E., 1987, Identification and synthesis of a recognition signal for the attachment of glycosaminoglycans to proteins, *Proc. Natl. Acad. Sci. USA* **84:**3194–3198.

Braell, W. A., and Lodish, H. F., 1981, Biosynthesis of the erythrocyte anion transport protein, *J. Biol. Chem.* **256:**11337–11344.

Brandt, A. E., Distler, J., and Jourdian, G. W., 1969, Biosynthesis of the chondroitin sulfate protein linkage region: Purification and properties of a glucuronosyltransferase from embryonic chick brain, *Proc. Natl. Acad. Sci. USA* **64:**374–378.

Brandt, A. E., Distler, J. J., and Jourdian, G. W., 1975, Biosynthesis of chondroitin sulfate proteoglycan, *J. Biol. Chem.* **250:**3996–4006.

Brennan, M. J., Oldberg, A., Ruoslahti, E., Brown, K., and Schwartz, N. B., 1983, Immunologic evidence for two distinct chondroitin sulfate proteoglycan core proteins: Differential expression in cmd mice, *Dev. Biol.* **98:**139–147.

Brennan, M. J., Oldberg, A., Pierschbacher, M. D., and Ruoslahti, E., 1984, Chondroitin/dermatan sulfate proteoglycan in human fetal membranes, *J. Biol. Chem.* **259:**13742–13750.

Bunge, R. P., and Bunge, M. B., 1983, Interrelationship between Schwann cell function and extracellular matrix production, *Trends Neurosci.* **6:**499–503.

Burkart, T., and Weismann, U. N., 1987, Sulfated glycosaminoglycans (GAG) in the developing mouse brain, *Dev. Biol.* **120:**447–456.

Burkart, T., Hofmann, K., Siegrist, H. P., and Herschkowitz, N. N., 1981, Quantitative measurement of in vivo sulfatide metabolism during development of the mouse brain: Evidence for a large rapid degradatable sulfatide pool, *Dev. Biol.* **83:**42–48.

Burkart, T., Caimi, L., Herschkowitz, N. N., and Weismann, U. N., 1983, Metabolism of sulfogalactosyl glycerolipids in the myelinating mouse brain, *Dev. Biol.* **98:**182–186.

Burnell, J. N., and Roy, A. B., 1978, Purification and properties of the ATP sulfurylase of rat liver, *Biochim. Biophys. Acta* **527:**239–248.

Campbell, S. C., and Schwartz, N. B., 1988, Kinetics of intracellular processing of chondroitin sulfate proteoglycan core protein and other matrix components, *J. Cell Biol.* **106:**2191–2202.

Carey, D. J., Rafferty, C., and Todd, M., 1987, Effects of inhibition of proteoglycan synthesis on the differentiation of cultured rat Schwann cells, *J. Cell Biol.* **105:**1013–1021.

Castejón, H. V., 1970, Histochemical demonstration of acid glycosaminoglycans in the nerve cell cytoplasm of mouse central nervous system, *Acta Histochem.* **35:**161–172.

Castellot, J. J., Jr., Choay, J., Lormeau, J.-C., Petitou, M., Sache, E., and Karnovsky, M., 1986, Structural determinants of the capacity of heparin to inhibit the proliferation of vascular smooth muscle cells. II. Evidence for a pentasaccharide sequence that contains a 3-O-sulfate group, *J. Cell Biol.* **102:**1979–1984.

Chiquet, M., and Fambrough, D., 1984, Chick myotendinous antigen. II. A novel extracellular glycoprotein complex consisting of large disulfide-linked subunits, *J. Cell Biol.* **98:**1937–1946.

Choay, J., Petitou, M., Lormeau, J., Sinay, P., Casu, B., and Gatti, G., 1983, Structure–activity relationship in heparin: A synthetic pentasaccharide with high affinity for antithrombin III and eliciting high anti-factor Xa activity, *Biochem. Biophys. Res. Commun.* **116:**492–499.

Chopra, R. K., Pearson, C. H., Pringle, G. A., Fackre, D. S., and Scott, P. G., 1985, Dermatan sulfate is located on serine-4 of bovine proteodermatan sulfate, *Biochem. J.* **232:**277–279.

Chun, J. J. M., and Shatz, C. J., 1988, A fibronectin-like molecule is present in the developing cat cerebral cortex and is correlated with subplate neurons, *J. Cell Biol.* **106:**857–872.

Cole, G., and Glaser, L., 1986, A heparin-binding domain from N-CAM is involved in neural cell-substratum adhesion, *J. Cell Biol.* **102:**403–412.

Cooper, A. R., and MacQueen, H. A., 1983, Subunits of laminin are differentially synthesized in mouse eggs and early embryos, *Dev. Biol.* **96:**467–471.

Crossin, K., Hoffmann, B., Grumet, S., Thiery, J.-P., and Edelman, G., 1986, Site-restricted expression of cytotactin during development of the chicken embryo, *J. Cell Biol.* **102:**1917–1930.

Davis, G., Varon, S., Engvall, E., and Manthorpe, M., 1985, Substratum-binding neurite-promoting factors: Relationships to laminin, *Trends Neurosci.* **8:**528–532.

Day, A., Ramis, C., Fisher, I., Gehron-Robey, P., Termine, J., and Young, M., 1986, Characterization of bone PGII cDNA and its relationship to PGII mRNA from other connective tissues, *Nucleic Acids Res.* **14:**9861–9876.

Day, A. A., McQuillan, C. I., Termine, J. O., and Young, M. R., 1987, Molecular cloning and sequence analysis of the cDNA for small proteoglycan II of bovine bone, *Biochem. J.* **248:**801–805.

De, K. K., Yamamoto, K., and Whistler, R. L., 1978, Enzymatic formation and hydrolysis of polysaccharide sulfates, *ACS Symp. Ser.* **77:**121–147.

Delpech, A., Girard, N., and Delpech, B., 1982, Location of hyaluronectin in the nervous system, *Brain Res.* **245:**251–257.

Doege, K., Fernandez, P., Hassell, J., Sasaki, M., and Yamada, Y., 1986, Partial cDNA sequence encoding a globular domain at the C terminus of the rat cartilage proteoglycan, *J. Biol. Chem.* **261:** 8108–8111.

Doege, K., Sasaki, M., Horigan, E., Hassell, J., and Yamada, Y., 1987, Complete primary structure of the rat cartilage proteoglycan core protein deduced from cDNA clones, *J. Biol. Chem.* **262:**17757–17767.

Dorfman, A., and Ho, P.-L., 1970, Synthesis of acid mucopolysaccharides by glial tumor cells in tissue culture, *Proc. Natl. Acad. Sci. USA* **66:**495–499.

Dziadek, M., and Timpl, R., 1985, Expression of nidogen and laminin in basement membranes during mouse embryogenesis and in teratocarcinoma cells, *Dev. Biol.* **111:**372–382.

Dziadek, M., Fujiwara, S., Paulsson, M., and Timpl, R., 1985, Immunological characterization of basement membrane types of heparan sulfate proteoglycan, *EMBO J.* **4:**905–912.

Easter, S., Bratton, B., and Scherer, S., 1984, Growth related order of the retinal fiber layer in goldfish, *J. Neurosci.* **4:**2173–2190.

Edgar, D., Timpl, R., and Thoenen, H., 1983, The heparin-binding domain of laminin is responsible for its effects on neurite outgrowth and neuronal survival, *EMBO J.* **3:**1463–1468.

Ekblom, P., Alitalo, K., Vaheri, A., Timpl, R., and Saxen, L., 1980, Induction of a basement membrane glycoprotein in embryonic kidney: Possible role of laminin in morphogenesis, *Proc. Natl. Acad. Sci. USA* **77:**485–489.

Esko, J. D., Stewart, T., and Taylor, W., 1985, Animal cell mutants defective in glycosaminoglycan biosynthesis, *Proc. Natl. Acad. Sci. USA* **82:**3197–3201.

Esko, J., Elgavish, A., Prasthofer, T., Taylor, W., and Weinke, J., 1986, Sulfate transport-deficient mutants of CHO cells, *J. Biol. Chem.* **261:**15725–15733.

Esko, J. D., Weinke, J., Taylor, W., Ekborg, G., Roden, L., Anantharamaiah, G., and Gawish, A., 1987, Inhibition of chondroitin and heparan sulfate biosynthesis in Chinese hamster ovary cell mutants defective in galactosyltransferase I, *J. Biol. Chem.* **262:**12189–12195.

Fahrig, T., Landa, C., Pesheva, P., Kuhn, K., and Schachner, M., 1987, Characterization of binding properties of the myelin-associated glycoprotein to extracellular matrix constituents, *EMBO J.* **6:**2875–2883.

Fedarko, N., and Conrad, H., 1986, A unique heparan sulfate in the nuclei of hepatocytes: Structural changes with the growth state of the cells, *J. Cell Biol.* **102:**587–599.

Fellini, S. A., Kimura, J. H., and Hascall, V. C., 1984, Localization of proteoglycan core protein in subcellular fractions isolated from rat chondrosarcoma chondrocytes, *J. Biol. Chem.* **259:**4634–4641.

Fransson, L.-Å., Havsmark, B., and Sheehan, J., 1981, Self-association of heparan sulfate, *J. Biol. Chem.* **256:**13039–13043.

Galligani, L., Hopwood, J., Schwartz, N. B., and Dorfman, A., 1975, Stimulation of synthesis of free chondroitin sulfate chains by β-D-xylosides in cultured cells, *J. Biol. Chem.* **250:**5400–5406.

Gallo, V., Bertolotto, A., and Levi, G., 1987, The proteoglycan chondroitin sulfate is present in a subpopulation of cultured astrocytes and in their precursors, *Dev. Biol.* **123:**282–285.

Geetha-Habib, M., Campbell, S., and Schwartz, N. B., 1984, Subcellular localization of the synthesis and glycosylation of chondroitin sulfate proteoglycan core protein, *J. Biol. Chem.* **259:**7300–7310.

Geller, D., Henry, J., Belch, J., and Schwartz, N. B., 1987, Co-purification and characterization of ATP-sulfurylase and adenosine-5'-phosphosulfate kinase from rat chondrosarcoma, *J. Biol. Chem.* **262:**7374–7382.

Glaser, L., and Brown, D. H., 1955, The enzymatic synthesis in vitro of hyaluronic acid chains, *Proc. Natl. Acad. Sci. USA* **41:**253–260.

Gloor, S., Odink, K., Guenther, J., Nick, H., and Monard, D., 1986, A glia-derived neurite promoting factor with protease inhibitory activity belongs to the protease nexins, *Cell* **47:**687–693.

Glössl, J., Beck, M., and Kresse, H., 1984, Biosynthesis of proteodermatan sulfate in cultured human fibroblasts, *J. Biol. Chem.* **259:**14144–14150.

Gordon, H., Sample, S., and Hall, Z., 1987, Genetic variants of the C2 muscle cell line defective in glycosaminoglycan biosynthesis, *J. Cell Biol.* **105:**199a.

Gospodarowicz, D., and Neufeld, G., 1987, Fibroblast growth factor: Molecular and biological properties, in: *Mesenchymal–Epithelial Interactions in Neural Development* (J. K. Wolff, J. Sievers, and M. Berry, eds.), pp. 191–222, Springer-Verlag, Berlin.

Gregory, J. D., and Lipmann, F., 1957, The transfer of sulfate among phenolic compounds with 3',5'-diphosphoadenosine as coenzyme, *J. Biol. Chem.* **229:**1081–1089.

Grey, H. M., and Chesnut, R., 1985, Antigen processing and presentation to T cells, *Immunol. Today* **6:**101–106.

Guenther, J., Nick, H., and Monard, D., 1985, A glia-derived neurite promoting factor with protease inhibitory activity, *EMBO J.* **4:**1963–1966.

Guha, A., Northover, B. J., and Bachhawat, B. K., 1960, Incorporation of radioactive sulfate into chondroitin sulfate in the developing brain of rats, *J. Sci. Ind. Res.* **C19:**287–289.

Gundersen, R., 1987, Response of sensory neurites and growth cones to patterned substrata of laminin and fibronectin in vitro, *Dev. Biol.* **121:**423–431.

Halfter, W., Reckhaus, W., and Kroger, S., 1987, Nondirected axonal growth on basal lamina from avian embryonic neural retina, *J. Neurosci.* **7:**3712–3722.

Hall, J., and Rosbash, M., 1988, Mutations and molecules influencing biological rhythms, *Annu. Rev. Neurosci.* **11:**373–393.

Hampson, I. N., Kumar, S., and Gallagher, J., 1983, Differences in the distribution of O-sulphate groups of cell-surface and secreted heparan sulphate produced by human neuroblastoma cells in culture, *Biochim. Biophys. Acta* **763:**183–190.

Hantaz-Ambroise, D., Vigny, M., and Koenig, J., 1987, Heparan sulfate proteoglycan and laminin mediate two different types of neurite outgrowth, *J. Neurosci.* **7:**2293–2304.

Hart, C. W., 1978, Sulfotransferase levels in developing cornea, *J. Biol. Chem.* **253:**347–353.

Hascall, V. C., and Riolo, R. L., 1972, Characteristics of the protein–keratan sulfate core and of keratan sulfate prepared from bovine nasal cartilage proteoglycan, *J. Biol. Chem.* **247:**4529–4538.

Hassell, J. R., Kimura, J. H., and Hascall, V. C., 1986, Proteoglycan core families, *Annu. Rev. Biochem.* **55:**539–568.

Hedgecock, E., Culotti, J., Hall, D., and Stern, B., 1987, Genetics of cell and axon migrations in C. elegans, *Development* **100:**365–382.

Henkart, P., Humphreys, S., and Humphreys, T., 1973, Characterization of sponge aggregation factor; a unique proteoglycan complex, *Biochemistry* 12:3045–3052.

Hirsch, M. R., Wietzerbin, J., Pierres, M., and Goridis, C., 1983, Expression of Ia antigens by cultured astrocytes treated with gamma-interferon, *Neurosci. Lett.* 41:199–204.

Ho, K.-L., Chang, C.-H., Yang, S., and Chason, J., 1984, Neuropathologic findings in thanatophoric dysplasia, *Acta Neuropathol.* 63:218–228.

Hoffman, H-P., Schwartz, N. B., Rodén, L., Prockop, D., 1984, Localization of xylosyltransferase in the cisterna of the rough endoplasmic reticulum, *Connect. Tissue Res.* 12:151–163.

Hoffman, S., and Edelman, G., 1987, A proteoglycan with HNK-1 antigenic determinants is a neuron-associated ligand for cytotactin, *Proc. Natl. Acad. Sci. USA* 84:2523–2527.

Inestrosa, N. C., Matthew, W., Reiness, C., Hall, Z., and Reichardt, L., 1985, Atypical distribution of asymmetric acetylcholinesterase in mutant PC12 pheochromocytoma cells lacking a cell surface heparan sulfate proteoglycan, *J. Neurochem.* 45:86–94.

Ishihara, M., Fedarko, N., and Conrad, H., 1986, Transport of heparan sulfate into the nuclei of hepatocytes, *J. Biol. Chem.* 261:13575–13580.

Ishihara, M., Fedarko, N., and Conrad, H., 1987, Involvement of phosphatidylinositol and insulin in the coordinate regulation of proteoheparan sulfate metabolism and hepatocyte growth, *J. Biol. Chem.* 262:4708–4716.

Jackson, F. R., Bargiello, T. A., Yun, S., and Young, M. W., 1986, Product of *per* locus of *Drosophila* shares homology with proteoglycans, *Nature* 320:185–188.

Jacobsson, I., Backstrom, G., Höök, M., Lindahl, U., Feingold, D. S., Malmström, A., and Rodén, L., 1979, Biosynthesis of heparin: Assay and properties of the microsomal uronosyl C-5 epimerase, *J. Biol. Chem.* 254:2975–2979.

Jalkanen, M., Rapraeger, A., and Bernfield, M., 1988, Mouse mammary epithelial cells produced basement membrane and cell surface proteoglycans containing distinct core proteins, *J. Cell Biol.* 106:953–962.

Jansson, L., Höök, M., Wasteson, A., and Lindahl, U., 1975, Biosynthesis of heparin. V. Solubilization and partial characterization of N- and O-sulphotransferases, *Biochem. J.* 149:49–55.

Jourdian, G. W., 1979, Biosynthesis of glycosaminoglycans, in: *Complex Carbohydrates of Nervous Tissue* (R. U. Margolis and R. K. Margolis, eds.), pp. 103–126, Plenum Press, New York.

Kanwar, Y. S., Rosenzweig, L., and Jakubowski, M., 1986, Xylosylated-proteoglycan-induced Golgi alterations, *Proc. Natl. Acad. Sci. USA* 83:6499–6503.

Keller, K. L., Keller, J., and Moy, J., 1980, Heparan sulfates from Swiss mouse 3T3 and SV3T3 cells: O-Sulfate difference, *Biochemistry* 19:2529–2536.

Keller, R., and Furthmayr, H., 1986, Isolation and characterization of basement membrane and cell proteoheparan sulphates from HR9 cells, *Eur. J. Biochem.* 161:707–714.

Kimata, K., Okayama, M., Oohira, A., and Suzuki, S., 1973, Cytodifferentiation and proteoglycan biosynthesis, *Mol. Cell. Biochem.* 1:211–228.

Kimata, K., Barrach, H., Brown, K. S., and Pennypacker, J. P., 1981, Absence of proteoglycan core protein in cartilage from the cmd/cmd (cartilage matrix deficiency) mouse, *J. Biol. Chem.* 256:6961–6968.

Kimura, J. H., Thomas, E. J., Hascall, V. C., Reiner, H., and Poole, A. R., 1981, Identification of core protein; an intermediate in proteoglycan biosynthesis in culture chondrocytes from the Swarm rat chondrosarcoma, *J. Biol. Chem.* 250:7890–7897.

Kimura, J. H., Lohmander, L. S., and Hascall, V. C., 1984, Studies on the biosynthesis of cartilage proteoglycan in a model system of cultured chondrocytes from the Swarm rat chondrosarcoma, *J. Cell Biochem.* 26:261–278.

Kleinman, H., McGarvey, M., Hassell, J., Star, V., Cannon, F., Laurie, G., and Martin, G., 1986, Basement membrane complexes with biological activity, *Biochemistry* 25:312–318.

Krayanek, S., 1980, Structure and orientation of extracellular matrix in developing chick optic tectum, *Anat. Rec.* 197:95–109.

Krayanek, S., and Goldberg, S., 1981, Oriented extracellular channels and axonal guidance in the embryonic chick retina, *Dev. Biol.* 84:41–50.

Krueger, R. C., and Schwartz, N. B., 1988, Investigation of a large chondroitin sulfate proteoglycan from embryonic chick brain, *4th Int. Congr. Cell Biol.*, accepted.

Krueger, R., Olson, C. A., and Schwartz, N. B., 1985, Deglycosylation of proteoglycan by hydrogen fluoride in pyridine, *Anal. Biochem.* **146:**232–237.

Krusius, T., and Ruoslahti, E., 1986, Primary structure for extracellular matrix proteoglycan core protein deduced from cloned cDNA, *Proc. Natl. Acad. Sci. USA* **83:**7683–7687.

Kunemund, V., Jungalwala, F. B., Fischer, G., Chou, D. K. H., Keilhauer, G., and Schachner, M., 1988, The L2/HNK-1 carbohydrate of neural cell adhesion molecules is involved in cell interactions, *J. Cell Biol.* **106:**213–223.

Lark, M., Laterra, J., and Culp, L., 1985, Close and focal contact adhesions of fibroblasts to a fibronectin-containing matrix, *Fed. Proc.* **44:**394–403.

Leivo, I., Vaheri, A., Timpl, R., and Wartiovaara, J., 1980, Appearance and distribution of collagens and laminin in the early mouse embryo, *Dev. Biol.* **76:**100–114.

Lindahl, U., Feingold, D., and Rodén, L., 1986, Biosynthesis of heparin, *Trends Biochem. Sci.* **11:**221–225.

Liu, X., Lorenz, L., Yu, Q., Hall, J. C., and Rosbash, M., 1988, Spatial and temporal expression of the *period* gene in *Drosophila melanogaster, Genes Dev.* **2:**228–238.

Majack, R. A., Cook, S., and Bornstein, P., 1986, Control of smooth muscle cell growth by components of the extracellular matrix: Autocrine role for thrombospondin, *Proc. Natl. Acad. Sci. USA* **83:**9050–9054.

Malmström, A., and Öberg, L., 1981, Biosynthesis of dermatan sulfate. Assay and properties of the uronosyl C-5 epimerase, *Biochem. J.* **201:**489–493.

Malmström, A., Fransson, L. Å., Höök, M., and Lindahl, U., 1975, Biosynthesis of dermatan sulfate. I. Formation of L-iduronic acid residues, *J. Biol. Chem.* **250:**3419–3425.

Malmström, A., Rodén, L., Feingold, D., Jacobsson, I., Backström, G., Höök, M., and Lindahl, U., 1980, Biosynthesis of heparin. Partial purification of the uronosyl C-5 epimerase, *J. Biol. Chem.* **255:**3878–3883.

Margolis, R. K., and Margolis, R. U., 1979, Structure and distribution of glycoproteins and glycosaminoglycans, in: *Complex Carbohydrates of Nervous Tissue* (R. U. Margolis and R. K. Margolis, eds.), pp. 45–73, Plenum Press, New York.

Margolis, R. K., Salton, S., and Margolis, R., 1987, Effects of nerve growth factor-induced differentiation on the heparan sulfate of PC12 pheochromocytoma cells and comparison with developing brain, *Arch. Biochem. Biophys.* **257:**107–114.

Margolis, R. U., Aquino, D. A., Klinger, M. M., Ripellino, J. A., and Margolis, R. K., 1986, Structure and localization of nervous tissue proteoglycans, *Ann. N.Y. Acad. Sci.* **481:**46–54.

Markovitz, A., Cifonelli, J. A., and Dorfman, A., 1959, The biosynthesis of hyaluronic acid by group A streptococcus. VI. Biosynthesis from uridine nucleotides in cell-free extracts, *J. Biol. Chem.* **234:**2343–2350.

Melvin, T., and Schwartz, N. B., 1988, *Pathol. Immunopathol. Res.* **7:**68–72.

Mian, N., 1986, Characterization of a high Mw plasma membrane bound protein and assessment of its role as a constituent of hyaluronate synthase complex, *Biochem. J.* **237:**343–357.

Miller, R. R., and Waechter, C. J., 1988, Partial purification and characterization of detergent solubilized N-sulfotransferase activity associated with calf brain microsomes, *J. Neurochem.* **51:**87–94.

Mitchell, D., and Hardingham, T., 1981, The effects of cycloheximide on the biosynthesis and secretion of proteoglycans by chondrocytes in culture, *Biochem. J.* **196:**521–529.

Nakanishi, S., 1983, Extracellular matrix during laminar pattern formation of neocortex in normal and reeler mutant mice, *Dev. Biol.* **95:**305–316.

Neufeld, E. F., and Hall, C. W., 1965, Inhibition of UDP-D-glucose dehydrogenase by UDP-D-xylose: a possible regulatory mechanism, *Biochem. Biophys. Res. Commun.* **19:**456–460.

Ng, K., and Schwartz, N. B., 1987, Solubilization of hyaluronate synthetase activity from oligo-dendroglioma, *IXth International Conference on Glycoconjugates* B55.

Noonan, D. M., Horigan, E. A., Ledbetter, S. P., Vogeli, G., Sasaki, M., Yamada, Y., and Hassell, J. R., 1988, Identification of cDNA clones encoding different domains of the basement membrane heparan sulfate proteoglycan, *J. Biol. Chem.* **263:**16379–16387.

Norling, B., Glimelius, B., and Wasteson, Å., 1984, A chondroitin sulphate proteoglycan from human cultured glial and glioma cells, *Biochem. J.* **221:**845–853.

Nuweyhid, N., Glaser, J. H., Johnson, J. C., Conrad, H. E., Hauser, S. C., and Hirschberg, C. B., 1986, Xylosylation and glucuronsylation reactions in rat liver Golgi apparatus and endoplasmic reticulum, *J. Biol. Chem.* **261**:12936–12941.

Oegema, T. R., Kraft, E. L., Jourdian, G. W., and van Valen, T. R., 1984, Phosphorylation of chondroitin sulfate in proteoglycans from the Swarm rat chondrosarcoma, *J. Biol. Chem.* **259**:1720–1726.

Ögren, S., and Lindahl, U., 1975, Cleavage of macromolecular heparin by enzymes from mouse mastocytoma, *J. Biol. Chem.* **250**:2690–2695.

Oldberg, A., Hayman, E. G., and Ruoslahti, E., 1981, Isolation of a chondroitin sulfate proteoglycan from a rat yolk sac tumor and immunochemical demonstration of its cell surface localization, *J. Biol. Chem.* **256**:10847–10852.

Oldberg, A., Antonsson, P., and Heinegård, D., 1987, The partial amino acid sequence from cartilage proteoglycan, deduced from a cDNA clone, contains numerous Ser-Gly sequences arranged in homologous repeats, *Biochem. J.* **243**:255–259.

Patterson, P. H., 1985, On the role of proteases, their inhibitors and the extracellular matrix in promoting neurite outgrowth, *J. Physiol. (Paris)* **80**:207–211.

Pearson, C. H., Winterbottom, N., Fachre, D. S., Scott, P. G., and Carpenter, M. R., 1983, The NH_2-terminal amino acid sequence of bovine skin proteodermatan sulfate, *J. Biol. Chem.* **258**:15101-15104.

Pejler, G., Backström, G., and Lindahl, U., 1987, Structure and affinity for antithrombin of heparan sulfate chains derived from basement membrane proteoglycans, *J. Biol. Chem.* **262**:5036–5043.

Philipson, L. H., and Schwartz, N. B., 1984, Subcellular localization of hyaluronate synthetase in oligodendroglioma cells, *J. Biol. Chem.* **259**:5017–5023.

Philipson, L. H., Westley, J., and Schwartz, N. B., 1985, The effect of hyaluronidase treatment of intact cells on hyaluronate synthetase activity, *Biochemistry* **24**:7899–7906.

Pixley, S. K. R., and Cotman, C., 1986, Laminin supports short-term survival of rat septal neurons in low density, serum-free cultures, *J. Neurosci. Res.* **15**:1–17.

Prehm, P., 1983a, Synthesis of hyaluronate in differentiated teratocarcinoma cells. Characterization of the synthease, *Biochem. J.* **220**:191–198.

Prehm, P., 1983b, Synthesis of hyaluronate in differentiated teratocarcinoma cells. Mechanism of chain growth, *Biochem. J.* **211**:181–189.

Prehm, P., 1984, Hyaluronate is synthesized at plasma membranes, *Biochem. J.* **220**:597–600.

Prehm, P., and Mausolf, A., 1986, Isolation of streptococcal hyaluronate synthase, *Biochem. J.* **235**:887–889.

Radoff, S., and Danishefsky, I., 1984, Location on heparin of the oligosaccharide section essential for anticoagulant activity, *J. Biol. Chem.* **259**:166–172.

Ramóm y Cajal, S., 1984, *The Neuron and the Glial Cell*, p. 265, Thomas, Springfield, Ill.

Rapraeger, A., Jalkanen, M., Endo, E., Koda, J., and Bernfield, M., 1985, The cell surface proteoglycan from mouse mammary epithelial cells bears chondroitin sulfate and heparin sulfate glycosaminoglycans, *J. Biol. Chem.* **260**:11046–11052.

Ratner, N., Bunge, R., and Glaser, L., 1985, A neuronal cell surface heparan sulfate proteoglycan is required for dorsal root ganglion neuron stimulation of Schwann cell proliferation, *J. Cell Biol.* **101**:744–754.

Reddy, R., Zehring, W., Wheeler, D., Pirrotta, V., Hadfield, C., Hall, J., and Rosbash, M., 1986, Molecular analysis of the *period* locus in Drosophila melanogaster and identification of a transcript involved in biological rhythms, *Cell* **38**:701–710.

Richardson, R. R., 1985, Congenital genetic murine (ch) hydrocephalus: A structural model of cellular dysplasia and disorganization with the molecular locus of deficient proteoglycan synthesis, *Child's Nerv. Syst.* **1**:87–99.

Rieger, F., Daniloff, J., Pincon-Raymond, M., Crossin, K., Grumet, M., and Edelman, G., 1986, Neuronal cell adhesion molecules and cytotactin are colocalized at the node of Ranvier, *J. Cell Biol.* **103**:379–391.

Riesenfeld, J., Höök, M., and Lindahl, U., 1982, Biosynthesis of heparin. Concerted action of early polymer-modification reactions, *J. Biol. Chem.* **257**:421–425.

Riggott, M., and Moody, S., 1987, Distribution of laminin and fibronectin along peripheral trigeminal axon pathways in the developing chick, *J. Comp. Neurol.* **258**:580–596.

Ripellino, J. A., Bailo, M., Margolis, R. U., and Margolis, R. K., 1988, Light and electronmicroscopic studies on the localization of hyaluronic acid in developing rat cerebellum, *J. Cell Biol.* **106**:845–855.

Robinson, H. C., Horner, A. A., Höök, M., Ögren, S., and Lindahl, U., 1978, A proteoglycan form of heparin and its degradation to single-chain molecules, *J. Biol. Chem.* **253**:6687–6693.

Robinson, J., Viti, M., and Höök, M., 1984, Structure and properties of an under-sulfated heparan sulfate proteoglycan synthesized by a rat hepatoma cell line, *J. Cell Biol.* **98**:946–953.

Rodén, L., 1980, Structure and metabolism of connective tissue proteoglycans, in: *The Biochemistry of Glycoproteins and Proteoglycans* (W. Lennarz, ed.), pp. 267–314, Plenum Press, New York.

Rodén, L., Baker, J. R., Helting, T., Schwartz, N. B., Stoolmiller, A., Yamagata, S., and Yamagata, T., 1972, Biosynthesis of chondroitin sulfate, *Methods Enzymol.* **28**:638–676.

Rodén, L., Koerner, T., Olsen, C., and Schwartz, N. B., 1985, Mechanisms of chain initiation in the biosynthesis of connective tissue polysaccharides, *Fed. Proc.* **44**:373–389.

Rogers, S., Edson, K., Letourneau, P., and McLoon, S., 1986, Distribution of laminin in the developing peripheral nervous system of the chick, *Dev. Biol.* **113**:429–435.

Rosamond, S., Brown, L., Gomez, C., Braciale, T. J., and Schwartz, B. D., 1987, Xyloside inhibits synthesis of the class II-associated chondroitin sulfate proteoglycan and antigen presentation events, *J. Immunol.* **139**:1946–1951.

Rosenberg, R. D., 1985, Role of heparin and heparinlike molecules in thrombosis and atherosclerosis, *Fed. Proc.* **44**:404–409.

Sai, S., Tanaka, T., Kosher, R. A., and Tanzer, M. C., 1986, Cloning and sequence analysis of a partial cDNA for chicken cartilage proteoglycan core protein, *Proc. Natl. Acad. Sci. USA* **83**:5081–5085.

Sanes, J., 1983, Roles of extracellular matrix in neural development, *Annu. Rev. Physiol.* **45**:581–600.

Schubert, D., 1984, *Developmental Biology of Cultured Nerve, Muscle, and Glia,* Academic Press, New York.

Schubert, D., and LaCorbiere, M., 1980a, Altered collagen and glycosaminoglycan secretion by a skeletal muscle myoblast variant, *J. Biol. Chem.* **255**:11557–11563.

Schubert, D., and LaCorbiere, M., 1980b, Role of a 16S glycoprotein complex in cellular adhesion, *Proc. Natl. Acad. Sci. USA* **77**:4137–4141.

Schubert, D., and LaCorbiere, M., 1982a, The specificity of extracellular glycoprotein complexes in mediating cellular adhesion, *J. Neurosci.* **2**:82–89.

Schubert, D., and LaCorbiere, M., 1982b, Properties of extracellular adhesion-mediating particles in myoblast clone and its adhesion-deficient variant, *J. Cell Biol.* **94**:108–114.

Schubert, D., and LaCorbiere, M., 1985, Isolation of a cell-surface receptor for chick neural retina adherons, *J. Cell Biol.* **100**:56–63.

Schubert, D., LaCorbiere, M., Klier, F., and Birdwell, C., 1983a, A role for adherons in neural retina cell adhesion, *J. Cell Biol.* **96**:990–998.

Schubert, D., LaCorbiere, M., Klier, F., and Birdwell, C., 1983b, The structure and function of myoblast adherons, *Cold Spring Harbor Symp. Quant. Biol.* **48**:539–549.

Schubert, D., LaCorbiere, M., and Esch, F., 1986, A chick neural retina adhesion and survival molecule is a retinol-binding protein, *J. Cell Biol.* **102**:2295–2301.

Schwartz, N. B., 1975, Biosynthesis of chondroitin sulfate: Immunoprecipitation of interacting xylosyltransferase and galactosyltransferase, *FEBS Lett.* **49**:342–345.

Schwartz, N. B., 1976, Biosynthesis of chondroitin sulfate: Role of phospholipids in the activity of UDP-D-galactose:D-xylose galactosyltransferase, *J. Biol. Chem.* **251**:285–292.

Schwartz, N. B., 1977, Regulation of chondroitin sulfate proteoglycan chondroitin sulfate chains and core protein, *J. Biol. Chem.* **252**:6316–6321.

Schwartz, N. B., 1979, Synthesis and secretion of an altered chondroitin sulfate proteoglycan, *J. Biol. Chem.* **254**:2272–2277.

Schwartz, N. B, 1982, Regulatory mechanisms in proteoglycan biosynthesis, in: *Glycosaminoglycans and Proteoglycans in Physiological and Pathologic Processes of Body Systems* (R. S. Varma and R. Varma, eds.), pp. 41–54, Karger, Basel.

Schwartz, N. B., 1986, Carbohydrate metabolism II: Special pathways, in: *Mammalian Biochemistry* (T. M. Devlin, ed.), pp. 406–437, Wiley, New York.

Schwartz, N. B., and Dorfman, A., 1975, Purification of rat chondrosarcoma xylosyltransferase, *Arch. Biochem.* **171**:136–144.

Schwartz, N. B., and Rodén, L., 1974a, Biosynthesis of chondroitin sulfate: Interaction between xylosyltransferase and galactosyltransferase, *Biochem. Biophys. Res. Commun.* **56**:717–724.

Schwartz, N. B., and Rodén, L., 1974b, Biosynthesis of chondroitin sulfate: Purification of UDP-D-xylose:core protein β-D-xylosyltransferase by affinity chromatography, *Carbohydr. Res.* **37**:167–180.

Schwartz, N. B., and Rodén, L., 1975, Biosynthesis of chondroitin sulfate: Solubilization of chondroitin sulfate glycosyltransferases and partial purification of UDP-D-galactose:D-xylose galactosyltransferase, *J. Biol. Chem.* **250**:5200–5207.

Schwartz, N. B., Habib, G., Campbell, S., D'Elvlyn, D., Gartner, M., Krueger, R., Olson, C., and Philipson, L., 1985, Synthesis and structure of proteoglycan core protein, *Fed. Proc.* **44**:369–372.

Shin, H.-S., Bargiello, T., Clark, B., Jackson, F., and Young, M., 1985, An unusual coding sequence from a Drosophila clock gene is conserved in vertebrates, *Nature* **317**:445–448.

Sievers, J., Hartmann, D., Gude, S., Pehlemann, F., and Berry, M., 1987, Influences of meningeal cells on the development of the brain, in: *Mesenchymal–Epithelial Interactions in Neural Development* (J. K. Wolff, J. Sievers, and M. Berry, eds.), pp. 171–188, Springer-Verlag, Berlin.

Siewert, J., and Strominger, J., 1967, Bacitracin: An inhibitor of the dephosphorylation of lipid pyrophosphate, an intermediate in biosynthesis of the peptidoglycan of bacterial cell walls, *Proc. Natl. Acad. Sci. USA* **57**:767–770.

Silver, J., and Sidman, R., 1980, A mechanism for the guidance and topographic patterning of retinal ganglion cell axons, *J. Comp. Neurol.* **189**:101–111.

Smalheiser, N., 1989, Morphologic plasticity of rapid-onset neurites in NG108-15 cells stimulated by substratum-bound laminin, *Devel. Brain Res.* **45**:39–47.

Smalheiser, N., and Schwartz, N., 1987, Kinetic analysis of 'rapid onset' neurite formation in NG108-15 cells reveals a dual role for substratum-bound laminin, *Dev. Brain Res.* **34**:111–121.

Smalheiser, N., Crain, S., and Reid, L., 1982, Retinal ganglion-cell outgrowth upon substrata derived from basement membrane-secreting tumor and CNS tissues, *Soc. Neurosci. Abstr.* **8**:927.

Smalheiser, N., Crain, S., and Reid, L., 1984, Laminin as a substrate for retinal axons in vitro, *Dev. Brain Res.* **12**:136–140.

Stamatoglou, S. C., and Keller, J., 1983, Correlation between cell substrate attachment in vitro and cell surface heparan sulfate affinity for fibronectin and collagen, *J. Cell Biol.* **96**:1820–1823.

Stewart, G. R., and Pearlman, A. L., 1987, Fibronectin-like immunoreactivity in the developing cerebral cortex, *J. Neurosci.* **7**:3325–3333.

Stoolmiller, A. C., and Dorfman, A., 1967, Mechanism of hyaluronic acid (HA) biosynthesis by group A streptococcus, *Fed. Proc.* **26**:2.

Stoolmiller, A., and Dorfman, A., 1969, The biosynthesis of hyaluronic acid by streptococcus, *J. Biol. Chem.* **244**:236–240.

Stoolmiller, A. C., Schwartz, N. B., and Dorfman, A., 1975, Biosynthesis of chondroitin 4-sulfate proteoglycan by a transplantable rat chondrosarcoma, *Arch. Biochem. Biophys.* **171**:124–135.

Sugahara, K., and Schwartz, N. B., 1979, Defect in phosphoadenosylphosphosulfate formation in brachymorphic mice, *Proc. Natl. Acad. Sci. USA* **76**:6615–6618.

Sugahara, K., and Schwartz, N. B., 1982a, A defect in 3'-phosphoadenosine 5'-phosphosulfate formation in brachymorphic mice, *Arch. Biochem. Biophys.* **214**:589–601.

Sugahara, K., and Schwartz, N. B., 1982b, Tissue distribution of the defect in PAPS synthesis in brachymorphic mice, *Arch. Biochem. Biophys.* **214**:602–609.

Sugahara, K., Schwartz, N. B., and Dorfman, A., 1979, Biosynthesis of hyaluronic acid by streptococcus, *J. Biol. Chem.* **254**:6252–6261.

Sugahara, K., Cifonelli, A. J., and Dorfman, A., 1981, Xylosylation of nascent peptides of chick cartilage chondroitin sulfate proteoglycan, *Fed. Proc.* **40**:1705.

Szakal, A. K., Kosco, M. H., and Tew, J. G., 1988, A novel *in vivo* follicular dendritic cell-dependent iccosome-mediated mechanism for delivery of antigen to antigen-processing cells, *J. Immunol.* **2**:341–353.

Torack, R. M., and Grawe, L., 1980, Subependymal glycosaminoglycan networks in adult and developing rat brain, *Histochemistry* **68**:55–65.

Treadwell, B. V., Mankin, D. P., Ho, P. K., and Mankin, H. J., 1980, Cell-free synthesis of cartilage proteins: Partial identification of proteoglycan core and link proteins, *Biochemistry* **19**:2269–2275.

Triscott, M. X., and van der Rijn, I., 1986, Solubilization of hyaluronic acid synthetic activity from streptococci and its activation with phospholipids, *J. Biol. Chem.* **261**:6004–6009.

Unanue, E. R., 1984, Antigen-presenting function of the macrophage, *Annu. Rev. Immunol.* **2**:395–428.

Upholt, W. B., Vertel, B. M., and Dorfman, A., 1979, Translation and characterization of messenger RNAs in differentiating chicken cartilage, *Proc. Natl. Acad. Sci. USA* **76**:4847–4851.

Vertel, B. M., Upholt, W. B., and Dorfman, A., 1984, Cell-free translation of messenger RNA for chondroitin sulfate proteoglycan core protein in rat cartilage, *Biochem. J.* **217**:259–263.

Waite, K. A., Mugnai, G., and Culp, L. A., 1987, A second cell-binding domain on fibronectin (RGDS-independent) for neurite extension of human neuroblastoma cells, *Exp. Cell Res.* **169**:311–327.

Wekerle, H., Linington, C., Lassmann, H., and Meyerann, R., 1986, Cellular immune reactivity within the CNS, *Trends Neurosci.* **9**:273–277.

Wewer, U. M., Taraboletti, G., Sobel, M., Albrechtsen, R., and Liotta, L., 1987, Role of laminin receptor in tumor cell migration, *Cancer Res.* **47**:5691–5698.

Woods, A., Couchman, J. R., Johansson, S., and Höök, M., 1986, Adhesion and cytoskeletal organisation of fibroblasts in response to fibronectin fragments, *EMBO J.* **5**:665–670.

Wu, T.-C., Wan, Y.-J., Chung, A., and Damjanov, I., 1983, Immunohistochemical localization of entactin and laminin in mouse embryos and fetuses, *Dev. Biol.* **100**:496–505.

Yamamoto, T., Iwasaki, Y., and Konno, H., 1987, Laminin: A messenger produced by central neurons? Immunohistochemical demonstration of its unique distribution, *Neurology* **37**(Suppl.1):234.

Zurn, A., Nick, H., and Monard, D., 1988, A glia-derived nexin promotes neurite outgrowth in cultured chick sympathetic neurons, *Dev. Neurosci.* **10**:17–24.

6

Lysosomal Degradation of Glycoproteins and Glycosaminoglycans

Larry W. Hancock and Glyn Dawson

1. INTRODUCTION

The catabolism of brain glycoconjugates is a complex process involving the interaction of endoglycosidases, exoglycosidases, and proteinases, as summarized in Table 1. Lysosomal catabolism of these glycoconjugates further requires the delivery of both the catabolic enzymes and their substrates to this organelle, and the maintenance of a functional milieu having the proper acidic pH and complement of cofactors (e.g., cations) to facilitate efficient catabolism.

Much of what is known about nervous system glycoconjugate catabolism has been derived from the biochemical analysis of inherited deficiencies of lysosomal enzymes (see Chapter 15). The description of specific enzyme deficiencies, in conjunction with the structural analysis of storage products in individual tissues and cell types, has provided enormous detail on the pathways of glycoconjugate catabolism and the relative importance of these pathways in specific tissues such as the nervous system. Identification and utilization of animals with inherited lysosomal enzyme deficiencies has provided further insight into the vital role which normal lysosomal catabolism plays in the development and maintenance of neural function. In addition, the recent utilization of inhibitors of lysosomal enzymes such as swainsonine in both animals and *in vitro* systems has shown exceptional promise for determining more precisely how glycoconjugate catabolism contributes to the survival and function of discrete neural cell populations.

Larry W. Hancock and Glyn Dawson • Departments of Pediatrics and Biochemistry and Molecular Biology, and the Kennedy Mental Retardation Research Center, University of Chicago, Chicago, Illinois 60637.

Table 1. Factors Necessary for
Glycoconjugate Catabolism

Catabolic enzymes	Subcellular compartmentalization	"Cofactors"
Endoglycosidases	Enzymes	pH
Exoglycosidases	Substrates	Ions
Proteases		"Activators"

In this chapter, we will address a number of areas related to the lysosomal catabolism of nervous system glycoconjugates, including discussions of the enzymes and ancillary proteins involved in catabolism, the interplay of catabolic enzymes during catabolism, the delivery of both enzymes and substrates to lysosomes, and possible mechanisms which may serve to regulate lysosomal catabolism. In addition, we will discuss the catabolism of specific neuroglycoconjugates, and address the possible importance of nonlysosomal glycoconjugate catabolism with particular emphasis on the nervous system.

Although much of the existing literature deals with *in vitro* studies carried out in nonneural cells or tissues, it seems clear that catabolism proceeds by similar mechanisms and employs analogous systems for regulation and targeting in both neural and nonneural cells. In cases where divergence has been shown to exist or is suspected of existing, we will attempt to delineate any tissue-specific differences.

2. GLYCOPROTEIN CATABOLISM

2.1. The Structure of Glycoprotein-Associated Oligosaccharides

The structures and biosynthesis of glycoprotein-associated oligosaccharides have been reviewed in detail elsewhere (see Chapters 3 and 4, and Kornfeld and Kornfeld, 1985). In general, two major classes of glycoprotein-associated oligosaccharides exist: those which are attached via *O*-glycosidic linkage to the serine or threonine residues of the polypeptide backbone (Figure 1), and those which are attached via *N*-glycosidic linkage to specific asparagine residues of the polypeptide backbone (Figure 2). As discussed below, distinct enzymes as well as distinct pathways of lysosomal catabolism may be involved in the degradation of *O*- and *N*-linked oligosaccharide chains. Chapter 15 includes details of glycosphingolipid catabolism, as derived from the analysis of inherited deficiencies of catabolic enzymes.

2.2. Catabolism of O-Linked Oligosaccharide Chains

As shown in Figure 1, *O*-linked oligosaccharides typically contain galactose, *N*-acetylgalactosamine, and sialic acid. Most are of the di- to tetrasaccharide size range, having no, one, or two sialic acid residues attached at the 3 and/or 6 positions of galactose. There is considerable structural similarity to nervous system sialoglycosphingolipids (gangliosides), and most of the hydrolases are able to degrade both glycolipids and glycoproteins (see Chapter 15). While no detailed studies of *O*-linked

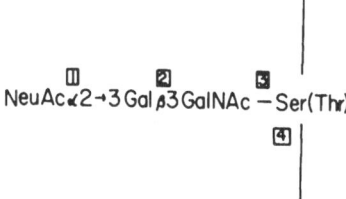

Figure 1. Typical *O*-glycosidically linked glycoprotein-associated oligosaccharide structure. Catabolism of this structure involves the action of exoglycosidases (Reactions 1–3) and proteases (Reaction 4) as described in the text.

oligosaccharide catabolism have been reported, catabolism presumably requires the action of three exoglycosidases (neuraminidase, β-galactosidase, and *N*-acetyl-α-hexosaminidase), as well as degradation of the polypeptide backbone by proteinases. By analogy with proteoglycan catabolism (discussed below), the catabolism of larger *O*-linked oligosaccharide structures such as those found on mucins may involve the action of endoglycosidase activities to produce oligosaccharide fragments, which would then be fully degraded by the action of exoglycosidases; an endo-α-galactosaminidase activity has, in fact, been isolated from *Diplococcus pneumoniae* (Umemoto *et al.*, 1977), but the role of such an enzyme in the mammalian nervous system is unknown. In contrast to the catabolism of *N*-linked oligosaccharides and proteoglycans, there is no evidence that proteolysis of the polypeptide backbone must precede catabolism of *O*-linked oligosaccharides, although such a mechanism has not been experimentally disproven.

The recent description of glycoproteins containing GlcNAc residues *O*-linked to serine or threonine (Torres and Hart, 1984; Holt and Hart, 1986; Schindler *et al.*, 1987; Holt *et al.*, 1987a,b; Hanover *et al.*, 1987; Abejon and Hirschberg, 1988) provides evidence for yet another class of *O*-glycosidic linkage. To date, little is known about the catabolism of these glycoproteins or the extent of their occurrence in nervous tissue, but the apparent cytoplasmic or nucleoplasmic orientation of at least some of the *O*-GlcNAc residues raises the intriguing possibility of a cytoplasmic site of glycosidase action for their catabolism.

In summary (see Figure 1), catabolism of mammalian *O*-linked glycoprotein oligosaccharide chains proceeds by the sequential action of exoglycosidases (Reactions 1, 2, and 3), which may be preceded by, concurrent with, or followed by degradation of the polypeptide backbone (Reaction 4).

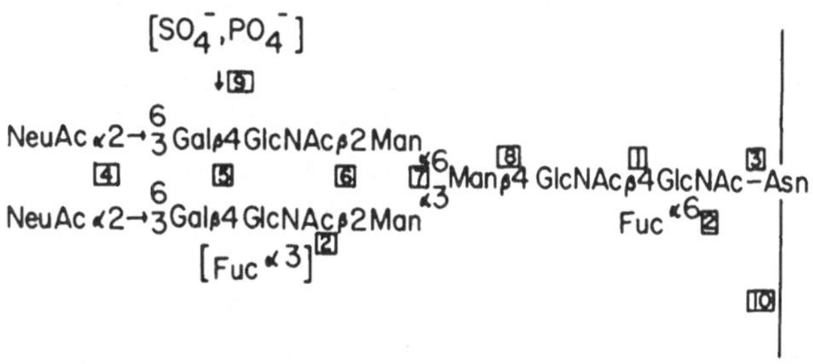

Figure 2. Typical *N*-glycosidically linked complex biantennary glycoprotein-associated oligosaccharide structure. Catabolism of this structure involves the action of endoglycosidases (Reactions 1 and 3), exoglycosidases (Reactions 2, 4–8), phosphatases and sulfatases (Reaction 9), and proteases (Reaction 10) as described in the text.

2.3. Catabolism of N-Linked Oligosaccharide Chains

N-Linked oligosaccharides have a branched trimannosyl "core" structure substituted with either mannose residues or N-acetylglucosamine, galactose, fucose, and sialic acid residues in bi-, tri-, or tetraantennary form. As discussed below, it is apparent that their catabolism involves the concerted action of endoglycosidases, proteases, and exoglycosidases.

2.3.1. Endoglycosidases

Since the culture medium of many microorganisms is rich in a number of endo-β-N-acetylglucosaminidase activities which are specific for various N-linked oligosaccharide structures found on mammalian glycoproteins, it was presumed that analogous endoglycosidase activities might function in mammalian glycoprotein catabolism. The accumulation in brain and other tissues of storage *oligosaccharides* rather than *glycopeptides* in individuals having inherited deficiencies of exoglycosidases involved in glycoprotein catabolism further suggested the existence of mammalian endoglycosidase activities. While a number of studies successfully demonstrated the presence of endo-β-N-acetylglucosaminidase activity in mammalian liver (Pierce *et al.*, 1979, 1980; Overdijk *et al.*, 1981; Tachibana *et al.*, 1982; Lisman *et al.*, 1985), in every case the activity was suggested to be cytoplasmic rather than lysosomal, based on subcellular fractionation and/or pH optimum. However, recent studies utilizing rat liver (Baussant *et al.*, 1986; Kuranda and Aronson, 1986; Brassart *et al.*, 1987) have described a lysosomal endo-β-N-acetylglucosaminidase activity capable of acting upon neutral or complex oligosaccharides having reducing-terminal di-N-acetylchitobiose residues. This suggests the sequential action of two lysosomal endoglycosidases, namely an aspartylglucosaminidase, which frees the N-linked oligosaccharide from the polypeptide backbone, and an endo-β-N-acetylglucosaminidase which acts on the free oligosaccharide to give, as products, GlcNAc and oligosaccharide having a single GlcNAc at its reducing terminus. Song *et al.* (1987) have demonstrated the absence of the latter activity in the tissues of certain species (e.g., sheep, cattle, and pigs); the absence of lysosomal endo-β-N-acetylglucosaminidase activity is consistent with the accumulation of di-N-acetylchitobiose-terminating oligosaccharides in certain species having genetic or inhibitor-induced deficiencies of lysosomal enzymes, including dogs (Warner and O'Brien, 1982), cows and cats (Abraham *et al.*, 1983), and sheep (Daniel *et al.*, 1984). In humans and rats, species which express lysosomal endo-β-N-acetylglucosaminidase activity, storage oligosaccharides having only a single GlcNAc at the reducing terminus accumulate in genetic and inhibitor-induced enzyme deficiencies. The identification of glycopeptide storage products in the tissue and urine of dogs suffering from lysosomal α-L-fucosidase deficiency (Abraham *et al.*, 1984) suggests that the initial removal of fucose residues from peripheral and/or internal GlcNAc residues may be requisite for the action of aspartylglucosaminidase. This implies that α-L-fucosidase acts early in N-linked oligosaccharide catabolism, and this is also true for fucoglycolipid catabolism.

The presence of significant amounts of storage oligosaccharides having both one and two GlcNAc residues at their reducing terminus in the tissue and urine of goats suffering from β-mannosidosis (Jones and Laine, 1981; Hancock *et al.*, 1986) suggested that both aspartylglucosaminidase and endo-β-*N*-acetylglucosaminidase could be acting on *N*-linked oligosaccharides in that species. Subsequent studies (Hancock and Dawson, 1987; Hancock, 1989) utilizing cultured goat fibroblasts have provided further evidence for the existence of both enzyme activities in goats, and have failed to show a precursor–product relationship between the two storage oligosaccharides. It is therefore probable that in goat brain and other tissues, aspartylglucosaminidase and endo-β-*N*-acetylglucosaminidase may act preferentially on discrete populations of *N*-linked oligosaccharide substrates, on the basis either of structure or of subcellular localization.

2.3.2. Exoglycosidases

Lysosomal exoglycosidases appear to act in a rather straightforward manner on glycoprotein-associated, glycopeptide, or free oligosaccharide substrates. Their action is discussed extensively in Chapter 15.

2.3.3. Proteinases

In all reported studies to date (except those originating from one laboratory, as discussed below), there has been no evidence to suggest that oligosaccharide catabolism is requisite for proteinase action. To the contrary, Kuranda and Aronson (1986) demonstrated the accumulation of the glycopeptide GlcNAcβ4GlcNAc-Asn in perfused rat liver treated with 5-diazo-4-oxo-L-norvaline (an inhibitor of aspartylglucosaminidase), suggesting the independent action of exoglycosidases and proteinases. In fibroblasts of goats suffering from the neurodegenerative disease β-mannosidosis, we showed the inhibition of Manβ4GlcNAcβ4GlcNAc storage oligosaccharide accumulation in cells treated with the cathepsin inhibitor leupeptin (Hancock and Dawson, 1987). In contrast, little effect was seen on the accumulation of Manβ4GlcNAc storage oligosaccharide in leupeptin-treated cells. These results imply that aspartylglucosaminidase action may be facilitated by at least limited proteolytic digestion of the glycoprotein substrate, while suggesting that endo-β-*N*-acetylglucosaminidase action may proceed independent of proteinase activity. They do not rule out, however, direct inhibition of aspartylglycosaminidase activity under the experimental conditions described.

2.3.4. Summary

As discussed above, *N*-linked oligosaccharide catabolism proceeds via the concerted action of endoglycosidases, proteinases, and exoglycosidases. Many of these exoglycosidases also act on various sialoglycosphingolipids, although activator proteins may also be required. Although species- and tissue-specific pathways may exist, it appears that a general pathway can be described as shown in Figure 2. In some species, the oligosaccharide may be released directly from the glycoprotein by an endo-β-*N*-acetylglucosaminidase activity (Reaction 1). In other species, the exoglycosidase α-L-fucosidase may first act on peripheral and/or internal fucose residues (Reaction 2), after which aspartylglucosaminidase (Reaction 3) effects release of the oligosaccharide from the glycoprotein; endo-β-N-acetylglucosaminidase may then act on the free oligosaccharide. The free oligosaccharide may then be fully degraded by the action of a number of lysosomal exoglycosidases, including neuraminidase (Reaction 4), β-galactosidase (Reaction 5), α-mannosidase (Reaction 7), and β-mannosidase (Reaction 8). Some controversies exist, e.g., the evidence for more than one lysosomal α-mannosidase. Studies carried out on fibroblasts and tissues from human patients with α-mannosidosis and on human skin fibroblasts pretreated with the lysosomal α-mannosidase inhibitor swainsonine (Jolly et al., 1981; Cenci di Bello et al., 1983; Hancock, 1989) have suggested the presence of at least two lysosomal α-mannosidase activities, which preferentially cleave terminal mannose residues in α1,3 or α1,6 linkage. Both activities appear to be inhibited by swainsonine, while only the α1,3 activity is absent in human α-mannosidosis and the 1,3-linked oligosaccharide preferentially accumulates in mannosidosis brain. However, the two putative α-mannosidases have not yet been separated biochemically. Sulfatases and phosphatases may remove sulfate and phosphate substituents on peripheral glycose residues prior to exoglycosidase action (Reaction 9). Finally, the polypeptide backbone is degraded by proteinase action (Reaction 10); limited proteinase digestion may be needed to facilitate aspartylglucosaminidase action.

2.4. Catabolism of the Polypeptide Backbone

As discussed above, at least limited proteolytic degradation of the polypeptide backbone of glycoproteins may be necessary for the efficient catabolism of N-linked oligosaccharide chains. Apart from the contribution to degradation of the oligosaccharide chains, it is apparent that proteolytic digestion of polypeptides is requisite for continued normal cellular function. Lysosomal catabolism of protein not only provides the cell with amino acids for de novo protein synthesis, but may also contribute to the remodeling of neural connections and modulation of cellular responsiveness via the turnover of cell-surface receptors and other constituents (Ivy, 1988). Recently, impaired proteolytic function has been implicated in the loss of neural plasticity observed in aging and in Alzheimer's disease (Ivy, 1988; Abraham et al., 1988).

In addition to catabolism, proteinases localized in lysosomes and/or prelysosomal compartments also appear to be involved in the proteolytic processing of the lysosomal enzymes themselves (Hasilik and von Figura, 1984). Although the significance of proteolytic processing of lysosomal enzymes remains somewhat unclear, the observation that virtually all soluble lysosomal enzymes undergo such processing events suggests an important function for this limited proteolysis. Our recent observation that such events may be partially impaired secondary to the accumulation of lysosomal storage products (Hancock et al., 1988) serves to further focus attention on this biochemical phenomenon; the implications of proteolytic processing with respect to the regulation of lysosomal enzyme activity in nervous and other tissues will be discussed later in this chapter.

2.5. Delivery of Catabolic Enzymes and Glycoprotein Substrates to the Lysosome

It is well established that the efficient delivery of most lysosomal enzymes to the lysosome is dependent upon the addition of a mannose 6-phosphate recognition signal to the lysosomal enzymes and their subsequent association with one of two mannose 6-phosphate receptors (reviewed by Kornfeld, 1987; von Figura and Hasilik, 1986). The recent observation that certain lysosomal enzymes such as glucocerebrosidase (Jonsson et al., 1987) and a number of lysosomal membrane glycoproteins (Lewis et al., 1985; Chen et al., 1985; D'Souza and August, 1986; Lippincott-Schwartz and Fambrough, 1986; Barriocanal et al., 1986; Green et al., 1987; Fambrough et al., 1988; Viitala et al., 1988) do not have the mannose 6-phosphate recognition marker, as well as the nearly normal levels of lysosomal enzymes in certain tissues (including nervous tissue) and cell types of patients suffering from I-cell disease (who are unable to synthesize the mannose 6-phosphate recognition marker; reviewed by Kornfeld, 1987), strongly suggests the existence of alternative mechanisms for delivery of lysosomal enzymes to the lysosome. It does appear clear, however, that nonlysosomal intracellular glycosidases do not represent "misrouted" lysosomal glycosidases, but are in fact distinct gene products; this will be discussed at greater length within the context of nonlysosomal catabolism.

Considerable data suggest that most glycoproteins are delivered to the lysosome for catabolism via the endocytic pathway; the actual mechanism for delivery may specifically involve receptor-mediated endocytosis (e.g., asialoglycoproteins and the hepatic asialoglycoprotein receptor; reviewed by Neufeld and Ashwell, 1980), or membrane components destined for lysosomal catabolism may be delivered to the lysosome as "passive" membrane components of the endocytic pathway, e.g., hormone and neurotransmitter receptors in nervous tissue.

Recent studies also suggest that a pentapeptide sequence found in certain soluble proteins (-Lys-Phe-Glu-Arg-Gln-), or a sequence closely related to the pentapeptide, may serve as a signal for translocation from the cytoplasm to the lysosome (McElligott et al., 1985; Dice et al., 1986; Backer and Dice, 1986; Dice, 1987; Chiang and Dice, 1988) under certain conditions such as serum deprivation of cultured cells. While the details of this translocation pathway are yet to be fully worked out, it would appear that translocation is mediated by a saturable receptor or binding protein. Taken together with other studies (Ahlberg et al., 1985), it seems that lysosomal catabolism must play a major role in the degradation of both membrane-associated and soluble cellular components of the nervous system.

2.6. Inhibitors of Glycoprotein Catabolism

During the past several years, a number of neurotoxic inhibitors (both synthetic and naturally occurring) of lysosomal glycosidases and proteinases have been described. In many instances, these inhibitors have been shown to produce phenotypes of inherited neuronal lysosomal enzyme deficiencies after their introduction into animals

Table 2. Inhibitors of Brain Glycoprotein Catabolism

Enzyme	Inhibitor
Neuraminidase	2,3-Dehydro-2-deoxy-*N*-acetylneuraminic acid (NeuAc-2-en)
β-Galactosidase	β-D-Galactopyranosylmethyl-*p*-nitrophenyl triazene (βGalMNT)
α-Mannosidase	Swainsonine
	α-D-Mannopyranosylmethyl-*p*-nitrophenyl triazene (αManMNT)
Aspartylglucosaminidase	5-Diazo-4-oxo-L-norvaline (DONva)

or cultured cells. They have also proven useful in studies designed to assess the interaction of hydrolytic enzymes in normal cells, since they allow for the accumulation of partially degraded substrates which can be identified and related to the pathways of catabolism. In view of the potential application of such inhibitors to the study of normal catabolic pathways for nervous system glycoproteins (and glycosphingolipids), as well as their possible utility in delineating the role of normal lysosomal function in a variety of biological systems, a number of inhibitors of glycoprotein catabolism will be discussed briefly. The inhibitors and their target enzymes are summarized in Table 2.

2.6.1. 2-Deoxy-2,3-dehydro-*N*-acetylneuraminic Acid (NeuAc-2-en)

NeuAc-2-en is a competitive inhibitor of both mammalian and bacterial neuraminidases (Sweeley and Usuki, 1987; Saito and Yu, 1986; Mendl and Tuppy, 1969) and is present in minor amounts in mammalian tissues and secretions. Although it is not clear that NeuAc-2-en can be experimentally introduced into the lysosomes of living cells, Sweeley and Usuki (1987) have recently demonstrated the inhibition of extracellular GM3 sialidase activity in cultured fibroblasts by NeuAc-2-en, and have suggested that the inhibition of extracellular GM3 sialidase activity may be an important mechanism for the modulation of cell growth *in vivo* (e.g., inhibition of the unrestricted growth of GM3-rich glioblastomas in the nervous system).

2.6.2. β-D-Galactopyranosylmethyl-*p*-nitrophenyl Triazene (βGalMNT)

βGalMNT is a chemically synthesized irreversible inhibitor of lysosomal β-galactosidase activity (as measured against artificial substrates, oligosaccharides, and GM1 ganglioside); it is ineffective against neutral glycolipid β-galactocerebrosidase. The compound has been used in cultured fibroblasts to determine the rate of turnover of normal and mutant GM1-β-galactosidase (van Diggelen *et al.*, 1980, 1981), and in perfused rat liver to assess the interaction of endoglycosidases, exoglycosidases, and proteases during glycoprotein catabolism (Kuranda and Aronson, 1985, 1986). More recently, it has been shown to induce a phenocopy of GM1 gangliosidosis in cultured neurotumor cells (Singer *et al.*, 1987). Given the potential importance of GM1 in neuritogenesis and *in vitro* differentiation (Facci *et al.*, 1984, 1988), this compound could prove valuable in the *in vitro* or *in situ* study of neuronal development and differentiation. The α-mannosyl and β-mannosyl analogues of βGalMNT have been synthesized, and αManMNT has been reported to be effective as an inhibitor of lysosomal α-mannosidases (Docherty *et al.*, 1986), although at much higher concentrations than βGalMNT in the same systems. Our own preliminary studies utilizing α- or βManMNT in cultured fibroblasts (L. Hancock *et al.*, unpublished observations) have suggested that these compounds are unlikely to be useful as inhibitors of lysosomal glycoprotein catabolism.

2.6.3. Swainsonine

Swainsonine is an indolizidine alkaloid which was originally isolated from the plant *Swainsonia canescens* (locoweed), and is a potent neurotoxin in cattle and sheep. It is a reversible inhibitor of lysosomal α-mannosidases (Dorling *et al.*, 1980), as well as the Golgi α-mannosidase II (Elbein *et al.*, 1981; Tulsiani *et al.*, 1982). As a result of this inhibition of both catabolic and biosynthetic (processing) α-mannosidases, swainsonine exerts effects both on the structures of glycoprotein-associated *N*-linked oligosaccharides (producing "hybrid" oligosaccharide chains which have structural elements of both high-mannose and complex *N*-linked oligosaccharides; see Chapter 4), and on the catabolism of α-mannose-containing oligosaccharide chains.

The effects of swainsonine on catabolic α-mannosidases have been demonstrated directly by analysis of swainsonine-treated cell or tissue extracts, and indirectly by the isolation and characterization of partially degraded oligosaccharides in tissue and urine of swainsonine-treated animals or from extracts of swainsonine-treated cultured cells (Cenci di Bello *et al.*, 1983; Daniel *et al.*, 1984; Hancock *et al.*, 1986; Tulsiani and Touster, 1987; Hancock, 1989). Winkler and Segal (1984a,b) have proposed that swainsonine may also inhibit the catabolism of the polypeptide chains of glycoproteins, perhaps by preventing removal of the carbohydrate components, although we and others have seen no effect of swainsonine on the catabolism of endocytosed glycoproteins (Kuranda and Aronson, 1985; Hancock, 1989). Tulsiani and Touster (1987) have also suggested that swainsonine may have selective inhibitory effects on cytoplasmic catabolic α-mannosidases in rat brain and other tissues, leading to the accumulation of storage oligosaccharides representative of impaired cytoplasmic catabolism. They have further speculated that such tissue-specific inhibition of nonlysosomal α-mannosidase activity might account for the apparent lack of toxic swainsonine effects in brain tissue of swainsonine-fed rats (Tulsiani and Touster, 1983).

It is clear that swainsonine treatment has no apparent effect on the efficient delivery of a number of soluble lysosomal enzymes to the lysosome (Hancock, 1989; Tropea *et al.*, 1988). This is to be expected, since swainsonine should have no effect on the synthesis of mannose 6-phosphate-containing oligosaccharide chains. Interestingly, Aerts *et al.*, (1986) have reported that swainsonine treatment of U937 human monocytes leads to the accumulation of the membrane-associated enzyme glucocerebrosidase in nonlysosomal compartments, implicating complex oligosaccharide chains in the proper lysosomal targeting of this enzyme.

To date, it is not clear what effects swainsonine might have on the targeting of heavily glycosylated lysosomal membrane glycoproteins, or what effects might be mediated by its action as a weak base, at least in cases of extreme swainsonine neuro-intoxication.

2.6.4. 5-Diazo-4-oxo-L-norvaline (DONva)

An animal model for the neurodegenerative disease aspartylglucosaminuria does not exist, but Kuranda and Aronson (1986) have recently demonstrated the inhibition of aspartylglucosaminidase activity, and consequent accumulation of partially degraded endocytosed glycoproteins, in perfused rat liver treated with DONva. Similar inhibition of aspartylglucosaminidase activity with DONva had previously been demonstrated by Tarentino and Maley (1969) in hen oviduct. The accumulation of GlcNAcβ4GlcNAc-Asn glycopeptide in DONva-treated perfused rat liver (Kuranda and Aronson, 1986) suggests that oligosaccharide removal is not requisite for degradation of the polypeptide backbone, and further supports the existence of a di-N-acetyl-chitobiose-cleaving glycosidase activity which acts on the core structure of N-linked oligosaccharide following cleavage of the GlcNAc-Asn bond. It is not clear from these studies, or from our own studies utilizing leupeptin-treated goat fibroblasts (Hancock and Dawson, 1987), to what extent polypeptide catabolism may precede cleavage of the GlcNAc-Asn bond under normal circumstances.

2.6.5. Other Glycosidase Inhibitors

In addition to its well-established effects on the processing α-glucosidases of the ER (see Chapter 4), the indolizidine alkaloid castanospermine has been shown to be an inhibitor of both lysosomal α-glucosidase and β-glucosidase (Saul et al., 1984; Chambers and Elbein, 1986); both enzymes are associated with neurovisceral storage diseases. In addition, conduritol B epoxide has been shown to be an effective inhibitor of acid β-glucosidase (the enzyme responsible for glucocerebroside catabolism), both after infusion into animals (the "Gaucher" mouse; Kanfer et al., 1975) and in vitro (Grabowski et al., 1986). In the latter study, this inhibitor proved useful in determining the effects of subtle mutations on the absolute amount and kinetic properties of lysosomal β-glucosidase in Gaucher patient-derived cells and tissue.

2.6.6. Proteinase Inhibitors

Few studies have dealt directly with the role of proteinases in glycoprotein catabolism by directly assessing the effects of proteinase inhibitors on glycoprotein-associated oligosaccharide catabolism. As mentioned above, our own studies (Hancock and Dawson, 1987) as well as those of Aronson and Docherty (1983) suggest that at least limited proteolytic cleavage facilitates hydrolysis of the GlcNAc-Asn bond by aspartylglucosaminidase. However, there is no evidence to suggest that proteolytic cleavage is requisite for endo-β-N-acetylglucosaminidase action. Recent studies have implicated impaired proteolytic action in the accumulation of the lipofuscin storage pigment observed in aging brain and in tissue derived from patients with neuronal ceroid lipofuscinosis (Abraham et al., 1988; Ivy, 1988; Dawson and Glaser, 1987), but there is no direct evidence to associate this storage material with partially degraded oligosaccharide chains or glycopeptides. Recent studies in our laboratory (L. W. Hancock, K. F. Johnson, and G. Dawson, unpublished observations) have, however, suggested that specific inhibitors of cysteine proteinases such as Ep459 (Hanada *et al.*, 1983) may, after addition to cultured human fibroblasts, have profound effects on the proteolytic processing of lysosomal glycosidases such as α-fucosidase and *N*-acetyl-β-hexosaminidase. It is not yet clear whether chronic proteinase inhibition might produce secondary effects on lysosomal glycosidase activities, although our preliminary results (Dawson, unpublished observations) suggest that activities of certain exoglycosidases (e.g., β-mannosidase) may be increased.

3. GLYCOSAMINOGLYCAN CATABOLISM

The synthesis and biological activity of nervous tissue glycosaminoglycans are covered in Chapter 5, and Rodén (1980) has provided an excellent review of the enzyme activities involved in the catabolism of connective tissue proteoglycans. *N*- and *O*-linked oligosaccharide chains of proteoglycans appear to be degraded in the same manner as those associated with glycoproteins. As shown in Figure 3, further proteoglycan catabolism involves the action of proteases, endoglycosidases, exoglycosidases, and sulfatases; in this section, we will concentrate on recently described enzyme activities, and on the interplay and subcellular sites of action of these classes of enzymes during the catabolism of glycosaminoglycans.

3.1. Catabolism of the Oligosaccharide Chains

3.1.1. Endoglycosidases

In addition to the endoglucuronidase and endo-β-*N*-acetylhexosaminidase activities which have previously been described (reviewed by Rodén, 1980), a number of additional catabolic endoglucuronidase activities have been documented during the past several years. Specifically, endoglucuronidase activities which may be involved in heparan sulfate proteoglycan catabolism have been found in rat hepatocytes (Bienkowski and Conrad, 1984; Kjellén *et al.*, 1985) and in cultured B16 melanoma cells (Nakajima et al., 1984), and are assumed to be active in nervous tissue. These

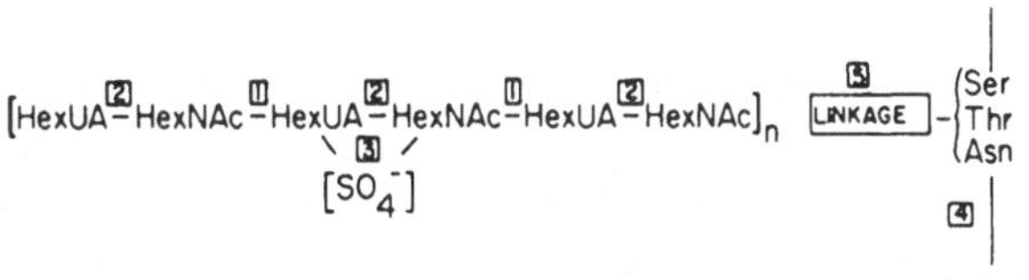

Figure 3. Stylized structure of a proteoglycan molecule. Catabolism of this structure involves the action of endo- and exoglycosidases (Reactions 1, 2, and 5), sulfatases (Reaction 3), and proteases (Reaction 4) as described in the text.

activities appear to be involved in fragmentation of the repeating disaccharide portion of the polysaccharide (Reaction 2, Figure 3), and the pH optimum of approximately 6 for the rat hepatocyte activity (Kjellén et al., 1985) suggests the possibility of a nonlysosomal site of action. It is thus theoretically possible that extracellular endoglucuronidase activity could be involved in the tissue-invasive, metastatic properties of B16 melanoma cells.

In contrast to the endoglucuronidase activities described above, Takagaki et al. (1988) have characterized a lysosomal endoglucuronidase (pH optimum 4.0) of rabbit liver which appears to be specific for the GlcUA-Gal bond in the linkage region (Reaction 5, Figure 3) of chondroitin sulfate proteoglycan. This activity is found in many tissues, and is probably involved in the latter stages of lysosomal chondroitin sulfate proteoglycan catabolism (i.e., catabolism of the linkage region glycopeptide fragment). It is possible that additional endoglycosidase activities, specific for the linkage regions of other proteoglycan molecules (Hascall, 1981), may also be involved in proteoglycan catabolism.

3.1.2. Exoglycosidases and Sulfatases

As discussed in greater detail below, proteinase and endoglycosidase action appears to precede the action of exoglycosidases and sulfatases in the catabolism of nervous system proteoglycans. In a variety of cultured cell types, metabolically labeled heparan sulfate proteoglycan has been shown to be degraded by limited proteolytic and endoglycosidase cleavage, followed by catabolism to monomeric constituents by exoglycosidases, sulfatases, and proteases. The evidence for the lysosomal localization of the enzymes responsible for the catabolism of proteoglycan fragments includes the kinetics of catabolism, the differential effects of inhibitors (including leupeptin and general inhibitors of lysosomal catabolism such as chloroquine), and the subcellular localization of fragments accumulating in inhibitor-treated cells (Yanagishita and Hascall, 1984; Bienkowski and Conrad, 1984; Kjellén et al., 1985; Iozzo, 1987). Hoppe et al. (1988) have suggested that proteolysis precedes lysosomal sulfate release during the catabolism of dermatan sulfate proteoglycan in cultured fibroblasts. Consistent with the findings of lysosomal sulfatase action, Fuchs et al. (1985) have reported the existence of $GlcNAc-6-SO_4$ as an intermediate in the catabolism of keratan sulfate. However, their results are consistent with the release of sulfate either from polymeric keratan sulfate or from GlcNAc-6-sulfate.

3.1.3. Proteinases

Studies in a number of cultured cell systems (Yanagishita and Hascall, 1984, 1985; Yanagishita, 1985; Bienkowski and Conrad, 1984; Kjellén et al., 1985; Hoppe et al., 1988), utilizing various inhibitors of proteinases in conjunction with metabolic labeling and subcellular fractionation, have clearly established the importance of proteolytic action in the catabolism of proteoglycans. In general, it appears that at least limited proteolytic and endoglycosidase digestion (perhaps in endosomes) (Reaction 4, Figure 3) precedes the action of lysosomal endoglycosidases, exoglycosidases, and sulfatases. It is clear that proteolytic cleavage is requisite for lysosomal degradation of the polysaccharide chains of proteoglycans, and there are further suggestions that proteolytic cleavage may be involved in the translocation of partially degraded proteoglycan molecules from prelysosomal to lysosomal compartments via an as yet unidentified mechanism.

3.2. Catabolism of the Polypeptide Backbone

As discussed above, proteolytic action on the polypeptide backbone is essential for proteoglycan catabolism. Leupeptin and other cysteine proteinase inhibitors at least partially inhibit proteoglycan catabolism, implicating cytosolic calcium-activated neutral proteinases and lysosomal cathepsins in this process. The involvement of cathepsins in what may be a nonlysosomal proteolytic cleavage is consistent with other studies which have demonstrated the nonlysosomal localization of the aspartyl proteinase cathepsin D (Diment and Stahl, 1985; Diment et al., 1988).

3.3. Summary

It appears that proteoglycan catabolism proceeds via proteolytic attack on the polypeptide backbone (Reaction 4, Figure 3) followed by endoglucuronidase or endohexosaminidase action (Reactions 1 and 2, Figure 3) on the repeating disaccharide portion of the polysaccharide chain; these enzymes may act in a prelysosomal compartment. Catabolism is completed in the lysosome, where sulfatases, β-glucuronidase, and N-acetyl-β-hexosaminidase act on the oligosaccharide fragments (Reactions 1–3, Figure 3), and the combined action of exoglycosidases, endoglycosidases, and proteases leads to degradation of the various linkage region (Rodén, 1980; Hascall, 1981) glycopeptides. Neural storage diseases (Chapter 15) attest to the fact that proteoglycans turn over rapidly in the nervous system.

3.4. Delivery of Substrate Proteoglycans to the Lysosome

Direct studies have demonstrated the binding of extracellular proteoglycans to both coated and uncoated areas of plasma membrane (Volker et al., 1984) and their subsequent association with intracellular organelles which could be identified as endosomes and lysosomes. Plasma membrane receptors for both hyaluronic acid and chondroitin sulfate proteoglycan have also been demonstrated in rat liver endothelial cells (Smedsrod et al., 1985; Laurent et al., 1986), and the acidic pH optimum of hyaluronidase and other catabolic enzymes suggests that delivery to the lysosomes might be mediated by receptor-mediated endocytosis. Synthesis, binding, and catabolism of hyaluronic acid and proteoglycans appear to be coordinately regulated in a number of cell types (Orkin et al., 1982; Morales and Hascall, 1988), suggesting a possibly complex mechanism for regulation of catabolism which may involve all of these processes. While recognition of hyaluronic acid and chondroitin sulfate appears to be related to chain length (and possibly negative charge), it is not precisely clear what signals may mediate proteoglycan association with cell-surface receptors. Preliminary studies suggesting a direct role for mannose 6-phosphate residues and the mannose 6-phosphate receptor in proteoglycan uptake (Brauker *et al.*, 1986) have been discounted on the basis of the normal uptake of extracellular proteoglycans observed in I-cell disease fibroblasts (Brauker and Wang, 1987).

4. SYNTHESIS AND TARGETING OF LYSOSOMAL ENZYMES

4.1. Mannose 6-Phosphate-Mediated Targeting

Virtually all soluble lysosomal enzymes are synthesized in the rough ER as prepropolypeptides; in the ER, they are subject to signal peptide cleavage, *N*-glycosylation, and carbohydrate modification. Subsequent to translocation to the Golgi, specific *N*-linked oligosaccharides of lysosomal enzymes are further modified by the action of two enzymes, UDP-GlcNAc:lysosomal enzyme *N*-acetylglucos-aminylphosphotransferase and *N*-acetylglucosaminylphosphoglycosidase, to give mannose 6-phosphate residues. These then serve as a recognition signal for interaction with one of two mannose 6-phosphate receptors, probably in the *trans*-Golgi. The 46-kDa cation-dependent receptor remains internal, whereas the 270-kDa cation-independent receptor (which also has binding sites for insulinlike growth factor II) is expressed on the surface of the cell. Their relative roles remain to be established. Receptor–ligand complexes are translocated to a prelysosomal acidic compartment, where a pH of 5.5 or less facilitates dissociation of the receptor and lysosomal enzymes. Lysosomal enzymes are subsequently localized to the lysosome, while the mannose 6-phosphate receptors may recycle to the plasma membrane and/or Golgi. The evidence for the participation of mannose 6-phosphate receptors in the targeting of lysosomal enzymes has recently been reviewed (von Figura and Hasilik, 1986; Kornfeld, 1987). While it has been clear for some time that the peptide conformation or some other structural feature of lysosomal enzyme polypeptides facilitates the formation of mannose 6-phosphate residues (Lang *et al.*, 1984), Faust *et al.* (1987a,b) have recently demonstrated that appropriate phosphorylation of cathepsin D can take place in *Xenopus* oocytes after the microinjection of human cathepsin D mRNA, suggesting the evolutionary conservation of this recognition signal.

4.2. Mannose 6-Phosphate-Independent Targeting

In spite of the well-established mannose 6-phosphate-dependent pathway for the targeting of many lysosomal enzymes, certain aspects of the targeting of lysosomal enzymes and membrane components suggest the existence of alternative signals which might mediate the intracellular segregation of lysosomal constituents. This may be especially important in nervous tissue. Among the observations which suggest alternative targeting mechanisms are the absence of mannose 6-phosphate residues on either lysosomal membrane components, or the lysosomal membrane-associated enzyme glucocerebrosidase, and the relatively high residual lysosomal enzyme activity in nervous tissue and other cell types of patients who are unable to synthesize mannose 6-

phosphate (I-cell disease; reviewed by Kornfeld, 1987). In addition, the recent development of the concept of a prelysosomal acidic compartment in which lysosomal enzymes dissociate from the mannose 6-phosphate receptors and from which the enzymes must be translocated to the lysosome (Griffiths *et al.*, 1988) suggests the necessity for additional biochemical signals which might mediate efficient translocation of lysosomal enzymes from the prelysosomal acidic compartment to the lysosome. It is also noteworthy that certain neural cell types, including rat oligodendrocytes (Hill *et al.*, 1985), appear to lack the cell surface 270-kDa cation-independent mannose 6-phosphate receptor. Although the recent discovery of the 46-kDa cation-dependent mannose 6-phosphate receptor (reviewed by von Figura and Hasilik, 1986; Kornfeld, 1987) may provide a mannose 6-phosphate-dependent mechanism for lysosomal enzyme targeting in nervous tissue, it is still not established that every cell type expresses one or the other mannose 6-phosphate receptor.

4.3. Proteolytic Processing

In addition to the extensive carbohydrate modifications discussed above, most if not all soluble lysosomal enzymes also undergo limited proteolytic processing in addition to signal peptide cleavage (reviewed by Hasilik and von Figura, 1984; Kornfeld, 1987; and in Chapter 15). Proteolytic processing may involve peptide cleavage at the amino- and carboxy-terminus, as well as internal proteolytic cleavage, and probably takes place in the lysosome or post-Golgi/prelysosomal compartments. The exact nature of these proteolytic cleavages, the proteolytic enzymes responsible for the cleavages, and the subcellular sites of cleavage remain to be determined. Recent studies (Oude Elferink *et al.*, 1986; Gabel and Foster, 1986b, 1987) have suggested that both proteolytic processing and limited dephosphorylation take place in endosomes, as well as in the Golgi and related compartments. Finally, it is not clear what role proteolytic processing may play in the activation or stabilization of lysosomal enzymes, since precursors of most enzymes have been shown to be active against both natural and artificial substrates (with the exception of cathepsin D). Our recent studies have, however, suggested that the proteolytic processing of lysosomal α-fucosidase may lower and broaden the pH optimum of the enzyme (Johnson *et al.*, 1988a); this observation would argue against the ''passive'' removal of nonessential portions of the polypeptide backbones of lysosomal enzymes after their delivery to the lysosome, and suggest a biological role for the proteolytic processing of lysosomal enzymes.

4.4. Mannose 6-Phosphate Receptor-Mediated Endocytosis

It is clear that the cell surface cation-independent mannose 6-phosphate receptor is capable of interacting with exogenous lysosomal enzymes, and of delivering lysosomal enzymes to the lysosomes via receptor-mediated endocytosis. The physiological significance of this pathway remains something of a mystery, since there is growing evidence that intracellular receptor-mediated targeting is responsible for virtually all delivery of lysosomal enzymes to the lysosome. The recent surprising discovery of identity between the cell-surface cation-independent mannose 6-phosphate receptor and the receptor for insulinlike growth factor II (IGF-II) (reviewed by Roth, 1988; Tong *et al.*, 1988) suggests the possibility that the primary function of the cell-surface 270-kDa mannose 6-phosphate receptor may involve the biological response to IGF-II. Phosphorylation of the mannose 6-phosphate receptors may serve as a means of regulating the delivery of lysosomal enzymes to the lysosome, analogous to the modulation of IGF-II receptor activity (Corvera *et al.*, 1988a,b) by phosphorylation. The recent demonstration that both mannose 6-phosphate receptors cycle from the cell surface to the Golgi (Duncan and Kornfeld, 1988) would suggest that modulation of rates or pathways of intracellular cycling for the receptors could effectively regulate delivery of newly synthesized lysosomal enzymes to the lysosome.

5. REGULATION OF LYSOSOMAL CATABOLISM

5.1. Lysosomal Activity and Stability of Catabolic Enzymes

It is well established that a number of lysosomal enzymes require divalent cations as cofactors in order to express full activity; for example, Zn^{2+} has been shown to activate lysosomal α-mannosidase (Jolly *et al.*, 1981) and Mg^{2+} appears to be essential for α-L-iduronidase activity (Schuchman and Desnick, 1988). It is also becoming increasingly apparent that an appropriate reducing environment is requisite for efficient lysosomal function (Gieselmann *et al.*, 1985; Lloyd, 1986; Schuchman & Desnick, 1988), and maintenance of an appropriate acidic pH is clearly a primary necessity for lysosomal function. While biological systems which carry out ion transport, proton transport, and other such necessary processes have been recognized for some time, it is only recently that evidence to support the concept of regulation of these phenomena has begun to emerge. Of particular interest, pH regulation along the endocytic pathway (perhaps extending to the lysosomal compartment) would now appear to be accomplished by the differential action of the Na^+/K^+-ATPase and the H^+ pump (Mellman *et al.*, 1986; Schmidt *et al.*, 1988; Cain and Murphy, 1988). It is apparent that intravesicular pH could then be regulated by overall membrane composition or activity of discrete membrane components.

In contrast to many of the enzymes responsible for complex glycolipid catabolism, there is no evidence suggesting that glycosidases responsible for the catabolism of brain glycoproteins and glycosaminoglycans require so-called "activator" proteins (Sandhoff *et al.*, 1987; Li *et al.*, 1988). This is not unexpected, based on the relative hydrophilicity of most glycoprotein and glycosaminoglycan substrates.

Other protein factors appear to be necessary for some aspects of efficient glycoprotein and glycosaminoglycan degradation, however. Specifically, the lysosomal membrane-associated enzyme acetyl CoA:α-glucosaminide *N*-acetyltransferase transfers an acetyl group from cytoplasmic acetyl CoA to terminal α-hexosamine groups of heparan sulfate, allowing for the action of catabolic α-*N*-acetylhexosaminidase (Bame and Rome, 1986). In addition, there is considerable evidence suggesting that lysosomal neuraminidase and β-galactosidase exist in the lysosome as a complex with a third component, "protective" protein. Maintenance of this complex is essential for stable lysosomal activity of both enzymes (Verheijen *et al.*, 1985). Our own recent studies (Johnson *et al.*, 1988b) have suggested the existence of a "protective" protein for lysosomal α-L-fucosidase; this protein appears to act in the lysosome to stabilize α-L-fucosidase against proteolytic degradation.

5.2. Synthesis of Lysosomal Enzymes

Traditionally, lysosomal enzymes have been viewed as "housekeeping" enzymes; consistent with that perspective, their synthesis has largely been considered to be constitutive (i.e., unregulated). A number of recent reports, while not providing definitive examples of regulated synthesis of lysosomal enzymes, at least provide evidence that synthesis of these enzymes *can* be regulated. Using cultured neurotumor cells, Fischer *et al.* (1986) showed a severalfold increase in activity of lysosomal enzymes in cultures following treatment with dibutyryl-cyclic AMP, with apparent concomitant expression of lysosomes in the neurites of "differentiated" cells.

Rochefort *et al.* (1987) reported the hyperexcretion of the "lysosomal" enzyme cathepsin D by cultured human breast carcinoma cells, and further demonstrated the induction of the enzyme (via increased mRNA) by estrogen treatment of the estrogen-responsive cell lines MCG7 and ZR75-1. Westley and May (1987) confirmed the induction of cathepsin D mRNA in estrogen-treated cell lines, and identified cathepsin D mRNA as one of several mRNAs induced in estrogen-responsive cell lines. Corticosteroids have also been reported to increase lysosomal hydrolase activities (especially arylsulfatase A) in cultured oligodendroglioma cells (Dawson and Kernes, 1979).

Frick *et al.* (1985) showed the induction of a number of mRNAs in BALB/c-3T3 cells after treatment with platelet-derived growth factor, and later one of the induced transcripts was identified as cathepsin L (Mason *et al.*, 1987). This later work identified MEP, the major excreted protein of certain transformed rodent cells, as cathepsin L. More recently (Troen *et al.*, 1988), MEP has been shown to be a secreted precursor of cathepsin L, the synthesis of which can be induced by transformation, growth factors, and tumor promoters. The identification and cloning of the promoter sequence for the rodent cathepsin L gene may provide considerable insight into the regulation of cathepsin L synthesis. This information should be critical in understanding the appropriate regulation of the synthesis of cathepsin L and other lysosomal enzymes.

In addition to secreted precursors of cathepsin D and cathepsin L described above, a precursor of cathepsin B has also been discovered in the ascitic fluid of certain patients with ovarian cancer (Mort *et al.*, 1983). While the discrete hypersecretion of individual lysosomal proteinases by certain tumor lines (in spite of the presence of mannose 6-phosphate residues on the secreted proenzymes, and the absence of hypersecretion of other lysosomal enzymes) provides interesting systems for the study of mannose 6-phosphate-independent targeting mechanisms for lysosomal enzymes as well as the regulation of lysosomal enzyme synthesis, the precise correlation of lysosomal proteinase secretion with the transformed phenotype remains a paradoxical problem.

5.3. Turnover of Lysosomal Enzymes

Based on the relatively long turnover times of lysosomal enzymes, especially in nondividing cells such as neurons, it is apparent that certain intrinsic properties of these molecules must regulate their lysosomal catabolism. Braulke *et al.* (1987) have recently reported the sulfation of lysosomal enzyme-associated *N*-linked oligosaccharides, but the rather rapid disappearance of the sulfate groups in the lysosome would suggest that their role, if any, probably involves targeting rather than stability. Gabel and Foster (1986a,b) have demonstrated the limited dephosphorylation of lysosomal enzymes in mouse L cells, and the steady-state accumulation of specific phosphorylated oligosaccharide structures on lysosomal enzymes could suggest a role for such specific oligosaccharide structures in the maintenance of active enzymes in the lysosomal compartment. In addition, the very high concentration of *N*-linked complex oligosaccharide chains on lysosomal membrane proteins (Lewis *et al.*, 1985; Chen *et al.*, 1985; D'Souza and August, 1986; Lippincott-Schwartz and Fambrough, 1986; Barriocanal *et al.*, 1986; Green *et al.*, 1987) has led to the suggestion that glycosylation may play a role in protecting the lysosomal membrane against inappropriate degradation, but no direct role of the oligosaccharide chains in lysosomal membrane stabilization has been demonstrated. While it is clear that lysosomal enzymes undergo a discrete series of proteolytic cleavages (at least some of which are lysosomal) during their maturation, it is not entirely clear how proteolytic processing may contribute to lysosomal stability of lysosomal enzymes. Our recent studies (Hancock *et al.*, 1988) in fibroblasts derived from patients suffering from infantile generalized *N*-acetylneuraminic acid storage disease (see Chapter 15) have clearly demonstrated impaired proteolytic processing of a number of lysosomal enzymes secondary to intralysosomal NeuAc accumulation, but as yet we have been unable to show any effect on lysosomal enzyme activities or turnover as a result of impaired proteolytic processing.

6. NONLYSOSOMAL CATABOLISM OF GLYCOPROTEINS AND GLYCOSAMINOGLYCANS

6.1. Cytoplasmic Catabolism

As previously noted, multiple pathways exist within the cell for the proteolytic degradation of cellular proteins. In addition to lysosomal catabolism, cytoplasmic pathways include the ATP-dependent ubiquitin-mediated pathway and calcium-dependent calpain-mediated proteolysis. Recent studies discussed earlier also suggest that certain cytoplasmic proteins may be transferred to the lysosome for catabolism by a peptide-mediated transport system, suggesting that there are present in the cell mechanisms for "communication" between cytoplasmic and lysosomal catabolic pathways (reviewed by Dice, 1987).

In contrast to the extensive literature on cytoplasmic proteolytic pathways, relatively little is known about the contribution which cytoplasmic pathways might make to the cellular catabolism of glycoconjugates. As discussed earlier, there is extensive work indicating the presence of endo-β-*N*-acetylglucosaminidase activity in extracts of tissues and cultured cells. Based on subcellular fractionation studies and the pH optimum of this endoglycosidase, it would appear that the activity is cytoplasmic; it is therefore theoretically possible that this enzyme activity could initiate the cytoplasmic catabolism of *N*-linked oligosaccharides. No cytoplasmic aspartylglucosaminidase activity has yet been described, but preliminary studies in cultured β-mannosidase-deficient goat fibroblasts which show storage oligosaccharides having both GlcNAc and di-*N*-acetylchitobiose residues at the reducing terminus in soluble and lysosomal subcellular fractions (Hancock, unpublished observations) do not rule out the existence of a cytoplasmic aspartylglucosaminidase activity (although lysosomal breakage could also account for the observed subcellular distribution of storage oligosaccharides).

As summarized in Table 3, both biochemical and genetic studies in a variety of tissue and cultured cells have pointed to the existence of distinct lysosomal and cytoplasmic forms of a number of exoglycosidases, including neuraminidase, β-galactosidase, α-mannosidase, and β-mannosidase. With the possible exception of *N*-acetyl-β-hexosaminidase, there appears to be sufficient evidence to support the existence of cytoplasmic glycosidases which would constitute a cytoplasmic pathway for *N*-linked oligosaccharide catabolism. Tulsiani and Touster (1987) have recently proposed such a pathway, based on their studies of glycoprotein catabolism in swainsonine-treated rat tissue. The recent observation of cytoplasmically oriented glycosylation sites for *O*-GlcNAc addition on integral membrane glycoproteins (Abejon and Hirschberg, 1988) provides further circumstantial evidence for the existence of cytoplasmic pathways for oligosaccharide catabolism, but it remains to be determined experimentally what role such pathways might play in overall cellular catabolism, and to what extent cytoplasmic and lysosomal pathways of glycoconjugate catabolism might interact.

In addition to the above-described studies involving glycoprotein catabolism, a number of recent reports also raise the possibility of cytoplasmic pathways for the catabolism of glycosaminoglycans. Specifically, Aquino *et al.* (1984) have used immunocytochemical techniques to demonstrate the intracellular distribution of a brain-specific chondroitin sulfate proteoglycan in tissue from adult rats. While this proteoglycan is clearly not preferentially associated with the extracellular matrix, it is not to this point completely clear whether or not it is associated with the cytoplasm or intracellular organelles. In contrast, Volker *et al.* (1984) have demonstrated the association of endocytosed proteoglycan with the endocytic/lysosomal compartments of bovine arterial tissue. Ripellino *et al.* (1988) have clearly shown the association of hyaluronic acid with the cytoplasm in developing and mature rat cerebellum tissue, but there is not yet any evidence as to whether the turnover of this cytoplasmically localized hyaluronic acid is mediated by cytoplasmic catabolic enzymes, or if it must be translocated to the lysosome for degradation. Although previously described brain hyaluronidase activity has a pH optimum (3.7) consistent with a lysosomal site of action (Margolis *et al.*, 1972), the existence of microsomal forms of β-glucuronidase which are distinct from lysosomal forms of the enzyme (Tulsiani *et al.*, 1978) at least raises the possibility of nonlysosomal glycosaminoglycan catabolic pathways. Even in the absence of nonlysosomal intracellular catabolic pathways for glycoproteins and

Table 3. Nonlysosomal Glycosidases of Mammalian Tissues

Enzyme/tissue source	Basis of distinction from lysosomal activity	Reference
Neuraminidase		
Rat liver	pH optimum; inhibition by Cu^{2+}, Hg^{2+}, Zn^{2+}; lack of activity against gangliosides and mucin	Tulsiani and Carubelli (1970)
	pH optimum; substrate specificity; subcellular fractionation	Miyagi and Tsuiki (1985)
Porcine brain	pH optimum; subcellular fractionation/localization	Venerando *et al.* (1978)
β-Galactosidase		
Human liver	Lack of immunological cross-reactivity; pH optimum; lack of binding to Con A	Ben-Yoseph *et al.* (1977)
α-Mannosidase		
Rat liver	pH optimum; ion sensitivity; subcellular fractionation/localization; kinetics of substrate hydrolysis	Marsh and Gourlay (1971)
	pH optimum; subcellular fractionation/localization (soluble); lack of immunological crossreactivity	Opheim and Touster (1978)
	pH optimum; immunological relatedness to ER activity; subcellular fractionation/localization (soluble); lack of binding to Con A; lack of inhibition by swainsonine and 1-deoxymannojirimycin	Bischoff and Kornfeld (1986)
Rat tissues	pH optimum; substrate specificity	Tulsiani and Touster (1987)
Human liver	pH optimum; binding to DEAE matrix; residual activity in α-mannosidosis	Carroll *et al.* (1972)
	pH optimum; lack of binding to Con A	Phillips *et al.* (1976)
Feline tissues	pH optimum; binding to DEAE matrix; lack of binding to Con A; residual activity in α-mannosidosis	Burditt *et al.* (1980)
β-Mannosidase		
Caprine tissues	pH optimum; lack of binding to Con A; residual activity in β-mannosidosis	Dawson (1982)
α-Fucosidase		
Human tissues	Degree of oligomerization; carbohydrate content; substrate specificity	Chien and Dawson (1980)
Monkey brain	Degree of oligomerization; heat inactivation susceptibility	Alam and Balasubramanian (1978)
β-Glucuronidase		
Rat liver	Microsomal activity; pH optimum; subcellular fractionation/localization; carbohydrate content; amino acid composition	Tulsiani *et al.* (1978)
Monkey brain	Microsomal activity; pH optimum; subcellular fractionation/localization; carbohydrate content	Alvares and Balasubramanian (1982)

glycosaminoglycans, the mechanisms by which cytoplasmic glycoconjugates are translocated to the lysosome for catabolism remain to be elucidated.

6.2. Extracellular Catabolism

Although it is generally accepted that extracellular catabolism makes a significant contribution to the turnover of the extracellular matrix, little is known about the mechanisms by which such extracellular catabolism might be accomplished. Recent studies by Sweeley and Usuki (1987) and Usuki *et al.* (1988) have suggested that the action of extracellular neuraminidase on GM3 may have significant effects on the growth of cells, and the inappropriate secretion of lysosomal glycosidases and proteinases by tumor cells such as glioblastomas (discussed earlier) has been proposed as a biochemical mechanism which might mediate tumor invasion and metastasis.

7. CATABOLISM OF NEURAL-SPECIFIC GLYCOCONJUGATES

Townsend *et al.* (1979) provided a review of the properties and occurrence of lysosomal glycosidases in nervous tissue; a selection of reports dealing with the properties of lysosomal hydrolases of nervous tissue is given in Table 4. In this section, we will deal at greater length with possible pathways for the catabolism of neural-specific glycoconjugates.

7.1. The HNK-1/L2 Epitope

The HNK-1/L2 epitope is a carbohydrate antigen which is distributed among a number of neuroglycoconjugates, including glycoproteins, glycolipids, and proteoglycans. Glycoproteins which bear this epitope, which has been identified as a glucuronic acid 3-sulfate substituent of the oligosaccharide chains (Chou *et al.*, 1986), include molecules which are involved in cell–cell adhesion (Kruse *et al.*, 1984; Poltorak *et al.*,

Table 4. Lysosomal Hydrolases of Nervous Tissue: A Partial Review of the Literature

Enzyme	References
Neuraminidase	Venerando *et al.* (1978), Wille and Trenker (1981), Saito and Yu (1986)
α-Fucosidase	Alam and Balasubramanian (1978), Chien and Dawson (1980)
β-Galactosidase	Farrell *et al.* (1973), Tanaka and Suzuki (1977)
β-Hexosaminidase	Overdijk *et al.* (1982)
α-Mannosidase	Burditt *et al.* (1980), Tulsiani and Touster (1983), Zanetta *et al.* (1983)
β-Glucuronidase	Alvares and Balasubramanian (1982)
Cathepsins	Ivy (1988)
Hyaluronidase	Margolis *et al.* (1972)
Endo-β-*N*-acetylglucosaminidase	Cook *et al.* (1984)

1987; Kunemund et al., 1988). The epitope has also been observed on other soluble and membrane-associated neural glycoproteins (Margolis et al., 1987; Shashoua et al., 1986). It is probable that previously described arylsulfatase A and β-glucuronidase activities are capable of effecting the catabolism of the HNK-1 oligosaccharide structure, but its catabolism could also involve extracellular as well as lysosomal pathways, based on the distribution of the epitope.

7.2. Neural Cell Adhesion Molecules (NCAMs)

NCAMs are cell-surface glycoproteins of neural cells which appear to be involved in the developmental regulation of cell–cell interactions (see Chapters 11 and 12). Of particular interest in the context of glycoconjugate catabolism is the high content of sialic acid in the NCAM molecules. It has been clearly shown that this sialic acid is expressed as α2-8-linked polysialic acid chains which are attached via the usual N-linked oligosaccharide core structures (Finne, 1982; Finne et al., 1983) to the polypeptide backbone. Early work implicated polysialic acid content as a modulator of NCAM function (Rutishauer et al., 1982; Hoffman and Edelman, 1983), and more recent studies (Rutishauser et al., 1985, 1988; Acheson and Rutishauser, 1988) have directly demonstrated the involvement of polysialic acid oligosaccharides in cell–cell adhesion and signal transduction of neural cells. To this point, it is not clear if developmental and other temporal alterations in cell-surface polysialic acid are mediated by NCAM turnover and replacement by molecules having altered polysialic acid content, or whether cell-surface polysialic acid content may be more directly modulated by the local action of extracellular exo- or endoneuraminidase activities. The resolution of this question is dependent upon future studies designed to identify and characterize catabolic enzymes and pathways involved in the turnover of NCAM molecules in mammalian nervous tissue.

7.3. Other Neural Glycoconjugates

Over the past several years, a number of distinct oligosaccharides have been isolated from brain chondroitin sulfate proteoglycans, all of which are attached via O-glycosidic linkage of the mannosyl-Ser/Thr type (Finne et al., 1979; Krusius et al., 1986, 1987). While most of the linkages present in these oligosaccharides should be susceptible to the "typical" spectrum of lysosomal glycosidases, it is possible that catabolism of the mannosyl-O-Ser/Thr linkage may require the action of a specific hydrolase. As reviewed elsewhere (Chapter 15), certain sialoglycosphingolipids such as GM2 (II^3NeuAcGgOse$_3$Cer) are essentially unique to the nervous system and require an activator protein for their lysosomal catabolism, although there is no evidence for neural-specific lysosomal hydrolases which may be involved in their degradation.

8. DEVELOPMENTAL MODULATION OF CATABOLIC ENZYMES

While it is clear that significant alterations in the glycoconjugate composition of most tissues occur during the course of development, there have been relatively few studies designed to assess the contribution which alterations in levels of catabolic enzymes might make to overall glycoconjugate composition. Shin-Buehring *et al.* (1980), in a survey of human fetal tissues (including brain) during development, reported alterations in the activity of a number of lysosomal enzymes (including α-glucosidase, β-galactosidase, α-galactosidase, and acid phosphatase), but were unable to draw any strong functional implications from their studies. In much the same manner, developmental studies in rodent cerebellum have shown alterations in the activity of neuraminidase (Wille and Trenker, 1981), α-mannosidase (Zanetta *et al.*, 1983), and endo-β-glucosaminidase (Cook *et al.*, 1984), although the significance of these changes in catabolic activity is not clearly established.

Schwarting *et al.* (1987) have assessed the developmental expression of glycolipids and glycoproteins bearing the HNK-1 epitope in rodent brain, and have suggested that the epitope may be concentrated in postmigratory cells of the embryonic nervous system; glycolipids and glycoproteins appear to be differentially regulated in their expression. These authors further suggest the possible correlation of expression with the activity of catabolic enzymes, which in turn may modulate the cell–cell interactions requisite for proper neural development.

Bourbon *et al.* (1987) have also correlated the expression of α-glucosidase with fetal rat lung development, although in this case increased α-glucosidase activity may facilitate the conversion of glycogen stores to substrates for surfactant phospholipid biosynthesis.

Developmental modulation of brain hyaluronic acid content (Oohira *et al.*, 1986; Margolis *et al.*, 1975) may be mediated by alterations in hyaluronidase activity (Polansky *et al.*, 1974), although it is possible that this could be related to absolute enzyme activity or to alterations in the relative distribution of enzyme activity between intracellular and extracellular compartments; similar mechanisms may exist for modulating the concentration and expression of other glycosaminoglycans in brain tissue.

In conclusion, it would appear that considerable additional work will be required in order to appreciate the relative contributions of synthesis and catabolism to the documented developmental alterations in glycoconjugate composition in the developing nervous system.

9. FUTURE DIRECTIONS

It is apparent from the preceding discussion that, while much is known about the enzymes involved in glycoprotein and glycosaminoglycan catabolism in the nervous system, relatively little is understood about many aspects of catabolism, including (1) the relative roles of lysosomal and nonlysosomal catabolism, (2) the interaction of catabolic enzymes, (3) the role of catabolism in nervous system development, and (4) the pathways of catabolism for specific neuroglycoconjugates.

In addition, the description over the past few years of inherited disorders which involve impaired lysosomal transport of catabolites including NeuAc (Hancock *et al.*, 1982, 1983; Hildreth *et al.*, 1986; see also Chapter 15), cystine (Gahl *et al.* 1982; Jones *et al.* 1982; Pisoni *et al.*, 1985), vitamin B_{12} (Rosenblatt *et al.*, 1985), and cholesterol (Sokol *et al.*, 1988) suggests that lysosomal catabolism of glycoconjugates and other constituents, and directed transport of catabolites to intracellular sites in neurons or glia where they may be available for synthetic processes, may be essential for normal cellular function. Clearly, the description of reutilization pathways for lysosomal catabolites is essential for appreciating the interaction of catabolic and synthetic processes.

Finally, much remains to be discovered concerning the regulation of catabolic activity, both at the level of enzyme synthesis and at the level of organelle biogenesis and turnover. The selective vulnerability of certain cell types (e.g., motor neurons) to specific deficiencies in lysosomal enzymes (e.g., N-acetyl-β-hexosaminidase iso-enzyme deficiencies) secondary to impaired glycoconjugate catabolism (e.g., GM2) emphasizes the vital role of lysosomal catabolism in normal neural cell function and survival. The biochemical correlation of glycoconjugate catabolism with other neural functions, such as memory consolidation and neural plasticity, may prove to be another fruitful area of investigation.

ACKNOWLEDGMENTS

The authors apologize for any inadvertent omissions of relevant studies, and take full responsibility for any misstatements. We thank Dr. Karl Johnson for his comments on the manuscript and sharing of preliminary data, and LaJoyce Safford for typing the manuscript. L.W.H. is supported by a USPHS FIRST AWARD, DK 38593, and G.D. by USPHS Grants HD-06426, NS-23131, and HD-09402.

10. REFERENCES

Abejon, C., and Hirschberg, C. H., 1988, Intrinsic membrane of glycoproteins with cytosol-oriented sugars in the endoplasmic reticulum, *Proc. Natl. Acad. Sci. USA* **85**:1010–1014.

Abraham, C. R., Selkoe, D. J., and Potter, H., 1988, Immunochemical identification of the serine protease inhibitor α1-antichymotrypsin in the brain amyloid deposits of Alzheimer's disease, *Cell* **52**:487–501.

Abraham, D., Blakemore, W. F., Jolly, R. D., Sidebotham, R., and Winchester, B., 1983, The catabolism of mammalian glycoproteins, *Biochem. J.* **215**:573–579.

Abraham, D., Blakemore, W. F., Dell, A., Herrtage, M. E., Jones, J., Littlewood, J. T., Oates, J., Palmer, A. C., Sidebotham, R., and Winchester, B., 1984, The enzymic defect and storage products in canine fucosidosis, *Biochem. J.* **221**:25–33.

Acheson, A., and Rutishauser, U., 1988, Neural cell adhesion molecule regulates cell contact-mediated changes in choline acetyltransferase activity of embryonic chick sympathetic neurons, *J. Cell Biol.* **106**: 479–486.

Aerts, J. M. F. G., Brul, S., Donker-Koopman, W. E., van Weely, S., Murray, G. J., Barranger, J. A., Tager, J. M., and Schram, A. W., 1986, Efficient routing of glucocerebrosidase to lysosomes requires complex oligosaccharide formation, *Biochem. Biophys. Res. Commun.* **141**:452–458.

Ahlberg, J., Berkenstam, A., Henell, F., and Glaumann, H., 1985, Degradation of short and long-lived proteins in isolated rat liver lysosomes, *J. Biol. Chem.* **260**:5847–5854.

Alam, T., and Balasubramanian, A. S., 1978, The purification, properties and characterization of three forms of α-L-fucosidase from monkey brain, *Biochim. Biophys. Acta* **524**:373–384.

Alvares, K., and Balasubramanian, A. S., 1982, Lysosomal and microsomal β-glucuronidase of monkey brain, *Biochim. Biophys. Acta* **708**:124–133.

Aquino, D. A., Margolis, R. U., and Margolis, R. K., 1984, Immunocytochemical localization of a chondroitin sulfate proteoglycan in nervous tissue, *J. Cell Biol.* **99**:1117–1129.

Aronson, N. N., Jr., and Docherty, P. A., 1983, Degradation of [6-^{3}H]- and [1-^{14}C]glucosamine-labeled asialo-α_1-acid glycoprotein by the perfused rat liver, *J. Biol. Chem.* **258**:4266–4271.

Backer, J. M., and Dice, J. F., 1986, Covalent linkage of ribonuclease S-peptide to microinjected proteins causes their intracellular degradation to be enhanced during serum withdrawal, *Proc. Natl. Acad. Sci. USA* **83**:5830–5834.

Bame, K. J., and Rome, L. H., 1986, Genetic evidence for transmembrane acetylation by lysosomes, *Science* **233**:1087–1089.

Barriocanal, J. G., Bonifacino, J. S., Yuan, L., and Sandoval, J., 1986, Biosynthesis, glycosylation, movement through the Golgi system and transport to the lysosomes by an N-linked carbohydrate-independent mechanism of three lysosomal integral membrane proteins (LIMPS), *J. Biol. Chem.* **261**: 16755–16763.

Baussant, T., Strecker, G., Wieruszeski, J.-M., Montreull, J., and Michalski, J.-C., 1986, Catabolism of glycoprotein glycans, *Eur. J. Biochem.* **159**:381–385.

Ben-Yoseph, Y., Shapira, E., Edelman, D., Burton, B. K., and Nadler, H. L., 1977, Purification and properties of neutral β-galactosidase activities from human liver, *Arch. Biochem. Biophys.* **184**:373–379.

Bienkowski, M. J., and Conrad, H. E., 1984, Kinetics of proteoheparan sulfate synthesis, secretion, endocytosis, and catabolism by a hepatocyte cell line, *J. Biol. Chem.* **259**:12989–12996.

Bischoff, J., and Kornfeld, R., 1986, The soluble form of rat liver α-mannosidase is immunologically related to the endoplasmic reticulum membrane α-mannosidase, *J. Biol. Chem.* **261**:4758–4765.

Bourbon, J. R., Doncet, E., and Rieutort, M., 1987, Role of α-glucosidase in fetal lung maturation, *Biochim. Biophys. Acta* **917**:203–210.

Brassart, D., Baussant, T., Wieruszeski, J.-M., Strecker, G., Montreuil, J., and Michalski, J.-C.., 1987, Catabolism of N-glycosylprotein glycans: Evidence for a degradation pathway of sialoglycoasparagines resulting from the combined action of the lysosomal aspartylglucosaminidase and endo-N-acetyl-β-D-glucosaminidase, *Eur. J. Biochem.* **169**:131–136.

Brauker, J. H., and Wang, J. L., 1987, Non-lysosomal processing of cell-surface heparan sulfate proteoglycans, *J. Biol. Chem.* **262**:13093–13101.

Brauker, J. H., Roff, C. F., and Wang, J. L., 1986, The effect of mannose 6-phosphate on the turnover of the proteoglycans in the extracellular matrix of human fibroblasts, *Exp. Cell Res.* **164**:115–126.

Braulke, T., Hille, A., Huttner, H. B., Hasilik, A., and von Figura, K., 1987, Sulfated oligosaccharides in human lysosomal enzymes, *Biochem. Biophys. Res. Commun.* **143**:178–185.

Burditt, L. J., Chotai, K., Hirani, S., Nugent, P. G., Winchester, B. G., and Blakemore, W. F., 1980, Biochemical studies on a case of feline mannosidosis, *Biochem. J.* **189**:467–473.

Cain, C. C., and Murphy, R. F., 1988, A chloroquine-resistant Swiss 3T3 cell line with a defect in late endocytic acidification, *J. Cell Biol.* **106**:269–277.

Carroll, M., Dance, N., Masson, P. K., Robinson, D., and Winchester, B. G., 1972, Human mannosidosis—The enzymic defect, *Biochem. Biophys. Res. Commun.* **49**:579–583.

Cenci di Bello, I., Dorling, P., and Winchester, B., 1983, The storage products in genetic and swainsonine-induced human mannosidosis, *Biochem. J.* **215**:693–696.

Chambers, J. P., and Elbein, A. D., 1986, Effects of castanospermine on purified lysosomal α-1,4-glucosidase, *Enzyme* **35**:53–56.

Chen, J. W., Murphy, T. L., Willingham, M. C., Pastan, I., and August, J. T., 1985, Identification of two lysosomal membrane glycoproteins, *J. Cell Biol.* **101**:85–95.

Chiang, H.-L., and Dice, J. F., 1988, Peptide sequences that target proteins for enhanced degradation during serum withdrawal, *J. Biol. Chem.* **263**:6797–6805.

Chien, S.-F., and Dawson, G., 1980, Purification and properties of two forms of human α-L-fucosidase, *Biochim. Biophys. Acta* **614**:476–488.

Chou, D. K. H., Ilyas, A. A., Evans, J. E., Costello, C., Quarles, R. H., and Jungalwala, F. B., 1986, Structure of sulfated glucuronyl glycolipids in the nervous system reacting with HNK-1 antibody and some IgM preparations in neuropathy, *J. Biol. Chem.* **261**:11717–11725.

Cook, N. J., Dontenwill, M., Meyer, A., Vincendon, G., and Zanetta, P., 1984, Postnatal modifications of endo-β-D-acetylglucosaminidase in the developing rat cerebellum, *Dev. Brain Res.* **15**:298–301.

Corvera, S., Roach, P. J., DePaoli-Roach, A. A., and Czech, M. P., 1988a, Insulin action inhibits insulin-like growth factor-II (IGF-II) receptor phosphorylation in H-35 hepatoma cells, *J. Biol. Chem.* **263**: 3116–3122.

Corvera, S., Yagaloff, K. A., Whitehead, R. E., and Czech, M. P., 1988b, Tyrosine phosphorylation of the receptor for insulin-like growth factor-II is inhibited in plasma membranes from insulin-treated rat adipocytes, *Biochem. J.* **250**:47–52.

Daniel, P. F., Warren, C. D., and James, L. F., 1984, Swainsonine-induced oligosaccharide excretion in sheep, *Biochem. J.* **221**:601–607.

Dawson, G., 1982, Evidence for two distinct forms of mammalian β-mannosidase, *J. Biol. Chem.* **257**: 3369–3371.

Dawson, G., and Glaser, P., 1987, Apparent cathepsin B deficiency in neuronal ceroid lipofuscinosis can be explained by peroxide inhibition, *Biochem. Biophys. Res. Commun.* **147**:267–274.

Dawson, G., and Kernes, S., 1979, Mechanism of action of hydrocortisone potentiation of sulfogalactosylceramide synthesis in mouse oligodendroglioma clonal cell line, *J. Biol. Chem.* **254**:163–167.

Dice, J. F., 1987, Molecular determinants of protein half-lives in eukaryotic cells, *FASEB J.* **1**:349–357.

Dice, J. F., Chiang, H.-L., Spencer, E. P., and Backer, J. M., 1986, Regulation of catabolism of microinjected ribonuclease A, *J. Biol. Chem.* **261**:6853–6859.

Diment, S., and Stahl, P., 1985, Macrophage endosomes contain proteases which degrade endocytosed protein ligands, *J. Biol. Chem.* **260**:15311–15317.

Diment, S., Leech, M. S., and Stahl, P., 1988, Cathepsin D is membrane-associated in macrophage endosomes, *J. Biol. Chem.* **263**:6901–6907.

Docherty, P. A., Kuranda, M. J., Aronson, N. N., Jr., BeMiller, J. N., Myers, R. W., and Bohn, J. A., 1986, Effect of α-D-mannopyranosyl-p-nitrophenyltriazene on hepatic degradation and processing of the N-linked oligosaccharide chains of α_1-acid glycoprotein, *J. Biol. Chem.* **261**:3457–3463.

Dorling, P. R., Huxtable, C. R., and Colegate, S. M., 1980, Inhibition of lysosomal α-mannosidase by swainsonine, an indolizidine alkaloid isolated from *Swainsonia canescens*, *Biochem. J.* **191**:649–651.

D'Souza, M. P., and August, J. T., 1986, A kinetic analysis of biosynthesis and localization of a lysosome-associated membrane glycoprotein, *Arch. Biochem. Biophys.* **249**:522–532.

Duncan, J. R., and Kornfeld, S., 1988, Intracellular movement of two mannose 6-phosphate receptors: Return to the Golgi apparatus, *J. Cell Biol.* **106**:617–628.

Elbein, A. D., Solf, R., Dorling, P. R., and Vosbeck, K., 1981, Swainsonine: An inhibitor of glycoprotein processing, *Proc. Natl. Acad. Sci. USA* **78**:7393–7397.

Facci, L., Leon, A., Toffano, G., Sonnino, S., Ghidoni, R., and Tettamanti, G., 1984, Promotion of neuritogenesis in mouse neuroblastoma cells by exogenous gangliosides. Relationship between the effect and the cell association of ganglioside GM1, *J. Neurochem.* **42**:299–305.

Facci, L., Skaper, S. D., Favaron, M., and Leon, A., 1988, A role for gangliosides in astroglial cell differentiation *in vitro*, *J. Cell Biol.* **106**:61–67.

Fambrough, D. M., Takeyasu, K., Lippincott-Schwartz, J., and Siegel, N. R., 1988, Structure of LEP100, a glycoprotein that shuttles between lysosomes and the plasma membrane, deduced from the nucleotide sequence of the encoding cDNA, *J. Cell Biol.* **106**:61–67.

Farrell, D. F., Baker, H. J., Herndon, R. M., Lindsey, J. R., and McKhann, G. M., 1973, Feline GM1 gangliosidosis: Biochemical and ultrastructural comparisons with the disease in man, *J. Neuropathol. Exp. Neurol.* **32**:1–18.

Faust, P. L., Wall, D. A., Perara, E., Lingappa, V. R., and Kornfeld, S., 1987a, Expression of human cathepsin D in *Xenopus* oocytes: Phosphorylation and intracellular targeting, *J. Cell Biol.* **105**:1937–1945.

Faust, P. L., Chirgwin, J. M., and Kornfeld, S., 1987b, Renin, a secretory glycoprotein, acquires phosphomannosyl residues, *J. Cell Biol.* **105**:1947–1955.

Finne, J., 1982, Occurrence of unique polysialosylcarbohydrate units in glycoproteins of developing brain, *J. Biol. Chem.* **257**:11966–11970.

Finne, J., Krusius, T., Margolis, R. K., and Margolis, R. U., 1979, Novel mannitol-containing oligosaccharides obtained by mild alkaline borohydride treatment of a chondroitin sulfate proteoglycan from brain, *J. Biol. Chem.* **254:**10295–10300.

Finne, J., Finne, U., Deagostini-Bazin, H., and Goridis, C., 1983, Occurrence of α,2-8 linked polysialosyl units in a neural cell molecule, *Biochem. Biophys. Res. Commun.* **112:**482–487.

Fischer, I., Shea, T. S., and Saperstein, V. S., 1986, Induction of lysosomal glycosidases by dibutyryl cAMP in neuroblastoma cells, *Neurochem. Res.* **11:**589–598.

Frick, K. K., Doherty, P. J., Gottesman, M. M., and Scher, C. D., 1985, Regulation of the transcript for a lysosomal protein: Evidence for a gene program modified by platelet-derived growth factor, *Mol. Cell. Biol.* **5:**2582–2589.

Fuchs, W., Beck, M., and Kresses, H., 1985, Intralysosomal formation and metabolic fate of N-acetyl-glucosamine-6-sulfate from keratan sulfate, *Eur. J. Biochem.* **151:**551–556.

Gabel, C. A., and Foster, S. A., 1986a, Lysosomal enzyme trafficking in mannose 6-phosphate receptor-positive mouse L cells: Demonstration of a steady state accumulation of phosphorylated acid hydrolases, *J. Cell Biol.* **102:**943–950.

Gabel, C. A., and Foster, S. A., 1986b, Mannose 6-phosphate receptor-mediated endocytosis of acid hydrolases: Internalization of β-glucuronidase is accompanied by a limited dephosphorylation, *J. Cell Biol.* **103:**1817–1827.

Gabel, C. A., and Foster, S. A., 1987, Postendocytic maturation of acid hydrolases: Evidence of pre-lysosomal processing, *J. Cell Biol.* **105:**1561–1570.

Gahl, W. A., Tietze, F., Bashan, N., Steinherz, R., and Schulman, J. D., 1982, Defective cystine exodus from isolated lysosome-rich fractions of cystinotic leucocytes, *J. Biol. Chem.* **257:**9570–9575.

Gieselmann, V., Hasilik, A., and von Figura, K., 1985, Processing of human cathepsin D in lysosomes *in vitro*, *J. Biol. Chem.* **260:**3215–3220.

Grabowski, G. A., Osieck-Newman, K., Dinur, T., Fabbro, D., Legler, G., Gatt, S., and Desnick, R. J., 1986, Human acid β-glucosidase, *J. Biol. Chem.* **261:**8263–8269.

Green, S. A., Zimmer, K.-P., Griffiths, G., and Mellman, I., 1987, Kinetics of intracellular transport and sorting of lysosomal membrane and plasma membrane proteins, *J. Cell Biol.* **105:**1227–1240.

Griffiths, G., Hoflack, B., Simons, K., Mellman, I., and Kornfeld, S., 1988, The mannose 6-phosphate receptor and the biogenesis of lysosomes, *Cell* **52:**329–341.

Hanada, K., Tamai, M., Adachi, T., Oguma, K., Kashiwagi, K., Ohmura, S., Kominami, E., Towatari, T., and Katunuma, N., 1983, Characterization of three new analogs of E-64 and their therapeutic application, in: *Proteinase Inhibitors: Medical and Biological Aspects* (N. Katunuma, ed.), pp. 25–36, Springer-Verlag, Berlin.

Hancock, L. W., 1989, Swainsonine effects on glycoprotein catabolism in cultured cells, in: *Proceedings of the International Symposium on Swainsonine and Related Glycosidase Inhibitors* (L. James, C. Warren, A. Elbein, and R. Molyneux, eds.), Iowa State University Press, Ames, in press.

Hancock, L. W., and Dawson, G., 1987, Evidence for two catabolic endoglycosidase activities in β-mannosidase-deficient goat fibroblasts, *Biochem. Biophys. Acta* **928:**13–21.

Hancock, L. W., Thaler, M. M., Horwitz, A. L., and Dawson, G., 1982, Generalized N-acetylneuraminic acid storage disease: Quantitation and identification of the monosaccharide accumulating in brain and other tissues, *J. Neurochem.* **38:**803–809.

Hancock, L. W., Horwitz, A. L., and Dawson, G., 1983, N-acetylneuraminic acid and sialoglycoconjugate metabolism in fibroblasts from a patient with generalized N-acetylneuraminic acid storage disease, *Biochim. Biophys. Acta* **760:**42–52.

Hancock, L. W., Jones, M. Z., and Dawson, G., 1986, Glycoprotein catabolism in normal and β-mannosidase-deficient goat skin fibroblasts, *Biochem. J.* **234:**175–183.

Hancock, L. W., Ricketts, J. P., and Hildreth, J., 1988, Impaired proteolytic processing of lysosomal N-acetyl-β-hexosaminidase in cultured fibroblasts from patients with infantile generalized N-acetyl-neuraminic acid storage disease, *Biochem. Biophys. Res. Commun.* **152:**83–92.

Hanover, J. A., Cohen, C. K., Willingham, M. C., and Park, M. K., 1987, O-linked N-acetylglucosamine is attached to protein of the nuclear pore. Evidence for cytoplasmic and nucleoplasmic glycoproteins, *J. Biol. Chem.* **262:**9887–9894.

Hascall, V. C., 1981, Proteoglycans: Structure and function, in: *Biology of Carbohydrates* (V. Ginsburg and P. Robbins, eds.), Vol. 1, pp. 1–50, Wiley, New York.

Hasilik, A., and von Figura, K., 1984, Processing of lysosomal enzymes in fibroblasts, in: *Lysosomes in Biology and Pathology* (J. T. Dingle, R. T. Dean, and W. Sly, eds.), Vol. 7, pp. 3–16, Elsevier, Amsterdam.

Hildreth, J., Sacks, L., and Hancock, L. W., 1986, N-acetylneuraminic acid accumulation in a buoyant lysosomal fraction of cultured fibroblasts from patients with infantile N-acetylneuraminic acid storage disease, *Biochem. Biophys. Res. Commun.* **139**:838–844.

Hill, D. F., Bullock, P. N., Chiapelli, F., and Rome, L. H., 1985, Binding and internalization of lysosomal enzymes by primary cultures of rat glia, *J. Neurochem. Res.* **14**:35–47.

Hoffman, S., and Edelman, G. M., 1983, Kinetics of homophilic binding by embryonic and adult forms of neural cell adhesion molecules, *Proc. Natl. Acad. Sci. USA* **80**:5762–5766.

Holt, G. D., and Hart, G. W., 1986, The subcellular distribution of terminal N-acetylglucosamine moieties, *J. Biol. Chem.* **261**:8049–8057.

Holt, G. D., Snow, C. M., Senior, A., Haltiwanger, R. S., and Gerace, L., and Hart, G. W., 1987a, Nuclear pore complex glycoproteins contain cytoplasmically disposed O-linked N-acetylglucosamine, *J. Cell Biol.* **104**:1157–1164.

Holt, G. D., Haltiwanger, R. S., Torres, C.-R., and Hart, G. W., 1987b, Erythrocytes contain cytoplasmic glycoproteins, *J. Biol. Chem.* **262**:14847–14850.

Hoppe, W., Rauch, U., and Kresse, H., 1988, Degradation of endocytosed dermatan sulfate proteoglycan in human fibroblasts, *J. Biol. Chem.* **263**:5926–5932.

Iozzo, R. V., 1987, Turnover of heparan sulfate proteoglycan in human colon carcinoma cells, *J. Biol. Chem.* **262**:1888–1900.

Ivy, G. O., 1988, Decreased neural plasticity in aging and certain pathologic conditions: Possible roles of protein turnover, in: *Neural Plasticity: A Lifespan Approach* (T. Petit and G. Ivy, eds.), pp. 351–371, Liss, New York.

Johnson, K. F., Hancock, L. W., and Dawson, G., 1988a, Post-synthetic processing of lysosomal α-fucosidase, *J. Cell Biol.* **107**:341a.

Johnson, K. F., Dawson, G., and Hancock, L. W., 1988b, Genetic diversity among α-L-fucosidase deficiencies: evidence for a "protective" factor, *Genome* **30**(Suppl. 1):226 (abstr.).

Jolly, R. D., Winchester, B. G., Gehler, J., Dorling, P. R., and Dawson, G., 1981, Mannosidosis: A comparative review of biochemical and related clinicopathological aspects of three forms of the disease, *J. Appl. Biochem.* **3**:273–291.

Jonas, A. J., Smith, M. L., and Schneider, J. A., 1982, ATP-dependent lysosomal cystine efflux is defective in cystinosis, *J. Biol. Chem.* **257**:13185–13188.

Jones, M. Z., and Laine, R. A., 1981, Caprine oligosaccharide storage disease, *J. Biol. Chem.* **256**:5181–5184.

Jonsson, L. M. V., Murray, G. J., Sorrell, S. H., Strijland, A., Aerts, J. M. F. G., Ginns, E. I., Barranger, J. A., Tager, J. M., and Schram, A. W., 1987, Biosynthesis and maturation of glucocerebrosidase in Gaucher fibroblasts, *Eur. J. Biochem.* **164**:171–179.

Kanfer, J. N., Legler, G., Sullivan, J., Raghavan, S. S., and Mumford, R. A., 1975, The Gaucher mouse, *Biochem. Biophys. Res. Commun.* **67**:85–90.

Kjellén, L., Pertoft, H., Oldberg, Å., and Höök, M., 1985, Oligosaccharides generated by an endo-glucuronidase are intermediates in the intracellular degradation of heparan sulfate proteoglycans, *J. Biol. Chem.* **260**:8416–8422.

Kornfeld, R., and Kornfeld, S., 1985, Assembly of asparagine-linked oligosaccharides, *Annu. Rev. Biochem.* **54**:631–644.

Kornfeld, S., 1987, Trafficking of lysosomal enzymes, *FASEB J.* **1**:462–468.

Kruse, T., Mailhammer, R., Wernecke, H., Faissner, A., Sommer, I., Goridis, C., and Schachner, M., 1984, Neural cell adhesion molecules and myelin-associated glycoprotein share a common carbohydrate moiety recognized by monoclonal antibodies L2 and HNK-1, *Nature* **311**:153–155.

Krusius, T., Finne, J., Margolis, R. K., and Margolis, R. U., 1986, Identification of an O-glycosidic mannose-linked sialylated tetrasaccharide and keratan sulfate oligosaccharides in the chondroitin sulfate proteoglycan of brain, *J. Biol. Chem.* **261**:8237–8242.

Krusius, T., Reinhold, V. N., Margolis, R. K., and Margolis, R. U., 1987, Structural studies on sialylated and sulphated O-glycosidic mannose-linked oligosaccharides in the chondroitin sulphate proteoglycan of brain, *Biochem. J.* **245**:229–234.

Kunemund, V., Jungalawala, F. B., Fischer, G., Chou, D. K. H., Keilhauer, G., and Schachner, M., 1988, The L2/HNK-1 carbohydrate of neural cell adhesion molecules is involved in cell interactions, *J. Cell Biol.* **106**:213-223.

Kuranda, M. J., and Aronson, N. N., Jr., 1985, Use of site-directed inhibitors to study *in situ* degradation of glycoproteins by the perfused rat liver, *J. Biol. Chem.* **260**:1858-1866.

Kuranda, M. J., and Aronson, N. N., Jr., 1986, A di-N-acetylchitobiase activity is involved in the lysosomal catabolism of asparagine-linked glycoproteins in rat liver, *J. Biol. Chem.* **261**:5803-5809.

Lang, L., Reitman, M. L., Tang, J., Roberts, R. M., and Kornfeld, S., 1984, Lysosomal enzyme phosphorylation: Recognition of a protein determinant allows specific phosphorylation of oligosaccharides present on lysosomal enzymes, *J. Biol. Chem.* **259**:14663-14671.

Laurent, T. C., Fraser, J. R. E., Pertoft, H., and Smedsrod, B., 1986, Binding of hyaluronate and chondroitin sulphate to liver endothelial cells, *Biochem. J.* **234**:653-658.

Lewis, V., Green, S. A., Marsh, M., Vihlco, P., Helenius, A., and Mellman, I., 1985, Glycoproteins of the lysosomal membrane, *J. Cell Biol.* **100**:1839-1847.

Li, S.-C., Sonnino, S., Tettamanti, G., and Li, Y.-T., 1988, Characterization of a non-specific activator protein for the enzymatic hydrolysis of glycolipids, *J. Biol. Chem.* **263**:6588-6591.

Lippincott-Schwartz, J., and Fambrough, D. M., 1986, Lysosomal membrane dynamics: Structure and interorganellar movement of a major lysosomal membrane glycoprotein, *J. Cell Biol.* **102**:1593-1605.

Lisman, J. J. W., vanderWal, C. J., and Overdijk, B., 1985, Endo-N-acetyl-β-D-glucosaminidase activity in rat liver, *Biochem. J.* **229**:379-385.

Lloyd, J. B., 1986, Disulphide reduction in lysosomes, *Biochem. J.* **237**:271-272.

Margolis, R. K., Ripellino, J. A., Goossen, B., Steinbrich, R., and Margolis, R. U., 1987, Occurrence of the HNK-1 epitope (3-sulfoglucuronic acid) in PC12 pheochromocytoma cells, chromaffin granule membranes, and chondroitin sulfate proteoglycans, *Biochem. Biophys. Res. Commun.* **145**:1142-1148.

Margolis, R. U., Margolis, R. K., Santella, R., and Atherton, D. M., 1972, The hyaluronidase of brain, *J. Neurochem.* **19**:2325-2332.

Margolis, R. U., Margolis, R. K., Chang, L. B., and Petri, C., 1975, Glycosaminoglycans of brain during development, *Biochemistry* **14**:85-88.

Marsh, C. A., and Gourlay, G. C., 1971, Evidence for a non-lysosomal α-mannosidase in rat liver homogenates, *Biochim. Biohys. Acta* **235**:142-148.

Mason, R. W., Gal, S., and Gottesman, M. M., 1987, The identification of the major excreted protein (MEP) from a transformed mouse fibroblast line as a catalytically active precursor form of cathepsin L, *Biochem. J.* **248**:448-454.

McElligott, M. A., Miao, P., and Dice, J. F., 1985, Lysosomal degradation of ribonuclease A and ribonuclease S-protein microinjected into human fibroblasts, *J. Biol. Chem.* **260**:11986-11993.

Mellman, I., Fuchs, R., and Helenius, A., 1986, Acidification of the endocytic and exocytic pathways, *Annu. Rev. Biochem.* **55**:663-700.

Mendl, P., and Tuppy, H., 1969, Kompetitive hemmung der *vibro cholerae* neuraminidase durch 2-deoxy-2,3-dehydro-N-acyl-neuraminnsauren, *Hoppe-Seyler's Z. Physiol. Chem.* **350**:1088-1092.

Miyagi, T., and Tsuiki, S., 1985, Purification and characterization of cytosolic sialidase from rat liver, *J. Biol. Chem.* **260**:6710-6716.

Morales, T. I., and Hascall, V. C., 1988, Correlated metabolism of proteoglycans and hyaluronic acid in bovine cartilage organ cultures, *J. Biol. Chem.* **263**:3632-3638.

Mort, J. S., Leduc, M. S., and Recklies, A. D., 1983, Characterization of a latent proteinase from ascitic fluid as a high molecular weight form of cathepsin B, *Biochim. Biophys. Acta* **755**:369-375.

Nakajima, M., Irimura, T., DiFerrante, N., and Nicholson, G. L., 1984, Metastatic melanoma cell heparanase, *J. Biol. Chem.* **259**:2283-2290.

Neufeld, E. F., and Ashwell, G., 1980, Carbohydrate recognition systems for receptor-mediated pinocytosis, in: *The Biochemistry of Glycoproteins and Proteoglycans* (W. J. Lennarz, ed.), pp. 241-266, Plenum Press, New York.

Oohira, A., Matsui, F., Matsuda, M., and Shoji, R., 1986, Developmental change in the glycosaminoglycan composition of the rat brain, *J. Neurochem.* **47**:588-593.

Opheim, D. J., and Touster, O., 1978, Lysosomal α-D-mannosidase of rat liver, *J. Biol. Chem.* **253**:1017-1023.

Orkin, R. W., Underhill, C. B., and Toole, B. P., 1982, Hyaluronate degradation by 3T3 and simian virus-transformed 3T3 cells, *J. Biol. Chem.* **257:**5821–5826.

Oude Elferink, R. P. j., van Doorn-van Wakeren, J., Hendriks, T., Strijland, A., and Tager, J. M., 1986, Transport and processing of endocytosed lysosomal α-glucosidase in cultured human skin fibroblasts, *Eur. J. Biochem.* **158:**339–344.

Overdijk, B., van der Kroef, W. M. J., Lisman, J. J. W., Pierce, R. J., Montreuil, J., and Spik, G., 1981, Demonstration and partial characterization of endo-N-acetyl-β-D-glucosaminidase in human tissues, *FEBS Lett.* **128:**364–366.

Overdijk, B., vanSteijn, G., Wolf, J. H., and Lisman, J. J. W., 1982, Purification and partial characterization of the carbohydrate structure of lysosomal N-acetyl-β-D-hexosaminidases from bovine brain, *Int. J. Biochem.* **14:**25–31.

Phillips, N. C., Robinson, D., and Winchester, B. G., 1976, Characterization of human liver α-D-mannosidase purified by affinity chromatography, *Biochem. J.* **153:**579–587.

Pierce, R. J., Spik, G., and Montreuil, J., 1979, Cytosolic location of an endo-N-acetyl-β-D-glucosaminidase activity in rat liver and kidney, *Biochem. J.* **180:**673–676.

Pierce, R. J., Spik, G., and Montreuil, J., 1980, Demonstration and cytosolic location of an endo-N-acetyl-β-D-glucosaminidase activity towards an asialo-N-acetyl-lactosamine-type substrate in rat liver, *Biochem. J.* **185:**261–264.

Pisoni, R., Thoene, J. G., and Christensen, H. N., 1985, Detection and characterization of carrier-mediated cationic amino acid transport in lysosomes of normal and cystinotic human fibroblasts, *J. Biol. Chem.* **260:**4791–4798.

Polansky, J. R., Toole, B. P., and Gross, J., 1974, Brain hyaluronidase: Changes in activity during chick development, *Science* **183:**862–864.

Poltorak, M., Sadoul, R., Keilhauer, G., Landa, C., and Schachner, M., 1987, The myelin-associated glycoprotein (MAG), a member of the L2/HNK-1 family of neural cell adhesion molecules, is involved in neuron–oligodendrocyte and oligodendrocyte–oligodendrocyte interaction, *J. Cell Biol.* **105:**1893–1899.

Ripellino, J. A., Margolis, R. U., and Margolis, R. K., 1988, Light and electron microscopic studies on the localization of hyaluronic acid in developing rat cerebellum, *J. Cell Biol.* **106:**845–855.

Rochefort, H., Capony, F., Garcia, M., Cavailles, V., Freiss, G., Chambron, M., Morriset, M., and Vignon, F., 1987, Estrogen-induced lysosomal proteases secreted by breast cancer cells: A role in carcinogenesis? *J. Cell Biochem.* **35:**17–29.

Rodén, L., 1980, Structure and metabolism of connective tissue proteoglycans, in: *The Biochemistry of Glycoproteins and Proteoglycans* (W. J. Lennarz, ed.), pp. 267–371, Plenum Press, New York.

Rosenblatt, D. S., Hosack, A., Matiaszule, N. V., Cooper, A. B., and Laframboise, R., 1985, Defect in vitamin B_{12} release from lysosomes: Newly described inborn error of vitamin B_{12} metabolism, *Science* **228:**1319–1321.

Roth, R. A., 1988, Structure of the receptor for insulin-like growth factor II: The puzzle amplified, *Science* **239:**1269–1271.

Rutishauser, U., Hoffman, S., and Edelman, G. M., 1982, Binding properties of a cell adhesion molecule from neural tissue, *Proc. Natl. Acad. Sci. USA* **79:**685–689.

Rutishauser, U., Watanabe, M., Silver, J., Troy, F. A., and Vimr, E. R., 1985, Specific alteration of N-CAM-mediated cell adhesion by an endoneuraminidase, *J. Cell Biol.* **101:**1842–1849.

Rutishauser, U., Acheson, A., Hall, A. K., Mann, D. M., and Sunshine, J., 1988, The neural cell adhesion molecule (NCAM) as a regulator of cell–cell interactions, *Science* **240:**53–57.

Saito, M., and Yu, R. K., 1986, Further characterization of a myelin-associated neuraminidase: Properties and substrate specificities, *J. Neurochem.* **47:**588–593.

Sandhoff, K., Schwarzmann, G., Sarmientos, F., and Conzelmann, E., 1987, Fundamentals of ganglioside catabolism, in: *Gangliosides and Modulation of Neuronal Functions* (H. Rahmann, ed.), pp. 231–250, Springer-Verlag, Berlin.

Saul, R., Molyneux, R. J., and Elbein, A. D., 1984, Studies on the mechanism of castanospermine inhibition of α- and β-glucosidase, *Arch. Biochem. Biophys.* **230:**668–675.

Schindler, M., Hogan, M., Miller, R., and DeGaetano, D., 1987, A nuclear-specific glycoprotein representative of a unique pattern of glycosylation, *J. Biol. Chem.* **262:**1254–1260.

Schmidt, S. L., Fuchs, R., Male, P., and Mellman, I., 1988, Two distinct subpopulations of endosomes involved in membrane recycling and transport to lysosomes, Cell 52:73–83.

Schuchman, E. H., and Desnick, R. J., 1988, Mucopolysaccharidosis Type I subtypes, J. Clin. Invest. 81: 98–105.

Schwarting, G. A., Jungalwala, F. B., Chou, D. K. H., Bayer, A. M., and Yamamoto, M., 1987, Sulfated glucuronic acid containing glycoconjugates are temporally and spatially regulated antigens in the developing mammalian nervous system, Dev. Biol. 120:60–76.

Shashoua, V. E., Daniel, P. F., Moore, M. E., and Jungalwala, F. B., 1986, Demonstration of glucuronic acid on brain glycoproteins which react with HNK-1 antibody, Biochem. Biophys. Res. Commun. 138: 902–909.

Shin-Buehring, Y. S., Dallinger, M., Osang, M., Rahm, P., and Schaub, J., 1980, Lysosomal enzyme activities of human fetal organs during development, Biol. Neonate 38:300–308.

Singer, H. S., Tiemeyer, M., Slesinger, P. A., and Sinnott, M. L., 1987, Inactivation of G_{M1}-ganglioside β-galactosidase by a specific inhibitor: A model for ganglioside storage disease, Ann. Neruol. 21:497–503.

Smedsrod, B., Kjellen, L., and Pertoft, H., 1985, Endocytosis and degradation of chondroitin sulfate by liver endothelial cells, Biochem. J. 229:63–71.

Sokol, J., Blanchette-Mackie, J., Kruth, H. S., Dwyer, N. K., Amende, L. M., Butler, J. D., Robinson, E., Patel, S., Brady, R. O., Comly, M. E., Vanier, M. T., and Pentchev, P. G., 1988, Type C Niemann–Pick disease, J. Biol. Chem. 263:3411–3417.

Song, S. Z., Li, S.-C., and Li, Y.-T., 1987, Absence of endo-β-N-acetylglucosaminidase activity in the kidneys of sheep, cattle, and pig, Biochem. J. 248:145–149.

Sweeley, C. C., and Usuki, S., 1987, The effect of a sialidase inhibitor on the cell cycle of cultured human fibroblasts, J. Cell Biol. 105:101a.

Tachibana, Y., Yamashita, K., and Kobata, A., 1982, Substrate specificity of mammalian endo-β-N-acetylglucosaminidase: Study with the enzyme of rat liver, Arch. Biochem. Biophys. 214:199–210.

Takagaki, K., Nakamura, T., Majima, M., and Endo, M., 1988, Isolation and characterization of a chondroitin sulfate-degrading β-glucuronidase from rabbit liver, J. Biol. Chem. 263:7000–7006.

Tanaka, H., and Suzuki, K., 1977, Substrate specificities of the two genetically distinct human brain β-galactosidases, Brain Res. 122:325–335.

Tarentino, A. L., and Maley, F., 1969, The purification and properties of a β-aspartyl N-acetylglucosamine amidohydrolase from hen oviduct, Arch. Biochem. Biophys. 130:295–303.

Tong, P. Y., Tollefsen, S. E., and Kornfeld, S., 1988, The cation-independent mannose 6-phosphate receptor binds insulin-like growth factor II, J. Biol. Chem. 263:2585–2588.

Torres, C.-R., and Hart, G. W., 1984, Topography and polypeptide distribution of terminal N-acetylglucosamine residues on the surfaces of intact lymphocytes, j. Biol. Chem. 259:3308–3317.

Townsend, R. R., Li, Y.-T., and Li, S.-C., 1979, Brain glycosidases, in: Complex Carbohydrates of Nervous Tissue (R. U. Margolis and R. K. Margolis, eds.), pp. 127–137, Plenum Press, New York.

Troen, B. R., Ascherman, D., Atlas, D., and Gottesman, M. M., 1988, Cloning and expression of the gene for the major excreted protein of transformed mouse fibroblasts, J. Biol. Chem. 263:254–261.

Tropea, J. E., Swank, R. T., and Segal, H. L., 1988, Effect of swainsonine on the processing and turnover of lysosomal β-galactosidase and β-glucuronidase from mouse peritoneal macrophages, J. Biol. Chem. 263:4309–4317.

Tulsiani, D. R. P., and Carubelli, R., 1970, Studies on the soluble and lysosomal neuraminidases of rat liver, J. Biol. Chem. 245:1821–1827.

Tulsiani, D. R. P., and Touster, O., 1983, Swainsonine, a potent mannosidase inhibitor elevates rat liver and brain lysosomal α-D-mannosidase, decreases Golgi α-D-mannosidase II, and increases the plasma levels of several acid hydrolases, Arch. Biochem. Biophys. 224:594–600.

Tulsiani, D. R. P., and Touster, O., 1987, Substrate specificities of rat kidney lysosomal and cytosolic α-D-mannosidases and effects of swainsonine suggest a role of the cytosolic enzyme in glycoprotein catabolism, J. Biol. Chem. 262:6506–6514.

Tulsiani, D. R. P., Six, H., and Touster, O., 1978, Rat liver microsomal and lysosomal β-glucuronidases differ in both carbohydrate and amino acid compositions, Proc. Natl. Acad. Sci. USA 75:3080–3084.

Tulsiani, D. R. P., Harris, T. M., and Touster, O., 1982, Swainsonine inhibits the biosynthesis of complex glycoproteins by inhibition of Golgi mannosidase II, *J. Biol. Chem.* **257**:7936–7939.

Umemoto, J., Bhavanandan, V. P., and Davidson, E. A., 1977, Purification and properties of an endo-α-D-galactosaminidase from *Diplococcus pneumoniae*, *J. Biol. Chem.* **252**:8609–8614.

Usuki, S., Lyu, S.-C., and Sweeley, C. C., 1988, Sialidase activities of cultured human fibroblasts and the metabolism of G_{M3} ganglioside, *J. Biol. Chem.* **263**:6847–6853.

van Diggelen, O. P., Galjaard, H., Sinnott, M. L., and Smith, P. J., 1980, Specific inactivation of lysosomal glycosidases in living fibroblasts by the corresponding glycosylmethyl-p-nitrophenyltriazenes, *Biochem. J.* **188**:337–343.

van Diggelen, O. P., Schram, A. W., Sinnott, M. L., Smith, P. J., Robinson, D., and Galjaard, H., 1981, Turnover of β-galactosidase in fibroblasts from patients with genetically different types of β-galactosidase deficiencies, *Biochem. J.* **200**:143–151.

Venerando, B., Preti, A., Lombardo, A., Cestaro, B., and Tettamanti, G., 1978, Studies on brain cytosol neuraminidase. II. Extractability, solubility and intraneuronal distribution of the enzyme in pig brain, *Biochim. Biophys. Acta* **527**:17–30.

Verheijen, F. W., Palmeri, Hoogeveen, A. T., and Galjaard, H., 1985, Human placental neuraminidase: Activation, stabilization, and association with β-galactosidase and its protective protein, *Eur. J. Biochem.* **149**:315–321.

Viitala, J., Carlsson, S. R., Siebert, P. D., and Fukuda, M., 1988, Molecular cloning of cDNA's encoding a human lysosomal membrane glycoprotein with apparent Mr 120,000, Lamp A, *Proc. Natl. Acad. Sci. USA* **85**:3743–3747.

Volker, W., Schmidt, A., Robenek, H., and Buddecke, E., 1984, Binding and degradation of proteoglycans by cultured arterial smooth muscle cells, *Eur. J. Cell Biol.* **34**:110–117.

von Figura, K., and Hasilik, A., 1986, Lysosomal enzymes and their receptors, *Annu. Rev. Biochem.* **56**:167–194.

Warner, T. G., and O'Brien, J. S., 1982, Structure analysis of the major oligosaccharides accumulating in canine G_{M1} gangliosidosis liver, *J. Biol. Chem.* **257**:224–232.

Westley, B. R., and May, F. E. B., 1987, Oestrogen regulates cathepsin D mRNA levels in oestrogen responsive human breast cancer cells, *Nucleic Acids Res.* **15**:3773–3786.

Wille, W., and Trenker, E., 1981, Changes in particulate neuraminidase activity during normal and staggerer mutant mouse development, *J. Neurochem.* **37**:443–446.

Winkler, J. R., and Segal, H. L., 1984a, Inhibition by swainsonine of the degradation of endocytosed glycoproteins in isolated rat liver parenchymal cells, *J. Biol. Chem.* **259**:1958–1962.

Winkler, J. R., and Segal, H. L., 1984b, Swainsonine inhibits glycoprotein degradation by isolated rat liver lysosomes, *J. Biol. Chem.* **259**:15369–15372.

Yanagishita, M., 1985, Inhibition of intracellular degradation of proteoglycans by leupeptin in rat ovarian granulosa cells, *J. Biol. Chem.* **260**:11075–11082.

Yanagishita, M., and Hascall, V. C., 1984, Metabolism of proteoglycans in rat ovarian granulosa cell culture, *J. Biol. Chem.* **259**:10270–10283.

Yanagishita, M., and Hascall, V. C., 1985, Effects of monensin on the synthesis, transport, and intracellular degradation of proteoglycans in rat ovarian granulosa cells in culture, *J. Biol. Chem.* **260**:5445–5455.

Zanetta, J. P., Roussel, G., Dontenwill, M., and Vincendon, G., 1983, Immunohistochemical localization of α-mannosidase during postnatal development of rat cerebellum, *J. Neurochem.* **40**:202–208.

Glycoproteins of the Synapse

James W. Gurd

1. INTRODUCTION

In 1967 Rambourg and Leblond published a series of observations on the carbohydrate-rich cell coat present at the surface of a variety of cells in the rat. Included within this survey were results obtained with nervous tissue which demonstrated an enrichment of carbohydrate-containing material in the region of the synaptic cleft, an observation which was subsequently confirmed and extended by Pfenninger (1973). Lectin-cytochemical studies demonstrated the presence of receptors for concanavalin A and *Ricinus communis* lectin on postsynaptic membranes and showed that these receptors did not aggregate under the influence of added lectin, suggesting that they might interact, either directly or indirectly, with the underlying postsynaptic density (Bittiger and Schnebli, 1974; Matus *et al.*, 1973; Cotman and Taylor, 1974; Kelly *et al.*, 1976). In parallel with these observations, biochemical analysis of synaptosomes (Brunn-graber *et al.*, 1967) and synaptic membranes (Gombos *et al.*, 1971; Breckenridge and Morgan, 1972; Breckenridge *et al.*, 1972; Margolis *et al.*, 1975; Morgan and Gombos, 1976; Churchill *et al.*, 1976) confirmed the association of glycoproteins with these structures and provided initial information concerning the identity of these synaptic components (reviewed in Mahler, 1979). These findings, in conjunction with the rapidly growing body of information concerning the structure, biosynthesis, and function of glycoproteins in general (Sharon and Lis, 1982; Olden *et al.*, 1982; Kornfeld and Kornfeld, 1985; Roth, 1987), stimulated a continuing interest in the identification and function of glycoproteins at the synapse. The present review will summarize current information relating to the identity, structure, and biochemical properties of glycoproteins associated with synaptic membranes, synaptic junctions, and postsynaptic densities. Of other organelles present at the synapse, mitochondria are deficient in glycoproteins (Zanetta *et al.*, 1977a; Krusius *et al.*, 1978), and the glycoprotein composition of synaptic vesicles is dealt with elsewhere in this volume (p. 309). A

James W. Gurd • Department of Biochemistry, Scarborough Campus, University of Toronto, West Hill, Ontario M1C 1A4, Canada.

number of synaptic membrane proteins of known function are glycosylated [e.g., the sodium channel (Casadei *et al.*, 1984), the 55-kDa subunit of the Na^+/K^+-ATPase (Dahl and Hokin, 1974), the muscarinic (Rauh *et al.*, 1986) and nicotinic (Hucho, 1986) acetylcholine receptors, a glutamate binding protein (Michaelis, 1975), and the opiate receptor (Gioannini *et al.*, 1982)] and their properties will not be discussed here.

2. CHARACTERIZATION OF SUBCELLULAR FRACTIONS USED FOR THE IDENTIFICATION OF SYNAPTIC GLYCOPROTEINS

The identification of synaptic glycoproteins has generally been achieved by the analysis of isolated subcellular fractions which are enriched in synaptic membranes (SMs), synaptic junctions (SJs), or postsynaptic densities (PSDs). The value of these fractions for the identification of synaptic components is clearly dependent upon the extent of contamination by membranes derived from other cellular organelles. The purity of synaptic membrane fractions is generally assumed to lie within the range of 70–80% (Gurd *et al.*, 1974; Morgan and Gombos, 1976). It is, therefore, apparent that while the presence of a particular component in SMs is indicative of a synaptic location, additional criteria are required in order to confirm a synaptic origin. Furthermore, synaptic membrane fractions consist predominantly of membranes derived from the presynaptic nerve terminal, with structures originating from the synaptic junctional complex (pre- and postsynaptic junctional membranes, PSDs) constituting only a small percentage of the total material. Thus, while useful in the initial identification of glycoproteins which may be located in the general region of the nerve terminal, SMs cannot be used for the identification of those species which are localized to the SJ complex itself and which correspond to glycoproteins identified in the synaptic cleft by cytochemical procedures.

The development of methods for the preparation of subcellular fractions enriched in SJs and PSDs made possible the identification of proteins which are specifically associated with these organelles. Under the appropriate conditions, extraction of SMs with detergent solubilizes the majority of extrajunctional membranes, leaving a residue which is enriched in structures derived from the SJ complex (Fiszer and de Robertis, 1967; de Robertis *et al.*, 1967; Cotman and Taylor, 1972; Davis and Bloom, 1973; Walters and Matus, 1975; Cohen *et al.*, 1977; Gurd *et al.*, 1982). The predominant material in these fractions generally consists of postsynaptic structures from which most of the presynaptic membrane and varying amounts of the postsynaptic membrane have been removed. Of the morphologically identifiable structures present in the SJ fraction prepared by the extraction of synaptic membranes with Triton X-100 (Cotman and Taylor, 1972), for example, approximately 55% correspond to postsynaptic membrane specializations (PSD with overlying membrane), with fewer (~ 10%) intact junctional complexes containing both pre- and postsynaptic membranes being present (Rostas *et al.*, 1979). PSDs prepared by a variety of procedures have similar morphological properties, consisting of PSDs from which all identifiable membranes have been removed. In spite of the similar morphological appearance, however, the protein composition of isolated PSDs may vary depending upon the rigorousness of the detergent extraction step employed (Cotman *et al.*, 1974; Cohen *et al.*, 1977; Matus and Taff-Jones, 1978; Gurd *et al.*, 1982). In addition to solubilizing extrajunctional SMs, the detergents used in these procedures will also solubilize membranes derived from other cellular organelles, such as the endoplasmic reticulum, Golgi apparatus, and surface membrane fragments of nonsynaptic origin, thereby reducing the problems of contamination which are associated with SM fractions.

Isolated SJ and PSD fractions have been noted to have "sticky" surfaces (Matus *et al.*, 1980) which may result in the adherence of contaminating proteins, and this possibility must be taken into account in considering the possible synaptic origin of proteins or glycoproteins present in these fractions. The detergents used in the preparation of PSDs and SJs may also solubilize proteins which are present as components of the synaptic apparatus *in situ* (e.g., several neurotransmitter receptors are extracted by detergents; Gioannini *et al.*, 1982; Salvaterra and Matthews, 1980; Rauh *et al.*, 1986; Bristow and Martin, 1987). The proteins and glycoproteins found in isolated SJ and PSD fractions therefore represent a minimum composition, consisting only of those components which resist the isolation procedure.

In view of the problems of interpretation arising from the possible contamination of SMs by other cellular membranes and of SJ and PSD fractions by proteins which adhere in a nonspecific fashion to the isolated organelle, the following guidelines are suggested to aid in assessing the synaptic origins of individual proteins and glycoproteins. (1) Synaptic constituents should copurify with and be enriched in isolated SM fractions. Although the degree of enrichment may vary with different components, typical enrichment values for synaptic proteins range from four- to sevenfold over the homogenate (Cotman and Matthews, 1971; Gurd *et al.*, 1974; Salvaterra and Matthews, 1980; deBlas and Mahler, 1976). (2) If specifically localized to the SJ complex, they should be further concentrated (two- to fourfold) in SJ and PSD fractions relative to SMs. (3) If integral components of the postsynaptic apparatus, they should resist extraction by reagents that do not substantially disrupt or solubilize SJs or PSDs. (4) If integral components of the SJ complex, they should be present in SM and PSD fractions prepared by more than one procedure. (5) Their concentration in the brain should increase in parallel with the process of synaptogenesis.

3. IDENTIFICATION AND CHARACTERIZATION OF SYNAPTIC GLYCOPROTEINS

3.1. Oligosaccharide Composition of Synaptic Glycoproteins

Analysis of the sugar composition of SM, SJ, and PSD fractions has confirmed at the biochemical level the cytochemical observations suggesting that the nerve terminal and synapse contain high concentrations of glycoproteins (Breckenridge *et al.*, 1972; Margolis *et al.*, 1975; Morgan and Gombos, 1976; Churchill *et al.*, 1976; Zanetta *et al.*, 1975, 1977b; reviewed in Mahler, 1979). Although most of the glycoprotein sugar in these fractions is associated with N-linked oligosaccharides, O-linked sugar chains are also present and account for 1–2% of the total glycoprotein carbohydrate present in

SMs (Krusius *et al.*, 1978). Analyses of the monosaccharide composition of SM, SJ, and PSD fractions have yielded variable results, and generalizations based on these reports are therefore difficult to make. Thus, while Churchill *et al.* (1976) reported that SMs, SJs, and PSDs each contained similar amounts of glycoprotein sugar (207, 212, and 215–258 nmoles/mg protein, respectively), other investigators (Morgan and Gombos, 1976; Margolis *et al.*, 1975; Webster and Klingman, 1980) have reported that SJs contained 30–40% less glycoprotein sugar than SMs. This variability presumably reflects, at least in part, the use of different subcellular fractionation procedures by different investigators and emphasizes the importance of extending these studies to the level of individual glycoproteins.

Relatively little information is available concerning the structure of the oligosaccharide groups associated with synaptic glycoproteins. Krusius *et al.* (1978) identified five different O-linked sugar chains on SM glycoproteins. The N-linked sugars have been resolved by Con A affinity chromatography into triantennary (or larger) complex chains (60–80% of the total sugar content), biantennary structures (21%), and high-mannose chains (13–16%) (Krusius *et al.*, 1978; Reeber *et al.*, 1984). The latter contain between five and eight Man residues, with Man_5 oligosaccharides being the predominant species (Gurd and Fu, 1982; Stanojev, 1987). Few, if any, hybrid oligosaccharides are associated with synaptic glycoproteins (Fu and Gurd, 1983).

Comparison of fucosyl-oligosaccharides associated with SM and SJ glycoproteins failed to reveal any major differences in composition in spite of the generally simplified glycoprotein composition of the latter fraction (Stanojev and Gurd, 1987; see also Section 3.2.2). SJs contain a slightly greater proportion of acidic fucosyl-oligosaccharides than SMs (Stanojev and Gurd, 1987), suggesting that more highly sialylated species may be concentrated in the region of the synaptic cleft.

Although the few available studies of the oligosaccharides associated with synaptic glycoproteins have not identified structures or properties which are unique or characteristic of these molecules, they have served to demonstrate the diversity of sugar structures which are present at the synapse. This diversity is not unexpected and presumably reflects the large number of different glycoproteins which are present in isolated synaptic fractions, and emphasizes the necessity of analyzing the sugar composition of individual glycoproteins if the role of these molecules at the synapse is to be clarified.

3.2. Identification of Synaptic Glycoproteins

3.2.1. Synaptic Membranes

The identification of individual glycoproteins present in SM and SJ fractions has served to differentiate between the composition of these fractions and to identify specific glycoproteins which appear to be uniquely associated with the synaptic apparatus. A variety of procedures have been employed to identify glycoproteins which are associated with the synapse. These include: (1) staining polyacrylamide gels with the periodic acid–Schiff procedure for carbohydrates (Waehneldt *et al.*, 1971; Breckenridge and Morgan, 1972; Gurd *et al.*, 1974; Goodrum and Tanaka, 1978); (2) reacting

Figure 1. Identification of Con A-binding glycoproteins of synaptic membranes, synaptic junctions, and postsynaptic densities. SMs, SJs, and PSDs were prepared as described by de Silva *et al.* (1979), Cotman and Taylor (1972), and Gurd *et al.* (1982), respectively. Equal amounts of protein (100 μg) were separated by poly-acrylamide gel electrophoresis in the presence of SDS. Glycoproteins were detected by reacting the gel with ^{125}I-labeled Con A and autoradiograms prepared. Additional experimental details as in Gurd (1980). a–f correspond to glycoproteins with apparent molecular sizes of 230, 180, 145, 116, 110, and 45–50 kDa, respectively.

gels or protein blots with radioiodinated or biotinylated lectins (Gurd, 1977a,b; Kelly and Cotman, 1977; Webster and Klingman, 1980; Mena *et al.*, 1981; Gordon-Weeks and Harding, 1983; Hampson and Poduslo, 1987; Figure 1); (3) radiolabeling oligosaccharides biosynthetically using monosaccharide precursors such as fucose (Zatz and Barondes, 1970; Gurd, 1979; Fu *et al.*, 1981), *N*-acetylglucosamine (Reeber *et al.*, 1984), and *N*-acetylmannosamine (Fu *et al.*, 1981); (4) chemically labeling sialic acid or galactose residues with NaB^3H_4 following oxidation with sodium periodate (Wang and Mahler, 1976; Cruz and Gurd, 1980, 1983) or galactose oxidase (Kelly and Cotman, 1977); and (5) isolating glycoproteins by lectin-affinity chromatography (Gurd and Mahler, 1974; Zanetta *et al.*, 1975, 1977b; Gurd, 1980). Although the relative proportions of individual components may vary with the method of detection used, presumably reflecting differences in the carbohydrate composition of different glycoproteins (Gurd, 1977b; Fu *et al.*, 1981), these procedures routinely identify eight to ten molecular weight classes of glycoproteins in SM fractions. A typical profile of Con A-binding glycoproteins present in SMs is shown in Figure 1.

It is clear that one-dimensional gel electrophoretic procedures underestimate the total number of glycoprotein species, since many molecular weight classes contain two or more glycoproteins which differ with respect to their sugar composition. Sequential lectin-affinity chromatography of SM glycoproteins radiolabeled with [³H]fucose on lentil lectin and WGA-Sepharose columns resolved each of the major molecular weight classes of glycoprotein into two or more subfractions, and 20 to 25 fucosylated glycoproteins were identified (Gurd and Mahler, 1974). Similarly, Zanetta *et al.*

(1977b) identified some 40 species following the fractionation of SM glycoproteins on Con A and *Ulex europeus* lectin-affinity columns. In the latter study, fucosyl-glycoproteins which bound to the *Ulex europeus* affinity column accounted for 26% of the total SM protein and 86% of the total glycoprotein sugar, indicating that the majority of SM glycoproteins contain at least one fucosylated oligosaccharide.

The significance of the diversity of glycoproteins present in SMs or of the heterogeneity of oligosaccharides associated with individual molecular weight classes is not known, but may reflect the occurrence of related glycoproteins with distinct sugar compositions in functionally different synaptic subtypes, the presence at individual synapses of a wide variety of structurally diverse glycoproteins, or a combination of these factors. Differentiation between these possibilities awaits the analysis of purified synaptic subtypes (e.g., see Docherty *et al.*, 1987).

3.2.2. Synaptic Junctions and Postsynaptic Densities

In general agreement with the cytochemical observations demonstrating the presence of lectin-receptor sites in the region of the synapse, and in particular on the postsynaptic membrane overlying the PSD, isolated SJ and PSD fractions react with Con A and other lectins (Gurd, 1977a,b; Kelly and Cotman, 1977; Gurd *et al.*, 1983b). Isolated SMs and SJs bind similar amounts of Con A ($3.5–4 \times 10^{-9}$ equivalents Con A bound/mg protein; Gurd, 1980; see also Cotman and Taylor, 1974), in contrast to morphological studies indicating that Con A binding sites are concentrated in the region of the synaptic cleft (Matus *et al.*, 1973; Cotman and Taylor, 1974). This apparent discrepancy between the cytochemical and biochemical observations may reflect, in part, the solubilization of glycoproteins during the preparation of SJs, as well as the high protein density of the PSD which would result in the same amount of protein being associated with a smaller surface area.

SJ fractions prepared by the extraction of SMs with Triton X-100 (Cotman and Taylor, 1972) are characterized by an increased concentration of a limited number of high-molecular-weight glycoproteins and a decrease in the relative amounts of the lower-molecular-weight components characteristic of SM fractions (Gurd, 1977a; Kelly and Cotman, 1977; Rostas *et al.*, 1979; Gurd, 1980; Mena *et al.*, 1981; Kelly and Montgomery, 1982; Groswald *et al.*, 1983; Figure 1). The major SJ glycoproteins identified on the basis of their reactivity with Con A have apparent molecular sizes of 100–110 (gp110), 116 (gp116),* 145 (gp145), 170–180 (gp180), and 230 kDa (gp230). These glycoproteins are also associated with PSDs prepared by several different procedures (Gurd, 1977b; Kelly and Cotman, 1977; Blomberg *et al.*, 1977; Gurd *et al.*, 1982, 1983b; Groswald *et al.*, 1983). The amount of gp110 present in PSDs prepared by the phase partitioning of SMs in the presence of *n*-octylglucoside is reduced (Gurd *et al.*, 1982; Figure 1), suggesting that it may be less tightly integrated into the structural framework of the PSD, or that it may be primarily associated with postsynaptic membranes rather than with the PSD.

---

*We have previously referred to gp116 as gp130 (Gurd, 1980). This glycoprotein has the same electrophoretic mobility as β-galactosidase indicating that it has an apparent M_r of 116,000. gp180, gp145, gp116, and gp110 correspond to glycoproteins i, ii, iii, and iv of Mena and Cotman (1982).

Several lines of experimental evidence suggest that the high-molecular-weight glycoproteins, and in particular gp116 and gp180, are exclusively associated with the postsynaptic apparatus and may constitute unique molecular markers for this structure. (1) gp110, gp116, and gp180 are present in SJs or PSDs isolated from a variety of brain regions and from a number of different species including rat, rabbit, chick, cow, hamster, monkey, human, mouse, lizard, frog, and shark (Rostas *et al.*, 1979; Freedman *et al.*, 1981; Nieto-Sampedro *et al.*, 1982; Gurd *et al.*, 1982). (2) They are not detected in myelin, mitochondria, nuclei, microsomes, axolemma membranes, or cytosol (Gurd, 1980; Mena *et al.*, 1981). (3) gp116 and gp180 copurify with the postsynaptic apparatus and are most highly concentrated in purified PSDs (Gurd *et al.*, 1983a; Figure 1), where they are present on the external, convex surface, consistent with a transmembrane orientation and localization of the carbohydrate-containing domains within the synaptic cleft (Gurd *et al.*, 1983b). (4) The major glycoproteins present in the SJ fraction, including gp110, gp116, and gp180, are resistant to extraction with a variety of reagents expected to solubilize extrinsic or weakly associated surface proteins, such as high or low salt concentrations, EGTA, chaotropic agents (KI, lithium diiodosalicylate), NaOH (0.05 M), and pyridine (33%, v/v; Fu and Gurd, 1983). (5) Each of the major SJ glycoproteins resists digestion with trypsin when present as part of intact SJs, suggesting that they are not superficially adsorbed to the SJ surface (Gurd, unpublished results). (6) Their concentration in SM fractions increases coincidentally with the time course of synaptogenesis in the forebrain, and in parallel with the concentration of SJs in the SM fraction (de Silva *et al.*, 1979; Kelly and Cotman, 1981). Together these results strongly support the contention that gp116 and gp180, and most probably gp110, are characteristic and possibly unique constituents of the postsynaptic apparatus, and as such are likely candidates for playing specific roles in one or more synaptic functions.

Comparison of tryptic peptide maps generated following the labeling of tyrosine residues with ^{125}I led Mena and Cotman (1982) to suggest that there was a high degree of sequence homology between gp180, gp145, and gp116, and to propose that these glycoproteins might represent differentially processed members of the same protein family. However, in contrast to these results, Kelly (1983) and Fu and Gurd (1983) found that each of the major SJ glycoproteins yielded unique ^{125}I-tryptic maps. Moreover, gp110, gp116, and gp180 each yielded characteristic and distinct [32P]phosphopeptide maps following the *in vitro* labeling of SJs with [32P]-ATP (Gurd *et al.*, 1983a). Clarification of the possible relationships between these postsynaptic glycoproteins remains a question of some interest, and awaits more detailed analysis of their structures through the application of peptide sequencing or gene cloning procedures.

The molecular composition of SJs isolated from the cerebellar cortex is distinct from that of forebrain SJs, the most notable difference being the reduced quantities of the 50-kDa subunit of Ca^{2+}/calmodulin-dependent protein kinase II in the former fraction (Flanagan *et al.*, 1982; Groswald *et al.*, 1983). The main M_r classes of glycoproteins which are characteristic of forebrain PSDs are also present in PSDs prepared from cerebellum (Gordon-Weeks and Harding, 1983). However, cerebellar SJs and PSDs are characterized by the presence of an additional prominent glycoprotein with an apparent M_r of 240 kDa (Groswald *et al.*, 1983; Gurd, unpublished observations). The concentration of gp240 in SJs was increased following the pretreatment of SMs with micromolar concentrations of Ca^{2+} (Groswald and Kelly, 1984). Peptide mapping experiments suggested that gp240 may be related to the 235-kDa fodrin polypeptide, although it did not react with antifodrin antibodies (Groswald and Kelly, 1984). A 240-kDa Con A-binding, brain-specific glycoprotein was recently partially purified from cerebellar membranes by Bezamahouta *et al.* (1988). The concentration of this glycoprotein in the cerebellum increased gradually between postnatal days 4 and 30 with the maximum rate of increase occurring between p12 and p18. The developmental expression of this glycoprotein is consistent with a synaptic localization and it was suggested that it may be related to the 240-kDa synaptic junctional glycoprotein described by Groswald and Kelly (1984).

The Con A binding activity of SJ glycoproteins is abolished following digestion with endoglycosidase H (Fu and Gurd, 1983; Groswald and Kelly, 1984; Figure 2), endoglycosidase H-sensitive oligosaccharides accounting for approximately 3, 13, and 1.8% of the apparent M_r of gp110, gp116, and gp180, respectively. Each of the major M_r classes of SJ and PSD glycoproteins reacts with the lectins from *Ulex europeus*, lentils, and wheat germ (Gurd, 1977b, 1980; Fu *et al.*, 1981; Figure 2), demonstrating the presence of complex as well as high-mannose sugar chains. The resolution of gp110 and gp180 into species which either did or did not bind to WGA (Gurd, 1980) indicated the occurrence of multiple glycoforms within these M_r classes.

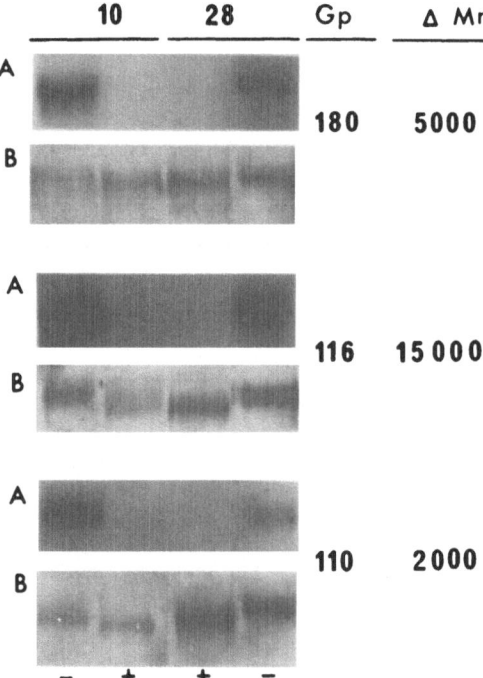

Figure 2. The effect of development on the Con A-binding oligosaccharides associated with the major synaptic junction glycoproteins. gp180, gp116, and gp110 were eluted from polyacrylamide gels of the Con A positive fraction obtained from synaptic junctions isolated from 10- and 28-day-old rat forebrains. Control glycoproteins ($-$) and glycoproteins digested with endoglycosidase H ($+$) were separated by gel electrophoresis, transferred to nitrocellulose sheets, and glycoproteins detected with WGA-biotin/HRP-avidin and O-dianisidine (B). Following staining with WGA, the blots were reacted with ^{125}I-labeled Con A and autoradiograms prepared (A). Digestion with Endo H abolishes all Con A-binding activity but does not alter WGA binding. ΔM_r: decrease in molecular weight following treatment with Endo H.

4. PHOSPHORYLATION OF SYNAPTIC GLYCOPROTEINS

The phosphorylation and dephosphorylation of proteins plays a central role in the regulation of a wide variety of cellular activities and there is abundant evidence that this enzymatic cycle is involved in the regulation and modulation of synaptic function (for recent reviews see Kennedy, 1983; Nairn et al., 1985; Nestler and Greengard, 1984). Because variations in phosphorylation level often reflect changes in the functional state of a protein, we investigated the ability of endogenous protein kinases to phosphorylate synaptic glycoproteins in order to gain some insight into the possible functional role(s) of the latter components.

Glycoproteins which bind to Con A accounted for 1–2% of the total protein-associated ^{32}P following the incubation of SMs, SJs, and PSDs with [γ-^{32}P]-ATP, or the in vivo administration of ^{32}PO$_4$ (Gurd et al., 1983a; Gurd, 1985a; Gurd and Bissoon, 1985; Kearney and Gurd, 1986 and unpublished results). Whereas several glycoproteins were phosphorylated following the incubation of SMs with [γ-^{32}P]-ATP (Kearney and Gurd, 1986), gp180 accounted for the majority (> 80%) of the ^{32}P transferred to PSD glycoproteins (Gurd, 1985a; Figure 3). Peptide mapping experiments demonstrated that gp180 was phosphorylated on multiple sites and that similar sites were labeled following both in vitro phosphorylation of SMs and PSDs and the in vivo phosphorylation of PSDs (Gurd, 1985a; Gurd and Bissoon, 1985; Kearney and Gurd, 1986). Phosphoamino acid analysis identified phosphotyrosine as well as phosphoserine and phosphothreonine as a product of the phosphorylation of gp180, under both in vivo and in vitro labeling conditions, thereby identifying tyrosine kinase as a constituent enzyme of the synaptic apparatus and gp180 as a potential endogenous

gp180

Figure 3. Phosphorylation of gp180 by calcium/calmodulin-dependent protein kinase. PSDs were incubated with [γ-^{32}P]-ATP in the presence (+) or absence (−) of calcium and calmodulin. Glycoproteins were isolated on Con A-sepharose, separated by gel electrophoresis, and autoradiograms prepared. Additional experimental details as in Gurd (1985a).

− +

substrate for this activity (Gurd, 1985a,b; Gurd and Bissoon, 1985; Kearney and Gurd, 1986; see also Ellis *et al.*, 1988).

PSDs contain high levels of calcium/calmodulin-dependent protein kinase activity (Grab *et al.*, 1981), and the phosphorylation of gp180 by both SMs and PSDs is enhanced in the presence of calcium and calmodulin (Gurd, 1985a; Kearney and Gurd, 1986; Figure 3). SM glycoproteins which were phosphorylated in a calcium- and calmodulin-dependent manner partitioned exclusively into the detergent-insoluble residue following extraction with Triton X-100, suggesting that the phosphorylation of glycoproteins by calcium/calmodulin-dependent protein kinase may occur primarily in the postsynaptic compartment (Kearney and Gurd, 1986). In view of the fact that calcium/calmodulin-dependent protein kinase II is a major protein component of PSDs (Kennedy *et al.*, 1983; Goldenring *et al.*, 1984; Kelly *et al.*, 1984) it seems probable that this enzyme mediates the phosphorylation of gp180, although this has yet to be demonstrated. The apparent immobility of postsynaptic glycoproteins in the plane of the membrane (e.g., see Matus *et al.*, 1973) suggests that gp180 and the protein kinase responsible for its phosphorylation may be located in close proximity, perhaps as a macromolecular complex within the structural framework of the PSD.

Synaptic glycoproteins are also dephosphorylated in a calcium- and calmodulin-dependent fashion (Kearney and Gurd, 1986). The identity of the protein phosphatase(s) responsible for this reaction is unknown but it is relevant that calcineurin, a calcium/calmodulin-dependent protein phosphatase of relatively broad substrate specificity (Pallen and Wang, 1984; Tallant and Cheung, 1984; King *et al.*, 1984), has been localized to the PSD by immunocytochemical procedures (Wood *et al.*, 1980). Indeed, the colocalization of calcineurin, calcium/calmodulin-dependent protein kinase II, and gp180 in the PSD suggests that the phosphorylation state of gp180 may be regulated by Ca^{2+} levels in the postsynaptic cell.

The functional sequelae to the phosphorylation of gp180 remain to be determined. However, the phosphorylation of this postsynaptic glycoprotein by both calcium/calmodulin-dependent protein kinase and tyrosine kinase identifies it as a potentially important modulator of synaptic activity.

5. BIOSYNTHESIS AND MODIFICATION

It is generally accepted that the majority, if not all, of the glycoproteins found at the nerve terminal are synthesized in the cell soma, processed in the membrane systems of the endoplasmic reticulum and Golgi apparatus, and transported in particulate form, most probably in combination with other proteins and lipids, by fast axoplasmic transport to the region of the synapse (see Chapter 9). There is, however, a growing body of evidence which suggests that the oligosaccharide moieties of glycoproteins may be altered either during transit down the axon or *in situ* at the synapse.

The hydrolysis of glycoprotein sialic acid by intrinsic SM sialidase(s) has been well documented (Yohe and Rosenberg, 1977; Cruz and Gurd, 1978). Glycoprotein sialidase is also associated with SJs, where it may act to release sialic acid from each of the major SJ glycoproteins (Cruz and Gurd, 1983). The presence of sialyltransferase activity in SM and SJ fractions has also been reported (Den *et al.*, 1975; Preti *et al.*, 1980; Rostas *et al.*, 1981; Durrie *et al.*, 1988; but see Ng and Dain, 1980) and it has been proposed that a sialylation–desialylation cycle may serve to regulate the negative charge density at the synapse (Tettamanti *et al.*, 1980). SJ fractions also express galactosyl- (Goodrum and Tanaka, 1978; Rostas *et al.*, 1981) and fucosyltransferase (Rostas *et al.*, 1981) activities. Although these enzymes are not enriched in SJs, the solubilization of membranes during the detergent extraction step used in the preparation of SJs reduces the likelihood that the measured activities represent impurities derived from nonsynaptic sources. In addition to these enzyme activities, α-mannosidase has also been detected in the region of the PSD by immunocytochemical procedures (Zanetta *et al.*, 1983).

The identification of glycosyltransferase and glycosidase activities in association with synaptic structures takes on added significance in light of reports indicating that the oligosaccharide and peptide portions of specific synaptic glycoproteins may turn over independently of each other (Goodrum and Morell, 1984), and that the oligosaccharide composition of glycoproteins undergoing axonal transport is distinct from that expressed by glycoproteins on the nerve terminal surface (Hart and Wood, 1986). These findings suggest that some post-Golgi processing of synaptic glycoproteins may occur either in the axon during transport or *in situ* at the synapse.

What might the functional consequences of the addition or removal of monosaccharides on synaptic glycoproteins be? The oligosaccharide groups of synaptic glycoproteins are located within the synaptic cleft, so that changes in sugar composition will alter the general molecular environment of the cleft. The physical dimensions of large, complex, *N*-linked oligosaccharides are such that they presumably cover a large part of the membrane surface area (a tetraantennary glycan may cover an area of 20 to 25 nm^2; Montreuil, 1984), and are capable of interacting with proteins or oligosaccharides on the opposing membrane. Moreover, the glycan moiety may directly interact with, and influence the conformation of, its associated polypeptide as occurs, for example, with the Fc fragment of IgG (Sutton and Phillips, 1983). Because the conformation of complex *N*-linked glycans is determined, at least in part, by the identity and linkages of the constituent monosaccharides (Montreuil, 1984; Homans *et al.*, 1987), the addition or removal of certain key sugars may change the oligosaccharide conformation and alter any molecular interactions in which it is involved. The modulation of the biological activity of N-CAM by its associated oligosaccharides (Hoffman and Edelman, 1983) exemplifies the potential influence which the sugar moiety may have on protein function.

In considering the possible significance of glycosidases and glycosyltransferases located at the synapse, it should be noted that most if not all, synaptic glycoproteins are oriented with their oligosaccharide moieties facing the synaptic cleft (Cotman and Taylor, 1974; Matus and Walters, 1976; Wang and Mahler, 1976; Chiu and Babitch, 1978; Gordon-Weeks *et al.*, 1981; Gurd *et al.*, 1983b). The structural modification of oligosaccharides would therefore necessitate that the requisite enzymes and substrates be present on the outer surface of the membrane. Although there is evidence for the occurrence of ectoglycosyltransferases in a variety of systems, including neurons (Shur and Roth, 1975; Cacan *et al.*, 1976; Matsui *et al.*, 1986), the lack of information that relates directly to the topography of glycosidases or glycosyltransferases at the synapse (see Durrie *et al.*, 1988) leaves the question of the function of these enzymes unresolved.

The recent finding of polyribosomes located at the base of dendritic spines has raised the possibility of some local synthesis of postsynaptic proteins (Steward, 1983; Steward and Falk, 1985, 1986). The occurrence of these polyribosomes is more frequent during postnatal and reactive synaptogenesis (Steward and Falk, 1985, 1986), and in the visual cortex of rats from enriched as compared to control environments (Greenenough *et al.*, 1985). These observations suggest that spine polyribosomes may be concerned with the construction of new synapses or in maintaining or altering the efficacy of active synapses. Although the identity of the protein(s) synthesized by these polyribosomes is currently unknown and there is no direct evidence to indicate that such proteins may be glycosylated, the location of many of the polyribosomes adjacent to membrane cisternae, together with the occurrence of glycosyltransferase activities in the postsynaptic apparatus, suggest the possibility that some glycosylation of proteins may occur at these sites (see also Steward and Reeves, 1988).

6. DEVELOPMENTAL ASPECTS

The development of a synapse may be regarded as a continuous process involving the initial contact and recognition between the pre- and postsynaptic cells, the formation of an adhesive membrane junction, the stabilization of the synaptic contact, and the subsequent morphological, biochemical, and functional maturation of the synapse. By analogy with their roles in other, generally simpler, systems, glycoproteins have been considered likely candidates for participation in these events, and in particular in the initial recognition and adhesion between putative pre- and postsynaptic cell membranes. Although there is little direct experimental evidence to support this proposal, suggestions that the cell adhesion molecule N-CAM may be involved in synapse formation (Chuong and Edelman, 1984; Jørgensen *et al.*, 1987), and evidence that the carbohydrate content of the growth cones of developing neurons is involved in the recognition between pre- and postsynaptic cells (Pfenninger and Maylié-Pfenninger, 1981; Pfenninger *et al.*, 1984; Dodd *et al.*, 1984; Dodd and Jessell, 1985) are in general accord with this hypothesis.

In a series of articles, Zanetta and his co-workers have described the transient accumulation of Con A-binding glycoproteins on the membranes of newly formed parallel fibers in the developing cerebellum (Zanetta *et al.*, 1978) and the simultaneous occurrence of a mannose-specific lectin (termed R1) on the surface of Purkinje cells (Zanetta *et al.*, 1985). These authors have proposed that interaction between the Con A-binding glycoproteins and R1 may serve a recognition function during the period of active synaptogenesis in the molecular layer (for additional references and a summary of this proposal, see Dontenwill *et al.*, 1985). Interestingly, granule cells grown in tissue culture demonstrate an increase in Con A-binding sites but not the subsequent disappearance of these glycoproteins (Webb *et al.*, 1985), generally supporting the suggestion that it is the interaction between granule cell axons and their target cells in the molecular layer which is important in regulating the levels of these Con A-binding glycoproteins (Dontenwill *et al.*, 1985).

A number of studies have investigated the effects of development on the protein and glycoprotein composition of synaptic fractions. In the absence of specific probes for individual synaptic proteins and glycoproteins, these studies are limited by the necessity of isolating the appropriate subcellular fraction prior to determining the level of expression of synaptic components at each developmental age analyzed. This requirement may bias the analyses toward those structures which have achieved a sufficient level of structural stability to withstand the stresses of the isolation procedures, while largely precluding the analysis of earlier, less well-developed synapses. In addition, because synaptic development is an asynchronous process, these studies are further complicated by the likelihood that isolated fractions will contain synapses at various stages of maturation. In spite of these methodological limitations, it is apparent that both the protein and glycoprotein composition of the synapse undergo developmentally related changes.

The most notable change in the protein composition of the synapse is a marked increase in the concentration of the 50-kDa subunit of calcium/calmodulin-dependent protein kinase II, which is present in low amounts in young animals and increases slowly to adult levels during the first 2–3 months of postnatal life (Rostas *et al.*, 1984; Kelly and Vernon, 1985; Weinberger and Rostas, 1986). Because this protein represents one of the most prominent constituents of the postsynaptic apparatus from adult brains (Kennedy *et al.*, 1983; Goldenring *et al.*, 1984; Kelly *et al.*, 1984), these results are indicative of a major structural reorganization of the PSD during synaptic maturation.

Analyses of the glycoprotein composition of synaptic fractions isolated from the brains of young and adult rats demonstrate both a general increase in the concentration of glycoproteins, and changes in the relative amounts of individual components (de Silva *et al.*, 1979; Kelly and Cotman, 1981; Fu *et al.*, 1981; Fu and Gurd, 1983; Rudge and Murphy, 1984; summarized in Table 1). In interpreting these studies, it should be noted that variations in the concentration of glycoproteins detected on the basis of the sugar portion of the molecule may reflect alterations in the monosaccharide composition of the glycan moiety rather than, or in addition to, changes in protein concentration. In the case of each of the major glycoproteins characteristic of SJs, gp110, gp116, and gp180, the amount of endoglycosidase H-sensitive, Con A-binding oligosaccharides associated with the polypeptide remains constant between 10 and 28 days (Figure 2) indicating that changes in Con A receptor activity accurately reflect developmentally related increases in the concentration of these proteins.

During development the oligosaccharides associated with SJ glycoproteins undergo some structural modifications. Between the second and third week of postnatal life, the high-mannose chains responsible for Con A binding shift from approximately equal proportions of Man_5 and Man_8 chains at 10 days to a preponderance of Man_5 chains at 28 days (Fu and Gurd, 1983), and there is a small increase in the sialylation of fucosyl-oligosaccharides (Stanojev, 1987). These developmentally related changes in the oligosaccharide composition of SJ glycoproteins may reflect variation in the activities of enzymes involved in the biosynthetic processing of glycoproteins (Quarles and Brady, 1970; Zanetta *et al.*, 1980; Harford and Waechter, 1980; Braulke and Biesold, 1981; Leon *et al.*, 1982; Volpe *et al.*, 1987), or in the local modification of oligosac-

Table 1. The Effect of Development on Synaptic Membrane Glycoproteins[a]

Glycoprotein M_r	Method of detection			
	Con A	FBP	WGA	Sialic acid
240	+			
230	+			
210			−	
200		−		
180	+ +	+ /0	+ +	−
145		−		−
116	+	+ +	+ /0	−
110	+	+ +	+	0/ −
88	−	+	+ +	− −
74		+ + +		+
65	+		+ +	+ +
50		+ +		+ +
40	+ +	+	+ +	−

[a]Glycoproteins were detected by reacting SM proteins which had been separated by polyacrylamide gel electrophoresis with ^{125}I-labeled lectins or by radiolabeling sialic acid residues as described by Cruz and Gurd (1980). FBP, fucose binding protein from *Lotus tetragonolobus;* WGA, wheat germ agglutinin; +, increase; −, decrease; 0, no change. A blank space indicates that no distinct component was detected. Adapted from Fu *et al.* (1981).

charides (e.g., synaptic sialoglycoprotein-sialidase activity increases three- to fourfold between the second and third week of postnatal development; Cruz and Gurd, 1981). Lectin cytochemical studies demonstrating changes in the lectin binding properties of the developing photoreceptor synapse (McLaughlin *et al.,* 1980) generally support the biochemical analyses indicating that synaptic development and maturation are accompanied by alterations in glycoprotein composition.

The relationship between glycoproteins present in growth cones and mature synapses was recently addressed by Greenberger and Pfenninger (1986), in a study comparing the composition of growth cones and synaptosomes from adult rat brain. A number of differences in the composition of the two fractions were noted indicating that the transition from growth cone to synapse is paralleled by changes in the level of expression of several glycoproteins. Of particular interest was the identification of several glycoproteins which were common to both structures, suggesting that some glycoproteins characteristic of the mature synapse may already be expressed, albeit at a lower level, in the growth cone.

7. IMMUNOLOGICAL APPROACHES TO THE STUDY OF SYNAPTIC GLYCOPROTEINS

In the absence of specific probes for synaptic glycoproteins, the preparation of subcellular fractions enriched in synaptic structures is required for the identification,

Table 2. Some Glycoprotein Antigens
Present in Neuronal Membranes

Antigen[a]	Reference[b]
N-CAM*	1
D2*	2
Thy-1*	3,4
CNSgp130*	5,6
gp50*	7
gp65*	8
gp58*,gp62*	9
BSP-2	10
BSP-3	11
F-10-44-2	12
NS-4	13
L1	14
Purkinje cell gp	15
A2B5	16
gp130	17
MRC OX-2	18

[a]*, present at synapses.
[b]References: 1, Persohn and Schachner (1987); 2, Jørgensen and Møller (1980); 3, Acton *et al.* (1978); 4, Bolin and Rouse (1986); 5, Lakin *et al.* (1983); 6, Gordon-Weeks (personal communication); 7, Beesley *et al.* (1987); 8, Hill *et al.* (1988); 9, Mahadik *et al.* (1982); 10, Rougon *et al.* (1982); 11, Hirn *et al.* (1982); 12, McKenzie *et al.* (1982); 13, Schnitzer and Schachner (1981); 14, Rathjen and Schachner (1984); 15, Reeber *et al.* (1981); 16, Eisenbarth *et al.* (1979); 17, Moss (1986); 18, Barclay and Ward (1982).

isolation, and analysis of these molecules. This necessity imposes strict limits upon the type of questions concerning these glycoproteins which are currently amenable to biochemical analysis, and it is apparent that more specific probes are required before a range of fundamental issues relating to tissue and cell specificity, developmental expression, biosynthetic processing, and function of individual glycoproteins can be addressed. Consequently, the production of antibodies to synaptic glycoproteins has been and remains an important, if elusive, goal to investigators interested in the characterization of these synaptic components. Although a number of antibodies which recognize neuronal membrane glycoproteins have been described, only a few of these react with synaptic glycoproteins (Table 2), and in no case has the immunoreactive species been shown to be uniquely associated with synaptic structures.

While the lack of antibodies which react with synapse-specific glycoproteins leaves the identity and occurrence of such molecules temporarily unresolved, it is becoming increasingly apparent that many of the glycoproteins which are present at the synapse are also found elsewhere in the nervous system. The glycoproteins D2, N-CAM, Thy-1, and CNSgp130, for example, have each been demonstrated by immunohistochemical and/or biochemical methodologies to be present at the synapse as well as elsewhere on the neuronal cell surface (for references see Table 2).

In collaboration with Dr. P. Beesley, we have recently initiated a program to produce monoclonal antibodies directed against synaptic glycoproteins. To this end, we have immunized mice with Con A-binding glycoproteins isolated from SMs or SJs and have obtained, to date, two monoclonal antibodies which recognize epitopes on different synaptic glycoproteins. The first of these, Mab-gp50, reacts with a prominent Con A-binding glycoprotein doublet of apparent M_r of 45,000 and 49,000, termed gp50 (Beesley et al., 1987). gp50 is neuron-specific and concentrated four- to fivefold in SMs. However, it is not unique to the synapse and is also expressed on the cell bodies as well as on the primary and proximal dendrites of granule cells in the cerebellum and hippocampus, and in pyramidal cells in the cerebral cortex. The second antibody, Mab-gp65, reacts with two immunologically related glycoproteins of apparent M_r of 55,000 and 65,000 (Hill et al., 1988). Both of these glycoproteins are neuron-specific and one of them, gp65, is moderately enriched in PSDs. However, neither glycoprotein is confined to the synapse, and gp65 is also expressed on a number of neurite subsets, including axons.

The difficulty in obtaining antibodies which react with glycoproteins which are uniquely associated with the synapse may indicate that such molecules do not exist, although the properties of gp116 and gp180 discussed above (Section 2), and the fact that many neurotransmitter receptors are glycosylated would suggest otherwise. However, these results may also indicate that such proteins are highly conserved and therefore constitute poor immunogens, or be simply due to a failure of immunization or detection procedures. Because of their potential value in elucidating the roles of glycoproteins in synaptic function, the acquisition of these antibodies must remain a high priority.

A novel approach to the identification of synaptic glycoprotein antigens based upon the retrograde axonal transport of antibodies following their specific uptake by adsorptive endocytosis at the nerve terminal has recently been described. Antibodies to glycoprotein fractions prepared from SM (Wenthold et al., 1984) or whole brain (Ritchie et al., 1985) were retrogradely transported in the hypoglossal or facial nerves following injection into the tongue or facial muscle, respectively. Because polyclonal antisera were used for these studies, it was not possible to identify the synaptic antigens responsible for the binding and uptake of antibodies. However, in combination with the production of monoclonal antibodies to individual synaptic components (Ritchie et al., 1986), this technique would appear to offer the potential of identifying individual synaptic glycoprotein antigens which are associated with specific classes of nerve terminals.

8. CONCLUDING REMARKS

It is evident from the experimental results reviewed in this chapter that considerable progress in the identification and characterization of glycoproteins at the synapse has been made since the initial observation of Rambourg and Leblond of elevated levels of carbohydrate in the region of the synaptic cleft. However, in spite of these advances, a number of fundamental questions concerning the structure and function of these synaptic components remain unanswered. We do not know, for example, the role of oligosaccharides in determining or modulating the function of individual synaptic glycoproteins, or the effect of the modification of these structures on synaptic activity. The involvement, if any, of the sugar moiety in the selection and packaging of glyco-proteins for transport from the site of synthesis in the cell body to the nerve terminal remains a subject of speculation. It is not known if all synapses have the same basic complement of glycoproteins or whether there is specificity with respect to synaptic type. The functional significance of the developmentally related changes in the amount and structure of glycoproteins at the synapse remains to be determined, as do the factors which regulate the expression of these synaptic molecules during ontogenesis of the nervous system. While the phosphorylation of the PSD glycoprotein, gp180, by both calcium/calmodulin-dependent protein kinase and tyrosine kinase identifies it as a potentially important regulator of synaptic activity, the functional consequences of these reactions remain unknown. It is apparent that the answers to these and similar questions related to the function of synaptic glycoproteins will only be obtained follow-ing detailed analysis of the structure, expression, and molecular interactions of indi-vidual glycoproteins. This, in turn, will necessitate the development and application of procedures which will facilitate the detection and isolation of these molecules. To this end, the successful production of antibodies is of primary importance. In addition, cloning of the genes for one or more synaptic glycoproteins, a goal which it is reason-able to expect will be achieved within the next few years, will provide new insights into the structure, organization, expression, and function of these important synaptic components.

ACKNOWLEDGMENTS

I thank Drs. P. Beesley, P. Gordon-Weeks, and O. Steward for providing me with as yet unpublished information.

9. REFERENCES

Acton, R. T., Addis, J., Carl, G. F., McLain, L. D., and Bridgers, W. F., 1978, Association of the Thy-1 differentiation alloantigen with synaptic complexes isolated from mouse brain, *Proc. Natl. Acad. Sci. USA* **75**:3283–3287.

Barclay, A. N., and Ward, H. A., 1982, Purification and characterization of membrane glycoproteins from rat thymocytes and brain, recognized by monoclonal antibody MRC-OX2, *Eur. J. Biochem.* **129**:447–458.

Beesley, P. W., Paladino, T., Gravel, C., Hawkes, R. A., and Gurd, J. W., 1987, Characterization of gp50, a major glycoprotein present in rat brain synaptic membranes, with a monoclonal antibody, *Brain Res.* **408**:65–78.

Benzamahouta, C., Zanetta, J.-P., Clos, J., Meyer, A., and Vincendon, G., 1988, Studies on the 240-kDa Con A-binding glycoprotein of rat cerebellum, a putative marker of synaptic junctions, *Dev. Brain Res.* **40**:193–200.

Bittiger, H., and Schnebli, H. P., 1974, Binding of concanavalin A and Ricin to synaptic junctions of rat brain, *Nature* **249**:370–371.

Blomberg, F., Cohen, R. S., and Seikevitz, P., 1977, The structure of postsynaptic densities isolated from dog cerebral cortex. II. Characterization and arrangement of some of the major proteins within the structure, *J. Cell Biol.* **74:**204–225.

Bolin, L. M., and Rouse, R. V., 1986, Localization of Thy-1 expression during postnatal development of the mouse cerebellar cortex, *J. Neurocytol.* **15:**29–36.

Braulke, T., and Biesold, D., 1981, Developmental patterns of galactosyltransferase activity in various regions of rat brain, *J. Neurochem.* **36:**1289–1291.

Breckenridge, W. C., and Morgan, I. G., 1972, Common glycoproteins of synaptic vesicles and the synaptosomal plasma membrane, *FEBS Lett.* **22:**253–256.

Breckenridge, W. C., Breckenridge, J. E., and Morgan, I. G., 1972, Glycoproteins of the synaptic region, *Adv. Exp. Med. Biol.* **32:**135–153.

Bristow, D. R., and Martin, I. L., 1987, Solubilization of the γ-aminobutyric acid/benzodiazepine receptor from rat cerebellum: Optimal preservation of the modulatory responses by natural brain lipids, *J. Neurochem.* **49:**1386–1393.

Brunngraber, E. G., Dekirjenjian, H., and Brown, B. D., 1967, The distribution of protein-bound N-acetylneuraminic acid in subcellular fractions of rat brain, *Biochem. J.* **103:**73–78.

Cacan, R., Verbert, A., and Montreuil, J., 1976, New evidence for cell surface galactosyltransferase, *FEBS Lett.* **64:**102–106.

Casadei, J. M., Gordon, R. D., Lampson, L. A., Schotland, D. L., and Barchi, R. L., 1984, Monoclonal antibodies against the voltage-sensitive Na$^+$ channel from mammalian skeletal muscle, *Proc. Natl. Acad. Sci. USA* **81:**6227–6231.

Chiu, T. C., and Babitch, J. A., 1978, Topography of glycoproteins in the chick synaptosomal plasma membrane, *Biochim. Biophys. Acta* **510:**112–123.

Chuong, C. M., and Edelman, G. M., 1984, Alterations in neural cell adhesion molecules during development of different regions of the nervous system, *J. Neurosci.* **4:**2354–2368.

Churchill, L., Cotman, C. W., Banker, G., Kelly, P., and Shannon, L., 1976, Carbohydrate composition of central nervous system synapses. Analysis of isolated synaptic junctional complexes and postsynaptic densities, *Biochim. Biophys. Acta* **448:**57–72.

Cohen, R. S., Blomberg, F., Berzins, K., and Seikevitz, P., 1977, The structure of postsynaptic densities isolated from dog cerebral cortex. Overall morphology and protein composition, *J. Cell Biol.* **74:**181–203.

Cotman, C. W., and Matthews, D. A., 1971, Synaptic plasma membranes from rat brain synaptosomes: Isolation and partial characterization, *Biochim. Biophys. Acta* **249:**380–394.

Cotman, C. W., and Taylor, D., 1972, Isolation and structural studies of synaptic complexes from rat brain, *J. Cell Biol.* **55:**696–711.

Cotman, C. W., and Taylor, D., 1974, Localization and characterization of concanavalin A receptors in the synaptic cleft, *J. Cell Biol.* **62:**236–242.

Cotman, C. W., Banker, G., Churchill, L., and Taylor, D., 1974, Isolation of postsynaptic densities from rat brain, *J. Cell Biol.* **63:**441–455.

Cruz, T. F., and Gurd, J. W., 1978, Reaction of synaptic plasma membrane sialoglycoproteins with intrinsic sialidase and wheat germ agglutinin, *J. Biol. Chem.* **253:**7314–7318.

Cruz, T. F., and Gurd, J. W., 1980, An adaptation of the sodium periodate/sodium borotritide labeling procedure which allows the unambiguous identification of synaptic sialylglycoproteins, *Anal. Biochem.* **108:**139–145.

Cruz, T. F., and Gurd, J. W., 1981, Synaptic membrane sialidase: The effects of development on activity, specificity and endogenous substrates, *Biochim. Biophys. Acta* **675:**201–208.

Cruz, T. F., and Gurd, J. W., 1983, Identification of intrinsic sialidase and sialylglycoprotein substrates in rat brain synaptic junctions, *J. Neurochem.* **40:**1599–1604.

Dahl, J. L., and Hokin, L. E., 1974, The sodium-potassium adenosine triphosphatase, *Annu. Rev. Biochem.* **43:**327–356.

Davis, G. A., and Bloom, F. E., 1973, Isolation of synaptic junctional complexes from rat brain, *Brain Res.* **62:**135–153.

deBlas, A., and Mahler, H. R., 1976, Studies of nicotinic acetylcholine receptors in mammalian brain. IV. Isolation of a membrane fraction enriched in receptor function for different neurotransmitters, *Biochem. Biophys. Res. Commun.* **72:**24–32.

Den, H. B., Kaufman, B., and Roseman, S., 1975, The sialic acids. XVIII. Subcellular distribution of seven glycosyltransferases in embryonic chick brain, *J. Biol. Chem.* **250**:739–746.

de Robertis, E., Azcurra, J. M., and Fiszer, S., 1967, Ultrastructure and cholinergic binding capacity of junctional complexes isolated from rat brain, *Brain Res.* **5**:45–56.

de Silva, N. S., Schwartz, C., and Gurd, J. W., 1979, Developmental alterations of rat brain synaptic membranes: Reaction of glycoproteins with plant lectins, *Brain Res.* **165**:283–293.

Docherty, M., Bradford, H. F., and Wu, J.-Y., 1987, The preparation of highly purified GABAergic and cholinergic synaptosomes from mammalian brain, *Neurosci. Lett.* **81**:232–238.

Dodd, J., and Jessell, T. M., 1985, Lactoseries carbohydrates specify subsets of dorsal root ganglion neurons projecting to the superficial dorsal horn of rat spinal cord, *J. Neurosci.* **5**:3278–3294.

Dodd, J., Solter, D., and Jessell, T. M., 1984, Monoclonal antibodies against carbohydrate differentiation antigens identify subsets of primary sensory neurons, *Nature* **311**:469–472.

Dontenwill, M., Roussel, G., and Zanetta, J. P., 1985, Immunohistochemical localization of a lectin-like molecule, R1, during the postnatal development of the rat cerebellum, *Dev. Brain Res.* **17**:245–253.

Durrie, R., Saito, M., and Rosenberg, A., 1988, Endogenous glycosphingolipid acceptor specificity of sialosyltransferase systems in intact Golgi membranes, synaptosomes and synaptic plasma membranes from rat brain, *Biochemistry* **27**:3759–3764.

Eisenbarth, G. S., Walsh, F. S., and Nirenberg, M., 1979, Monoclonal antibody to a plasma membrane antigen of neurons, *Proc. Natl. Acad. Sci. USA* **76**:4913–4917.

Ellis, P. D., Bissoon, N., and Gurd, J. W., 1988, Characterization of tyrosine protein kinase associated with synaptic membranes and postsynaptic densities from rat brain, *J. Neurochem.* **51**:611–620.

Fiszer, S., and de Robertis, E., 1967, Action of Triton-X100 on ultrastructure and membrane bound enzymes of isolated nerve endings from rat brain, *Brain Res.* **5**:31–44.

Flanagan, S. D., Yost, B., and Crawford, G., 1982, Putative 51,000-Mr protein marker for postsynaptic densities is virtually absent in cerebellum, *J. Cell Biol.* **94**:743–748.

Freedman, M. S., Clark, B. D., Cruz, T. F., Gurd, J. W., and Brown, I. R., 1981, Selective effects of LSD and hypothermia on the synthesis of synaptic proteins and glycoproteins, *Brain Res.* **207**:129–145.

Fu, S. C., and Gurd, J. W., 1983, Developmental changes in the oligosaccharide composition of synaptic junctional glycoproteins, *J. Neurochem.* **41**:1726–1734.

Fu, S. C., Cruz, T. F., and Gurd, J. W., 1981, Development of synaptic glycoproteins. Effect of postnatal age on the synthesis and concentration of synaptic membrane and synaptic junctional fucosyl- and sialyl-glycoproteins, *J. Neurochem.* **36**:1338–1351.

Gioannini, T., Foucaud, B., Hiller, J. M., Hatten, M. E., and Simon, E. J., 1982, Lectin binding of solubilized opiate receptors: Evidence for their glycoprotein nature, *Biochem. Biophys. Res. Commun.* **105**:1128–1134.

Goldenring, J. W., McGuire, J. S., Jr., and DeLorenzo, R. J., 1984, Identification of the major postsynaptic density protein as homologous with the major calmodulin-binding subunit of a calmodulin-dependent protein kinase, *J. Neurochem.* **42**:1077–1084.

Gombos, G., Morgan, I. G., Waehneldt, T. V., Vincendon, G., and Breckenridge, W. C., 1971, Glycoproteins of the synaptosomal plasma membrane, *Adv. Exp. Med. Biol.* **25**:101–113.

Goodrum, J. F., and Morell, P., 1984, Analysis of the apparent biphasic axonal transport kinetics of fucosylated glycoproteins, *J. Neurosci.* **4**:1830–1839.

Goodrum, J. F., and Tanaka, R., 1978, Proteins and glycoproteins of subsynaptosomal fractions. Implications for exocytosis and recycling of synaptic vesicles, *Neurochem. Res.* **3**:599–617.

Gordon-Weeks, P. R., and Harding, S., 1983, Major differences in the concanavalin A binding glycoproteins of postsynaptic densities from rat forebrain and cerebellum, *Brain Res.* **277**:380–385.

Gordon-Weeks, P. R., Jones, D. H., Gray, E. G., and Barron, J., 1981, Trypsin separates synaptic junctions to reveal pre- and post-synaptic concanavalin A receptors, *Brain Res.* **219**:224–230.

Grab, D. J., Carlin, R. K., and Seikevitz, P., 1981, Function of calmodulin in postsynaptic densities. II. Presence of a calmodulin-activatable protein kinase activity, *J. Cell Biol.* **89**:440–448.

Greenberger, L. M., and Pfenninger, K. H., 1986, Membrane glycoproteins of the nerve growth cone: Diversity and growth regulation of oligosaccharides, *J. Cell Biol.* **103**:1369–1382.

Greenenough, W. T., Hwang, H.-N. S., and Gorman, C., 1985, Evidence for active synapse formation or altered postsynaptic metabolism in the visual cortex of rats reared in complex environments, *Proc. Natl. Acad. Sci. USA* **82**:4549–4552.

Groswald, D. E., and Kelly, P. T., 1984, Evidence that a cerebellum-enriched, synaptic junction glycoprotein is related to fodrin and restricts extraction with Triton in a calcium-dependent manner, *J. Neurochem.* **42**:534–546.

Groswald, D. E., Montgomery, P. R., and Kelly, P. T., 1983, Synaptic junctions isolated from cerebellum and forebrain: Comparisons of morphological and molecular properties, *Brain Res.* **278**:63–80.

Gurd, J. W., 1977a, Synaptic plasma membrane glycoproteins: Molecular identification of lectin receptors, *Biochemistry* **16**:369–374.

Gurd, J. W., 1977b, Identification of lectin receptors associated with rat brain postsynaptic densities, *Brain Res.* **126**:154–159.

Gurd, J. W., 1979, Molecular and biosynthetic heterogeneity of fucosyl-glycoproteins associated with rat brain synaptic junctions, *Biochim. Biophys. Acta* **555**:221–229.

Gurd, J. W., 1980, Subcellular distribution and partial characterization of the major concanavalin A receptors associated with rat brain synaptic junctions, *Can. J. Biochem.* **58**:941–951.

Gurd, J. W., 1985a, Phosphorylation of the postsynaptic density glycoprotein gp180 by calcium/calmodulin-dependent protein kinase, *J. Neurochem.* **45**:1128–1135.

Gurd, J. W., 1985b, Phosphorylation of the postsynaptic density glycoprotein gp180 by endogenous tyrosine kinase, *Brain Res.* **333**:385–388.

Gurd, J. W., and Bissoon, N., 1985, In vivo phosphorylation of the postsynaptic density glycoprotein, gp180, *J. Neurochem.* **45**:1136–1140.

Gurd, J. W., and Fu, S. C., 1982, Concanavalin A receptors associated with rat brain synaptic junctions are high mannose-type oligosaccharides, *J. Neurochem.* **39**:719–725.

Gurd, J. W., and Mahler, H. R., 1974, Fractionation of synaptic membrane glycoproteins by lectin-affinity chromatography, *Biochemistry* **13**:5193–5198.

Gurd, J. W., Jones, L. R., Mahler, H. R., and Moore, W. J., 1974, Isolation and partial characterization of rat brain synaptic plasma membranes, *J. Neurochem.* **22**:281–290.

Gurd, J. W., Gordon-Weeks, P., and Evans, W. H., 1982, Biochemical and morphological comparison of postsynaptic densities prepared from rat, hamster and monkey brains by phase partitioning, *J. Neurochem.* **39**:1117–1124.

Gurd, J. W., Bissoon, N., and Kelly, P. T., 1983a, Synaptic junctional glycoproteins are phosphorylated by cAMP-dependent protein kinase, *Brain Res.* **269**;287–296.

Gurd, J. W., Gordon-Weeks, P. R., and Evans, W. H., 1983b, Identification and localization of concanavalin A binding sites on isolated postsynaptic densities, *Brain Res.* **276**:141–146.

Hampson, D. R., and Poduslo, S. E., 1987, Comparison of proteins and glycoproteins of neural plasma membranes, axolemma, synaptic membranes and oligodendroglial membranes, *J. Neurosci.* **17**:277–284.

Harford, J. B., and Waechter, C. J., 1980, A developmental change in dolichol phosphate mannose synthase activity in pig brain, *Biochem. J.* **188**:481–490.

Hart, C. E., and Wood, J. G., 1986, Analysis of the carbohydrate composition of axonally transported glycoconjugates in sciatic nerve, *J. Neurosci.* **6**:1365–1371.

Hill, I. E., Selkirk, C. P., Hawkes, R. B., and Beesley, P. W., 1988, Characterization of novel glycoprotein components of synaptic membranes and postsynaptic densities, gp55 and gp65, with a monoclonal antibody, *Brain Res.* **461**:27–43.

Hirn, M., Pierres, M., Deagostini-Bazin, H., Hirsch, M. R., Goridis, C., Ghandour, M. S., Langley, O. K., and Gombos, G., 1982, A new brain cell surface glycoprotein identified by monoclonal antibody, *Neuroscience* **7**:239–250.

Hoffman, S., and Edelman, G. M., 1983, Kinetics of homophilic binding by embryonic and adult forms of the neural cell adhesion molecule, *Proc. Natl. Acad. Sci. USA* **80**:5762–5766.

Homans, S. W., Dwek, R. A., and Rademacher, T. W., 1987, Solution conformations of N-linked oligosaccharides, *Biochemistry* **26**:6572–6578.

Hucho, F., 1986, The nicotinic acetylcholine receptor and its ion channel, *Eur. J. Biochem.* **158**:211–226.

Jørgensen, O. S., and Møller, M., 1980, Immunocytochemical demonstration of the D2 protein in the presynaptic complex, *Brain Res.* **194**:419–429.

Jørgensen, O. S., Mogensen, J., and Divac, I., 1987, The N-CAM D2-protein as marker for synaptic remodelling in the red nucleus, *Brain Res.* **405**:39–45.

Kearney, K., and Gurd, J. W., 1986, Phosphorylation of synaptic membrane glycoproteins: The effects of Ca^{2+} and calmodulin, *J. Neurochem.* **46**:1683–1691.

Kelly, P. T., 1983, Nervous system glycoproteins: Molecular properties and possible functions, in: *The Biology of Glycoproteins* (R. J. Ivatt, ed.), pp. 323–369, Plenum Press, New York.

Kelly, P. T., and Cotman, C. W., 1977, Identification of proteins and glycoproteins at synapses of the central nervous system, *J. Biol. Chem.* **252**:786–793.

Kelly, P. T., and Cotman, C. W., 1981, Developmental changes in morphology and molecular composition of synaptic junctional structures, *Brain Res.* **206**:251–271.

Kelly, P. T., and Montgomery, P. R., 1982, Subcellular localization of the 52,000 molecular weight major postsynaptic density protein, *Brain Res.* **233**:265–286.

Kelly, P. T., and Vernon, P., 1985, Changes in the subcellular distribution of calmodulin kinase II during brain development, *Dev. Brain Res.* **18**:211–224.

Kelly, P. T., Cotman, C. W., Gentry, C., and Nicolson, G., 1976, Distribution and mobility of lectin receptors on synaptic membranes of identified neurons in the central nervous system, *J. Cell Biol.* **71**: 487–496.

Kelly, P. T., McGuiness, T. L., and Greengard, P., 1984, Evidence that the major postsynaptic density protein is a component of a Ca^{2+}/calmodulin-dependent protein kinase, *Proc. Natl. Acad. Sci. USA* **81**:945–949.

Kennedy, M. B., 1983, Experimental approaches to understanding the role of protein phosphorylation in the regulation of neuronal function, *Annu. Rev. Neurosci.* **6**:493–525.

Kennedy, M. B., Bennet, M. K., and Erondu, N. E., 1983, Biochemical and immunochemical evidence that the "major postsynaptic density protein" is a subunit of a calmodulin-dependent protein kinase, *Proc. Natl. Acad. Sci. USA* **80**:7357–7361.

King, M. M., Huang, C. Y., Chock, P. B., Nairn, A. C., Hemmings, H. C., Jr., Chan, K.-F. J., and Greengard, P., 1984, Mammalian brain phosphoproteins as substrates for calcineurin, *J. Biol. Chem.* **259**:8080–8083.

Kornfeld, R., and Kornfeld, S., 1985, Assembly of asparagine-linked oligosaccharides, *Annu. Rev. Biochem.* **54**:631–634.

Krusius, T., Finne, J., Margolis, R. U., and Margolis, R. K., 1978, Structural features of synaptosomal, mitochondrial, and soluble glycoproteins of rat brain, *Biochemistry* **17**:3849–3854.

Lakin, K. H., Allen, A. K., and Farbre, J. W., 1983, Purification and preliminary biochemical characterization of the human and rat forms of the central nervous system-specific molecule F3-87-8, *J. Neurochem.* **42**:385–394.

Leon, M., Broquet, P., Morélis, R., and Louisot, P., 1982, Postnatal developmental changes in fucosyltransferase activity in rat brain, *J. Neurochem.* **39**:1770–1773.

Mahadik, S. P., Korenovsky, V., Ciccarone, V., and Laev, H., 1982, Synaptic membrane antigens in developing rat brain cerebral cortex and cerebellum, *J. Neurochem.* **39**:1340–1347.

Mahler, H. R., 1979, Glycoproteins of the synapse, in: *Complex Carbohydrates of the Nervous System* (R. U. Margolis and R. K. Margolis, eds.), pp. 165–184, Plenum Press, New York.

Margolis, R. K., Margolis, R. U., and Lai, D., 1975, Distribution and metabolism of glycoproteins and glycosaminoglycans in subcellular fractions of brain, *Biochemistry* **22**:4797–4804.

Matsui, Y., Lombard, D., Massarelli, R., Mandel, P., and Dreyfus, H., 1986, Surface glycosyltransferase activities during development of neuronal cell cultures, *J. Neurochem.* **46**:144–150.

Matus, A. I., and Taff-Jones, D. H., 1978, Morphology and molecular composition of isolated postsynaptic junctional structures, *Proc. R. Soc. London Ser. B* **203**:135–151.

Matus, A. I., and Walters, B. B., 1976, Type 1 and type 2 synaptic junctions: Differences in distribution of concanavalin A binding sites and stability of the junctional adhesion, *Brain Res.* **108**:249–256.

Matus, A. I., DePetris, S., and Raff, M. C., 1973, Mobility of concanavalin A receptors in myelin and synaptic membranes, *Nature New Biol.* **244**:278–280.

Matus, A. I., Pehling, G., Akerman, M., and Maeder, J., 1980, Brain postsynaptic densities: Their relationship to glial and neuronal filaments, *J. Cell Biol.* **87**:346–359.

McKenzie, J. L., Dalchau, R., and Fabre, J. W., 1982, Biochemical characterization and localization in brain of a human brain-leucocyte membrane glycoprotein recognised by a monoclonal antibody, *J. Neurochem.* **39**:1461–1466.

McLaughlin, B. J., Wood, J. G., and Gurd, J. W., 1980, The localization of lectin binding sites during photoreceptor synaptogenesis in the chick retina, *Brain Res.* **191**:345–357.

Mena, E. E., and Cotman, C. W., 1982, Synaptic cleft glycoproteins contain homologous amino acid sequences, *Science* **216**:422–423.

Mena, E. E., Foster, A. C., Fagg, G. E., and Cotman, C. W., 1981, Identification of synapse specific components: Synaptic glycoproteins, proteins, and transmitter binding sites, *J. Neurochem.* **37**:1557–1566.

Michaelis, E. K., 1975, Partial purification and characterization of a glutamate-binding protein from rat brain, *Biochem. Biophys. Res. Commun.* **65**:1004–1012.

Montreuil, J., 1984, Spatial conformation of glycans and glycoproteins, *Biol. Cell* **51**:115–131.

Morgan, I. G., and Gombos, G., 1976, Biochemical studies of synaptic macromolecules: Are there specific synaptic components? in: *Neuronal Recognition* (S. H. Barondes, eds.), pp. 179–204, Plenum Press, New York.

Moss, D. J., 1986, Characterization of a soluble trypsin fragment of gp130: A neuronal glycoprotein associated with the cytoskeleton, *J. Cell. Biochem.* **32**:97–112.

Nairn, A. C., Hemmings, H. C., and Greengard, P., 1985, Protein kinases in the brain, *Annu. Rev. Biochem.* **54**:931–976.

Nestler, E. J., and Greengard, P., 1984, *Protein Phosphorylation in the Nervous System*, Wiley, New York.

Ng, S. S., and Dain, J. A., 1980, Specificity and properties of young rat brain sialyl transferases, in: *Cell Surface Glycolipids* (C. Sweeley, ed.), pp. 345–358, Washington, D.C.

Nieto-Sampedro, M., Bussineau, C. M., and Cotman, C. W., 1982, Isolation, morphology and protein and glycoprotein composition of synaptic junctions from the brain of lower vertebrates. Antigen PSD-95 as a junctional marker, *J. Neurosci.* **2**:722–734.

Olden, K., Parent, J. B., and White, S. L., 1982, Carbohydrate moieties of glycoproteins: A re-evaluation of their function, *Biochim. Biophys. Acta* **650**:209–232.

Pallen, C. J., and Wang, J. H., 1984, A multifunctional calmodulin-stimulated phosphatase, *Arch. Biochem. Biophys.* **237**:281–291.

Persohn, E., and Schachner, M., 1987, Immunoelectron microscopic localization of the neural cell adhesion molecules L1 and N-CAM during postnatal development of the mouse cerebellum, *J. Cell Biol.* **105**:569–576.

Pfenninger, K. H., 1973, Synaptic morphology and cytochemistry, *Prog. Histochem. Cytochem.* **5**:1–85.

Pfenninger, K. H., and Maylié-Pfenninger, M.-F., 1981, Lectin labeling of sprouting neurons. 1. Regional distribution of surface glycoconjugates, *J. Cell Biol.* **89**:536–546.

Pfenninger, K. H., Maylié-Pfenninger, M.-F., Friedman, L. B., and Simkowitz, P., 1984, Lectin labeling of sprouting neurons. III. Type-specific glycoconjugates on growth cones of different origins, *Dev. Biol.* **106**:97–108.

Preti, A., Fiorilli, A., Lombardo, A., Caimi, L., and Tettamanti, G., 1980, Occurence of sialyltransferase activity in the synaptosomal membranes prepared from calf brain cortex, *J. Neurochem.* **35**:281–296.

Quarles, R., and Brady, R. O., 1970, Sialoglycoproteins and several glycosidases in developing rat brain, *J. Neurochem.* **17**:801–807.

Rambourg, A., and Leblond, C. P., 1967, Electron microscope observations on the carbohydrate-rich cell coat present at the surface of cells in the rat, *J. Cell Biol.* **32**:27–53.

Rathjen, F. G., and Schachner, M., 1984, Immunocytological and biochemical characterization of a new neuronal cell surface component (L1 antigen) which is involved in cell adhesion, *EMBO J.* **3**:1–10.

Rauh, J. J., Lambert, M. P., Cho, N. J., Chin, H., and Klein, W. L., 1986, Glycoprotein properties of muscarinic acetylcholine receptors from bovine cerebral cortex, *J. Neurochem.* **46**:23–32.

Reeber, A., Vincendon, G., and Zanetta, J. P., 1981, Isolation and immunohistochemical localization of a "Purkinje cell specific glycoprotein subunit" from rat cerebellum, *Brain Res.* **229**:53–65.

Reeber, A., Zanetta, J. P., and Vincendon, G., 1984, Glycans of synaptosomal membrane glycoproteins from adult rat forebrain. Characterization after fractionation by concanavalin A affinity chromatography, *Biochim. Biophys. Acta* **801**:444–455.

Ritchie, T. C., Fabian, R. F., and Coulter, J. D., 1985, Axonal transport of antibodies to subcellular and protein fractions of rat brain, *Brain Res.* **343**:252–261.

Ritchie, T. C., Fabian, R. H., Choate, J. V. A., and Coulter, J. D., 1986, Axonal transport of monoclonal antibodies, *J. Neurosci.* **6**:1177–1184.

Rostas, J. A. P., Kelly, P. T., Pesin, H. R., and Cotman, C. W., 1979, Protein and glycoprotein composition of synaptic junctions prepared from discrete synaptic regions and different species, *Brain Res.* **168:** 151–167.

Rostas, J. A. P., Leung, W. N., and Jeffrey, P., 1981, Glycosyltransferase activities in chick brain synaptic junctions, *Neurosci. Lett.* **24:**155–160.

Rostas, J. A. P., Brent, V. A., and Güldner, F. H., 1984, The maturation of postsynaptic densities in chicken forebrain, *Neurosci. Lett.* **45:**297–304.

Roth, J., 1987, Subcellular organization of glycosylation in mammalian cells, *Biochim. Biophys. Acta* **906:** 405–436.

Rougon, G., Deagostini-Bazin, H., Hirn, M., and Goridis, C., 1982, Tissue- and developmental stage-specific forms of a neural cell surface antigen linked to differences in glycosylation of a common polypeptide, *EMBO J.* **1:**1239–1244.

Rudge, J. S., and Murphy, S., 1984, Concanavalin A binding glycoproteins in subcellular fractions from the developing rat cerebral cortex, *J. Neurochem.* **43:**891–894.

Salvaterra, P. M., and Matthews, D. A., 1980, Isolation of rat brain subcellular fraction enriched in putative neurotransmitter receptors and synaptic junctions, *Neurochem. Res.* **5:**181–195.

Schnitzer, J., and Schachner, M., 1981, Expression of Thy-1, H-2 and NS-4 cell surface antigens and tetanus toxin receptors in early perinatal and adult mouse cerebellum, *J. Neuroimmunol.* **1:**429–456.

Sharon, N., and Lis, H., 1982, Glycoproteins, *Proteins* **5:**1–144.

Shur, B. D., and Roth, S., 1975, Cell surface glycosyltransferases, *Biochim. Biophys. Acta* **415:**473–512.

Stanojev, D., 1987, N-Glycans of synaptic glycoproteins in rat forebrain, M.Sc. thesis, University of Toronto.

Stanojev, D., and Gurd, J. W., 1987, Characterization of fucosyl-oligosaccharides associated with synaptic membrane and synaptic junctional glycoproteins, *J. Neurochem.* **48:**1604–1611.

Steward, O., 1983, Polyribosomes at the base of dendritic spines of central nervous system neurons—Their possible role in synapse construction and modification, *Cold Spring Harbor Symp. Quant. Biol.* **48:** 745–759.

Steward, O., and Falk, P. M., 1985, Polyribosomes under developing spine synapses: Growth specializations of dendrites at sites of synaptogenesis, *J. Neurosci. Res.* **13:**75–88.

Steward, O., and Falk, P. M., 1986, Protein-synthetic machinery at postsynaptic sites during synaptogenesis: A quantitative study of the association between polyribosomes and developing synapses, *J. Neurosci.* **6:**412–423.

Steward, O., and Reeves, T. M., 1988, Protein-synthetic machinery beneath postsynaptic sites on CNS neurons: Association between polyribosomes and other organelles at the synaptic site, *J. Neurosci.* **8:** 176–184.

Sutton, B. J., and Phillips, D. C., 1983, The three dimensional structure of the carbohydrate within the Fc fragment of immunoglobulin G, *Biochem. Soc. Trans.* **11:**130–132.

Tallant, E. A., and Cheung, W. Y., 1984, Characterization of bovine brain calmodulin-dependent protein phosphatase, *Arch. Biochem. Biophys.* **232:**269–279.

Tettamanti, G., Preti, A., Cestaro, B., Masserimi, M., Sonnino, S., and Ghidoni, R., 1980, Gangliosides and associated enzymes at the nerve ending membranes, in: *Cell Surface Glycolipids* (C. S. Sweeley, ed.), pp. 321–343, American Chemical Society, Washington, D.C.

Volpe, J. L., Sakakihara, Y., and Ishii, S., 1987, Dolichol-linked glycoprotein synthesis in developing mammalian brain: Maturational changes of the N-acetylglucosaminyl phosphotransferase, *Dev. Brain Res.* **33:**277–284.

Waehneldt, T. V., Morgan, I. G., and Gombos, G., 1971, The synaptosomal plasma membrane: Protein and glycoprotein composition, *Brain Res.* **34:**403–406.

Walters, B. B., and Matus, A. I., 1975, Proteins of the synaptic junction, *Biochem. Soc. Trans.* **3:**109–112.

Wang, Y.-J., and Mahler, H. R., 1976, Topography of the synaptosomal plasma membrane, *J. Cell Biol.* **71:**639–658.

Webb, M., Gallo, V., Schneider, A., and Balazs, R., 1985, The expression of concanavalin A binding glycoproteins during the development of cerebellar granule neurons *in vitro, Int. J. Dev. Neurosci.* **3:** 199–208.

Webster, J. C., and Klingman, J. D., 1980, Synaptic junctional glycoconjugates from chick brain. Glycoprotein identification and carbohydrate composition, *Neurochem. Res.* **5:**401–414.

Weinberger, R. P., and Rostas, J. A. P., 1986, Subcellular distribution of a calmodulin-dependent protein kinase activity in rat cerebral cortex during development, *Dev. Brain Res.* **29**:37–50.

Wenthold, R. J., Skaggs, K. K., and Reale, R. R., 1984, Retrograde axonal transport of antibodies to synaptic membrane components, *Brain Res.* **304**:162–165.

Wood, J. G., Wallace, R. W., Whitaker, J. N., and Cheung, W. Y., 1980, Immunocytochemical localization of calmodulin and a heat-labile calmodulin binding protein (CaM-BP80) in basal ganglia of mouse brain, *J. Cell Biol.* **84**:66–76.

Yohe, H. C., and Rosenberg, A., 1977, Action of intrinsic sialidase of rat brain synaptic membranes on sialolipid and sialoprotein components in situ, *J. Biol. Chem.* **252**:2412–2418.

Zanetta, J. P., Morgan, I. G., and Gombos, G., 1975, Synaptosomal plasma membrane glycoproteins: Fraction by affinity chromatography on concanavalin A, *Brain Res.* **83**:337–348.

Zanetta, J. P., Reeber, A., Ghandour, M. S., Vincendon, G., and Gombos, G., 1977a, Glycoproteins of intrasynaptosomal mitochondria, *Brain Res.* **83**:337–348.

Zanetta, J. P., Reeber, A., Vincendon, G., and Gombos, G., 1977b, Synaptosomal plasma membrane glycoproteins. ii. Isolation of fucosyl-glycoproteins by affinity chromatography on the Ulex europeus lectin specific for L-fucose, *Brain Res.* **138**:317–328.

Zanetta, J. P., Roussel, G., Ghandour, M. S., Vincendon, G., and Gombos, G., 1978, Postnatal development of rat cerebellum: Massive and transient accumulation of concanavalin A binding glycoproteins in parallel fiber axolemma, *Brain Res.* **142**:301–319.

Zanetta, J. P., Federico, A., and Vincendon, G., 1980, Glycosidases and cerebellar ontogenesis in the rat, *J. Neurochem.* **34**:831–834.

Zanetta, J. P., Roussel, G., Dontenwill, M., and Vincendon, G., 1983, Immunohistochemical localization of α-mannosidase during postnatal development of the rat cerebellum, *J. Neurochem.* **40**:202–208.

Zanetta, J. P., Dontenwill, M., Meyer, A., and Roussel, G., 1985, Isolation and immunohistochemical localization of a lectin-like molecule from the rat cerebellum, *Dev. Brain Res.* **17**:233–243.

Zatz, M., and Barondes, S. H., 1970, Fucose incorporation into glycoproteins of mouse brain, *J. Neurochem.* **17**:157–163.

Glycoproteins of Myelin and Myelin-Forming Cells

Richard H. Quarles

1. INTRODUCTION AND OVERVIEW OF THE CHEMICAL STRUCTURE OF PNS AND CNS MYELIN

Myelin is formed as an extension and modification of the surface membrane of the oligodendrocyte in the central nervous system (CNS) and the Schwann cell in the peripheral nervous system (PNS). The extended plasma membranes are wrapped around the axons in a spiral fashion, leading to a structure that is seen in the electron microscope to consist of alternating major dense and intraperiod lines. The relationship of the myelin sheath to the myelin-forming cells and the axon, as well as the morphological and biochemical differences between CNS and PNS myelin, are described in detail elsewhere (Morell *et al.*, 1989). However, the basic structures of CNS and PNS myelin sheaths are similar, with the major dense lines consisting of the structure formed by the junction of the cytoplasmic surfaces of the plasma membrane of myelin-forming cells and the intraperiod lines corresponding to the point at which the extracellular surfaces come together. The membranes composing the compacted myelin sheaths are characterized by a high lipid content and only a few major proteins. CNS myelin has two major proteins: the hydrophobic 30-kDa proteolipid protein (PLP) accounting for about half of the total, and the highly positively charged, 14- to 21-kDa myelin basic proteins (MBPs) accounting for over one-third of the total. More than half of the total protein in PNS myelin is the integral, 30-kDa P0 protein, while about one-fourth to one-third is accounted for by lower-molecular-weight positively charged proteins including the same MBPs found in CNS myelin and another 14-kDa component called the P2 protein. Current concepts about how these major myelin proteins are incorporated into the myelin sheath are summarized in Figure 1.

Richard H. Quarles • Section on Myelin and Brain Development, Laboratory of Molecular and Cellular Neurobiology, National Institute of Neurological Disorders and Stroke, National Institutes of Health, Bethesda, Maryland 20892.

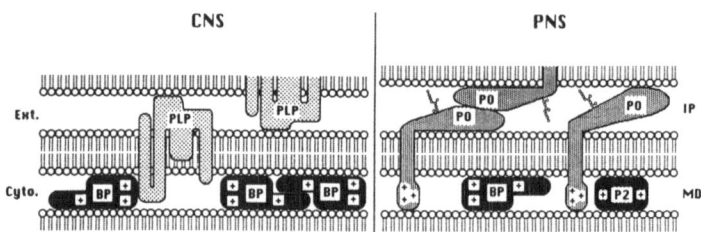

Figure 1. Diagrammatic representation of current concepts of the molecular organization of compact CNS and PNS myelin. The apposition of extracellular (Ext.) surfaces of the oligodendrocyte or Schwann cell membranes to form the intraperiod (IP) line is shown in the upper part of the figure. The apposition of the cytoplasmic (Cyto.) surfaces of the membranes to form the major dense (MD) line is shown in the lower part. The extrinsic myelin basic protein (BP) is believed to play a role in stabilizing the major dense lines of both CNS and PNS myelin. Proteolipid protein (PLP) is a very hydrophobic integral membrane protein that traverses the bilayer of CNS myelin several times and is believed to help stabilize the intraperiod line. The P0 glycoprotein of PNS myelin crosses the bilayer one time and may play a role in stabilizing both the intraperiod and major dense lines (see text). Taken from Morell *et al.* (1989) with permission.

Since glycoproteins are well-known constituents of cell surface membranes, it is not surprising that they are important in the formation of myelin. However, the P0 protein of the PNS is the only one of the major myelin proteins of mammals that is glycosylated. The other glycoproteins associated with myelin and myelin-forming cells appear to be quantitatively rather minor constituents. A historical review of the identification and early characterization of myelin-related glycoproteins has been presented (Quarles, 1979) and will be treated only briefly here. This chapter will emphasize the very rapid recent advances in our knowledge of these components derived from the techniques of molecular biology and immunology. The chapter will deal both with how glycoproteins are believed to be involved in glia–axon interactions and myelinogenesis during development, and with their possible roles in demyelinating diseases.

2. P0 GLYCOPROTEIN

2.1. Introduction

P0 is a 30-kDa integral membrane glycoprotein that accounts for more than half of the total protein in PNS myelin. It is specific for myelin of the PNS, as it is not found in the CNS.

2.2. Chemistry

About 6% by weight of P0 glycoprotein is carbohydrate that appears to exist primarily as a nonasaccharide containing three mannoses, three *N*-acetylglucosamines, one fucose, one galactose, and one sialic acid (Kitamura *et al.*, 1976; Roomi *et al.*, 1978). Amino acid analyses showed a relatively high content of hydrophobic amino acids (Ishaque *et al.*, 1980; Kitamura *et al.*, 1976). In addition to being glycosylated, P0 is also phosphorylated (Singh and Spritz, 1976; Wiggins and Morell, 1980), sulfated (Matthieu *et al.*, 1975b; Wiggins and Morell, 1980), and acylated (Agrawal *et al.*, 1983; Sakamoto *et al.*, 1986).

The complete amino acid sequence of rat P0 was determined from cDNA cloning experiments (Lemke and Axel, 1985), and that for bovine P0 was obtained by automatic Edman degradation of peptide fragments (Sakamoto *et al.*, 1987). Both proteins consist of 219 amino acid residues with the same amino- and carboxy-terminal residues. The protein appears to be highly conserved since there were only 15 amino acid differences between the two species, and most of those could be explained by single base changes in the codons. The initiator ATG methionine determined from the cDNA is followed by 28 uncharged amino acids comprising a signal sequence that is not in the mature protein, and P0 is the only one of the major myelin proteins synthesized as a precursor with a terminal signal sequence. The sequence of the mature protein reveals a single membrane-spanning domain separating a 124-amino-acid extracellular domain from a 68-amino-acid intracellular domain. The extracellular domain contained one canonical acceptor sequence for *N*-glycosylation at Asn-93 and several uncharged stretches. Interestingly, computer analysis revealed that P0 is a member of the immunoglobulin superfamily based on amino acid sequence homologies of its extracellular domain with variable-region folds of immunoglobulins and with other members of the family (Lai *et al.*, 1987, 1988; Lemke *et al.*, 1988; Uyemura *et al.*, 1987). It carries all the essential features of prototypical Ig domains including two cysteine residues thought to be disulfide bonded separated by 77 amino acids, as well as reiterated stretches of alternating hydrophobic residues. This finding has important implications for the function and evolutionary history of P0 and will be discussed in more detail below.

2.3. Biosynthesis and Metabolism

It has long been known that the accumulation of P0 during development generally correlates with the laying down of myelin sheaths (see Quarles, 1979, for a review). A recent correlative biochemical and morphometric investigation of developing cat nerve has shown that P0 accumulates synchronously with the increase in myelin area (Willison *et al.*, 1987). Northern blot analyses indicate that the P0 gene is transcribed as a single mRNA species approximately 1.9 kb in length, and that the level of message is high during active myelination and falls to a lower steady-state level in adults (Lemke and Axel, 1985). There do not appear to be alternative splicing events as are found with other myelin proteins. Genomic analysis indicates a relatively small gene without the large introns as are found for the MBP gene (Lemke, 1986; Lemke *et al.*, 1988). It is divided into multiple exons with one encoding the 5′ untranslated end and most of the signal sequence, two encoding the extracellular domain, one the transmembrane domain, and two the cytoplasmic domain and the untranslated 3′ sequences. Recent studies indicate that Schwann cell-specific expression of the P0 gene is controlled by *cis*-acting elements localized upstream of exon 1 (Lemke *et al.*, 1988). Interestingly, there is a similar 5′ untranslated element shared by the P0 and

MBP genes (19 identities out of 25 nucleotides) that is not in the PLP gene and is a good candidate for conveying PNS specificity (Lemke *et al.*, 1988).

Autoradiographic studies with [^{3}H]fucose have shown that P0 passes through the Golgi apparatus of Schwann cells, after which it is transferred to the outermost regions of the myelin sheath and then moves into the myelin at a rate that is only about 1% of the rate of the phospholipid movement (Gould, 1977). P0 has also been observed in the Golgi apparatus immunocytochemically (Trapp *et al.*, 1981), and metabolic experiments with the ionophore monensin have revealed a key role for the Golgi apparatus in the posttranslational processing and the directing of P0 to myelin (Rapaport *et al.*, 1982; Linington and Waehneldt, 1983). *In vitro* incubation of sliced nerve preparations with radioactive precursors was shown to be useful for studying the metabolism of P0 and other myelin proteins (Smith, 1980). Incubation of such a preparation with radioactive amino acids and sugars in the presence of tunicamycin demonstrated the synthesis of a slightly smaller, nonglycosylated form of P0 which appeared to be incorporated into myelin based on subcellular fractionation experiments (Smith and Sternberger, 1982; Rapaport *et al.*, 1982). The slice system was also used for pulse–chase experiments in the presence and absence of cycloheximide to determine that about 33 min elapses between the synthesis of P0 and its appearance in myelin, and that 20 min of this lag occurs after fucosylation in the Golgi (Rapaport and Benjamins, 1981).

Early studies, primarily in tissue culture systems from the rat, indicated that P0 and other PNS myelin proteins are not synthesized by Schwann cells in the absence of axonal contact (e.g., see Mirsky *et al.*, 1980). However, cultured mouse Schwann cells may be different since recent studies demonstrate expression of basal levels of P0, P1, and P2 in the absence of neurons (Burroni *et al.*, 1988). Also, using the model of the permanently transected sciatic nerve, Poduslo and co-workers have found evidence of P0 synthesis in the absence of axons (Poduslo *et al.*, 1985a,b). The amount of P0 present under these circumstances is very low, being calculated at less than 0.1% of the normal adult level in a recent quantitative investigation of this model (Willison *et al.*, 1988a). The P0 that is synthesized in the absence of axons appears to have predominantly high-mannose oligosaccharides rather than complex oligosaccharides (Poduslo *et al.*, 1985a), and to be abnormally routed to lysosomes rather than myelin (Brunden and Poduslo, 1987).

2.4. Localization and Structural Function in Compact PNS Myelin

Since P0 is quantitatively the major protein in PNS myelin and has been shown immunocytochemically to be distributed throughout the layered myelin sheaths (Trapp *et al.*, 1979), it must play a major role along with the lipids in the structure of PNS myelin. As schematically indicated in Figure 1, it seems likely that the extracellular domain of P0 plays a role in stabilizing the intraperiod line of the myelin while the cytoplasmic domain may be important in stabilizing the major dense line. The chemical structure based on sequence analysis as described in Section 2.2 is consistent with this hypothesis.

Histochemical evidence and surface probe studies indicating that part of the P0 molecule, and particularly the glycosylated domain, is in the intraperiod line of PNS myelin were summarized previously (Quarles, 1979). Based on the hydrophobic stretches that occur in the extracellular domain of P0, Lemke and Axel (1985) suggested that it might bind to itself and stabilize the intraperiod line by homophilic interactions as illustrated to the left in the schematic representation of PNS myelin in Fig. 1. The fact that P0 is a member of the Ig superfamily and the well-characterized ability of Ig-like domains to interact with other Ig-like domains add strength to this hypothesis. In fact, two of the three molecules in the superfamily with which P0 shows the strongest sequence homology (the polyimmunoglobulin receptor and the T4 protein) exhibit binding to other Ig domains as part of their function (Mostov *et al.*, 1984; Gay *et al.*, 1987). Since P0 is not in the CNS, the nonglycosylated and more hydrophobic PLP which has greater capacity for penetrating the lipid bilayers presumably plays an analogous role in stabilizing the intraperiod line of central myelin. It is well known that the periodicity of peripheral myelin is slightly greater than that of central myelin (as indicated in Figure 1), and electron microscopic studies show that a double membrane can be more readily visualized at the intraperiod line of PNS myelin than CNS myelin (Peters and Vaughn, 1970). It is likely that this structural difference between the PNS and CNS is accounted for by the different proteins at this location. The large, glycosylated, more hydrophilic extracellular domain of P0 and possibly its homophilic interactions would prevent fusion of the extracellular Schwann cell surfaces in PNS myelin, while the more hydrophobic PLP would allow greater fusion of these surfaces in central myelin.

P0 glycoprotein also has a positively charged cytoplasmic domain which appears to play an important role in stabilizing the major dense line of PNS myelin. P0 and the positively charged MBPs and P2 protein are likely to do this by interacting with acidic lipids in the bilayer. MBP is believed to play a major role in stabilizing the major dense line in the CNS, based in part on the finding that Shiverer murine mutants in which MBP is undetectable exhibit very little compaction of the cytoplasmic membrane surfaces (Hogan and Greenfield, 1984). By contrast, the structure of PNS myelin in this mutant appears normal (Kirschner and Ganser, 1980). Furthermore, rodents contain very little P2 protein. Therefore, it seems likely that the positive domain of P0 is capable of stabilizing the major dense line in the PNS. That it may perform this role in normal animals is indicated by the much lower content of MBP in PNS myelin (5–18%) than in CNS myelin (~ 30%).

The function of P0 glycoprotein described in the preceding paragraphs is a static one in maintaining the compacted, spiraled structure of the already formed myelin sheath. P0 may also play a more dynamic role during the elaboration and spiraling of sheaths during myelination that occurs in developing animals. It has been proposed that the hydrophobic extracellular domain of P0 creates a self-adhesive extracellular membrane face which guides the wrapping process and ultimately compacts adjacent lamellae (Lemke and Axel, 1985). This more dynamic role during myelinogenesis in the PNS is likely to occur in conjunction with another glycoprotein constituent of myelin sheaths called the myelin-associated glycoprotein (MAG) which will be described in detail in Section 3.

2.5. Phylogeny

As indicated previously, P0 glycoprotein is specific for the PNS in mammals and other higher vertebrates. However, intriguing studies in fish reviewed by Waehneldt *et al.* (1986) have indicated that hydrophobic P0-like glycoproteins are important in the structure of both PNS and CNS myelin in this species. This was shown by the finding that fish CNS and PNS myelins have proteins in the range of 22 to 32 kDa that react with antisera to mammalian P0 protein but not with antisera to PLP. The relationship of the fish CNS myelin proteins to P0 and not PLP was also indicated by amino acid analysis and peptide mapping studies. Overall, the species investigations indicate a discontinuity of CNS myelin protein composition at the transition from fish to tetrapods expressed as an alteration from several P0-like glycoproteins in the myelin of aquatic vertebrates to a nonglycosylated PLP-like component in terrestrial animals. Of special interest in the phylogenetic sense are lungfish, with a glycosylated PLP-like protein but no P0, and the simultaneous presence of P0 and PLP in the bichir. The results indicate that myelin proteins could be useful molecular markers in establishing vertebrate phylogenetic relationships, particularly at the fish–tetrapod transition, and that investigation of fish may provide useful clues about the function of glycoproteins during myelination.

An interesting proposal with regard to the Ig superfamily is that it arose during evolution from a common precursor whose function was to mediate interactions between cells (Williams *et al.*, 1985). A molecule with properties such as those of P0, containing a single Ig domain that binds to itself, has been proposed as the evolutionary antecedent of the Ig superfamily (Williams *et al.*, 1985; Williams, 1987). Furthermore, the similarity to a primordial recognition molecule is also supported by the finding that the genomic sequences encoding the Ig-like extracellular domain of P0 are split between two exons. This is in contrast to the prototypical Ig-like domains in the superfamily, but in keeping with the theory that Ig-like domains evolved from ancestral half domains (reviewed in Lemke *et al.*, 1988). Neural cell adhesion molecule (N-CAM), another member of the family involved in adhesion in the nervous system, also has been found to have Ig-like domains with coding split between two exons (Owens *et al.*, 1987). The complex immunoglobulins themselves may be the last and most sophisticated development in this diverse protein family.

2.6. Involvement in Peripheral Nerve Diseases

Since P0 is the major protein of PNS myelin, it is not surprising that it is lost in demyelinating diseases of the PNS. This has been demonstrated, for example, immunocytochemically in Guillain–Barré syndrome (Schober *et al.*, 1981) and biochemically in Wallerian degeneration (Wood and Dawson, 1974; Willison *et al.*, 1988a). Since P0 is particularly susceptible to degradation by plasmin, it has been suggested that macrophage plasminogen activator could participate in inflammatory demyelination in the PNS (Cammer *et al.*, 1981).

In spite of the known loss of P0 in demyelinating diseases, there is little evidence to suggest that it is the principal target of an immunological response or other pathological process in these diseases. Attempts to induce experimental allergic neuritis (EAN) with purified P0 have generally been negative (reviewed in Quarles, 1979), although guinea pigs with EAN induced with whole myelin have cells sensitized to P0 (Carlo et al., 1975). Injection of polyclonal anti-P0 antisera into rat sciatic nerves produced demyelination, while injection of anti-MBP or anti-P2 antisera did not (Hughes et al., 1985). The positive effects obtained with anti-P0 antisera, as well as with antigalactocerebroside antisera, were attributed to the surface exposure of these antigens on Schwann cells and myelin sheaths. However, there is little evidence of anti-P0 antibodies in human inflammatory polyneuropathies or other demyelinating diseases (reviewed in Ilyas et al., 1988b).

3. MYELIN-ASSOCIATED GLYCOPROTEIN (MAG)

3.1. Introduction

MAG is another extensively studied glycosylated protein that is related to myelin sheaths. Like P0, MAG appears to be a nervous system-specific protein. However, unlike P0, MAG is a quantitatively minor component of isolated myelin, is found in both the CNS and PNS, and is present in myelin-related membranes of oligodendrocytes and Schwann cells rather than in the compact myelin itself. Some quantitative data obtained by radioimmunoassay on the distribution of MAG in CNS and PNS tissues are shown in Table 1. Since MAG is quantitatively a minor component, it is generally not seen on polyacrylamide gels of whole myelin stained for protein and was

Table 1. Quantitation of MAG in Rat and Human Tissues
by Radioimmunoassay[a]

	MAG (ng/μg total protein)
Rat	
Whole brain	2.7 ± 0.14 (8)
Brain stem	4.1 ± 0.70 (8)
Optic nerve	4.4 ± 0.42 (4)
Purified brain myelin	7.4 ± 0.60 (6)
Sciatic nerve	0.7 ± 0.11 (6)
Spleen, liver, heart, kidney, and lymph node	not detected (<0.01)
Human	
Frontal white matter	5.9 ± 0.62 (5)
Occipital white matter	4.6 ± 0.75 (5)
Corpus callosum	4.5 ± 0.62 (5)
Optic nerve	3.2 ± 0.50 (4)
Cortical gray matter	1.3 ± 0.55 (10)
Tibial nerve	0.28 ± 0.08 (3)
Sciatic nerve	0.12 (2)

[a]Radioimmunoassay was done as described by Johnson et al. (1982), and data are from Johnson et al. (1982, 1986) and Johnson and Quarles (1986) with permission.

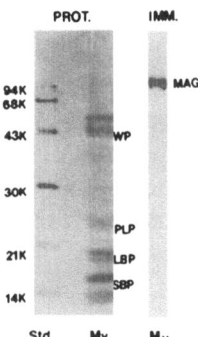

Figure 2. Western blot of proteins in purified myelin showing the 100-kDa myelin-associated glycoprotein (MAG). The panel on the left is a nitrocellulose strip stained for protein with amido black. Std, molecular weight standards; My, purified rat brain myelin (20 μg protein). The strip on the right also contained 20 μg of rat myelin protein and was immunostained with polyclonal anti-MAG antiserum by the peroxidase–antiperoxidase procedure. WP, Wolfgram protein; PLP, proteolipid protein; LBP, large myelin basic protein; SBP, small myelin basic protein. PLP is the major protein of CNS myelin, but appears minor on the blot because it transfers to the nitrocellulose poorly. Reproduced from Quarles et al. (1985) with permission.

originally detected in rat brain myelin by sensitive radiolabeling experiments with [^{3}H]fucose (Quarles et al., 1973a). A sensitive and useful means for detecting MAG after electrophoresis of whole homogenates or purified myelin fractions is by immunostaining on Western blots (Figure 2).

3.2. Chemical Properties

MAG is a 100-kDa glycoprotein and nearly one-third of its molecular mass is accounted for by carbohydrate (Quarles et al., 1983). The carbohydrate composition is typical of an N-linked glycoprotein containing mannose, N-acetylglucosamine, galactose, fucose, and sialic acid. Nearly one-fifth of the total carbohydrate is sialic acid, and radiolabeling studies indicate that rat brain MAG is sulfated (Matthieu et al., 1975a). The sialic acid and sulfate contribute to a low pI for the molecule in the range of 3 to 4, and MAG divides into multiple components in the electrofocusing direction on two-dimensional gels probably because of microheterogeneity in the oligosaccharides (Noronha et al., 1984b). MAG is an integral membrane glycoprotein and detergents are required for its solubilization. The most widely used method for purification involves selective extraction with lithium diiodosalicylate (LIS)–phenol followed by gel filtration (Quarles et al., 1985), although recent techniques involving immunoaffinity chromatography appear useful (Poltorak et al., 1987; Noronha and Quarles, unpublished results).

A major advance in our chemical knowledge of the MAG polypeptide has come in the past year as a result of the independent molecular cloning of MAG from rat brain by three separate groups (Arquint et al., 1987; Lai et al., 1987; Salzer et al., 1987). Sequencing of the cDNA clones has demonstrated that rat brain MAG exists in two forms, containing 610 and 566 amino acids, that are identical except for different C-terminal domains (Figure 3). MAG contains a single membrane-spanning domain and a large extracellular N-terminus that consists of five domains that are related in sequence to each other and to members of the Ig superfamily (Williams, 1987). The implications of MAG being Ig-like will be considered further in Sections 3.5 and 3.6.

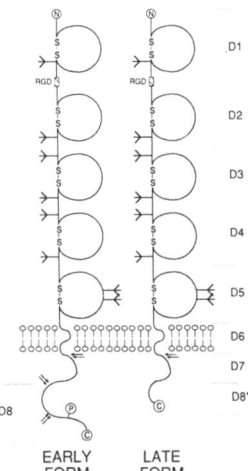

Figure 3. Models of early and late expressed forms of MAG. Five Ig-like domains are indicated as disulfide-bonded loops and the oligosaccharide side chains by branched structures. The Arg-Gly-Asp (RGD) sequence that may function in binding to cell surfaces is in domain D1. The transmembrane domain D6 anchors the protein in the bilayer. The two forms are generated by alternative splicing of the mRNA and differ in the C-terminal, cytoplasmic domains, D8 or D8*. P indicates the potential site for tyrosine kinase phosphorylation on the early form. Arrows represent potential sites for proteolysis. Reproduced from Lai *et al.* (1987) with permission.

Overall, the primary structure of MAG as a transmembrane protein with a cytoplasmic domain and a large glycosylated N-terminus is similar to that previously postulated (Trapp *et al.*, 1984), and is well suited to mediate glia–axon interactions in myelin formation and maintenance.

The two forms of MAG are coded for by a single gene on chromosome 7 in mice (Barton *et al.*, 1987; D'Eustachio *et al.*, 1988) and arise by alternative splicing of the primary mRNA transcript (Salzer *et al.*, 1987; Lai *et al.*, 1987). The genome includes 13 exons, 10 of which are protein coding and correspond to domains arbitrarily designated D0 to D8 and D8* (see Figure 3). D0 is a peptide at the N-terminus which resembles a signal peptide sequence and is presumably removed in the final processed protein. D1 to D5 are Ig-like extracellular domains that are related in sequence to each other and to other members of the Ig superfamily. There are eight potential sites for *N*-linked glycosylation in these domains. D6 contains 21 hydrophobic amino acids that make up the transmembrane domain separating the glycosylated N-terminus from the cytoplasmic domains that consist of D7 and D8 in the larger form and D7 and D8* in the smaller form. The larger mRNA codes for the smaller form of MAG because it contains an extra exon (No. 12) that codes for 10 amino acids of D8* followed by a termination codon. The result is that the 54 C-terminal amino acids of D8 in the larger form of MAG are replaced by a different sequence of 10 amino acids (D8*) in the smaller form. As discussed in Section 3.3, the two forms of MAG are developmentally regulated.

Much less is known about the chemistry of MAG in the PNS. Its presence in peripheral nerve was first detected by immunological methods (Sternberger *et al.*, 1979; Figlewicz *et al.*, 1981) and subsequent studies have revealed extensive immunological cross-reactivity between CNS and PNS MAG. *In vitro* translation experiments with rat sciatic nerve mRNA (Frail *et al.*, 1985) and experiments with antibodies that discriminate between the two alternatively spliced forms of CNS MAG (Noronha *et al.*, 1988) suggest that only the smaller form of MAG is present in peripheral nerve. On the other hand, earlier biochemical experiments had indicated that MAG in rat sciatic nerve has a slightly higher apparent M_r than that in brain (Figlewicz *et al.*, 1981). A possible explanation for this discrepancy would be that PNS contains only the smaller MAG polypeptide but it is more heavily glycosylated *in vivo* leading to a protein with slightly larger apparent M_r. However, definitive knowledge about the structure of the PNS MAG polypeptide must await characterization of MAG cDNAs from the PNS.

Limited biochemical studies on human MAG suggest that its structure is similar to that of rat MAG (Noronha *et al.*, 1984a,b). A difference is that the C-terminus of human MAG appears to be particularly susceptible to a myelin-related Ca^{2+}-activated, neutral protease (CANP) that cleaves it to a slightly smaller derivative (90 kDa) called dMAG (Sato *et al.*, 1982, 1984a). This proteolytic modification of MAG may be important in multiple sclerosis as discussed in more detail in Section 3.7.1. Another difference between rat MAG and human MAG occurs in the oligosaccharide moieties of the molecule and is revealed by a wide variety of monoclonal antibodies that have attained a great deal of prominence in the recent literature and are discussed in Section 3.4.

3.3. Development, Biosynthesis, and Metabolism

During development, MAG is undetectable in brain or nerve prior to myelination and increases dramatically as myelin sheaths are laid down (Johnson *et al.*, 1982; Johnson and Quarles, 1986; Willison *et al.*, 1987). In the PNS, MAG accumulation occurs substantially earlier than the accumulation of the proteins of compact myelin such as P0 (Figure 4) (Willison *et al.*, 1987), suggesting a function in the early stages of myelin formation. Similarly, in the CNS, MAG accumulation is slightly earlier than that of MBP although the difference from the proteins of compact myelin is not as dramatic as in the PNS (Johnson and Quarles, 1986).

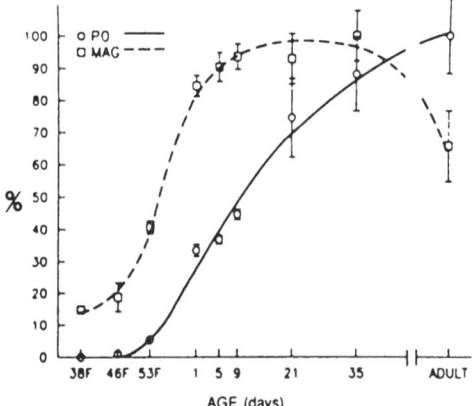

Figure 4. Comparison of the developmental accumulation of P0 and MAG in cat sciatic nerve. Both components are expressed as a percentage of their maximum levels. Maximum levels of P0 and MAG are 238 and 1.4 ng/μg total protein, respectively. MAG is at 15% of its maximum before any P0 is detected, and at 85% of maximum when P0 is only 30% of maximum. Reproduced from Willison *et al.* (1987) with permission.

In vivo labeling with radioactive sugars such as fucose and *N*-acetylglucosamine was used extensively in the early studies on MAG (see Quarles, 1979). Double labeling experiments with ^{3}H- and ^{14}C-labeled sugars demonstrated that MAG synthesized in rat brain during the early period of myelinogenesis has a higher apparent M_r than that synthesized in more mature rats (Quarles *et al.*, 1973b). This developmental decrease in apparent M_r now appears to be due at least in part to developmental regulation of the two forms of MAG formed by alternative splicing of the mRNA discussed in Section 3.2. Frail and Braun (1984) detected two mRNAs for MAG by *in vitro* translation experiments that coded for 67- and 72-kDa MAG polypeptides. During active myelination the principal message was for the 72-kDa peptide, but the proportion of this fell off with age and the message for the 67-kDa form gradually increased. More recent RNA protection experiments using complete cDNA clones for the two alternatively spliced MAG mRNAs gave similar results (Lai *et al.*, 1987). The message for the larger MAG peaks in rat brain between 17 and 29 days of age and then decreases, while that for the smaller form is low early in development and becomes the predominant adult species. Thus, it appears that the principal form of MAG synthesized during active myelinogenesis in the rodent CNS has the larger C-terminal cytoplasmic domain, while the form with the smaller cytoplasmic domain is the principal form synthesized in mature CNS.

An interesting aspect of amino acid sequences in the cytoplasmic domains of MAG is the presence of potential phosphorylation sites for cAMP-dependent kinases, calcium–calmodulin-dependent protein kinase, protein kinase C, and tyrosine kinase (Arquint *et al.*, 1987; Salzer *et al.*, 1987; Lai *et al.*, 1987). Interestingly, some of these potential phosphorylation sites are in the early, larger form of MAG but not in the smaller form, suggesting developmental changes in phosphorylation of MAG. The possible phosphorylation of MAG *in vitro* and *in vivo* is now under active investigation in a number of laboratories.

3.4. Monoclonal Antibodies Binding to a Carbohydrate Determinant in MAG and Other Glycoconjugates

In the past 5 years, interest in MAG has escalated as a result of reports that a variety of monoclonal antibodies of diverse origin react with this glycoprotein. One of the best-known examples is the antibody HNK-1, which was raised to a human lymphoblastoma and is widely used to identify a subset of human lymphocytes with natural killer function (Abo and Balch, 1981). McGarry *et al.* (1983) reported that this antibody reacted with MAG, and this was also demonstrated by other groups (Sato *et al.*, 1983; Murray and Steck, 1984). Thus, in addition to the amino acid sequence homologies shared between MAG and Ig-like molecules discussed in Section 3.2, MAG shares an epitope with cells of the immune system. Another group of antibodies with similar specificity that have received considerable attention are human IgM paraproteins that react with MAG in patients with polyneuropathy occurring in association with IgM gammopathy (Braun *et al.*, 1982; Steck *et al.*, 1983a; Stefansson *et al.*, 1983; Ilyas *et al.*, 1984a; Mendell *et al.*, 1985). The relationship of these human antibodies to the neuropathy is discussed in Section 3.7.2. Studies on HNK-1 and the human IgM paraproteins revealed that they were reacting with a carbohydrate determinant on MAG (Ilyas *et al.*, 1984a; Inuzuka *et al.*, 1984; Frail *et al.*, 1984; Nobile-Orazio *et al.*, 1984; Kruse *et al.*, 1984) that is also expressed on a variety of other glycoconjugates. In addition, there is a pronounced species variation with regard to expression of the carbohydrate epitope on MAG, with MAG from humans and most other large mammals reacting strongly with the antibodies, while MAG from rats and other small experimental animals reacts weakly or not at all (Ilyas *et al.*, 1984a; Inuzuka *et al.*, 1984; O'Shannessy *et al.*, 1985).

In adult human CNS, MAG has received a lot of attention because quantitatively it is the most prominent component reacting with these antibodies. However, it soon became apparent that the reactive epitope(s) is also expressed on other glycoconjugates, especially in the PNS. The PNS antigens included some 19- to 28-kDa glycoproteins of myelin (Inuzuka *et al.*, 1984; Nobile-Orazio *et al.*, 1984; O'Shannessy *et al.*, 1985) (see Section 4.1) and some novel, newly identified sphingoglycolipids (Ilyas *et al.*, 1984a; Freddo *et al.*, 1985). The principal reactive glycolipid in human nerve is glucuronyl 3-sulfate paragloboside (Chou *et al.*, 1985, 1986; Ariga *et al.*, 1987). A detailed discussion of the glycolipid antigens is beyond the scope of this chapter and is reviewed elsewhere (Quarles *et al.*, 1986). However, it is relevant to mention that experiments with the glycolipid antigen and chemically modified derivatives indicate that the sulfated glucuronic acid is an important part of the epitope (Ilyas *et al.*, 1986b, 1988b) suggesting that this or a similar structure is present in MAG. However, these experiments also indicated that the precise requirements for binding vary somewhat from antibody to antibody. For example, HNK-1 absolutely requires the sulfate moiety but not the carboxyl group of the glucuronic acid for specificity, and some human IgM antibodies will react with the desulfated lipid as long as the carboxyl group is present. Therefore, the carbohydrate epitope(s) on MAG is not necessarily sulfated glucuronic acid, and unpublished work in our laboratory suggests that the epitope is probably somewhat different. It has been reported that the carbohydrate structure reacting with the antibodies is on complex tri- or tetraantennary oligosaccharides of MAG (Shy *et al.*, 1986).

In addition to HNK-1 and the human IgM paraproteins, there is now a whole family of antibodies raised to a diverse group of immunogens that react with the same spectrum of glycoconjugate antigens and appear to have similar specificities. All of these antibodies (summarized in Table 2) react with a carbohydrate epitope on MAG, sulfated glucuronyl paragloboside, and other glycoproteins including the 19- to 28-kDa glycoproteins of PNS myelin. The frequency with which antibodies of this specificity are turning up suggests that the reactive carbohydrate structure is highly immunogenic. An interesting subgroup in this family are the "L2 antibodies" which, along with HNK-1, have been shown by Schachner and co-workers (Kruse *et al.*, 1984, 1985; Keilhauer *et al.*, 1985; Noronha *et al.*, 1986b) to react with a number of adhesion proteins of the nervous system including N-CAM, L1 protein (also called NILE glycoprotein and Ng-CAM) and J-1 (also called cytotactin). Schachner and colleagues have hypothesized that all proteins expressing the L2/HNK-1 epitope are involved in adhesion and provided some evidence that the reactive carbohydrate structure may be functionally involved (Keilhauer *et al.*, 1985). As we will see in the next section, there is good reason to believe that MAG is an adhesion protein, and the presence of the reactive carbohydrate structure in MAG adds further credence to this hypothesis.

Table 2. Antibodies Reacting with a Carbohydrate Epitope in MAG
and Other Glycoconjugates

Antibody	Antigen	References
HNK-1 (anti-leu 7)	Human lymphoblastoma	See text
Human IgM paraproteins associated with neuropathy	Naturally occurring; unknown	See text
L2 monoclonal antibodies	Glycoprotein fraction of mouse brain	See text
Monoclonal (G7C8, F7F7, etc.) and polyclonal anti-MAG antibodies	Human MAG	Dobersen et al. (1985), O'Shannessy et al. (1985), Ilyas et al. (1986a)
10C5, 11B5	Melanoma	Noronha et al. (1986a)
NC-1	Quail ciliary ganglia	Tucker et al. (1984)
4F4	Embryonic rat forebrain	Schwarting et al. (1987)

Recently another carbohydrate epitope called L3 has been identified on MAG and some other adhesion molecules of the nervous system (Kucherer et al., 1987), but nothing has been reported about identification of the sugars involved.

3.5. Function

3.5.1. Localization

An important clue about the function of MAG came from immunocytochemical studies at the light microscopic level showing that MAG was localized at the inner rim of myelin sheaths in the CNS and PNS and that it appeared to be absent from compact myelin (Sternberger et al., 1979). This selective periaxonal localization suggested that the MAG could be involved in forming and maintaining the junction between the myelin-forming oligodendrocytes or Schwann cells and axons. Although it was subsequently reported that MAG was present in compact CNS myelin based on postembedding immunostaining at the electron microscopic level (Webster et al., 1983), Trapp and Quarles (1984) concluded that the bulk of the evidence favors a selective periaxonal localization for MAG and an absence from compact myelin in both the CNS and PNS, as originally reported. The periaxonal localization in the PNS has been confirmed by the immunogold procedure using ultrathin cryosections of peripheral nerve (Trapp et al., 1986; Martini and Schachner, 1986). In addition to the presence of MAG in periaxonal membranes, strong correlative evidence suggesting a functional role for MAG in maintaining the Schwann cell–axon junction was provided by an investigation of this glycoprotein in the ventral roots of quaking mice (Trapp et al., 1984). In normal mice there is a constant 12- to 14-nm periaxonal space between the inner Schwann cell membrane and the axolemma (Figure 5A), but in aging quaking mice a pathological change occurs in which this periaxonal space breaks down (Figure 5B). In addition, normal myelin sheaths always have a cytoplasmic collar between the periaxonal Schwann cell membrane and compact myelin (Figure 5A), whereas in quaking mice this space was often absent, with the periaxonal membrane fusing with the inner layer

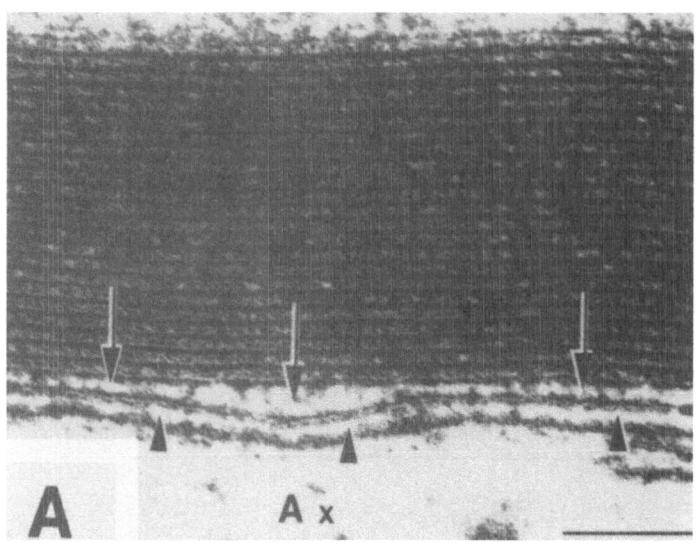

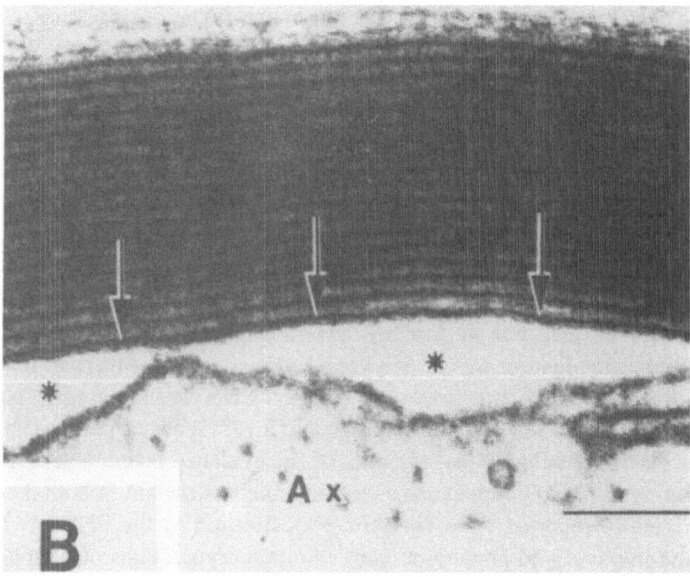

Figure 5. Electron micrographs of myelinated fibers in the L4 ventral root of an 11-month-old quaking mouse. (A) Normally myelinated fiber in which the extracellular leaflet of the periaxonal Schwann membrane is separated from the axolemma by a 12- to 14-nm periaxonal space (arrowheads). Also, a Schwann cell cytoplasmic collar (arrows) separates the cytoplasmic leaflet of the inner compact myelin lamella from the cytoplasmic leaflet of the Schwann cell periaxonal membrane. (B) Abnormal myelin fiber with a dilation of the periaxonal space (asterisks). The cytoplasmic leaflet of the periaxonal membrane has fused with the cytoplasmic side of the inner compact myelin lamella to form a major dense line (arrows). Normally spaced membranes (A) contain MAG in the periaxonal Schwann cell membrane, while MAG is absent in the abnormal situation (B) (see text). Ax, axon. Bars = 0.1 μm. Reproduced from Trapp *et al.* (1984) with permission.

of myelin to form an additional major dense line (Figure 5B). Using the "thick–thin" approach in which semithin sections are examined immunocytochemically and compared to adjacent thin sections examined electron microscopically, it was demonstrated that in all sheaths exhibiting the normal structure (Figure 5A), MAG was present in the periaxonal membrane, whereas MAG was absent from the membranes exhibiting the pathological changes shown in Figure 5B. Based on these findings, it was hypothesized that MAG could be functionally involved in maintaining the normal 12- to 14-nm periaxonal space and the Schwann cell cytoplasmic collar (Trapp et al., 1984). A molecular model was proposed (Trapp et al., 1984) for MAG very much resembling the structure that has now been determined from sequencing of cDNA clones (see Figure 3). It was suggested that the large glycosylated N-terminal domains maintain the extracellular periaxonal space by interaction with a component on the axolemma and the C-terminal domain helps to maintain the cytoplasmic collar possibly by interacting with cytoskeletal elements.

In addition to the periaxonal membrane, the "thick–thin" immunocytochemical approach revealed that MAG was also present in Schmidt–Lanterman incisures, paranodal loops, and outer mesaxons of PNS myelin sheaths (Trapp and Quarles, 1982). All of these regions are characterized by 12- to 14-nm spacing of the extracellular surfaces of adjacent Schwann cell membranes and the presence of cytoplasm at the inner surface of the membranes. Therefore, it was also hypothesized that MAG could function in maintaining the spacing of these adjacent Schwann cell membranes much as it maintains the spacing in the periaxonal region. Consistent with the possibility of maintaining the periaxonal cytoplasmic collar and the other cytoplasmic pockets by interacting with cytoskeletal elements, it is of interest that microfilament components, F-actin and spectrin, colocalize with MAG in the periaxonal region, Schmidt–Lanterman incisures, paranodal loops, and inner and outer mesaxon of myelinating Schwann cells (Trapp et al., 1985).

The localization of MAG in membranes of myelin-forming oligodendrocytes and Schwann cells as discussed above is now well established, but in recent years there have been indications that MAG may also be present in some previously unexpected locations. Some, but not all, of these reports can probably be explained by the use of antibodies such as HNK-1 and others described in Section 3.4 that identify a carbohydrate determinant that is on MAG but is also expressed on other glycoconjugates. Observations made with these antibodies must be interpreted with caution. Nevertheless, there are reports of unexpected localizations with antibodies that react with polypeptide epitopes that appear to be specific for MAG. For example, recent immunocytochemical observations indicate the presence of MAG in the extracellular matrix of myelinated peripheral nerves in the region of interstitial collagens and basement membranes of Schwann cells (Martini and Schachner, 1986; B. D. Trapp, unpublished observation). In this regard, biochemical studies revealed that MAG binds to several types of collagen and to heparin, but not to laminin or fibronectin (Fahrig et al., 1987). The physiological significance of MAG in the extracellular matrix remains to be determined.

There is a report of MAG in small neurons of chicken dorsal root ganglion using a polyclonal antiserum that is believed to be MAG-specific (Omlin et al., 1985). Another indication that MAG may be expressed in some neurons resulted from one of the cloning studies discussed in Section 3.2 (Lai *et al.*, 1987). One of the groups involved in the cloning had previously identified a brain-specific partial cDNA coding for a protein that they called 1B236 (Milner and Sutcliffe, 1983; Sutcliffe *et al.*, 1983). Immunocytochemical studies with antisera raised to synthetic peptides corresponding to amino acid sequences in 1B236 indicated that this protein was present in some neuronal tracts in adult rat brain (Milner and Sutcliffe, 1983; Malfroy *et al.*, 1985; Bloom *et al.*, 1985), but more recent immunocytochemical and *in situ* hybridization studies demonstrated that it is also synthesized in oligodendrocytes during active myelination (Higgins *et al.*, 1986). Extensive immunological experiments utilizing antibodies raised to synthetic peptides of 1B236 and to purified MAG demonstrate that these proteins cannot be distinguished immunologically and are almost certainly identical (Lai *et al.*, 1987; Noronha *et al.*, 1987). These experiments also indicate that quantitatively most of the MAG in the CNS is associated with myelin and myelin-forming cells in both developing and adult animals (Noronha *et al.*, 1987). The possibility that some MAG is expressed in neurons and its possible function in this location deserve further investigation.

3.5.2. Adhesion Properties of MAG

The immunocytochemical studies described in the previous section provide circumstantial evidence that MAG may be involved in the interaction between myelinating cells and axons, and in the interactions between adjacent uncompacted Schwann cell membranes. However, in themselves they give no direct demonstration that MAG possesses adhesive properties that would enable it to perform such a function. Direct evidence that MAG is capable of mediating cell–cell interactions was recently reported utilizing an anti-MAG monoclonal antibody (Poltorak *et al.*, 1987). In experiments involving monolayer cultures as target cells and single cell suspensions as probe cells, it was shown that Fab fragments of the anti-MAG monoclonal antibody inhibited oligodendrocyte to neuron and oligodendrocyte to oligodendrocyte adhesion, but not oligodendrocyte to astrocyte adhesion. In addition, MAG-containing liposomes bound to the cell surfaces of appropriate target cells and the binding was inhibited by the Fab fragments, establishing the capacity of MAG to function as an adhesion molecule.

Also, a clue about a possible molecular mechanism by which MAG may mediate cell–cell or membrane–membrane interactions came out of the cDNA cloning experiments described in Section 3.2. As shown in Figure 3, domain 1 of MAG contains an Arg-Gly-Asp sequence which has been demonstrated to be a cell attachment site for other glycoproteins (Ruoslahti and Pierschbacher, 1986). For example, this is the crucial sequence in the domain of fibronectin that enables it to attach to cell surfaces. Experiments to probe the possible function of this sequence in myelination will be important.

3.5.3. Ig-like Properties

Computer analyses of amino acid sequences in the five extracellular domains of MAG revealed significant sequence homologies among themselves, with sequences of

Ig variable and constant regions, and with sequences of other proteins that have been defined as belonging to the Ig superfamily (Arquint *et al.*, 1987; Lai *et al.*, 1987, 1988; Salzer *et al.*, 1987; Williams *et al.*, 1985; Williams, 1987). The strongest internal homologies were between domains 3 and 4 of MAG, with 45% of the residues being identical, and the other domains showing significant homology with at least one other domain. Particularly characteristic and Ig-like were two cysteines in each domain separated by sequences of 43 to 62 residues (Figure 1). The Ig superfamily includes molecules of the immune system such as the immunoglobulins themselves, histocompatibility antigens, the polyimmunoglobulin receptor, etc., as well as molecules in the nervous system including Thy-1 and N-CAM. The strongest amino acid sequence homologies were between MAG and N-CAM. Like MAG, N-CAM has five Ig-like domains, and domains 3, 4, and 5 of MAG gave higher alignment scores with N-CAM domains than with each other. Interestingly, there is a modestly significant relationship between domain 3 of MAG and the Ig variable-region-like sequence of the P0 glycoprotein of PNS myelin (Lai *et al.*, 1987; Lemke *et al.*, 1988).

Since members of the Ig superfamily are involved in recognition and cell–cell interactions, the identification of MAG as a member of this family increases the likelihood that it functions in cell–cell or membrane–membrane interactions during myelinogenesis and suggests mechanisms by which it might perform this function. Ig-like proteins have the property of interacting with or binding to other Ig-like proteins. An example of this in the immune system is the binding of circulating IgA and IgM to the polyimmunoglobulin receptor on the surface of epithelial cells (Mostov *et al.*, 1984). In the nervous system, N-CAM is believed to mediate adhesion between cells by binding to itself (Edelman, 1986). Since MAG does not appear to be expressed on the surface of myelinated axons, the role of MAG in maintaining the 12- to 14-nm periaxonal space may involve heterophilic interaction with another Ig-like molecule. However, homophilic interactions with other MAG molecules may take place in locations where there are adjacent Schwann cell membranes such as the mesaxons and Schmidt-Lanterman incisures.

3.5.4. Dysmyelinating Mutants

The dysmyelinating murine mutants have been studied extensively by biochemists with the hope of obtaining clues about the molecular mechanisms of myelination occurring in normal animals. In the myelin-deficient mouse (mld) with a defect in the synthesis of MBP and a hypomyelination of the CNS, there are abnormally rapid metabolism of MAG and other myelin constituents, a delay in the developmental switch between the two forms of MAG, and an accumulation of MAG-rich endoplasmic reticulum in oligodendrocytes (Matthieu *et al.*, 1986a).

The quaking mutant exhibits a severe hypomyelination in the CNS and a less severe hypomyelination of the PNS. Quantitative measurements of MAG in quaking mice have revealed that MAG is not decreased as much as the proteins of compact myelin such as MBP and P0 glycoprotein (Inuzuka *et al.*, 1987). Similar observations have been made in other hypomyelinating mutants (Inuzuka *et al.*, 1985, 1986; Yanagisawa and Quarles, 1986; Yanagisawa *et al.*, 1986), and presumably reflect the fact that these mutants have a greater deficiency of compact myelin than of the myelin-

1   2   3

Figure 6. Western blot showing anomalously high apparent M_r for MAG in quaking CNS. Brain-stem homogenates from 36-day-old control (lane 1) and quaking (lanes 2 and 3) mice were electrophoresed. Total quantities of protein in lanes 1, 2, and 3 were 20, 60, and 120 μg respectively, and the blot was immunostained with polyclonal anti-MAG antiserum. Reproduced from Inuzuka *et al.* (1987) with permission.

related membranes in which MAG is concentrated. *In vivo* double labeling experiments with radioactive fucose (Matthieu *et al.*, 1974, 1978) and more recent immunoblotting experiments (Inuzuka *et al.*, 1987) have demonstrated that MAG in both the CNS and PNS of quaking mice has an anomalously high apparent M_r (Figure 6). This higher-than-normal M_r does not appear to be due to an excess of the larger MAG polypeptide generated by alternative splicing of the mRNA and characteristic of immature brain (Section 3.3). Frail and Braun (1985) found an overabundance of the message coding for the smaller form of MAG in this mutant by *in vitro* translation experiments, and MAG could not be detected in the mutant brains with an antiserum specific for the larger form (Noronha *et al.*, 1988). This suggests that the larger form of MAG found in the quaking nervous system *In vivo* may be due to abnormal glycosylation.

Similarly, a larger-than-normal MAG was found in peripheral nerve of trembler mice (Inuzuka *et al.*, 1985), a mutant with a hypomyelination of the PNS. The CNS in these mutants myelinates normally and MAG in the CNS is normal, so the presence of abnormal MAG coincided with the specific dysmyelination of the PNS in this mutant. Immunocytochemical studies have revealed an abundance of spiraled, uncompacted Schwann cell membranes in both the trembler and quaking mutants (Heath and Trapp, unpublished results; Trapp, 1988). Whether the abnormal MAG found in these mutants contributes to this, to the breakdown of the periaxonal space in quaking mice (Trapp *et al.*, 1984), or to other aspects of the pathologies requires further investigation which may yield important information about the function of this molecule.

3.6. Phylogeny

As discussed in Section 2.5 with regard to P0 glycoprotein, it is possible that molecules in the Ig superfamily arose during evolution from a primordial molecule involved in cell–cell interactions (Williams *et al.*, 1985). Lai *et al.* (1988) have proposed that MAG is a prototype for a subgroup of the Ig superfamily called the C-2 set (Williams, 1987) that is characterized by sequences similar to variable domains but with Cys–Cys distances similar in size or shorter than those commonly found in constant domains. N-CAM is also a member of this subgroup and appears to be involved in many morphogenetic events outside of the nervous system during embryogenesis (Edelman, 1986). It may be that adhesion molecules such as MAG and N-CAM most closely resemble ancestors of the Ig family.

The typical ultrastructure of myelin with alternating major and less dense lines seems to have appeared first among the vertebrates in the cartilaginous fish (reviewed in Waehneldt *et al.*, 1986, and Matthieu *et al.*, 1986b). A phylogenetic study by Matthieu *et al.* (1986b) using immunological criteria for identification indicated that MAG and MBP were present in all vertebrates tested from fish to human. Furthermore, unlike MBP, the apparent M_r of MAG varied little among the species, suggesting that conservation of molecular structure may be important for its function throughout the vertebrates. However, in another species study using polyclonal and monoclonal anti-MAG polypeptide antibodies, little or no reactivity was detected with some of the antibodies in some of the lower vertebrates (O'Shannessy *et al.*, 1985). Further chemical information will be needed before a final evaluation about the degree of conservation of MAG structure during evolution can be made. Interestingly, the HNK-1/L2 carbohydrate epitope that has been suggested to be characteristic of adhesion molecules (see Section 3.4) is expressed very strongly on glycoproteins (some in the M_r range of MAG) in some of the lower vertebrates, especially fish (O'Shannessy *et al.*, 1985).

3.7. Clinical Aspects

3.7.1. Multiple Sclerosis

Multiple sclerosis (MS) a disease characterized by focal loss of myelin in regions of the CNS, is generally believed to be autoimmune in nature and possibly instigated by a viral infection. Since glycoproteins are known to be cell surface antigens, receptors for some viruses, and sometimes altered by viral infections, a possible role for MAG in the autoimmune or viral aspects of MS is a reasonable suggestion. An involvement of MAG in MS was suggested by immunocytochemical and biochemical studies demonstrating greater loss of MAG than other myelin proteins at the periphery of many MS plaques (Itoyama *et al.*, 1980; Prineas *et al.*, 1984; Gendelman *et al.*, 1985; Johnson *et al.*, 1986; Moller *et al.*, 1987). Since plaques expand outwards, this is the region where the earliest pathological changes are likely to occur and the results suggest a key role for MAG in the molecular pathogenesis of this disease. These results and their possible significance have been reviewed in more detail elsewhere (Quarles, 1979, 1984, 1988), and this chapter will emphasize our current hypothesis about the reason for these observations.

Although there is evidence for a low level of humoral (Wajgt and Gorny, 1983; Sato *et al.*, 1986; Moller and Quarles, unpublished results) and cellular (Johnson *et al.*, 1986) immunity to MAG in MS, the response appears weak and probably not sufficient to account for the preferential MAG loss described above. We consider it likely that the MAG loss observed at the periphery of MS plaques may be caused by the high susceptibility of human MAG to a calcium-activated neutral protease that cleaves the C-terminus of MAG to produce a slightly lower M_r derivative called dMAG (Sato *et al.*, 1982, 1984a). dMAG is not as tightly associated with membranes as intact MAG and is released into the supernatant after being formed by the *in vitro* incubation of myelin (Sato *et al.*, 1982; Steck *et al.*, 1983b). The similar release of dMAG *in vivo* is suggested by the finding of substantial amounts of this derivative but no intact MAG in

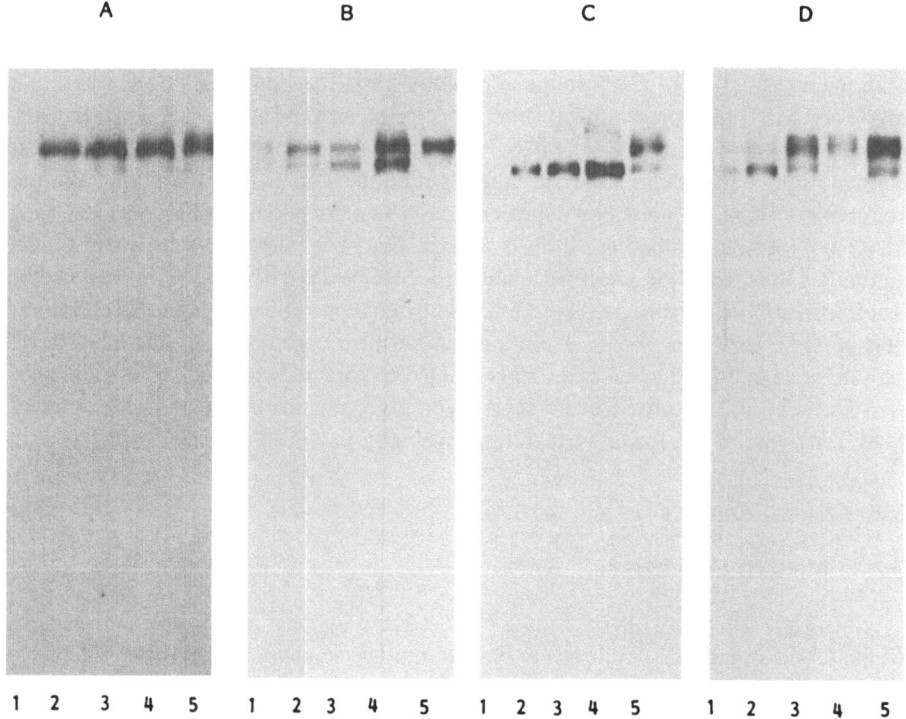

Figure 7. Western blots of homogenates of brain tissue from multiple sclerosis (MS) patients and control subjects showing a greater proportion of MAG in the form of dMAG in MS. Panels A to D show samples from different MS brains and the lanes represent the following regions: 1, plaque; 2, inner periplaque; 3, outer periplaque; 4, macroscopically normal-appearing white matter; and 5, control matched for age, region, and postmortem time. The blots were immunostained with an anti-MAG monoclonal antibody, D7E10. The two closely spaced bands are MAG and its proteolytic derivative, dMAG. Control samples contained very little dMAG, the highest being about 25% of the total as seen in panel D, lane 5. Panel B is representative of six of nine plaque regions examined with about half the total MAG in the form of dMAG. One plaque region had all intact MAG (A), one had all dMAG (C), and a third had progressively more dMAG toward the inner part of the plaque (D). Reproduced from Moller *et al.* (1987) with permission.

the cerebrospinal fluid of normal subjects and patients with neurological disease (Yanagisawa et al., 1985). The production of dMAG may be accelerated in MS, since incubation of myelin from patients with this disease leads to the formation of significantly more dMAG than incubation of control myelin (Sato et al., 1984b). Furthermore, immunoblotting experiments demonstrate that MS white matter contains a greater proportion of the MAG in the form of dMAG than control white matter (Figure 7) (Moller et al., 1987). Taken as a whole, these results suggest that increased activity of calcium-activated neutral protease in affected MS white matter converts MAG to dMAG, leading to its release from oligodendroglia membranes and a preferential loss from the tissue. Whether this loss of MAG contributes to the cellular pathology in MS requires further investigation.

3.7.2. Peripheral Neuropathy Associated with IgM Gammopathy

An IgM gammopathy or paraproteinemia is a condition caused by the neoplastic expansion of a B-cell clone leading to the appearance of large amounts of a monoclonal antibody in the circulation. The high frequency with which gammopathy occurs in association with neuropathy in comparison to its frequency in control subjects clearly establishes a relationship between these two conditions. A high proportion of patients with neuropathy and IgM gammopathy have a serum monoclonal antibody reacting with MAG (Braun et al., 1982; Steck et al., 1983a; Stefansson et al., 1983; Ilyas et al., 1984a; Mendell et al., 1985; Quarles et al., 1986). As discussed in Section 3.4, these antibodies react with a carbohydrate epitope that is on MAG and other protein and lipid glycoconjugates of peripheral nerve. In our experience, more than half of the patients with neuropathy occurring in association with IgM gammopathy have a monoclonal antibody with this specificity (Quarles et al., 1986). At this time, more than 40 patients with an anti-MAG IgM paraprotein have been identified in our laboratory, and numerous others have been detected by other groups. It seems likely that the neuropathy in these patients is caused by binding of the monoclonal IgM to MAG or the other glycoconjugates in the nerves.

The capacity of the human anti-MAG antibodies to cause demyelination under appropriate conditions has been demonstrated by experiments involving the injection of the human antibody into cat nerves (Hays et al., 1987; Willison et al., 1988b). However, as discussed in more detail elsewhere (Mendell et al., 1985; Hays et al., 1987; Willison et al., 1988a; Quarles, 1988), the pathology observed in this experimental model is in some respects different from that seen in biopsies from patients with the human disease. These differences may relate to the acute exposure of nerve tissue to antibody in the intraneural injection model in contrast to chronic exposure in patients. Nevertheless, additional work is needed to fully document that the antibody causes the neuropathy by binding to MAG and/or the other reactive glycoconjugates. Other ways in which the antibody could be related to the neuropathy include causing the neuropathy in a nonspecific way, being the result of an underlying neuropathy of unknown cause, or preventing remyelination and recovery from a neuropathy with a different cause.

4. OTHER GLYCOPROTEINS ASSOCIATED WITH MYELIN AND MYELIN-FORMING CELLS

4.1. 19- to 28-kDa Glycoproteins of PNS Myelin

It has been known for many years that purified PNS myelin contains a number of protein components that electrophorese between P0 and myelin basic protein (P1) in the apparent M_r range of 19 to 28 kDa (Greenfield *et al.*, 1973) and that some proteins of this size are glycosylated (reviewed in Quarles, 1979). The components of this size account for as much as 10–15% of PNS myelin protein (Greenfield *et al.*, 1973). Some of these glycosylated components may be proteolytic degradation products of P0 (Ishaque *et al.*, 1980) or incompletely reduced P0 (Cammer *et al.*, 1980). However, whereas P0 is both fucosylated and sulfated, Matthieu *et al.* (1975b) demonstrated a component in this size range that is fucosylated but not sulfated. Kitamura *et al.* (1976) characterized a component called PAS-II of approximately this size that had an amino acid and carbohydrate composition different from P0. Smith and Perret (1986) reported that a 19-kDa glycoprotein in PNS myelin is immunologically distinct from P0, and similar unpublished observations have been made by O'Shannessy and Quarles. Thus, there appear to be one or more glycoproteins of PNS myelin in this size range that are distinct from P0.

Several glycoproteins in the PNS between 19 and 28 kDa expressing the carbohydrate epitope recognized by HNK-1 and the human IgM paraproteins were discussed in Section 3.4. These glycoproteins appear to be myelin components since they are enriched in purified myelin (Inuzuka *et al.*, 1984) and their deposition during development closely parallels that of P0 (Willison *et al.*, 1987). However, they appear to be distinct from P0, since P0 expresses little or none of the HNK-1 epitope (Inuzuka *et al.*, 1984), and at least some of the HNK-1-reactive glycoproteins do not react with anti-P0 antiserum (O'Shannessy and Quarles, unpublished results). The relationship of these glycoproteins to those of the same size discussed in the previous paragraph is uncertain. An interesting phylogenetic consideration is the possible relationship of these glycoproteins to the P0-related proteins in fish myelin (see Section 3.4), since fish myelin has a number of HNK-1-reactive proteins of about the same size (O'Shannessy *et al.*, 1985; O'Shannessy and Quarles, unpublished results). Clearly, additional work is required to characterize the glycoproteins of lower M_r than P0 in PNS myelin.

4.2. 170-kDa Glycoprotein of PNS Myelin

Coomassie blue staining and lectin binding to the proteins of rat PNS myelin identified a 170-kDa glycoprotein that appeared to account for about 5% of the total myelin protein (Shuman *et al.*, 1983, 1986). It appears to be specific for PNS myelin since it is not in CNS myelin or the nonneural tissues examined. Immunocytochemical studies indicate that it is in compact rat, bovine, and human myelin and that a neuronal influence is necessary for its synthesis by Schwann cells. Recent studies show that synthesis of this glycoprotein and galactocerebroside by Schwann cells can be induced by cAMP, but the synthesis of P0 and MBP are not induced by this treatment (Shuman *et al.*, 1988). The significance of this observation for the control of myelination and chemical characterization of the 170-kDa glycoprotein await further studies.

4.3. Myelin/Oligodendrocyte Glycoprotein

A monoclonal antibody (8-18C5) raised to Thy-1-depleted, rat cerebellar glycoproteins was used to identify a novel glycoprotein antigen in CNS myelin. The purified glycoprotein contains a 51- to 54-kDa component as well as 20- to 29-kDa components which may be degradation products (Linington *et al.*, 1984; Linington and Lassman, 1987). The glycoprotein appears to be in compact CNS myelin and is absent from the PNS. There is experimental evidence indicating that it may be an important target for antibody-mediated demyelination in diseases of the CNS (Linington and Lassman, 1987).

4.4. Adhesion Proteins Produced by Myelin-Forming Cells

Glycoproteins such as N-CAM, L1 (also called Ng-CAM and NILE), and cytotactin (probably the same as the J1 adhesion protein), which have been studied with regard to adhesion between various types of neural cells, are expressed by oligodendrocytes and/or Schwann cells and may play a role in myelination. A detailed review of these adhesion proteins is beyond the scope of this chapter and only a few particularly pertinent considerations will be mentioned here. Interestingly, although brain N-CAM occurs in three forms with sizes of 120, 140, and 180 kDa due to alternative splicing of the mRNA, the principal form expressed in oligodendrocytes and myelin appears to be the 120 kDa form (Bhatt and Silberburg, 1986). It appears to be involved in Ca^{2+}-dependent oligodendrocyte aggregation. The J1 adhesion protein has also been detected on oligodendrocytes (Kruse *et al.*, 1985), but the L1 protein has only been detected on postmitotic neurons in the CNS (Fushiki and Schachner, 1986). As discussed in Section 3.4, all of these adhesion proteins including MAG have been shown to express the HNK-1/L2 carbohydrate epitope. Recently, two large glycosylated polypeptides (150 and 225 kDa) expressing this epitope have been detected in myelinating human oligodendrocytes but not in myelin (Gulcher *et al.*, 1986). These glycoproteins did not appear to correspond to the other adhesion molecules.

Immunocytochemical studies in the PNS by Martini and Schachner (1986) have demonstrated the presence of N-CAM and L1 in Schwann cells during the early stages of myelination. After the Schwann cell membrane has formed about 1.5 loops, L1 disappears and is replaced by MAG. N-CAM is also reduced when myelination proceeds, but it is still detected periaxonally in mature fibers. Based on these observations, Martini and Schachner (1986) suggested that N-CAM and/or L1 mediate the initial cell surface interactions during myelination, and that MAG takes over these functions during the later stages. On the other hand, biochemical studies described in Section 3.3 indicate that MAG appears well in advance of the proteins of compact myelin during peripheral nerve development (Willison *et al.*, 1987). Also, MAG has been detected on Schwann cell surface membranes at the time of initial contact with axons during remyelination in the PNS of quaking mice (Trapp *et al.*, 1984). Immunocytochemical studies by Rieger *et al.* (1986) on the developing PNS have shown a concentration of N-CAM, Ng-CAM, and cytotactin at the nodes of Ranvier in myelinated fibers as well as abnormalities in the distribution of these three proteins in the dysmyelinating trembler mutant.

4.5. Other Little-Characterized Glycoproteins of Myelin and Myelin-Forming Cells

It is clear from experiments involving the incorporation of radioactive sugar precursors (Quarles *et al.*, 1973a), lectin binding (McIntyre *et al.*, 1979; Lane and Fagg, 1980; Poduslo *et al.*, 1980; Poduslo, 1981, 1983), and sodium [^{3}H]borohydride reduction (Poduslo *et al.*, 1976; Mena *et al.*, 1981) that purified CNS myelin fractions contain a very heterogeneous population of glycoproteins. Similarly, there are numerous glycoproteins in isolated oligodendroglial plasma membranes (Poduslo, 1983). Similar procedures demonstrate many glycoproteins associated with PNS myelin (Figlewicz *et al.*, 1981; Linington and Waehneldt, 1981; Pleasure *et al.*, 1982) and with Schwann cell plasma membranes (Pleasure *et al.*, 1982). Some of the glycoproteins revealed by these general detection techniques correspond to the extensively or partially characterized glycoproteins discussed elsewhere in this chapter, but essentially nothing is known about many of the others. The latter category may include some that are in membranes contaminating the myelin or plasma membrane fractions, but others, whose properties and function in myelin formation and maintenance have yet to be investigated, are probably genuine constituents of these membranes.

5. CONCLUSIONS AND PERSPECTIVES

Recent applications of the techniques of recombinant DNA technology have led to major advances in our knowledge of the well-established myelin and myelin-related proteins including MBP, PLP, P0 glycoprotein, and MAG. As described in this chapter, these studies have revealed the complete primary structure of two glycoproteins involved in myelination, P0 and MAG, and given important clues as to how these proteins may function in cell–cell interactions and membrane layering during myelin formation. The molecular mechanisms suggested are open to experimental testing.

As transmembrane proteins, both P0 and MAG appear well suited to mediate interactions between intracellular cytoskeletal elements and adjacent axonal or glial membrane surfaces. MAG in particular could play a key role in the chemomechanical forces involved in generating the spiraled myelin sheaths. The immunocytochemical studies reviewed earlier as well as very recent studies indicate that conversion of Schwann cell mesaxon to compact myelin involves the insertion of P0 into the membrane and removal of MAG (Trapp, 1988). Furthermore, there appear to be intermediate stages in which P0 and MAG coexist in the same membranes, raising the possibility of interactions between these two members of the Ig superfamily in the process of myelin formation.

It is becoming apparent that other glycoproteins such as N-CAM, about which much is known but which have not been greatly studied in the context of myelination, may also be important in the interactions of myelin-forming cells with axons. Also, there are many glycoproteins associated with myelin and myelin-forming cells that have not yet been characterized but which may play important roles in myelin formation as well as in demyelinating diseases. Clearly, in the coming years the techniques of molecular biology will continue to be applied to gain important information about the genetic control of myelination and how the coordinated expression of proteins and lipids in the surface membranes of myelin-forming cells leads to the generation of spiraled and compacted myelin sheaths. Cell surface glycoproteins undoubtedly play key roles in these complex cellular processes. Furthermore, the potential involvement of glycoproteins in the pathogenesis of demyelinating diseases, as cell surface antigens or in other ways, continues to be a promising area of investigation.

ACKNOWLEDGMENTS

The author is grateful to the following recent members of the Section on Myelin and Brain Development who contributed to our research on MAG and related glycoconjugates: Hiroko Baba, Shuichiro Goda, Jeffrey Hammer, Amjad Ilyas, Carl Lauter, Eileen Madden, Johanna Moller, Antonio Noronha, Daniel O'Shannessy, Nancy Slattery, Hugh Willison, and Katsuhiko Yanagisawa. The many contributions of Roscoe Brady to our research program over the years and the biochemistry–morphology collaboration with Bruce Trapp are greatly appreciated. Also, I am grateful to Remmari McDonald for assistance in preparation of the manuscript.

6. REFERENCES

Abo, T., and Balch, C. M., 1981, A differentiation antigen of human NK and K cells identified by a monoclonal antibody (HNK-1), J. Immunol. 127:1024–1029.

Agrawal, H. C., Schmidt, R. E., and Agrawal, D., 1983, In vivo incorporation of [³H]palmitic acid into P0 protein, the major intrinsic protein of rat sciatic nerve myelin. Evidence for covalent linkage of fatty acid to P0, J. Biol. Chem. 258:6556–6560.

Ariga, T., Kohriyama, T., Freddo, L., Latov, N., Saito, M., Kon, K., Ando, S., Suzuki, M., Hemling, M., Rinehart, K. L., Jr., Kusonki, S., and Yu, R. K., 1987, Characterization of sulfated glucuronic acid-containing glycolipids reacting with IgM paraproteins in patients with neuropathy, J. Biol. Chem. 262: 848–853.

Arquint, M., Roder, J., Chia, L., Down, J., Wilkerson, D., Bayley, H., Braun, P., and Dunn, R., 1987, Molecular cloning and the primary structure of the myelin-associated glycoprotein, Proc. Natl. Acad. Sci. USA 84:600–604.

Barton, D. E., Arquint, M., Roder, J., Dunn, R., and Franke, U., 1987, The myelin-associated glycoprotein gene: Mapping to human chromosome 19 and mouse chromosome 7 and expression in quivering mice, Genomics 1:107–112.

Bhatt, S., and Silberburg, D. H., 1986, Oligodendrocyte cell adhesion molecules are related to neural cell adhesion molecule (N-CAM), J. Neurosci. 6:3348–3354.

Bloom, F. E., Battenberg, E. L. F., Milner, R. J., and Sutcliffe, J. G., 1985, Immunocytochemical mapping of 1B236: A brain specific neuronal protein deduced from the sequence of its mRNA, J. Neurosci. 5:1781–1802.

Braun, P. E., Frail, D. E., and Latov, N., 1982, Myelin-associated glycoprotein is the antigen for a monoclonal IgM in polyneuropathy, *J. Neurochem.* **39:**1261–1265.

Brunden, K. R., and Poduslo, J. F., 1987, Lysosomal delivery of the major myelin glycoprotein in the absence of myelin assembly: Posttranslational regulation of the level of expression by Schwann cells, *J. Cell Biol.* **104:**661–669.

Burroni, D., White, F. V., Ceccarini, C., Matthieu, J. M., and Constantino-Ceccarini, E., 1988, Expression of myelin components in mouse Schwann cells in culture, *J. Neurochem.* **50:**331–336.

Cammer, W., Sirota, S. R., and Norton, W. T., 1980, The effect of reducing agents on the apparent molecular weight of the myelin P0 protein and the possible identity of P0 and "Y" proteins, *J. Neurochem.* **34:**404–409.

Cammer, W., Brosnan, C. F., Bloom, B. R., and Norton, W. T., 1981, Degradation of P0, P1 and Pr proteins in peripheral nervous system myelin by plasmin: Implications regarding the role of macrophages in demyelinating diseases, *J. Neurochem.* **36:**1506–1514.

Carlo, D. J., Karkhanis, Y. D., Bailey, P. J., Wisniewski, H. M., and Brostoff, S. W., 1975, Allergic neuritis: Evidence for the involvement of P2 and P0 proteins, *Brain Res.* **76:**423–430.

Chou, D. K. H., Ilyas, A. A., Evans, J. E., Quarles, R. H., and Jungalwala, F. B., 1985, Structure of a glycolipid reacting with monoclonal IgM in neuropathy and with HNK-1, *Biochem. Biophys. Res. Commun.* **128:**383–388.

Chou, D. K. H., Ilyas, A. A., Evans, J. E., Costello, C., Quarles, R. H., and Jungalwala, F. B., 1986, Structure of sulfated glucuronyl glycolipids in the nervous system reacting with HNK-1 antibody and some IgM paraproteins in neuropathy, *J. Biol. Chem.* **261:**11717–11725.

D'Eustachio, P., Colman, D. R., and Salzer, J. L., 1988, Chromosomal location of the mouse gene that encodes the myelin-associated glycoproteins, *J. Neurochem.* **50:**589–593.

Dobersen, M. J., Hammer, J. A., Noronha, A. B., MacIntosh, T. D., Trapp, B. D., Brady, R. O., and Quarles, R. H., 1985, Generation and characterization of mouse monoclonal antibodies to the myelin-associated glycoprotein (MAG), *Neurochem. Res.* **10:**423–437.

Edelman, G. M., 1986, Cell adhesion molecules in the regulation of animal form and tissue pattern, *Annu. Rev. Cell Biol.* **259:**14857–14862.

Fahrig, T., Landa, C., Pesheva, P., Kuhn, K., and Schachner, M., 1987, Characterization of binding properties of the myelin-associated glycoprotein to extracellular matrix constituents, *EMBO J.* **6:**2875–2883.

Figlewicz, D. A., Quarles, R. H., Johnson, D., Barbarash, G. R., and Sternberger, N. H., 1981, Biochemical demonstration of the myelin-associated glycoprotein in the peripheral nervous system, *J. Neurochem.* **37:**749–758.

Frail, D. E., and Braun, P. E., 1984, Two developmentally regulated messenger RNAs differing in their coding region may exist for the myelin-associated glycoprotein, *J. Biol. Chem.* **259:**14857–14862.

Frail, D. E., and Braun, P. E., 1985, Abnormal expression of the myelin-associated glycoprotein in the central nervous system of dysmyelinating mutant mice, *J. Neurochem.* **45:**1071–1075.

Frail, D. E., Edwards, A. M., and Braun, P. E., 1984, Molecular characteristics of the myelin-associated glycoprotein that is recognized by a monoclonal IgM in human neuropathy patients, *Mol. Immunol.* **21:**721–725.

Frail, D. E., Webster, H. d., and Braun, P. E., 1985, Developmental expression of the myelin-associated glycoprotein in the peripheral nervous system is different from that in the central nervous system, *J. Neurochem.* **45:**1308–1310.

Freddo, L., Ariga, T., Saito, M., Macala, M.-C., Yu, R. K., and Latov, N., 1985, The neuropathy of plasma cell dyscrasia: Binding of IgM M-proteins with peripheral nerve glycolipids, *Neurology* **35:**1420–1424.

Fushiki, S., and Schachner, M., 1986, Immunocytological localization of cell adhesion molecules L1 and N-CAM and the shared carbohydrate epitope L2 during development of the mouse neocortex, *Dev. Brain Res.* **24:**153–167.

Gay, D., Maddion, P., Sekaly, R., Talle, M. A., Godfrey, M., Long, E., Goldstein, G., Chess, L., Axel, R., Kappler, J., and Merrack, P., 1987, Functional interaction between human T-cell protein CD4 and the major histocompatibility complex HLA-DR antigen, *Nature* **328:**626–629.

Gendelman, H. E., Pezeshkpour, G. H., Pressman, N. J., Wolinsky, J. S., Quarles, R. H., Dobersen,

M. J., Trapp, B. D., Kitt, C. A., Aksamit, A., and Johnson, R. T., 1985, A quantitation of myelin-associated glycoprotein and myelin basic protein loss in different demyelinating diseases, *Ann. Neurol.* **18**:324–328.

Gould, R. M., 1977, Incorporation of glycoproteins into peripheral nerve myelin, *J. Cell Biol.* **75**:326–339.

Greenfield, S., Brostoff, S., Eylar, E. H., and Morell, P., 1973, Protein composition of myelin of the peripheral nervous system, *J. Neurochem.* **20**:1207–1216.

Gulcher, J. R., Marton, L. S., and Stefansson, K., 1986, Two large glycosylated polypeptides found in myelinating oligodendrocytes but not in myelin, *Proc. Natl. Acad. Sci. USA* **83**:2118–2122.

Hays, A. P., Latov, N., Takatsu, M., and Sherman, W. H., 1987, Experimental demyelination of nerve induced by serum of patients with neuropathy and an anti-MAG IgM protein, *Neurology* **37**:242–256.

Higgins, G. A., Schmale, H., Bloom, F. E., Wilson, M. C., and Milner, R. J., 1986, Developmental shift in the cellular expression of the brain-specific gene 1B236: Localization to oligodendrocytes revealed by *in situ* hybridization, *Soc. Neurosci. Abstr.* **12**:213.

Hogan, E. L., and Greenfield, S., 1984, Animal models of genetic disorders of myelin, in: *Myelin* (P. Morell, ed.), pp. 489–534, Plenum Press, New York.

Hughes, R. A. C., Powell, H. C., Braheny, S. L., and Brostoff, S., 1985, Endoneural injection of antisera to myelin antigens, *Muscle Nerve* **8**:516–522.

Ilyas, A. A., Quarles, R. H., MacIntosh, T. D., Dobersen, M. J., Trapp, B. D., Dalakas, M. C., and Brady, R. O., 1984a, IgM in a human neuropathy related to paraproteinemia binds to a carbohydrate determinant in the myelin-associated glycoprotein and to a ganglioside, *Proc. Natl. Acad. Sci. USA* **81**:1225–1229.

Ilyas, A. A., Quarles, R. H., and Brady, R. O., 1984b, Monoclonal antibody HNK-1 reacts with a peripheral nerve ganglioside, *Biochem. Biophys. Res. Commun.* **122**:1206–1211.

Ilyas, A. A., Dobersen, M. J., Willison, H. J., and Quarles, R. H., 1986a, Mouse monoclonal and rabbit polyclonal antibodies prepared to human myelin-associated glycoprotein also react with glycosphingo-lipids of peripheral nerve, *J. Neuroimmunol.* **12**:99–106.

Ilyas, A. A., Dalakas, M. C., Brady, R. O., and Quarles, R. H., 1986b, Sulfated glucuronyl glycolipids reacting with anti-myelin-associated glycoprotein antibodies including IgM paraproteins in neuropathy: Species distribution and partial characterization of epitopes, *Brain Res.* **385**:1–9.

Ilyas, A. A., Chou, D. K. H., Jungalwala, F. B., and Quarles, R. H., 1988a, Characterization of SGPG epitopes reacting with IgM paraproteins and HNK-1, *Trans. Am. Soc. Neurochem.*, **19**:96 (abstr.).

Ilyas, A. A., Willison, H. J., Quarles, R. H., Jungalwala, F. B., Cornblath, D. R., Trapp, B. D., Griffin, D. E., Griffin, J. W., and McKhann, G. M., 1988b, Serum antibodies to gangliosides in Guillain-Barré syndrome, *Ann. Neurol.*, **23**:440–447.

Inuzuka, T., Quarles, R. H., Noronha, A. B., Dobersen, M. J., and Brady, R. O., 1984, A human lymphocyte antigen is shared with a group of glycoproteins in peripheral nerve, *Neurosci. Lett.* **51**:105–111.

Inuzuka, T., Quarles, R. H., Health, J., and Trapp, B. D., 1985, Myelin-associated-glycoprotein and other proteins in Trembler mice, *J. Neurochem.* **44**:793–797.

Inuzuka, T., Duncan, I. A., and Quarles, R. H., 1986, Myelin proteins in the CNS of "shaking pups," *Dev. Brain Res.* **27**:43–50.

Inuzuka, T., Johnson, D., and Quarles, R. H., 1987, Myelin-associated glycoprotein in the central and peripheral nervous system of quaking mice, *J. Neurochem.* **49**:597–602.

Ishaque, A., Roomi, M. W., Szymanska, I., Kowalski, S., and Eylar, E. H., 1980, The P0 glycoprotein of peripheral nerve myelin, *Can. J. Biochem.* **58**:913–921.

Itoyama, Y., Sternberger, N. H., Webster, H. d., Quarles, R. H., Cohen, S. R., and Richardson, E. P., Jr., 1980, Immunocytochemical observations on the distribution of myelin-associated glycoprotein and basic protein in multiple sclerosis brain, *Ann. Neurol.* **7**:167–177.

Johnson, D., and Quarles, R. H., 1986, Deposition of myelin-associated glycoprotein in specific regions of the developing rat central nervous system, *Dev. Brain Res.* **28**:263–266.

Johnson, D., Quarles, R. H., and Brady, R. O., 1982, A radioimmunoassay for the myelin-associated glycoprotein, *J. Neurochem.* **39**:1356–1362.

Johnson, D., Sato, S., Quarles, R. H., Inuzuka, T., Brady, R. O., and Tourtellotte, W., 1986, Quantitation of the myelin-associated glycoprotein in human nervous tissue from controls and multiple sclerosis patients, *J. Neurochem.* **46:**1086–1093.

Keilhauer, G., Faissner, A., and Schachner, M., 1985, Differential inhibition of neurone–neurone, neurone–astrocyte and astrocyte–astrocyte adhesion by L1, L2 and N-CAM antibodies, *Nature* **316:**728–730.

Kirschner, D. A., and Ganser, A. L., 1980, Compact myelin exists in the absence of basic protein in the shiverer mutant mouse, *Nature* **283:**207–209.

Kitamura, K., Suzuki, M., and Uyemura, K., 1976, Purification and partial characterization of two glycoproteins in bovine peripheral nerve myelin membrane, *Biochim. Biophys. Acta* **455:**806–816.

Kruse, J., Mailhammer, R., Wernecke, A., Faissner, I., Sommer, C., Goridis, C., and Schachner, M., 1984, Neural cell adhesion molecules and myelin-associated glycoprotein share a common carbohydrate moiety recognized by monoclonal antibodies L2 and HNK-1, *Nature* **311:**153–155.

Kruse, J., Keilhauer, G., Faissner, A., Timpl, R., and Schachner, M., 1985, The J1 glycoprotein—A novel nervous system adhesion molecule of the L2/HNK-1 family, *Nature* **316:**728–730.

Kucherer, A., Faissner, A., and Schachner, M., 1987, The novel carbohydrate epitope L3 is shared by some neural cell adhesion molecules, *J. Cell Biol.* **104:**1597–1602.

Lai, C., Brow, M. A., Nave, K. A., Noronha, A. B., Quarles, R. H., Bloom, F. E., Milner, R. J., and Sutcliffe, J. G., 1987, Two forms of 1B236/myelin-associated glycoprotein, a cell adhesion molecule for postnatal neural development, are produced by alternative splicing, *Proc. Natl. Acad. Sci. USA* **84:**4337–4341.

Lai, C., Watson, J. B., Bloom, F. E., Sutcliffe, J. G., and Milner, R. J., 1988, Neural protein 1B236/MAG defines a subgroup of the immunoglobulin superfamily, *Immunol. Rev.* **100:**127–149.

Lane, J. D., and Fagg, G. E., 1980, Protein and glycoprotein composition of myelin subfractions from the developing rat optic nerve and tract, *J. Neurochem.* **34:**163–171.

Lemke, G., 1986, Molecular biology of the major myelin genes, *Trends Neurosci.* **9:**266–270.

Lemke, G., and Axel, R., 1985, Isolation and sequence of a cDNA encoding the major structural protein of peripheral myelin, *Cell* **40:**501–508.

Lemke, G., Lamar, E., and Patterson, J., 1988, Isolation and analysis of the gene encoding peripheral myelin protein zero, *Neuron,* **1:**73–83.

Linington, C., and Lassman, H., 1987, Antibody responses in chronic experimental allergic encephalomyelitis: Correlation of serum demyelinating activity with antibody titre to the myelin/oligodendrocyte glycoprotein (MOG), *J. Neuroimmunol.* **17:**61–69.

Linington, C., and Waehneldt, T. V., 1981, The glycoprotein composition of peripheral nervous system myelin subfractions, *J. Neurochem.* **36:**1528–1535.

Linington, C., and Waehneldt, T. V., 1983, Peripheral nervous system myelin assembly in vitro: Perturbation by the ionophore monensin, *J. Neurochem.* **41:**426–433.

Linington, C., Webb, M., and Woodhams, P. L., 1984, A novel myelin-associated glycoprotein defined by a mouse monoclonal antibody, *J. Neuroimmunol.* **6:**387–396.

Malfroy, B., Bakhit, C., Bloom, F. E., Sutcliffe, J. G., and Milner, R. J., 1985, Brain-specific polypeptide 1B236 exists in multiple molecular forms, *Proc. Natl. Acad. Sci. USA* **82:**2009–2013.

Martini, R., and Schachner, M., 1986, Immunoelectron microscopic localization of neural cell adhesion molecules (L1, N-CAM, and MAG) and their shared carbohydrate epitope and myelin basic protein in developing sciatic nerve, *J. Cell Biol.* **103:**2439–2448.

Matthieu, J. M., Brady, R. O., and Quarles, R. H., 1974, Anomalies of myelin-associated glycoprotein in quaking mice, *J. Neurochem.* **22:**291–296.

Matthieu, J. M., Quarles, R. H., Poduslo, J., and Brady, R. O., 1975a, ^{35}S sulfate incorporation into myelin glycoproteins. I. Central nervous system, *Biochim. Biophys. Acta* **392:**159–166.

Matthieu, J. M., Everly, J. L., Brady, R. O., and Quarles, R. H., 1975b, ^{35}S sulfate incorporation into myelin glycoproteins. II. Peripheral nervous tissue, *Biochim. Biophys. Acta* **392:**167–174.

Matthieu, J. M., Koellreutter, B., and Joyet, M. L., 1978, Protein and glycoprotein composition of myelin and myelin subfractions from brains of Quaking mice, *J. Neurochem.* **30:**783–790.

Matthieu, J. M., Roch, J. M., Omlin, F. X., Rambaldi, I., Almazan, G., and Braun, P. E., 1986a, Myelin instability and oligodendrocyte metabolism in myelin-deficient mutant mice, *J. Cell Biol.* **103:**2673–2682.

Matthieu, J. M., Waehneldt, T. V., and Eschmann, N., 1986b, Myelin-associated glycoprotein and myelin basic protein are present in central and peripheral myelin throughout phylogeny, *Neurochem. Int.* **8:** 521–526.

McGarry, R. C., Helfand, S. L., Quarles, R. H., and Roder, J. C., 1983, Recognition of myelin-associated glycoprotein by the monoclonal antibody HNK-1, *Nature* **306:** 376–378.

McIntyre, L. J., Quarles, R. H., and Brady, R. O., 1979, Lectin-binding proteins in central-nervous-system myelin: Detection of glycoproteins of purified myelin on polyacrylamide gels by ^{3}H-concanavalin A binding, *Biochem. J.* **183:**205–212.

Mena, E. E., Moore, B. W., Hagen, S., and Agrawal, H. C., 1981, Demonstration of five major glycoproteins in myelin and myelin subfractions, *Biochem. J.* **195:**525–528.

Mendell, J. R., Sahenk, Z., Whitaker, J. N., Trapp, B. D., Yates, A. J., Griggs, R. C., and Quarles, R. H., 1985, Polyneuropathy and IgM monoclonal gammopathy; studies on the pathogenetic role of antiMAG antibodies, *Ann. Neurol.* **17:**243–254.

Milner, R. J., and Sutcliffe, J. G., 1983, Gene expression in rat brain, *Nucleic Acids Res.* **11:**5497–5520.

Mirsky, R., Winter, J., Abney, E. R., Pruss, R. M., Gavrilovic, J., and Raff, M. C., 1980, Myelin-specific proteins and glycolipids in rat Schwann cells and oligodendrocytes in culture, *J. Cell Biol.* **84:**483–494.

Moller, J. R., Yanagisawa, K., Brady, R. O., Tourtellotte, W. W., and Quarles, R. H., 1987, Myelin-associated glycoprotein in multiple sclerosis lesions: A quantitative and qualitative analysis, *Ann. Neurol.* **22:**469–474.

Morell, P., Quarles, R. H., and Norton, W. T., 1989, Myelin formation, structure, and biochemistry, in: *Basic Neurochemistry*, 4th ed. (G. Siegel, B. Agranoff, W. Albers, and P. Molioff, eds.), pp. 109–136, Raven Press, New York.

Mostov, K. E., Friedlander, M., and Blobel, G., 1984, The receptor for transepithelial transport of IgA and IgM contains multiple immunoglobulin domains, *Nature* **308:**37–43.

Murray, N., and Steck, A. J., 1984, Indication of a possible role in a demyelinating neuropathy for an antigen shared between myelin and NK cells, *Lancet* **1:**711–713.

Nobile-Orazio, E., Hays, A. P., Latov, N., Perman, G., Golier, J., Shy, M. E., and Freddo, L., 1984, Specificity of mouse and human monoclonal antibodies to myelin-associated glycoprotein, *Neurology* **34:**1336–1342.

Noronha, A., Tolliver, T., Dobersen, M., Hammer, J., and Quarles, R. H., 1984a, Tryptic fragments of human and rat myelin-associated glycoprotein, *Trans. Am. Soc. Neurochem.* **15:**233 (Abstract).

Noronha, A., Tolliver, T. J., Grojec, P. L., Curtis, M. A., and Quarles, R. H., 1984b, Biochemical and immunochemical characterization of the myelin-associated glycoprotein from human tissue, *Soc. Neurosci. Abstr.* **10:**81.

Noronha, A. B., Harper, J., Ilyas, A. A., Reisfeld, R. A., and Quarles, R. H., 1986a, Myelin-associated glycoprotein shares an antigenic determinant with a glycoprotein of human melanoma cells, *J. Neurochem.* **47:**1558–1565.

Noronha, A. B., Ilyas, A. A., Antonicek, H., Schachner, M., and Quarles, R. H., 1986b, Molecular specificity of L2 monoclonal antibodies that bind to carbohydrate determinants of neural cell adhesion molecules and their resemblance to other monoclonal antibodies recognizing the myelin-associated glycoprotein, *Brain Res.* **385:**237–244.

Noronha, A. B., Hammer, J. A., Lai, C., Brow, M. A., Watson, J. B., Bloom, F. E., Milner, R. J., Sutcliffe, J. G., and Quarles, R. H., 1987, Relationship of the myelin-associated glycoprotein (MAG) and the brain 1B236 protein, *J. Neurochem.* **48S:**33 (Abstract).

Noronha, A., Hammer, J., Milner, R., Sutcliffe, J. G., and Quarles, R., 1988, Immunological characterization of MAG in normal and mutant CNS and PNS, *Trans. Am. Soc. Neurochem.*, **19:**118 (abstr.).

Omlin, F. X., Matthieu, J. M., Philippe, E., Roch, J. M., and Droz, B., 1985, Expression of myelin-associated glycoprotein by small neurons of the dorsal root ganglion in chickens, *Science* **227:**1359–1360.

O'Shannessy, D. J., Willison, H. J., Inuzuka, T., Dobersen, M. J., and Quarles, R. H., 1985, The species distribution of nervous system antigens that react with anti-myelin-associated glycoprotein antibodies, *J. Neuroimmunol.* **9:**255–268.

Owens, G. C., Edelman, G. M., and Cunningham, B. A., 1987, Organization of the neural cell adhesion molecule gene, *Proc. Natl. Acad. Sci. USA* **84:**294–298.

Peters, A., and Vaughn, J. E., 1970, Morphology and development of the myelin sheath, in: *Myelination* (A. N. Davison and A. Peters, eds.), pp. 3–79, Thomas, Springfield, Ill.

Pleasure, D., Hardy, M., Kreider, B., Stern, J., Doan, H., Shuman, S., and Brown, S., 1982, Schwann cell surface proteins and glycoproteins, *J. Neurochem.* **39:**486–492.

Poduslo, J. F., 1981, Developmental regulation of the carbohydrate composition of glycoproteins associated with central nervous system myelin, *J. Neurochem.* **36:**1924–1931.

Poduslo, J. F., Quarles, R. H., and Brady, R. O., 1976, External labeling of galactose in surface membrane glycoproteins of the intact myelin sheath, *J. Biol. Chem.* **251:**153–158.

Poduslo, J. F., Harman, J. L., and McFarlin, D. E., 1980, Lectin receptors in central nervous system myelin, *J. Neurochem.* **34:**1733–1744.

Poduslo, J. F., Berg, C. T., Ross, S. M., and Spencer, P. S., 1985a, Regulation of myelination: Axons not required for the biosynthesis of basal levels of the major myelin glycoprotein by Schwann cells in denervated distal segments of the adult cat sciatic nerve, *J. Neurosci. Res.* **14:**177–185.

Poduslo, J. F., Dyck, P. J., and Berg, C. T., 1985b, Regulation of myelination: Schwann cell transition from a myelin-maintaining state to a quiescent state after permanent nerve transection, *J. Neurochem.* **44:**388–400.

Poduslo, S. E., 1983, Proteins and glycoproteins in plasma membranes and in membrane lamellae produced by purified oligodendroglia in culture, *Biochim. Biophys. Acta* **728:**59–65.

Poltorak, M., Sadoul, R., Keilhauer, G., Landa, C., Fahrig, T., and Schachner, M., 1987, Myelin-associated glycoprotein, a member of the L2/HNK-1 family of neural cell adhesion molecules, is involved in neuron–oligodendrocyte and oligodendrocyte–oligodendrocyte interaction, *J. Cell Biol.* **105:**1893–1899.

Prineas, J. W., Kwon, E. E., Sternberger, N. H., and Lennon, V. A., 1984, The distribution of myelin-associated glycoprotein in actively demyelinating multiple sclerosis lesions, *J. Neuroimmunol.* **6:**251–264.

Quarles, R. H., 1979, Glycoproteins in myelin and myelin-related membranes, in: *Complex Carbohydrates of Nervous Tissue* (R. U. Margolis and R. K. Margolis, eds.), pp. 209–233, Plenum Press, New York.

Quarles, R. H., 1984, Myelin-associated glycoprotein in development and disease, *Dev. Neurosci.* **6:**285–303.

Quarles, R. H., 1988, Myelin-associated glycoprotein: Functional and clinical aspects, in: *Neuronal and Glial Proteins: Structure, Function and Clinical Applications* (P. J. Marangos, I. Campbell, and R. H. Cohen, eds.), pp. 295–320, Academic Press, New York.

Quarles, R. H., Everly, J. L., and Brady, R. O., 1973a, Evidence for the close association of a glycoprotein with myelin in rat brain, *J. Neurochem.* **21:**1177–1191.

Quarles, R. H., Everly, J. L., and Brady, R. O., 1973b, Myelin-associated glycoprotein: A developmental change, *Brain Res.* **58:**506–509.

Quarles, R. H., Barbarash, G. R., Figlewicz, D. A., and McIntyre, L. J., 1983, Purification and partial characterization of the myelin-associated glycoprotein from adult rat brain, *Biochim. Biophys. Acta* **757:**140–143.

Quarles, R. H., Barbarash, G. R., and MacIntosh, T. D., 1985, Methods for the identification and characterization of glycoproteins in central and peripheral myelin, *Res. Methods Neurochem.* **6:**303–357.

Quarles, R. H., Ilyas, A. A., and Willison, H. J., 1986, Antibodies to glycolipids in demyelinating diseases of the human peripheral nervous system, *Chem. Phys. Lipids* **42:**235–248.

Rapaport, R. N., and Benjamins, J. A., 1981, Kinetics of entry of P_0 protein into peripheral nerve myelin, *J. Neurochem.* **37:**164–171.

Rapaport, R. N., and Benjamins, J. A., and Skoff, R. P., 1982, Effects of monensin on assembly of P0 protein into peripheral nerve myelin, *J. Neurochem.* **39:**1101–1110.

Rieger, F., Daniloff, J. K., Pincon-Raymond, M., Crossin, K. L., Grumet, M., and Edelman, G. M., 1986, Neuronal cell adhesion molecules and cytotactin are colocalized at the node of Ranvier, *J. Cell Biol.* **103:**379–391.

Roomi, M. W., Ishaque, A., Kahn, N. R., and Eylar, E. H., 1978, The P0 protein: The major glycoprotein of peripheral nerve myelin, *Biochim. Biophys. Acta* **536:**112–121.

Ruoslahti, E., and Pierschbacher, M. D., 1986, Arg-Gly-Asp: A versatile cell recognition signal, *Cell* **44:**517–518.

Sakamoto, Y., Kitamura, K., Yoshimura, K., Nishijima, T., and Uyemura, K., 1986, Fatty acid-linked peptides from bovine P0 protein in peripheral nerve myelin, *Biomed. Res.* 7:261–266.

Sakamoto, Y., Kitamura, K., Yoshimura, K., Nishijima, T., and Uyemura, K., 1987, Complete amino acid sequence of P0 protein in bovine peripheral nerve myelin, *J. Biol. Chem.* 262:4208–4214.

Salzer, J. L., Holmes, W. P., and Colman, D. R., 1987, The amino acid sequences of the myelin-associated glycoproteins: Homology to the immunoglobulin gene superfamily, *J. Cell Biol.* 104:957–965.

Sato, S., Quarles, R. H., and Brady, R. O., 1982, Susceptibility of the myelin-associated glycoprotein and basic protein to a neutral protease in highly purified myelin from human and rat brain, *J. Neurochem.* 39:97–105.

Sato, S., Baba, H., Tanaka, M., Yanagisawa, K., and Miyatake, T., 1983, Antigenic determinant shared between myelin-associated glycoprotein from human brain and natural killer cells, *Biomed. Res.* 4: 489–493.

Sato, S., Yanagisawa, K., and Miyatake, T., 1984a, Conversion of myelin-associated glycoprotein (MAG) to a smaller derivative by calcium activated neutral protease (CANP)-like enzyme in myelin and inhibition by E-64 analogue, *Neurochem. Res.* 9:629–635.

Sato, S., Quarles, R. H., Brady, R. O., and Tourtellotte, W. W., 1984b, Elevated neutral protease activity in myelin from brains of patients with multiple sclerosis, *Ann. Neurol.* 15:264–267.

Sato, S., Baba, T., Inuzuka, T., and Miyatake, T., 1986, Anti-myelin-associated glycoprotein antibody in sera from patients with demyelinating diseases, *Acta Neurol. Scand.* 74:115–120.

Schober, R., Itoyama, Y., Sternberger, N. H., Trapp, B. D., Richardson, E. P., Asbury, A. K., Quarles, R. H., and Webster, H. d., 1981, Immunocytochemical study of P0 glycoprotein, P1 and P2 basic proteins, and myelin-associated glycoprotein (MAG) in lesions of idiopathic polyneuritis, *Neuropathol. Appl. Neurobiol.* 7:437–451.

Schwarting, G. A., Jungalwala, F., Chou, D. K. H., Boyer, A. M., and Yamoto, M., 1987, Sulfated glucuronic-acid containing glycoconjugates are temporally and spatially regulated antigens in the developing mammalian nervous system, *Dev. Biol.* 120:65–76.

Shuman, S., Hardy, M., and Pleasure, D., 1983, Peripheral nervous system myelin and Schwann cell glycoproteins: Identification by lectin binding and partial purification of a peripheral nervous system myelin-specific 170,000 molecular weight glycoprotein, *J. Neurochem.* 41:1277–1285.

Shuman, S., Hardy, M., and Pleasure, D., 1986, Immunochemical characterization of peripheral nervous system myelin 170,000-M_r glycoprotein, *J. Neurochem.* 47:811–818.

Shuman, S., Hardy, M., Sobue, G., and Pleasure, D., 1988, A cyclic AMP analogue induces synthesis of a myelin-specific glycoprotein by cultured Schwann cells, *J. Neurochem.* 50:190–194.

Shy, M. E., Gabel, C. A., Vietorisz, E., and Latov, N., 1986, Characterization of oligosaccharides that bind to human anti-MAG antibodies and to the mouse monoclonal antibody, HNK-1, *J. Neuroimmunol.* 12:291–298.

Singh, H., and Spritz, N., 1976, Protein kinases associated with peripheral nerve myelin. I. Phosphorylation of endogenous myelin proteins and exogenous substrates, *Biochim. Biophys. Acta* 448:325–337.

Smith, M. E., 1980, Biosynthesis of peripheral nervous system myelin proteins in vitro, *J. Neurochem.* 35: 1183–1189.

Smith, M. E., and Perret, V., 1986, Immunological nonidentity of 19K protein and TP_0 in peripheral nervous system myelin, *J. Neurochem.* 47:924–929.

Smith, M. E., and Sternberger, N. H., 1982, Glycoprotein biosynthesis in peripheral nervous system myelin: Effect of tunicamycin, *J. Neurochem.* 38:1044–1049.

Steck, A., Murray, N., Meier, C., Page, N., and Perruisseaku, G., 1983a, Demyelinating neuropathy and monoclonal IgM antibody to myelin-associated glycoprotein, *Neurology* 33:19–23.

Steck, A. J., Murray, N., Vandevelde, M., and Zurbriggen, A., 1983b, Human monoclonal antibodies to myelin-associated glycoprotein. Comparison of specificity and use for immunocytochemical localisation of antigen, *J. Neuroimmunol.* 5:145–156.

Stefansson, K., Marton, L., Antel, J. P., Wollman, R. L., Roos, R. P., Chejfec, G., and Arnason, B. H. G. W., 1983, Neuropathy accompanying monoclonal IgM gammopathy, *Acta Neuropathol.* 59:255–261.

Sternberger, N. H., Quarles, R. H., Itoyama, Y., and Webster, H. d., 1979, Myelin-associated glycoprotein demonstrated immunocytochemically in myelin and myelin forming cells of developing rat, *Proc. Natl. Acad. Sci. USA* 76:1510–1514.

Sutcliffe, J. G., Milner, R. J., Shinnick, T. M., and Bloom, F. E., 1983, Identifying the protein products of brain-specific genes with antibodies to chemically synthesized peptides, *Cell* **33**:671–682.

Trapp, B. D., 1988, Distribution of MAG and P0 protein during myelin compaction in quaking mouse peripheral nerve *J. Cell Biol.* **107**:675–685.

Trapp, B. D., and Quarles, R. H., 1982, Presence of myelin-associated glycoprotein correlates with alterations in the periodicity of peripheral myelin, *J. Cell Biol.* **92**:877–882.

Trapp, B. D., and Quarles, R. H., 1984, Immunocytochemical localization of the myelin-associated glycoprotein: Fact or artifact? *J. Neuroimmunol.* **6**:231–249.

Trapp, B. D., McIntyre, L. J., Quarles, R. H., Sternberger, N. H., and Webster, H. d., 1979, Immunocytochemical localization of rat peripheral nervous system myelin proteins: P_2 protein is not a component of all peripheral nervous system myelin sheaths, *Proc. Natl. Acad. Sci. USA* **76**:3552–3556.

Trapp, B. D., Itoyama, Y., Sternberger, N. H., Quarles, R. H., and Webster, H. d., 1981, Immunocytochemical localization of P_0 protein in Golgi complex membranes and myelin of developing rat Schwann cells, *J. Cell Biol.* **90**:1–6.

Trapp, B. D., Quarles, R. H., and Suzuki, K., 1984, Immunocytochemical studies of quaking mice support a role for the myelin-associated glycoprotein in forming and maintaining the periaxonal space and periaxonal cytoplasmic collar in myelinating Schwann cells, *J. Cell Biol.* **99**:595–606.

Trapp, B. D., Griffin, J. W., Wong, M., O'Connell, M., and Andrews, S. B., 1985, Localization of F-actin in one micron frozen sections of myelinated peripheral nerve, *J. Cell Biol.* **101**(2):33a.

Trapp, B. D., O'Connell, M. F., and Andrews, S. B., 1986, Ultrastructural immunolocalization of MAG and P0 proteins in cryosections of peripheral nerve, *J. Cell Biol.* **103**(2):228a.

Tucker, G. C., Aoyama, H., Lipinski, M., Tursz, T., and Thiery, J. P., 1984, Identical reactivity of monoclonal antibodies HNK-1 and NC-1: Conservation in vertebrates on cells derived from neural primordium and on some leukocytes, *Cell Differ.* **14**:223–230.

Uyemura, K., Suzuki, M., Sakamoto, Y., and Tanaka, S., 1987, Structure of P0 protein: Homology to immunoglobulin superfamily, *Biomed. Res.* **8**:353–357.

Waehneldt, T. V., Matthieu, J. M., and Jeserich, G., 1986, Commentary: Appearance of myelin proteins during vertebrate evolution, *Neurochem. Int.* **4**:463–474.

Wajgt, A., and Gorny, M., 1983, CSF antibodies to myelin basic protein and to myelin-associated glycoprotein in multiple sclerosis. Evidence of the intrathecal production, *Acta Neurol. Scand.* **68**:337–343.

Webster, H. d., Palkovits, C. G., Stoner, G. L., Favilla, J. T., Frail, D. E., and Braun, P. E., 1983, Myelin-associated glycoprotein—Electron microscopic immunocytochemical localization in compact developing and adult central nervous system myelin, *J. Neurochem.* **41**:1469–1479.

Wiggins, R. C., and Morell, P., 1980, Phosphorylation and fucosylation of myelin proteins in vitro by sciatic nerve from developing rats, *J. Neurochem.* **34**:627–634.

Williams, A. F., 1987, A year in the life of the immunoglobulin superfamily, *Immunol. Today* **8**:298–303.

Williams, A. F., Barclay, A. N., Clark, M. J., and Gagnon, J., 1985, Cell surface glycoproteins and the origin of immunity, in: *Gene Expression during Normal and Malignant Differentiation* (L. C. Anderson, C. G. Ghamberg, and P. Ekblon, eds.), pp. 125–138, Academic Press, New York.

Willison, H. J., Ilyas, A. A., O'Shannessy, D. J., Pulley, M., Trapp, B. D., and Quarles, R. H., 1987, Myelin-associated glycoprotein and related glycoconjugates in developing cat peripheral nerve: A correlative biochemical and morphometric study, *J. Neurochem.* **49**:1853–1862.

Willison, H. J., Trapp, B. D., Bacher, J. T., and Quarles, R. H., 1988a, The expression of myelin-associated glycoprotein in regenerating cat sciatic nerve, *Brain Res.* **444**:10–16.

Willison, H. J., Trapp, B. D., Bacher, J. D., Dalkas, M. C., Griffin, J. W., and Quarles, R. H., 1988b, Demyelination induced by intraneural injection of human anti-myelin associated glycoprotein antibodies, *Muscle Nerve* **11**:1169–1176.

Wood, J. G., and Dawson, R. M. C., 1974, Lipid and protein changes in sciatic nerve during Wallerian degeneration, *J. Neurochem.* **22**:631–635.

Yanagisawa, K., and Quarles, R. H., 1986, Jimpy mice: Quantitation of myelin-associated glycoprotein and other proteins, *J. Neurochem.* **47**:322–325.

Yanagisawa, K., Quarles, R. H., Johnson, D., Brady, R. O., and Whitaker, J., 1985, A derivative of myelin-associated glycoprotein (dMAG) in cerebrospinal fluid of normals and patients with neurological disease, *Ann. Neurol.* **18**:464–469.

Yanagisawa, K., Duncan, I. D., Hammang, J. P., and Quarles, R. H., 1986, Myelin-deficient rat: Analysis of myelin proteins, *J. Neurochem.* **47**:1901–1907.

9

Axonal Transport and Intracellular Sorting of Glycoconjugates

Jeffry F. Goodrum, George C. Stone, and Pierre Morell

1. INTRODUCTION TO AXONAL TRANSPORT

The movement of material along axons was first described in the now classic studies of Weiss and Hiscoe (1948). These investigators observed that when a ligature was placed on a nerve, axonal material accumulated proximal to the tie, resulting in a swelling of the nerve. Following removal of the ligature the swollen region was displaced down the nerve at 1–2 mm/day. Various adaptations of this technique, combined with the application of radioisotope methodology introduced to the field in the early 1960s, have led to the elucidation of a complex system of intracellular traffic within axons. While a full understanding of the mechanisms of axonal transport remains to be achieved, much has been learned as summarized below.

The neuronal cell body contains almost all the metabolic machinery for the synthesis and degradation of macromolecules and organelles. Therefore, the maintenance and normal functioning of the axon and its terminals are critically dependent on the process of axonal transport for the continued delivery of materials to and from the cell body. The initiation of axonal transport within the neuronal cell body involves a complex process of sorting and routing of materials, analogous to that described in many other cell types. These transported materials, which include the various glycoconjugates, are predominantly macromolecular, and almost all appear to be transported as components of organelles and other cytological structures. Material moving anterogradely (away from the cell body) is transported at one of at least five rates ranging from a fraction of a millimeter per day (slow axonal transport) to several hundred

Jeffry F. Goodrum • Biological Sciences Research Center and Department of Pathology, University of North Carolina, Chapel Hill, North Carolina 27599.   *George C. Stone* • Division of Molecular Biology, Nathan Kline Institute, Orangeburg, New York 10962.   *Pierre Morell* • Biological Sciences Research Center and Department of Biochemistry, University of North Carolina, Chapel Hill, North Carolina 27599.

Table 1. Axonal Transport Components and Their Compositions[a]

Transport rate component		Rate (mm/day) in mammals	Identified components	Postulated cytological structures involved
Anterograde				
Fast	I	200–400	Proteins, glycoproteins, phospholipids, glycolipids, neurotransmitters and associated enzymes	Vesicles, agranular ER
Intermediate	II	50–200	F_1ATPase, diphosphatidylglycerol	Mitochondria
	III	15	Myosinlike actin-binding protein	?
Slow	SCb/IV	2–5	Actin, clathrin, calmodulin, enolase	Microfilaments, axoplasmic matrix
	SCa/V	0.2–1	Tubulin, MAPS, neurofilament triplet	Microtubules, neurofilaments
Retrograde		100–200	Similar to fast component and including lysosomal hydrolases	Prelysosomal structures

[a] Adapted from Lasek (1981).

millimeters per day (fast axonal transport). The macromolecular composition of material transported at each of these rates appears to be compositionally distinct although
not necessarily unique. All the glycoconjugates appear to be restricted to the fast
transport component. There is also a retrograde (toward the cell body) axonal transport
which moves at about one half the most rapid anterograde rate. The major characteristics of these different rate components of axonal transport are summarized in Table
1. It is known that axonal transport is temperature sensitive, requires a local energy
supply, and at least the most rapid component appears to be dependent on intact
microtubules. Recently, some details of the mechanisms involved in the transduction
of the energy from phosphate bonds to physical displacement of particles have emerged
(see Schnapp and Reese, 1986). For recent reviews on various aspects of axonal
transport, see Bisby (1980), Grafstein and Forman (1980), Ochs (1982), Schwab and
Thoenen (1983), Elam and Cancalon (1984), and Smith and Bisby (1987).

2. AXONAL TRANSPORT METHODOLOGY

2.1. Radioisotope Methodology Is Used to Study Glycoconjugate Transport

The most common method for studying axonal transport has been the labeling of
biosynthesized macromolecules by applying radioactive precursors in the vicinity of
neuronal cell bodies and subsequently monitoring the movement of these newly synthesized radioactive molecules into the axons and terminal fields of these neurons. For
glycoconjugates a variety of sugar precursors have been used, including ^{3}H- and ^{14}C-
labeled fucose, galactose, mannose, glucosamine, N-acetylglucosamine, N-acetylgalactosamine, and N-acetylmannosamine. $^{35}SO_4$ is also frequently used for studies of
glycoprotein and glycosaminoglycan transport. Radioactive fucose is highly specific
for labeling glycoproteins (Margolis and Margolis, 1972; Chan et al., 1979), and thus
is a commonly used precursor for studies of glycoprotein transport. One can simply
determine the acid-precipitable radioactivity incorporated from labeled fucose in a
tissue and assume that it is present in glycoprotein. In contrast, glucosamine labeling is
much less specific; it will label gangliosides and glycosaminoglycans as well as
glycoproteins. Thus, transport of different glycoconjugates can be studied simultaneously if the classes are separated (e.g., see Gammon et al., 1985).

2.2. A Specific Example: Glycoprotein Transport in the Optic System

By far the majority of studies of glycoconjugate transport have focused on
glycoproteins, and the most commonly used neural pathway for such studies has been
the optic system. A radioactive precursor such as [^{3}H]fucose is injected into the
vitreous chamber of one eye, bringing the precursor into direct contact with the ganglion cells of the retina (Figure 1A). The time course of arrival of radioactive macromolecules into successive structures of the optic pathway is then measured. Making
the simplifying assumption of a pulse label, an idealized version of such a transport
time course is shown in Figure 1B. Following somal uptake, incorporation, and routing

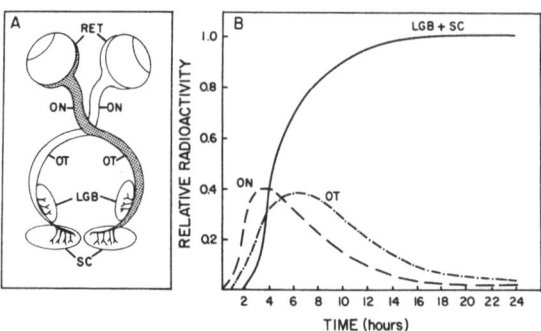

Figure 1. Axonal transport of glyco-proteins in the optic system. (A) Diagram of the rat optic system, which is > 95% crossed. RET, retina; ON, optic nerve; OT, optic tract; LGB, lateral geniculate body; SC, superior colliculus. (B) Idealized transport time course of a pulse of labeled glycoproteins through the optic system. Radioactivity appears successively in optic structures, passing through the axonal structures (ON and OT) and accumulating in the terminals (LGB and SC).

into the transport mechanism, the labeled material passes sequentially through the ipsilateral optic nerve and the contralateral optic tract and accumulates in nerve terminal regions (optic tectum in fish and birds, lateral geniculate body and superior colliculus in mammals). In actuality, the shape of a transport time-course curve is highly influenced by the uptake and incorporation rates for the precursor (see Section 3.1.2), any posttranslational modifications of the labeled material, accumulation of somal pools of the labeled macromolecules (see Section 3.1.2), and the metabolic turnover of the labeled molecule (see Section 5.4).

In order to demonstrate that the radioactivity that appears in these optic structures is due to the axonal transport of macromolecules, three related possibilities must first be ruled out. These possibilities are systemic labeling, diffusion of precursors, and axonal transport of precursors. Systemic labeling is presumably due to leakage of radioactive precursor from the site of injection and delivery to other tissues by the circulation. For most systems, including the optic pathway, such labeling is expected to be a uniform phenomenon, and thus levels of incorporated radioactivity in the corresponding structures contralateral to the ones of the injected pathway are taken to represent systemic labeling. The difference in amount of radioactivity between corresponding left and right structures is taken to represent transported material. The extent of systemic labeling is usually represented as the ratio of the radioactivity in the corresponding left and right structures and varies greatly with the glycoconjugate precursor as well as the species used (e.g., see Forman *et al.*, 1971; Ledeen *et al.*, 1976; Karlsson and Linde, 1977; Goodrum *et al.*, 1979b).

Related to the problem of systemic labeling is the diffusional spread of labeled precursor from the vitreous chamber to the optic nerve and subsequent local incorporation. The extent of such diffusion is usually determined by the use of microtubule-disrupting drugs such as colchicine, which block axonal transport (for a review see Hanson and Edstrom, 1978). In the presence of such compounds, radioactivity above that from systemic labeling can be attributed to diffusion, and usually extends only a few millimeters down the optic nerve.

Controls to rule out transport of free sugars and subsequent local incorporation of these precursors into glycoprotein usually involve demonstrating that inhibition of protein synthesis in the cell bodies greatly diminishes the amount of labeled glycoprotein that appears in the axons and nerve terminals (e.g., McEwen *et al.*, 1971; Forman *et al.*, 1971, 1972; Ambron *et al.*, 1975). Such demonstrations are not completely convincing because local application of inhibitors may have widespread effects. In most cases this control is not taken too seriously because the capacity of axons for local synthesis is very restricted and the amount of free precursor available in the axons and terminals is generally far too low to account for the extent of labeled glycoconjugates which eventually appear in these structures (e.g., Karlsson and Sjostrand, 1971; Forman *et al.*, 1972; Bennett *et al.*, 1973; McLean *et al.*, 1975). There is, however, some evidence for local axonal and nerve ending incorporation of sugars into glycoconjugates, which will be discussed in Sections 4.1.6 and 5.1.

3. INTRACELLULAR SORTING OF GLYCOCONJUGATES AND THE INITIATION OF AXONAL TRANSPORT

3.1. The Initiation Phase of Axonal Transport

From site(s) of synthesis confined to the perikaryal region of the neuron, protein and lipid components must consistently reach appropriate site(s) of utilization, via axonal transport, in order to maintain the exaggerated cytoarchitecture, as well as support diverse afferent and efferent functional domains. In practical terms, distinct subsets of the at least 10^4 proteins being synthesized at any given time must be continually exported from the cell body to enter one of the five rate components of axonal transport (Table 1). The cellular and molecular events within the perikaryon that are responsible for segregating and compartmentalizing newly synthesized proteins and glycoconjugates in preparation for axonal transport comprise the initiation phase of axonal transport (for additional reviews see Hammerschlag and Stone, 1986; Stone and Hammerschlag, 1987).

The glycoconjugates are transported predominantly in the fastest rate component (see Table 1 and Section 4). The initiation phase of this fastest rate component is well studied and is in certain ways analogous to the synthesis, sorting, and routing of proteins and glycoproteins in nonneuronal systems such as exocrine pancreas (Palade, 1983) and transformed epithelial cells (Sabatini *et al.*, 1982).

3.1.1. Initiation of Fast Axonal Transport May Be Studied *in Vitro*

Events that occur during the initiation of fast transport may be studied by radiolabeling of neuronal perikarya with various protein and glycoconjugate precursors, coupled with administration of pharmacological agents that have specific effects on intracellular transport. A useful system for this work is an *in vitro* preparation of bullfrog primary afferent neurons: the dorsal root ganglion with associated axons that comprise the spinal nerve and contribute to the sciatic nerve (Stone *et al.*, 1978; Stone and Hammerschlag, 1981). The advantage of such a preparation is that virtually all anabolic reactions, as well as intracellular transport and fast axonal transport processes can be maintained in minimal organ culture conditions (i.e., oxygenated Ringer's at

18–22°C). Accordingly, labeled precursors (amino acids or sugars) as well as pharmacological agents can be administered selectively to sensory neuron cell bodies, maintained in one compartment of a small Lucite chamber, while newly synthesized fast-transported proteins or glycoconjugates can be followed in the respective spinal nerves, maintained in a second compartment separated from the first by a Vaseline barrier. Thus, the specific consequences of manipulating molecular events that occur in the cell body during the initiation of axonal transport can be assessed by examining how the population of transported macromolecules found in the nerve(s) is modified.

3.1.2. The Initiation Phase for Transported Glycoprotein Is Protracted

An examination of the kinetics of intracellular transport in nonneuronal cells has provided evidence for differential routing (Gumbiner and Kelly, 1982; Williams *et al.*, 1985), as well as offered a prospective functional role for glycoconjugates as sorting signals (Olden *et al.*, 1982, 1985). Similar studies examining the kinetics of the initiation phase of axonal transport are technically more difficult, due partly to the limited synthesis of species destined to undergo fast axonal transport, and have not been performed successfully. However, examination of transport downflow profiles has yielded useful data regarding the duration of the initiation phase for proteins and glycoproteins. Following pulse-labeling of perikarya, transport of newly synthesized species is permitted to proceed for selected periods. The distribution of radioactivity along the axons is then used to derive a graphic relationship of distance traveled by the wavefront of labeled transported material over time: an extension of the approach used to calculate rates of transport (see Section 4.1.1 and Figure 3). Values for the duration of the somal phase are then computed by extrapolating this function to zero distance. This method has been used successfully to derive estimates of the time course of initiation in mammals (Bisby, 1976, 1978; Toews *et al.*, 1982; Harry *et al.*, 1987) and poikilotherms (Gross, 1973; Gross and Kreutzberg, 1978; Nichols *et al.*, 1982), as well as to demonstrate the high degree of temperature dependence of initiation events (Gross, 1973).

Because sugars are added to preexisting polypeptides while amino acids are incorporated into nascent peptides, one would expect that the somal processing times for sugar-labeled glycoproteins would be shorter than for proteins labeled with an amino acid precursor. Experiments comparing the incorporation and transit of [^{3}H]fucose- and [^{35}S]methionine-labeled fast-transported proteins in bullfrog primary afferent neurons indicate that at 18°C the time course of initiation of [^{3}H]fucose-containing glycoprotein species is longer than [^{35}S]methionine-labeled proteins (Stone and Dougher, unpublished). Such findings are explained by the slow uptake and kinetics of incorporation into glycoproteins of most sugars as compared to amino acids (e.g., Karlsson and Sjostrand, 1971; Bennett *et al.*, 1973; Goodrum *et al.*, 1979b; Toews *et al.*, 1982; Gammon *et al.*, 1985). Another posttranslational modifying group, sulfate, is taken up and incorporated more rapidly than sugars, resulting in a shorter initiation period (Goodrum *et al.*, 1979b). The possible existence of an intrasomal storage pool(s) through which newly synthesized fast-transported proteins pass prior to entering the axon (Berry, 1980; Goodrum and Morell, 1982, 1984; Hammerschlag and Stone, 1987) may also contribute to these differences in initiation times.

Recent experiments are focusing on resolution of the magnitude of these effects on the kinetics of initiation. For example, recent evidence suggests that the functional site of intrasomal storage pool(s) appears to be after newly synthesized fast-transported proteins exit the Golgi. Monensin and Co^{2+} block entry of newly synthesized [^{3}H]leucine-labeled proteins into such pools (see Section 3.3), but have no effect on protracted release of proteins into the axon (Hammerschlag and Stone, 1987). Models have been proposed for morphological correlates of intrasomal storage pools (Berry, 1980; Goodrum and Morell, 1982) but evidence for the existence of unique subcellular components has yet to appear.

3.2. Routing of Proteins and Glycoproteins during the Initiation of Fast Axonal Transport

3.2.1. Monensin Defines a Common Intrasomal Route for Fast-Transported Glycoproteins and Nonglycosylated Proteins

The cationic ionophore monensin has the ability to inhibit protein traffic through the Golgi (Tartakoff, 1980; Tartakoff et al., 1980). When newly synthesized proteins destined for fast axonal transport are labeled with [^{35}S]methionine and [^{3}H]mannose in the presence of monensin, the relative abundance of both ^{35}S- and ^{3}H-labeled fast-transported proteins exported from the cell body is depressed to the same extent. These data reflect a monensin-induced blockage of Golgi function that inhibits the passage of fast-transported [^{3}H]glycoproteins as well as ^{35}S-labeled nonglycosylated fast-transported proteins. Thus, it appears that all fast-transported proteins, whether glycosylated or not, proceed along a common intra-Golgi route (Hammerschlag et al., 1982) (Figure 2).

Studies of the relative effects of monensin on incorporation of various ^{3}H-labeled sugars have further resolved the site of action of monensin within the Golgi, and thus confirm the route common to all fast-transported proteins. For example, monensin does not significantly effect the somal incorporation of sugars such as [^{3}H]mannose that are added to proteins cotranslationally (i.e., prior to the monensin site of action). By contrast, monensin reduces the incorporation of galactose or fucose to an extent

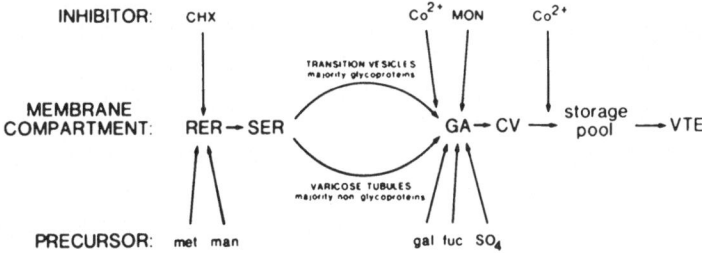

Figure 2. Model describing the initiation phase of fast axonal transport. The routing of fast-transported proteins through subcellular membrane compartments has been defined using a variety of precursors known to be incorporated at specific sites, as well as inhibitors that act at unique locations along the intracellular transport pathway. RER, rough endoplasmic reticulum; SER, smooth endoplasmic reticulum; GA, Golgi apparatus; CV, coated vesicles; VTE, vesiculotubular elements; CHX, cycloheximide; MON, monensin.

comparable to the reduction in export of fast-transported proteins. This is expected because these sugars are added via reactions that occur in the Golgi distal to the site of action of monensin (Hammerschlag and Stone, 1982).

3.2.2. Co^{2+} Distinguishes Differential Routing of Fast-Transported Glycoproteins and Nonglycosylated Proteins

Co^{2+} is known to antagonize Ca^{2+}-mediated membrane fusion events at two sites, the transfer of membrane-associated and secretory proteins between the smooth ER and Golgi complex, and between the Golgi complex and post-Golgi compartments (Kern and Kern, 1969; Douglas, 1974; Beiger et al., 1975). Application of Co^{2+} to neuronal perikarya reduces the abundance of all proteins that exit the cell body via fast axonal transport. This inhibition by Co^{2+} preferentially affects the intrasomal transit of fast-transported proteins of molecular weight greater than 35,000; movement of lower-molecular-weight proteins is less affected. This differential inhibition may be accounted for by the existence of two Ca^{2+}-mediated fusion events, with the route of proteins of molecular weight greater than 35,000 involved in both steps (Stone and Hammerschlag, 1981). Though the sorting signals that generate such differential routing have yet to be determined, it is relevant that 89% of the fast-transported proteins of molecular weight greater than 35,000 are glycosylated, while only 19% of lower-molecular-weight fast-transported proteins contain detectable amount of sugars (Stone and Hammerschlag, 1983). Thus, the likelihood exists that glycosylated fast-transported proteins are representative of a subpopulation of species that enter the Golgi via a Ca^{2+}-mediated fusion of transition vesicles, while nonglycosylated fast-transported proteins are transferred via a non-Ca^{2+}-requiring process, through a system of tubules continuous between the ER and Golgi (Teichberg and Holtzman, 1973; Morre et al., 1974; Stone and Hammerschlag, 1981; Hammerschlag and Stone, 1982; Lindsey and Ellisman, 1983) (Figure 2).

3.2.3. Are There Glycoconjugate Sorting Signals for Fast-Transported Proteins?

Specific carbohydrate moieties can confer molecular identity that mediates protein sorting in nonneural systems. Mannose 6-phosphate has been shown to be involved in the proper delivery of hydrolases to lysosomes (Sly et al., 1981; Neufeld and McKusick, 1983), while a similar mechanism utilizing glucose-6-phosphate appears to direct certain matrix proteins to the extracellular face of the plasma membrane in chick retina (Marchase et al., 1982; Koro and Marchase, 1982). The presence of complex N-linked, or O-linked glycoconjugates can also influence intracellular transport. Inhibition of the formation of such side chains with tunicamycin or swainsonine results, for certain proteins, in impaired export or altered kinetics of passage along the secretory pathway (Olden et al., 1985; Yeo et al., 1985a,b). Attempts have been made to use these drugs as probes of glycoconjugate function in studies of axonal transport using the in vitro bullfrog preparation (Stone and Hammerschlag, unpublished). Clear-cut results were not obtained, and it appears that, at least in this system, these drugs do not easily penetrate the tissue.

Nevertheless, as noted above (Sections 3.2.1 and 3.2.2), the fact that not all fast-transported proteins possess carbohydrate side chains, as well as the observation that all fast-transported proteins appear to be routed through the Golgi apparatus, argues against the need for sugars to determine whether or not a particular protein undergoes fast transport. A more likely candidate for a feature potentially common to all fast-transported proteins is a unique portion of the primary sequence representing a "signal" sequence. The presence of "signal" sequences appears to account for numerous examples of protein sorting in nonneuronal cells (Walter *et al.*, 1985; Friedlander and Blobel, 1985; Audigier *et al.*, 1987).

A more plausible role for carbohydrates on rapidly transported glycoproteins is in specifying the final intracellular destination of fast-transported proteins. By a mechanism similar to hydrolases being routed to lysosomes via interaction between mannose-6-phosphate and its receptor in the Golgi (Brown and Farquhar, 1984), fast-transported proteins could be segregated into specific regions of the Golgi by recognition of sugars with appropriate lectin-type molecules (Yeo *et al.*, 1985a,b; Olden *et al.*, 1982). Such segregation would thereby result in compartmentalization of these species within a similar population of transport organelles, and subsequent delivery to similar intracellular domains.

4. AXONAL TRANSPORT OF GLYCOCONJUGATES

4.1. Transport of Glycoproteins

4.1.1. Glycoproteins Are Transported at a Single (?) Fast Rate

Axonal transport rates can most accurately be determined in long peripheral nerves where the appearance of labeled material in many consecutive nerve segments can be assayed over time. From such data a series of downflow curves is generated and the rate calculated from the displacement of either the foot or the crest of these curves (see Ochs, 1982). An example of such curves from rat sciatic nerve following [^{3}H]fucose labeling of the dorsal root ganglion is shown in Figure 3. In such studies it has been found that glycoprotein transport in peripheral nerves proceeds at the same rate (300–450 mm/day) as the rapidly transported proteins labeled with an amino acid precursor (Toews *et al.*, 1982; Aquino *et al.*, 1985; Harry *et al.*, 1987).

In CNS nerves and tracts the reported rates of glycoprotein transport (and rapid transport in general) have usually been much slower than for sciatic nerve, between 40 and 200 mm/day, and variable from nerve to nerve (e.g., Karlsson and Sjostrand, 1971; Bennett *et al.*, 1973; Frizell and Sjostrand, 1974a; Specht and Grafstein, 1977; Levin, 1977; Goodrum *et al.*, 1979b). In most cases, however, these apparent differences relate to methodological considerations. For example, in the case of studies using the optic system, measurements are frequently based only on the time of initial arrival of transported material in the nerve terminal regions, and therefore include the incorporation and processing time in the cell bodies prior to the initiation of transport (see Section 3.1.2). Such measurements set the lower limit on transport rate, not the actual rate. When measurements are made between two or more positions along the

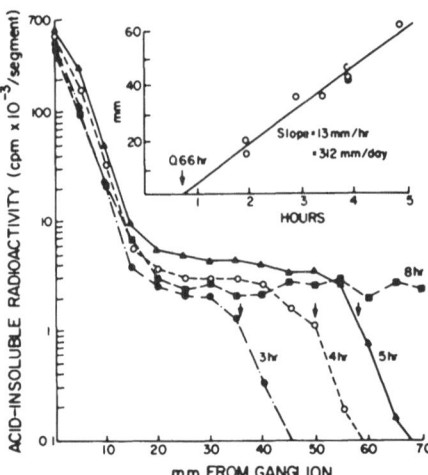

Figure 3. Distribution of acid-insoluble radioactivity in the dorsal root ganglion and sciatic nerve at various times following injection of [³H]fucose into the L5 ganglia. The position of the initial crest of transported radioactivity in the nerve at 3, 4, and 5 hr after injection is indicated by the arrows. Crest locations from all experiments of 5 hr or less are plotted against the length of time between injection and sacrifice in the inset. The transport rate was determined from the slope of the best straight line through these points, as determined by linear regression. The intercept of the abscissa represents the time interval between precursor injection and the appearance of newly synthesized glycoproteins in the nerve. From Toews *et al.* (1982).

optic system, the rapid transport rate for glycoprotein (Goodrum and Morell, 1982), and protein in general (Blaker *et al.*, 1981; Crossland, 1985), is found to be similar to that in peripheral nerves. These results suggest that the rate for fast transport of glycoproteins in similar in both the CNS and PNS of homeothermic animals. In poikilothermic animals the rapid transport rate for glycoprotein is slower, but when corrected for the temperature differences (assuming a Q_{10} of 2–3), they are comparable to that in mammals (e.g., Elam and Agranoff, 1971a; McEwen *et al.*, 1971; Forman *et al.*, 1972; Edstrom and Mattsson, 1972; Ambron *et al.*, 1974a; Gross and Beidler, 1975).

When proteins in neuronal cell bodies are labeled with an amino acid precursor, the passage of rapidly transported proteins through the axons of these cells is followed over the next several days and weeks by more slowly moving material which makes up the bulk of the total transported protein (see Table 1). Glycoproteins appear to be restricted almost exclusively to the rapidly moving component (fast transport), although a few studies have suggested multiple rates in certain systems. Ambron and associates have reported data from the invertebrate *Aplysia* giant neuron R2 which they have interpreted as indicating either preferential sites of deposition of labeled material within the axon (Thompson *et al.*, 1976) or peaks of migrating glycoprotein with different rapid transport rates (Ambron *et al.*, 1974a; Goldberg and Ambron, 1981). In another study, Levin (1977) reported two waves of transported glycoprotein moving through the hypothalamus at 96 and 48 mm/day, respectively, following stereotaxic injection of [³H]fucose into the locus coeruleus in rats. However, the large amount of systemic labeling which occurred (right to left hypothalamus ratios of only 1.2–1.3) preclude a firm conclusion regarding these data.

4.1.2. Much Newly Synthesized Glycoprotein is Committed to Axonal Transport

Given the large amount of membrane area in the axon and terminal fields of most neurons, one would expect a considerable percentage of the glycoprotein labeled in

Table 2. Retinal Export and Cellular Distribution of Axonally Transported Glycoproteins

Preparation	Precursor	Percent exported[a]	Percent distribution of transported label[b]		Tissues used		Reference[c]
			Axons	Terminals	Axons	Terminals	
Goldfish optic system	[3H]fucose	—	30	70[d]	ON, OT[e]	Tectum	1
Scardinius erythrophthalmus (teleost fish) optic system	[3H]N-acetylglucosamine	—	35	65[d]	ON, OT	Tectum	2
	[3H]N-acetylmannosamine	—	35	65[d]	ON, OT	Tectum	2
Garfish olfactory system	35SO4	—	83	17[f]	Olf. nerve	Olf. bulb	3
Chicken optic system	[3H]N-acetylmannosamine	—	25	75	ON	OL	4
Rabbit optic system	[3H]fucose	12	5	95[d]	OT	OL	5
	[3H]fucose	5	10	90	OT	LGB, SC	6
	35SO4	1	13	87[f]	OT	LGB, SC	7
Rat optic system	[3H]fucose	15	10	90	OT	SC	8
	[3H]fucose	—	20	80	ON, OT	LGB, SC	9
	[3H]glucosamine	12	5	95	ON, OT	LGB, SC	10
	35SO4	1	10	90	OT	LGB, SC	11

[a]Percent of total retinal incorporation into glycoprotein.
[b]Estimated from values at the peak of accumulation in the nerve terminals except where noted.
[c]References: (1) Forman et al. (1972); (2) Rosner et al. (1973); (3) Elam and Peterson (1976); (4) Rosner (1975); (5) Gremo and Marchisio (1975); (6) Karlsson and Sjostrand (1971); (7) Karlsson and Linde (1977); (8) Goodrum and Morell (1982); (9) Blaker et al. (1981); (10) Gammon et al. (1985); (11) Goodrum et al. (1979b).
[d]Measured before peak accumulation in the terminals.
[e]ON, optic nerve; OT, optic tract; LGB, lateral geniculate body; SC, superior colliculus.
[f]Label in both glycoproteins and glycosaminoglycans.

neuronal cell bodies to be committed to axonal transport. Consistent with this expectation, Ambron et al. (1974b) estimated that in the single R2 neuron of Aplysia, 45% of the incorporated [^{3}H]fucose was exported into the axon within 10 hr after labeling. In the optic system, where the retinal ganglion cells are only one of many cell types, from 5 to 12% of the total retinal glycoprotein labeling with fucose and glucosamine is subsequently found in ganglion cell axonal and terminal regions (Table 2). In contrast, when ^{35}SO$_4$ is used to label retinal macromolecules, only 1% of the incorporated label is exported (see Table 2). It may be that relatively more SO$_4$ than fucose is incorporated into cells other than the ganglion cells of the retina.

4.1.3. Glycoproteins Are Preferentially Transported to Nerve Terminals

It is a common observation that the majority of label in transported glycoprotein appears in the nerve terminal regions (see Table 2). When the relative labeling of the axonal and terminal regions of the optic system are compared at the peak of accumulation, between 70 and 95% of the label is found in the terminal regions, depending on the precursor and species. In contrast, just 40–60% of transported [^{3}H]glycerol-labeled phospholipid appears in the nerve terminal regions (Blaker et al., 1981). Autoradiographic observations with [^{3}H]fucose confirm this preferential terminal distribution of transported glycoprotein (Bennett et al., 1973; and see below). It should be noted that the relative distribution of label between axons and terminals is dependent on axonal length. For example, in the garfish olfactory nerve, which has very long axons, much more ^{35}SO$_4$-labeled material is found in axons than terminals (Elam and Peterson, 1976).

This preferential transport of glycoproteins to nerve endings has also been examined at the level of the transported peptides. One-dimensional gel electrophoretic analysis of [^{3}H]fucose-labeled glycoproteins in the optic system of rat revealed that most of the transported glycoproteins moved through the axons to the terminal regions. Two peptides, with approximate molecular weights of 49,000 and 28,000, were exceptions, being preferentially deposited along the axons (Goodrum and Morell, 1982). Similar results were obtained in the sciatic nerve of rat where it appeared that although most glycoproteins moved toward the terminals, a peptide with an approximate molecular weight of 49,000 was left behind (Toews et al., 1982).

4.1.4. Most Transported Glycoproteins Are Localized to Membranous Structures

Glycoproteins are predominantly membranous, and axonally transported glycoproteins would be expected to be components of membranes destined to be integrated into either the axonal or terminal membranes. Quantitative autoradiographic studies have provided the best indications of the cellular and subcellular localization of transported glycoprotein. In the chicken ciliary ganglion, at all times examined following injection of either [^{3}H]fucose or [^{3}H]glucosamine into the midbrain, the concentration of radioactivity in the axons was much lower than that in the terminal regions (Bennett et al., 1973). Silver grains in the axons were localized primarily over the axolemma, but also over vesicular structures and smooth ER (Markov *et al.*, 1976). Grains in the terminals were localized primarily in regions of synaptic vesicles (55%) and synaptic plasma membrane (22%) (Bennett *et al.*, 1973). In the axons of the R2 neuron of *Aplysia*, vesicles were the major labeled structure, having a specific activity 4–10 times that of any other structure (Thompson *et al.*, 1976).

Traditional subcellular fractionation techniques have also demonstrated the primarily membranous localization of transported glycoproteins; between 65 and 95% of transported glycoproteins are found in membranous fractions (Karlsson and Sjostrand, 1971; Forman *et al.*, 1971, 1972; Di Giamberadino *et al.*, 1973; Goodrum *et al.*, 1979b; Gammon *et al.*, 1985). Beyond this observation, however, subcellular fractionation has provided little information on the specific subcellular localization of transported glycoproteins. Transport-labeled axonal and synaptic constituents tend to become widely distributed throughout the standard subcellular fractions (e.g., see Zatz and Barondes, 1971; Elam and Agranoff, 1971a,b: Karlsson and Sjostrand, 1971; Edstrom and Mattsson, 1972; Di Giamberadino *et al.*, 1973; Marko and Cuenod, 1973; Frizell and Sjostrand, 1974a). Usually it is the microsomal fractions which not only contain most of the label but also have the highest specific activity. That transported glycoproteins are not substantially enriched in fractions presumed to represent nerve endings reflects the fragility of nerve endings and the low yield and purity of ''synaptosome'' fractions, rather than necessarily implying a nonterminal localization of transported glycoproteins.

4.1.5. Some Transported Glycoproteins May Be Secreted

Following subcellular fractionation, from 5 to 30% of axonally transported glycoprotein is found in the soluble fractions, but it has not been determined how much of this soluble glycoprotein is in fact transported in soluble form. When the compositions of the soluble and particulate fractions were examined by one-dimensional gel electrophoresis, only a few qualitative differences were seen (Goodrum *et al.*, 1979b; Elam, 1982), suggesting that many of these glycoproteins may have been artifactually solubilized by the fractionation procedure. However, differences in metabolism and carbohydrate complexity between the soluble and particulate fractions have been reported (Ripellino and Elam, 1980; Elam, 1982), suggesting that there may be some truly soluble transported glycoproteins. There is, in addition, evidence suggesting that some transported glycoproteins are secreted from axons and their terminals (Caroni *et al.*, 1985; Tedeschi and Wilson, 1987; Stone *et al.*, 1987). These data suggest that there may be a small population of soluble transported glycoproteins, perhaps confined to the lumen of transported vesicles and secreted at nerve terminals.

4.1.6. Transported Glycoproteins Have Some Distinctive Characteristics

The polypeptide composition of axonally transported glycoproteins has been examined in many studies, by both one- and two-dimensional gel electrophoresis. As expected, a large number of labeled polypeptides are found (Karlsson and Sjostrand, 1971; Edstrom and Mattsson, 1973; Stone *et al.*, 1978; Goodrum *et al.*, 1979b; Padilla and Morell, 1980; Roger *et al.*, 1980; Tytell *et al.*, 1980; Goodrum and Morell, 1982, 1984; Elam, 1982; Toews *et al.*, 1982; Wenthold and McGarvey, 1982; Gammon *et al.*, 1985). Two-dimensional gel electrophoresis has revealed that a majority of the fast-transported proteins contain carbohydrate side chains, and that most of these species are of relatively higher molecular weight. In one study, gel spots corresponding to 92 individual fast-transported proteins were examined; 61% were found to contain incorporated galactose, or galactose and fucose. An analysis of molecular weights of the glycosylated proteins further revealed that 89% of 55 species of greater than 35,000 were glycoconjugates, while only 19% of 37 species between 20,000 and 35,000 incorporated detectable quantities of these sugars (Stone and Hammerschlag, 1983). Prominent labeling is frequently reported at molecular weights of 25,000–30,000, 45,000–50,000, 110,000–140,000, and > 200,000. Although none of these glycoproteins has been functionally identified, it has been suggested that the 45,000–50,000 gel band may contain the glycoprotein subunit of the Na^+,K^+-ATPase (Goodrum and Morell, 1982). Transported glycoproteins labeled with $^{35}SO_4$ show an overlapping but different distribution of molecular weights with more of the label in the high-molecular-weight region of the gels (Goodrum *et al.*, 1979b; Stone *et al.*, 1983).

In the single R2 neuron of *Aplysia*, Ambron *et al.* (1974a, 1980, 1981a,b) have shown that some glycoproteins remain within the cell body while others are preferentially exported into the axon, suggesting that there may be some glycoproteins which serve functions specific to axons and their terminals. In other systems, the glycoprotein labeling pattern seen in the nerve terminal regions is generally somewhat less complex than that seen in the region containing the cell bodies (e.g., Goodrum and Morell, 1982; Gammon *et al.*, 1985), but little can be said about preferential export in such heterogeneous systems.

Examination of the carbohydrate moieties of transported glycoproteins has also revealed distinctions between these and the general population of nervous system glycoproteins. Karlsson (1979, 1980) found that [^{3}H]fucose-labeled transported glycoproteins in the rabbit visual system have lectin affinities somewhat different from the general population, binding strongly to concanavalin A, wheat germ agglutinin, and the lectin from *Lens culinaris*, but not to several other lectins, indicating many exposed *N*-acetylglucosamine and mannopyranoside residues. In addition, many transported glycoproteins were found to bind to the fucose-specific lectin from *Aleuria aurantia* but not other fucose-specific lectins (Gustavsson *et al.*, 1982; Ohlson and Karlsson, 1983a,b). Hart and Wood (1986) also found that glycoproteins that accumulated proximal to a cold block within the axon have lectin affinities indicative of a high mannose content. This is in contrast to glycoproteins on the nerve terminal membranes which exhibit lectin binding indicative of a more complex carbohydrate composition, and suggests that some local processing of transported glycoproteins may take place (see Section 5.1).

The complexity of glycopeptides generated by proteolytic digestion of transported glycoproteins in fish has been examined by Elam and collaborators (Elam *et al.*, 1970; Elam and Agranoff, 1971b; Elam and Peterson, 1976, 1979). They found these glycopeptides to have a broad size distribution and a typical carbohydrate composition in terms of sugar types, indicative of a large general population of transported glyco-proteins. In the garfish, but not the goldfish, a fraction of very large, [^{3}H]fucose-labeled glycopeptides was found that was not present when systemically labeled glycoproteins were examined.

4.1.7. Glycoprotein Transport Changes during Development and Regeneration

It seems reasonable to expect that different populations of glycoconjugates might be required by neurons during the periods of axon elongation, synapse formation, and maturation, respectively. Because axonal transport provides all the materials necessary for axonal growth and maintenance, such differences should be detectable by examining axonally transported glycoconjugates either in developing or regenerating systems.

Developmental studies of axonal transport have been most frequently carried out in the chick embryo optic system. Axonal transport of glycoproteins has been demonstrated autoradiographically as early as embryonic day 7 (Gremo et al., 1974), a time before the growing axons have reached the tectum. Rapidly transported glycoproteins can be detected in the optic tectum by embryonic days 10–14, the period of fiber ingrowth and synaptogenesis (Bondy and Madsen, 1971; Marchisio et al., 1973; Panzetta et al., 1983). These data demonstrate that rapid axonal transport of glycoconjugates is occurring throughout the growth of axons. The amount of transported glycoproteins in the developing chick optic system, as a fraction of that synthesized in the retina, increases dramatically at times corresponding to the periods of synaptogenesis and functional maturation; this fraction later declines to adults levels (Bondy and Madsen, 1971, 1974; Marchisio et al., 1973; Gremo and Marchisio, 1975). This correlation is not found for nonglycosylated, rapidly transported proteins. These results suggest that glycoproteins play specific roles in the formation and functioning of synapses.

It has also been reported that the rate of rapid transport for proteins and glycoproteins increases during development (Hendrickson and Cowan, 1971; Marchisio et al., 1973; Specht, 1977; Matthews et al., 1982b), but these data are only suggestive. These estimates of rate were based on either the determination of time of arrival of transported label in the terminal region, which is subject to error (see Section 4.1.1), or based on insufficiently short time points to make estimates of differences in rate. One determination of rapid transport rate in a developing animal found no difference from the adult value (Ochs, 1973).

Changes in composition of transported glycoprotein during development have been reported. Skene and Willard (1981a,b,c) have identified several proteins, termed growth-associated proteins, which are transported in much higher quantities in growing axons and postulated that these proteins have specific functions in axonal elongation. In contrast to these growth-associated proteins, Specht (1982) described a fucosylated, 50,000-dalton glycoprotein which appears to be transported in the developing hamster optic system only after eye opening, and suggested it is a specific component of mature synapses.

Results similar to those described above are also found in regenerating axons. The amount of transported glycoproteins is reported to increase severalfold during regeneration in frog, chick, and goldfish optic systems (Frizell and Sjostrand, 1974a,b; Skene and Willard, 1981a,b,c; Whitnall *et al.*, 1982; Sbaschnig-Agler *et al.*, 1984), but the transport rate does not change (Frizell and Sjostrand, 1974a). Griffin *et al.* (1981) found that the transported glycoprotein in regenerating rat sciatic nerve is preferentially deposited in the regenerating portion of the nerve. In the regenerating toad optic system and rabbit hypoglossal nerve, the same growth-associated proteins are present as described above in developing nerves (Skene and Willard, 1981a,b,c). Cole and Elam (1981, 1983) have reported that in regenerating garfish olfactory nerve there is an increase in the amount of mannose-rich, Con A-binding glycopeptides being transported, and postulated that these glycopeptides are involved in cell–cell interactions necessary for successful regeneration. Finally, Tedeschi and Wilson (1987) have reported an increase in *in vitro* secretion of transported glycoproteins from regenerating sciatic nerve and suggest that these secreted proteins play a role in regeneration.

4.2. Transport of Glycosaminoglycans

Despite increasing evidence of the functional importance of proteoglycans in nervous tissue (see Chapters 3, 5, 13, and 14), relatively little attention has been given to the axonal transport of the glycosaminoglycans. Nearly ten years ago, Margolis and Margolis (1979) stressed the need to define the form of axonally transported glycosaminoglycans and their fate at the nerve endings. However, progress in this area has been slow.

Glycosaminoglycans can be labeled either with a sugar precursor, usually glucosamine, or with SO_4 (see Chapter 5). Elam *et al.* (1970), studying axonal transport in the goldfish, first demonstrated that $^{35}SO_4$ would label transported glycosaminoglycan as well as glycoprotein, and that the glycosaminoglycan was transported at about the same rate as the glycoprotein. Transport of glycosaminoglycan has since been demonstrated in the sciatic nerve of frog (Edstrom and Mattsson, 1972) and cat (Held and Young, 1972), in garfish olfactory nerve (Elam and Peterson, 1976), and in the optic system of rabbit (Karlsson and Linde, 1977) and rat (Goodrum *et al.*, 1979b; Gammon *et al.*, 1985). In all cases where it has been examined, the glycosaminoglycan transport rate was comparable to that for glycoproteins. No evidence for any slower rates was found.

In comparison to the other glycoconjugates, very little of the glycosaminoglycan synthesized in cell body regions is transported (see Table 3). For example, in the rat only 1–2% of the [^{3}H]glucosamine incorporated into retinal glycosaminoglycan was found to be exported (Gammon *et al.*, 1985). This relatively small export of glycosaminoglycan is consistent with the very low levels of glycosaminoglycan reported to be present in axon and nerve terminal regions (see Chapter 3). In most cases, as with the glycoproteins, almost all of the glycosaminoglycan that is transported appears to be preferentially destined for the nerve terminals (see Table 3).

In the goldfish and garfish, transported glycosaminoglycan has a subcellular localization similar to that for glycoprotein, less than 20% being soluble and the

Table 3. Retinal Export and Cellular Distribution of Axonally Transported Glycosaminoglycans

Preparation	Precursor	Percent exported[a]	Percent distribution of transported label[b]		Tissues used		References[c]
			Axons	Terminals	Axons	Terminals	
Garfish olfactory system	$^{35}SO_4$	—	83	17[d]	Olf. nerve	Olf. bulb	1
Rabbit optic system	$^{35}SO_4$	1	13	87[d]	OT	LGB, SC	2
Rat optic system	[^{3}H]glucosamine	1	5	95	ON, OT	LGB, SC	3
	$^{35}SO_4$	1	10	90	OT	LGB, SC	4

[a]Percent of total retinal incorporation into glycosaminoglycans.
[b]Estimated from values at the peak of accumulation in the nerve terminals.
[c]References: (1) Elam and Peterson (1976); (2) Karlsson and Linde (1977); (3) Gammon et al. (1985); (4) Goodrum et al. (1979b).
[d]Label in both glycoproteins and glycosaminoglycans.

remainder being in membrane fractions (Elam and Agranoff, 1971b; Elam, 1982). Unlike the fish, in the rat optic system well over 50% of the [^{3}H]glucosamine-labeled glycosaminoglycan is soluble, and the remainder is associated predominantly with light membranes (sedimenting at < 0.85 M sucrose) (Gammon et al., 1985). These values are consistent with those found for whole rat brain glycosaminoglycan (see Chapter 3). As in the case of soluble transported glycoproteins, the localization of this soluble glycosaminoglycan during transport is not known. Glycosaminoglycan has been found as a constituent of secretory vesicles such as chromaffin granules and is released from these vesicles during exocytosis (see Salton et al., 1983). It may be that this soluble fraction of glycosaminoglycan is transported within membrane vesicles and secreted at the nerve terminal, as proposed above for soluble glycoproteins.

Characterization of the transported glycosaminoglycan indicates that all the major classes are transported, and roughly in proportion to their synthesis in the cell body region, i.e., there is no preferential transport of any class (Elam et al., 1970; Held and Young, 1972; Elam and Peterson, 1976; Karlsson and Linde, 1977; Gammon et al., 1985).

4.3. Transport of Glycolipids

4.3.1. Gangliosides Are Rapidly Transported

As is the case with glycosaminoglycans, until recently little attention was paid to ganglioside transport. The growing interest in the role of gangliosides in the processes of development and regeneration (see Chapter 2) has made knowledge of their transport characteristics an important issue. Labeling studies commonly utilize either N-acetylmannosamine, an immediate and specific precursor of sialic acid, or the more general precursor glucosamine, which is more efficiently incorporated into gangliosides but labels other glycoconjugates as well (see Ledeen et al., 1976). Lipid precursors such as acetate can also been used.

Axonally transported, [^{3}H]glucosamine-labeled material soluble in organic solvents was first reported by Grafstein and co-workers in the goldfish optic system (Forman et al., 1971; McEwen et al., 1971) and postulated to be rapidly transported gangliosides. Subsequently, Forman and Ledeen (1972) and Rosner et al. (1973), using both [^{3}H]glucosamine and [^{3}H]-N-acetylmannosamine, demonstrated this material to be transported ganglioside. The rapid axonal transport of gangliosides has also been demonstrated in the visual systems of the chicken (Bondy and Madsen, 1974; Rosner, 1975; Maccioni et al., 1977; Landa et al., 1979; Rosner and Merz, 1982), rabbit (Ledeen et al., 1976, 1981), and rat (Gammon et al., 1985), and in rat sciatic nerve (Aquino et al., 1985, 1987; Harry et al., 1987). The studies in sciatic nerve demonstrated the gangliosides to be transported at the same rate as the glycoproteins. However, in several of these studies, especially those in fish, there was evidence of transport of soluble precursor when N-acetylmannosamine was used, suggesting that some of the labeled ganglioside in terminal regions could have been synthesized locally rather than axonally transported (see Section 5.1). In no case was there any evidence of more slowly transported gangliosides. There is one contrary report (Igarashi et al.,

Table 4. Retinal Export and Cellular Distribution of Axonally Transported Gangliosides

Preparation	Precursor	Percent exported[a]	Percent distribution of transported label[b]		Tissues used		Reference[c]
			Axons	Terminals	Axons	Terminals	
Scardinius erythrophthalamus (teleost fish) optic system	[³H]N-acetylglucosamine	—	43	57[d]	ON, OT	Tectum	1
	[³H]N-acetylmannosamine	—	25	75[d]	ON, OT	Tectum	1
Chicken optic system	[³H]N-acetylmannosamine	—	29	71	ON	OL	2
Rabbit optic system	[³H]N-acetylmannosamine	8	25	75[d]	ON, OT	LGB, SC	3
Rat optic system	[³H]glucosamine	40	25	75	ON, OT	LGB, SC	4

[a]Percent of total retinal incorporation into gangliosides.
[b]Estimated from values at the peak of accumulation in the nerve terminals except where noted.
[c]References: (1) Rosner et al. (1973); (2) Rosner (1975); (3) Ledeen et al. (1976); (4) Gammon et al. (1985).
[d]Measured before peak accumulation in the terminals.

1985, 1986) in which it is suggested that the major frog ganglioside is transported at only 3 mm/day and, furthermore, that precursor CMP-sialic acid is axonally transported at 8 mm/day. These results cannot be reconciled with previous studies and it should be noted that the latter investigation did not control adequately for extraaxonal diffusion.

The cellular and subcellular distribution of transported ganglioside (Table 4) is somewhat different from both the glycoproteins and the glycosaminoglycans (compare to Tables 2 and 3). Ledeen *et al.* (1976) reported that at least 8% of rabbit retinal ganglioside was exported. In the rat over 40% of the glucosamine-labeled retinal ganglioside is committed to transport; considerably more than the fraction of newly synthesized retinal glycoprotein which is exported (Gammon *et al.*, 1985). In both the rabbit and the rat optic systems, about 75% of the transported ganglioside appeared in the terminal regions, somewhat less than for glycoproteins. Analysis of downflow curves for gangliosides in the rat sciatic nerve indicates they are more attenuated than those for glycoprotein; this has been interpreted as due to extensive deposition of gangliosides in the axons (Harry *et al.*, 1987).

In mammals all the major gangliosides appear to be axonally transported (Ledeen *et al.*, 1976, 1981; Gammon *et al.*, 1985; Harry *et al.*, 1987). GT1b, GD1b, and GM1 are the major transported gangliosides in rat sciatic nerve. LM1, a putative myelin-specific ganglioside (Chou *et al.*, 1985), is not transported (Harry *et al.*, 1987). In the chicken, ganglioside transport appears more specific in that while systemically applied precursor labels many gangliosides, transported gangliosides consist predominantly of GD1a (Rosner, 1975; Landa *et al.*, 1979; Rosner and Merz, 1982; Panzetta *et al.*, 1983).

The axonal transport of gangliosides in developing and regenerating axons has not been extensively studied. Transport of gangliosides has been demonstrated in developing chick optic system (Bondy and Madsen, 1974; Landa *et al.*, 1979) and the amount of transported ganglioside is reported to increase many fold in the regenerating goldfish optic system (Sbaschnig-Agler *et al.*, 1984).

4.3.2. Some Other Glycolipids May Also Be Transported

Aquino *et al.* (1987) have reported that following [^{3}H]glucosamine labeling of the rate dorsal root ganglion sensory neurons, about half the transported lipid label is in neutral glycosphingolipids, the remainder being in gangliosides. This is in contrast to labeling the motor neurons via injection of the ventral spinal cord where these authors found little transport of neutral glycosphingolipids. This is an interesting demonstration of a class of macromolecules whose transport is restricted to certain neuronal cell types.

5. FATE OF TRANSPORTED GLYCOCONJUGATES

5.1. Transported Glycoconjugates May Be Modified at Their Destinations

The occasional reports of transported precursors in nerves and terminal regions following injection into cell body regions (see Section 4.3) raise the possibility that local incorporation of precursor may contribute to the labeling of axonal and nerve terminal glycoconjugates. There is a controversial body of literature reporting the presence of a variety of glycosyltransferases specific for glycoproteins, gangliosides, and glycosaminoglycans in nerve terminals (see Chapters 2, 4, 5, and Bosmann, 1972; Brandt *et al.*, 1975; Den *et al.*, 1975; Goodrum *et al.*, 1979a), and on the ability of synaptosomal and subsynaptosomal fractions to rapidly incorporate some sugar precursors following either *in vivo* or *in vitro* labeling (Barondes, 1968; Barondes and Dutton, 1969; Bosmann and Hemsworth, 1970a; Festoff *et al.*, 1971; Dutton *et al.*, 1973; Margolis *et al.*, 1975). The major objection to most of these studies has been that the subcellular fractions used are far too contaminated with microsomal membranes to justify such conclusions (e.g., see Raghupathy *et al.*, 1972). However, it can also be argued that microsome fractions are too contaminated with synaptic membranes to justify this objection (see Section 4.1.4). The most direct evidence for the local modification of transported glycoproteins comes from studies in *Aplysia*. Ambron *et al.* (1974a) observed a glycoprotein in the axon of the R2 neuron that was absent from the cell body, and suggested that this new molecule was formed by glycosylation within the axon. In support of this hypothesis, Ambron and Treistman (1977) found that an intraaxonal injection of [^{3}H]-N-acetylgalactosamine resulted in a labeling of axonal glycoproteins and glycolipids which was insensitive to protein synthesis inhibitors.

The possibility of local glycosylation of transported glycoproteins is also suggested by lectin-binding studies. Glycoproteins undergoing axonal transport have lectin binding affinities indicative of high-mannose type oligosaccharides, whereas neuronal cell surface lectin receptors have affinities indicative of the complex oligosaccharide type (see Section 4.1.6). Presuming that most axonally transported glycoproteins are destined for insertion into the plasma membrane, these results suggest that at least some transported glycoproteins are glycosylated at their destinations.

Glycoconjugates may be modified by removal as well as addition of sugars. Metabolic studies have shown that at least some fucose and SO$_4$ residues on transported glycoproteins turn over much faster than the peptide backbones (see Section 5.4), suggesting that the carbohydrate composition of transported glycoproteins can change both during and following transport. Consistent with these data, several glycosidases which could modify transported glycoconjugates have also been reported to be present in synaptosomal preparations (Bosmann and Hemsworth, 1970b; Schengrund and Rosenberg, 1970; see Townsend *et al.*, 1979, for a review).

The only evidence of local modification of glycosaminoglycans is that for the rapid turnover of SO$_4$ residues relative to the carbohydrate backbone (see Section 5.4). There is little evidence for any such local modifications occurring in transported gangliosides (Ledeen *et al.*, 1976; Aquino *et al.*, 1985).

5.2. Transported Glycoconjugates May Be Redistributed to New Locations

Following the initial deposition of transported glycoconjugates within the axon or terminals, these macromolecules may be redistributed to new locations within the cell. Marchisio *et al.* (1975) observed a continuing increase in labeling of chick optic lobe following administration of colchicine, and concluded that this newly arriving label must be from previously deposited material in the axons. However, a redistribution of deposited label in axons was not observed in regenerating rat sciatic nerve (Griffin *et al.*, 1981).

In addition to redistribution within the neuron, some transported label finds its way to adjacent cells. There have been several reports that following axonal transport of labeled glycoproteins, myelin slowly becomes labeled (Di Giamberadino *et al.*, 1973; Monticone and Elam, 1975; Autilio-Gambetti *et al.*, 1975), but the direct transfer of intact macromolecules has been ruled out. Labeling of myelin-enriched fractions has been attributed to axolemmal contamination (Matthieu *et al.*, 1978), and the labeling of myelin-specific glycoproteins has been attributed to turnover of transported glycoprotein followed by reincorporation of the labeled sugar into myelin (Toews *et al.*, 1982).

Postsynaptic cells also can become labeled from transported glycoconjugates. Some of this label is subsequently axonally transported to the terminals of these postsynaptic cells (Specht and Grafstein, 1973, 1977; Drager, 1974; Casagrande and Harting, 1975; Matthews *et al.*, 1982a). The extent of labeling of postsynaptic cells varies considerably. Bennett *et al.* (1973) found minimal labeling of postsynaptic cells in chick ciliary ganglion, while Specht and Grafstein (1977) found 10–15% of the transported label in the rat lateral geniculate body subsequently transported to the visual cortex. It is not known whether this phenomenon is due solely to the uptake and reincorporation of labeled sugar following metabolic turnover of transported glycoconjugates, or if the transfer of intact macromolecules also occurs (see Grafstein and Forman, 1980).

5.3. Glycoconjugates Are Components of Retrograde Transport

Membranous material, in the form of prelysosomal structures, is axonally transported toward the cell body (retrograde transport, see Table 1). Retrograde transport presumably serves to return macromolecules to the cell body for degradation (see Bisby, 1980). As would be expected for the transport of membranous material, proteins and lipids have been shown to be constituents of retrograde transport (see Grafstein and Forman, 1980).

The retrograde transport of glycoproteins was first demonstrated in regenerating hypoglossal and vagus nerves where [^{3}H]fucose-labeled material was shown to accumulate distal to ligatures on the nerves (Frizell *et al.*, 1976). Using ligatures or nerve crushes, the retrograde transport of glycoproteins has subsequently been described in goldfish optic nerve (Whitnall *et al.*, 1982), and the motor (Aquino *et al.*, 1985) and sensory (Harry *et al.*, 1987) components of rat sciatic nerve. The subcellular localization and composition of glycoproteins in the retrograde transport component have not been examined.

The retrograde transport of gangliosides has also been demonstrated in rat sciatic nerve (Aquino *et al.*, 1985, 1987; Harry *et al.*, 1987). Upon fractionation of the transported gangliosides the anterograde and retrograde patterns were found to be similar (Aquino *et al.*, 1985). The retrograde transport of glycosaminoglycans has not been reported.

Table 5. Turnover of Axonally Transported Glycoproteins

Precursor	Tissue	$t_{1/2}$ (days)		Reference[a]
		Fast	Slow	
[³H]fucose	Chick ciliary ganglion		7	1
	Pigeon optic lobe	3	38	2
	Pigeon optic tract	7	69	2
	Goldfish optic tectum		20	3
	Rabbit OT, LGB, and SC		10	4
	Mouse LGB and SC	2–4	7–10	5
	Rat SC	1	7	6
	Rat sciatic nerve		7	7
[³H]glucosamine	Goldfish optic tectum		20	8
	Rat SC	2	15	9
³⁵SO₄[b]	Goldfish optic tectum	1	7	10
	Rabbit LGB and SC		2–3	11
	Rat SC	0.5	2–3	12

[a]References: (1) Bennett et al. (1973); (2) Marko and Cuenod (1973); (3) Forman et al. (1972), Monticone and Elam (1975), Ripellino and Elam (1980); (4) Karlsson and Sjostrand (1971); (5) Specht and Grafstein (1977); (6) Goodrum et al. (1979b), Goodrum annd Morell (1982); (7) Toews et al. (1982); (8) Forman et al. (1971); (9) Gammon et al. (1985); (10) Elam and Agranoff (1971a); (11) Karlsson and Linde (1977); (12) Goodrum et al. (1979b).
[b]Label in both glycoproteins and glycosaminoglycans.

5.4. Many Glycoconjugates Exhibit an Apparent Biphasic Metabolic Turnover at Nerve Endings

Labeling of axons and nerve terminals via axonal transport provides an opportunity to study the metabolic properties of glycoconjugates from a single class of neurons. A number of such studies have been reported and are summarized in Table 5. Most investigators report the loss of incorporated radioactivity from sugars to be biphasic, with a brief period of rapid loss followed by a prolonged slow loss. This biphasic loss of radioactivity is most prominent for [³H]fucose-labeled glycoproteins, for which the rapid phase has a half-time of around 1 day and the half-time for the slow phase is usually around 7–10 days, although much longer times have been reported. [³H]glucosamine label in glycoproteins appears to turn over somewhat more slowly than fucose. In the rat optic system there is a suggestion of a small rapid phase, with a half-time of around 2 days, and a slow phase with a half-time of about 15 days (Gammon et al., 1985).

A factor which complicates interpretation of studies of metabolic turnover at nerve endings is the removal of material by retrograde transport. For [³H]glucosamine-labeled glycoproteins transported to the goldfish optic tectum, Whitnall et al. (1982) estimated that 40% of the initial loss of label from the tectum was due to retrograde transport.

This biphasic loss of radioactivity from transported glycoconjugates may complicate interpretation of the time course of axonal transport. Several studies present results

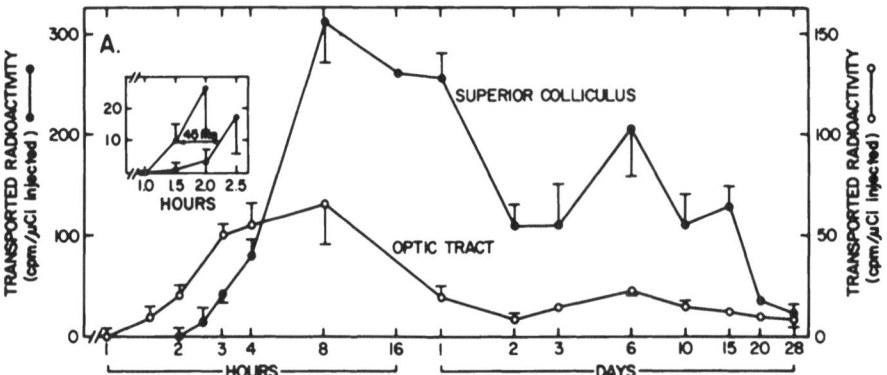

Figure 4. Time course of transported [³H]fucose-labeled glycoproteins in the rat optic tract and superior colliculus. The inset shows the time delay in initial arrival of transported radioactivity in these structures, from which a transport rate of > 300 mm/day can be estimated. Note that the second rise in radioactivity occurs simultaneously in both structures, indicating that it is not due to slowly transported glycoproteins. From Goodrum and Morell (1982).

interpreted as indicating that [³H]fucose labels populations of glycoproteins which are exported at different times, resulting in temporally separated waves of rapid transport (Figure 4; Specht and Grafstein, 1977; Goodrum et al., 1979b; Padilla and Morell, 1980). A more recent study from our laboratory suggests that the apparent arrival of two waves of [³H]fucose radioactivity at nerve endings is the result of the summation of a monophasic time course of radioactive polypeptide arrival and the biphasic metabolic turnover of the fucose residues on these polypeptides (Goodrum and Morell, 1984). Such rapid, independent turnover of a part of the fucose on glycoproteins can presumably occur due to its terminal location on carbohydrate chains. In contrast, when glucosamine (which labels predominantly nonterminal sugars on glycoproteins) is used as a precursor, the rapid phase of turnover and biphasic transport time courses are much less significant (e.g., Gammon et al., 1985).

This turnover of sugar residues independently of the peptide backbone is consistent with the possible presence of glycosidases in axons and terminals (Section 5.1). This rapid loss of radioactivity may represent either an actual loss of sugar residues from glycoproteins via glycosidases, or the substitution of unlabeled for labeled sugar residues via the concerted action of a glycosidase and a glycosyltransferase. Presumably, the slower phase of radioactivity loss represents the actual metabolic turnover of the glycoproteins.

$^{35}SO_4$ label in transported proteins is also lost in a biphasic fashion, but much more rapidly than is labeled fucose or glucosamine (Karlsson and Linde, 1977; Goodrum et al., 1979b; Gammon et al., 1985). The fast and slow phases of turnover have half-times of around 0.5 and 2–3 days, respectively. Such rapid loss of label suggests that the SO_4 residues turn over independently of the carbohydrate and peptide backbones.

The turnover of transported glycosaminoglycans and gangliosides has not been extensively examined. Loss of SO_4 label in transported glycosaminoglycans is similar to that in glycoproteins, with fast and slow phases having half-times of a few hours and a few days, respectively (Goodrum *et al.*, 1979b; Gammon *et al.*, 1985). Glycosaminoglycans labeled with [^{3}H]glucosamine also appear to have a biphasic decay, with a small rapid phase and a slow phase that decays with a half-time of around 20 days (Gammon *et al.*, 1985). Labeling experiments involving any of several different precursors suggest that transported gangliosides have a metabolic turnover time on the order of weeks (Ledeen *et al.*, 1976; Gammon *et al.*, 1985).

6. FUTURE PROSPECTS

Studies of the axonal transport of glycoconjugates have provided much basic information about glycoconjugates in the nervous system as well as about axonal transport itself, but many important questions remain. The issues of cellular sorting and routing of materials into the axon are being pursued successfully but the problems of sorting and routing of glycoconjugates to their specific axonal or terminal destinations is only beginning to be addressed. Included in this issue is the question of the possible secretion of transported glycoconjugates by the nerve terminal.

The question of local modification of glycoconjugates by the addition or removal of carbohydrate moieties is in need of renewed examination. Such modifications could be involved in the functional activity of the axon or terminal, or perhaps could signal entry into the retrograde transport compartment. The nature of glycoconjugate material carried by retrograde transport and the mechanisms involved in the routing of material into this compartment are poorly understood.

Development and regeneration encompass another important area for further work. Surprisingly little research has been done in developing systems to characterize glycoconjugate transport, especially in mammals. The importance of both gangliosides and proteoglycans in the development as well as adult functioning of the nervous system is becoming more and more recognized, but very little is known about the details of their intracellular life, including their axonal transport. The elucidation of these issues will provide a better understanding both of the function of glycoconjugates in nervous tissue and of axonal transport.

7. REFERENCES

Ambron, R. T., and Treistman, S. N., 1977, Glycoproteins are modified in the axon of R2, the giant neuron of Aplysia californica, after intra-axonal injection of [^{3}H]N-acetylgalactosamine, *Brain Res.* **121**:287–309.

Ambron, R. T., Goldman, J. E., and Schwartz, J. H., 1974a, Axonal transport of newly synthesized glycoproteins in a single identified neuron of Aplysia californica, *J. Cell Biol.* **61**:665–675.

Ambron, R. T., Goldman, J. E., Thompson, E. B., and Schwartz, J. H., 1974b, Synthesis of glycoproteins in a single identified neuron of Aplysia californica, *J. Cell Biol.* **61**:649–664.

Ambron, R. T., Goldman, J. E., and Schwartz, J. H., 1975, Effect of inhibiting protein synthesis on axonal transport of membrane glycoproteins in an identified neuron of Aplysia, *Brain Res.* **94**:307–323.

Ambron, R. T., Goldman, J. E., Shkolnik, L. J., and Schwartz, J. H., 1980, Synthesis and axonal transport of membrane glycoproteins in an identified serotonergic neuron of Aplysia, *J. Neurophysiol.* **43**:929–944.

Ambron, R. T., Sherbany, A. A., and Schwartz, J. H., 1981a, Distribution of membrane glycoproteins among the organelles of a single identified neuron of Aplysia. II. Isolation and characterization of a glycoprotein associated with vesicles, *Brain Res.* **207**:33–48.

Ambron, R. T., Sherbany, A. A., Shkolnik, L. J., and Schwartz, J. H., 1981b, Distribution of membrane glycoproteins among the organelles of a single identified neuron of Aplysia. I. Association of a [^{3}H]glycoprotein with vesicles, *Brain Res.* **207**:17–32.

Aquino, D. A., Bisby, M. A., and Ledeen, R. W., 1985, Retrograde axonal transport of gangliosides and glycoproteins in the motoneurons of rat sciatic nerve, *J. Neurochem.* **45**:1262–1267.

Aquino, D. A., Bisby, M. A., and Ledeen, R. W., 1987, Bidirectional transport of gangliosides, glycoproteins and neutral glycosphingolipids in the sensory neurons of rat sciatic nerve, *Neuroscience* **20**:1023–1029.

Audigier, Y., Friedlander, M., and Blobel, G., 1987, Multiple topogenic sequences in bovine opsin, *Proc. Natl. Acad. Sci. USA* **84**:5783–5787.

Autilio-Gambetti, L., Gambetti, P., and Shafer, B., 1975, Glial and neuronal contribution to proteins and glycoproteins recovered in myelin fractions, *Brain Res.* **84**:336–340.

Barondes, S. H., 1968, Incorporation of radioactive glucosamine into macromolecules at nerve endings, *J. Neurochem.* **15**:699–706.

Barondes, S. H., and Dutton, G. R., 1969, Acetoxycyclohexamide effect on synthesis and metabolism of glucosamine-containing macromolecules in brain and in nerve endings, *J. Neurobiol.* **1**:99–110.

Beiger, W., Seybold, J., and Kern, H. F., 1975, Studies on intracellular transport of secretory proteins in the rat exocrine pancreas. III. Effect of cobalt, lanthanum, and antimycin A, *Virchows Arch. A* **368**:329–345.

Bennett, G., Di Giamberadino, L., Koenig, H. L., and Droz, B., 1973, Axonal migration of protein and glycoprotein to nerve endings. II. Radioautographic analysis of the renewal of glycoproteins in nerve endings of chicken ciliary ganglion after intracerebral injection of [^{3}H]fucose and [^{3}H]glucosamine, *Brain Res.* **60**:129–146.

Berry, R. W., 1980, Evidence for multiple somatic pools of individual axonally transported proteins, *J. Cell Biol.* **87**:379–385.

Bisby, M. A., 1976, Orthograde and retrograde axonal transport of labeled protein in motoneurons, *Exp. Neurol.* **50**:628–640.

Bisby, M. A., 1978, Fast axonal transport of labeled protein in sensory axons during regeneration, *Exp. Neurol.* **61**:281–300.

Bisby, M. A., 1980, Retrograde axonal transport, in: *Advances in Cellular Neurobiology* (S. Fedoroff and L. Hertz, eds.), pp. 69–117, Academic Press, New York.

Blaker, W. D., Goodrum, J. F., and Morell, P., 1981, Axonal transport of the mitochondria-specific lipid, diphosphatidylglycerol, in the rat visual system, *J. Cell Biol.* **89**:579–584.

Bondy, S. C., and Madsen, C. J., 1971, Development of rapid axonal flow in the chick embryo, *J. Neurobiol.* **2**:279–286.

Bondy, S. C., and Madsen, C. J., 1974, The extent of axoplasmic transport during development, determined by migration of various radioactively-labeled materials, *J. Neurochem.* **23**:905–910.

Bosmann, H. B., 1972, Synthesis of glycoproteins in brain. Identification, purification, and properties of glycosyltransferases from purified synaptosomes of guinea pig cerebral cortex, *J. Neurochem.* **19**:763–778.

Bosmann, H. B., and Hemsworth, B. A., 1970a, Incorporation of amino acids and monosaccharides into macromolecules by isolated synaptosomes and synaptosomal mitochondria, *J. Biol. Chem.* **245**:363–371.

Bosmann, H. B., and Hemsworth, B. A., 1970b, Intraneural glycosidases. I. Glycosidase, β-glucuronidase and acid phosphatase activity in rat and guinea pig cerebral cortical synaptosomes, *Physiol. Chem. Phys.* **2**:249–262.

Brandt, A. E., Distler, J. J., and Jourdian, G. W., 1975, Biosynthesis of chondroitin sulfate proteoglycan: Subcellular distribution of glycosyl transferases in embryonic chick brain, *J. Biol. Chem.* **250**:3996–4006.

Brown, W. J., and Farquhar, M. G., 1984, The mannose-6-phosphate receptor for lysosomal enzymes is concentrated in cis Golgi cisternae, *Cell* **36**:295–307.

Caroni, P., Carlson, S. S., Schweitzer, E., and Kelly, R. B., 1985, Presynaptic neurons may contribute a unique glycoprotein to the extracellular matrix at the synapse, *Nature* **314:**441–443.

Casagrande, V. A., and Harting, J. K., 1975, Transneuronal transport of tritiated fucose and proline in the visual pathways of tree shrew Tupaia glis, *Brain Res.* **96:**367–372.

Chan, J. Y., Nwokoro, N. A., and Schachter, H., 1979, L-fucose metabolism in mammals, *J. Biol. Chem.* **254:**7060–7068.

Chou, K. H., Nolan, C. E., and Jungalwala, F. B., 1985, Subcellular fractionation of rat sciatic nerve and specific localization of ganglioside LM1 in rat nerve myelin, *J. Neurochem.* **44:**1898–1912.

Cole, G. J., and Elam, J. S., 1981, Axonal transport of glycoproteins in regenerating olfactory nerve: Enhanced glycopeptide concanavalin A-binding, *Brain Res.* **222:**437–441.

Cole, G. J., and Elam, J. S., 1983, Characterization of axonally transported glycoproteins in regenerating garfish olfactory nerve, *J. Neurochem.* **41:**691–702.

Crossland, W. J., 1985, Fast axonal transport in the visual pathway of the chick and rat, *Brain Res.* **340:** 373–377.

Den, H., Kaufman, B., McGuire, E. J., and Roseman, S., 1975, The sialic acids. XVIII. Subcellular distribution of seven glycosyltransferases in embryonic chicken brain, *J. Biol. Chem.* **250:**739–746.

Di Giamberadino, L., Bennett, G., Koenig, H., and Droz, B., 1973, Axonal migration of protein and glycoprotein to nerve endings. III. Cell fraction analysis of chicken ciliary ganglion after intracerebral injection of labeled precursors of proteins and glycoproteins, *Brain Res.* **60:**147–159.

Douglas, W. W., 1974, Involvement of calcium in exocytosis and the exocytosis vesiculation sequence, in: *Calcium and Cell Regulation* (R. M. S. Smellie, ed.), pp. 1–28, Biochemical Society, London.

Drager, U. C., 1974, Autoradiography of tritiated proline and fucose transported transneuronally from the eye to the visual cortex in pigmented and albino mice, *Brain Res.* **82:**284–292.

Dutton, G. R., Haywood, P., and Barondes, S. H., 1973, ^{14}C glucosamine incorporation into specific products in the nerve ending fraction in vivo and in vitro, *Brain Res.* **57:**397–408.

Edstrom, A., and Mattsson, H., 1972, Rapid axonal transport in vitro in the sciatic system of the frog of fucose-, glucosamine- and sulphate-containing material, *J. Neurochem.* **19:**1717–1729.

Edstrom, A., and Mattsson, H., 1973, Electrophoretic characterization of leucine-, glucosamine- and fucose-labeled proteins rapidly transported in frog sciatic nerve, *J. Neurochem.* **21:**1499–1507.

Elam, J. S., 1982, Composition and subcellular distribution of glycoproteins and glycosaminoglycans undergoing axonal transport in garfish olfactory nerves, *J. Neurochem.* **39:**1220–1229.

Elam, J. S., and Agranoff, B. W., 1971a, Rapid transport of protein in the optic system of the goldfish, *J. Neurochem.* **18:**375–387.

Elam, J. S., and Agranoff, B. W., 1971b, Transport of proteins and sulfated mucopolysaccharides in the goldfish visual system, *J. Neurobiol.* **2:**379–390.

Elam, J. S., and Cancalon, P., 1984, *Axonal Transport in Neuronal Growth and Regeneration*, Plenum Press, New York.

Elam, J. S., and Peterson, N. W., 1976, Axonal transport of sulfated glycoproteins and mucopolysaccharides in the garfish olfactory nerve, *J. Neurochem.* **26:**845–850.

Elam, J. S., and Peterson, N. W., 1979, Axonal transport of glycoproteins in the garfish olfactory nerve: Isolation of high molecular weight glycopeptides labeled with [^{3}H]fucose and [^{3}H]glucosamine, *J. Neurochem.* **33:**571–573.

Elam, J. S., Goldberg, J. M., Radin, N. S., and Agranoff, B. W., 1970, Rapid axonal transport of sulfated mucopolysaccharide proteins, *Science* **170:**458–460.

Festoff, B. W., Appel, S. H., and Day, E., 1971, Incorporation of ^{14}C glucosamine into synaptosomes in vitro, *J. Neurochem.* **18:**1871–1886.

Forman, D. S., and Ledeen, R. W., 1972, Axonal transport of gangliosides in the goldfish optic nerve, *Science* **177:**630–633.

Forman, D. S., McEwen, B. S., and Grafstein, B., 1971, Rapid transport of radioactivity in goldfish optic nerve following injections of labeled glucosamine, *Brain Res.* **28:**119–130.

Forman, D. S., Grafstein, B., and McEwen, B. S., 1972, Rapid axonal transport of [^{3}H]fucosyl glycoproteins in the goldfish optic system, *Brain Res.* **48:**327–342.

Friedlander, M., and Blobel, G., 1985, Bovine opsin has more than one signal sequence, *Nature* **318:**338–343.

Frizell, M., and Sjostrand, J., 1974a, The axonal transport of [³H]fucose labeled glycoproteins in normal and regenerating peripheral nerves, *Brain Res.* **78**:109–123.

Frizell, M., and Sjostrand, J., 1974b, Transport of proteins, glycoproteins and cholinergic enzymes in regenerating hypoglossal neurons, *J. Neurochem.* **22**:845–850.

Frizell, M., McLean, W. G., and Sjostrand, J., 1976, Retrograde axonal transport of rapidly migrating labeled proteins and glycoproteins in regenerating peripheral nerves, *J. Neurochem.* **27**:191–196.

Gammon, C. M., Goodrum, J. F., Toews, A. D., Okabe, A., and Morell, P., 1985, Axonal transport of glycoconjugates in the rat visual system, *J. Neurochem.* **44**:376–387.

Goldberg, D. J., and Ambron, R. T., 1981, Two rates of fast axonal transport of [³H]glycoprotein in an identified invertebrate neuron, *Brain Res.* **229**:445–455.

Goodrum, J. F., and Morell, P., 1982, Axonal transport, deposition and metabolic turnover of glycoproteins in the rat optic pathway, *J. Neurochem.* **38**:696–704.

Goodrum, J. F., and Morell, P., 1984, Analysis of the apparent biphasic axonal transport kinetics of fucosylated glycoproteins, *J. Neurosci.* **4**:1830–1839.

Goodrum, J. F., Bosmann, H. B., and Tanaka, R., 1979a, Glycoprotein galactosyltransferase activity in synaptic junctional complexes isolated from rat forebrain, *Neurochem. Res.* **4**:331–337.

Goodrum, J. F., Toews, A. D., and Morell, P., 1979b, Axonal transport and metabolism of [³H]-fucose- and [³⁵S]sulfate-labeled macromolecules in the rat visual system, *Brain Res.* **176**:255–272.

Grafstein, B., and Forman, D. S., 1980, Intracellular transport in neurons, *Physiol. Rev.* **60**:1167–1283.

Gremo, F., and Marchisio, P. C., 1975, Dynamic properties of axonal transport of proteins and glycoproteins: A study based on the effects of metaphase blocking drugs in the developing optic pathway of chick embryos, *Cell Tissue Res.* **161**:303–316.

Gremo, F., Sjostrand, J., and Marchisio, P. C., 1974, Radioautographic analysis of ³H-fucose labeled glycoproteins transported along the optic pathway of chick embryos, *Cell Tissue Res.* **153**:465–476.

Griffin, J. W., Price, D. L., Drachman, D. B., and Morris, J., 1981, Incorporation of axonally transported glycoproteins into axolemma during nerve regeneration, *J. Cell Biol.* **88**:205–214.

Gross, G. W., 1973, The effect of temperature on the rapid axoplasmic transport in C-fibers, *Brain Res.* **56**:359–363.

Gross, G. W., and Beidler, L. M., 1975, A quantitative analysis of isotope concentration and rapid transport velocities in the C-fibers of the garfish olfactory nerve, *J. Neurobiol.* **6**:213–232.

Gross, G. W., and Kreutzberg, G. W., 1978, Rapid axoplasmic transport in the olfactory nerve of the pike. I. Basic transport parameters for proteins and amino acids, *Brain Res.* **139**:65–76.

Gumbiner, B., and Kelly, R. B., 1982, Two distinct intracellular pathways transport secretory and membrane glycoproteins to the surface of pituitary tumor cells, *Cell* **28**:51–59.

Gustavsson, S., Ohlson, C., and Karlsson, J. O., 1982, Glycoproteins of axonal transport: Affinity chromatography on fucose-specific lectins, *J. Neurochem.* **38**:852–855.

Hammerschlag, R., and Stone, G. C., 1982, Membrane delivery by fast axonal transport, *Trends Neurosci.* **5**:12–15.

Hammerschlag, R., and Stone, G. C., 1986, Prelude to fast axonal transport: Sequence of events in the cell body, in: *Axoplasmic Transport* (Z. Iqbal, ed.), pp. 21–34, CRC Press, Boca Raton, Fla.

Hammerschlag, R., and Stone, G. C., 1987, Further studies on the initiation of fast axonal transport, in: *Axonal Transport* (R. S. Smith and M. A. Bisby, eds.), pp. 37–51, Liss, New York.

Hammerschlag, R., Stone, G. C., Bolen, F. A., Lindsey, J. D., and Ellisman, M. H., 1982, Evidence that all newly synthesized proteins destined for fast axonal transport pass through the Golgi apparatus, *J. Cell Biol.* **93**:568–575.

Hanson, M., and Edstrom, A., 1978, Mitosis inhibitors and axonal transport, *Int. Rev. Cytol. Suppl.* **7**:373–402.

Harry, G. J., Goodrum, J. F., Toews, A. D., and Morell, P., 1987, Axonal transport characteristics of gangliosides in sensory axons of rat sciatic nerve, *J. Neurochem.* **48**:1529–1536.

Hart, C. E., and Wood, J. G., 1986, Analysis of the carbohydrate composition of axonally transported glycoconjugates in sciatic nerve, *J. Neurosci.* **6**:1365–1371.

Held, I., and Young, I. J., 1972, Transport of radioactivity derived from labeled N-acetylglucosamine in mammalian motor axons, *J. Neurobiol.* **3**:153–161.

Hendrickson, A. E., and Cowan, W. M., 1971, Changes in the rate of axoplasmic transport during postnatal development of the rabbit's optic nerve and tract, *Exp. Neurol.* **30**:403–422.

Igarashi, M., Komiya, Y., and Kurokawa, M., 1985, CMP-sialic acid, the sole sialosyl donor, is intra-axonally transported, *FEBS Lett.* **192**:239–242.

Igarashi, M., Komiya, Y., and Kurokawa, M., 1986, A ganglioside species (GD1a) migrates at a slow rate and CMP-sialic acid several fold faster in Xenopus sciatic nerve: Fluorographic demonstration, *J. Neurochem.* **47**:1720–1727.

Karlsson, J. O., 1979, Proteins of axonal transport: Interaction of rapidly transported proteins with lectins, *J. Neurochem.* **32**:491–494.

Karlsson, J. O., 1980, Proteins of rapid axonal transport: Polypeptides interacting with the lectin from Lens culinaris, *J. Neurochem.* **34**:1184–1190.

Karlsson, J. O., and Linde, A., 1977, Axonal transport of [^{35}S]sulphate in retinal ganglion cells of the rabbit, *J. Neurochem.* **28**:293–297.

Karlsson, J. O., and Sjostrand, J., 1971, Rapid intracellular transport of fucose-containing glycoproteins in retinal ganglion cells, *J. Neurochem.* **18**:2209–2216.

Kern, H. F., and Kern, D., 1969, Electronenmikroskopische untersuchungen uber die wirkung von kobaltchlorid auf das exokrine pankreasgewebe des meerschweinchens, *Virchows Arch. Abt. B Zellpath.* **4**:54–70.

Koro, L. A., and Marchase, R. B., 1982, A UDP-glucose: glycoprotein glucose-1-phosphotransferase in embryonic chicken neural retina, *Cell* **31**:739–748.

Landa, C. A., Maccioni, H. J. F., and Caputto, R., 1979, The site of synthesis of gangliosides in the chick optic system, *J. Neurochem.* **33**:825–838.

Lasek, R. J., 1981, Cytoskeletons and cell motility in the nervous system, in: *Basic Neurochemistry* (G. J. Seigel, R. W. Albers, B. W. Agranoff, and R. Katzman, eds.), Little, Brown, Boston.

Ledeen, R. W., Skrivanek, J. A., Tirri, L. J., Margolis, R. K., and Margolis, R. U., 1976, Gangliosides of the neuron: Localization and origin, *Adv. Exp. Med. Biol.* **71**:83–103.

Ledeen, R. W., Skrivanek, J. A., Nunez, J., Sclafani, J. R., Norton, W. T., and Farooq, M., 1981, Implications of the distribution and transport of gangliosides in the nervous system, in: *Gangliosides in Neurological and Neuromuscular Function* (M. M. Rapport and A. Gorio, eds.), pp. 211–223, Raven Press, New York.

Levin, B. E., 1977, Axonal transport of [^{3}H]fucosyl glycoproteins in noradrenergic neurons in the rat brain, *Brain Res.* **130**:421–432.

Lindsey, J. D., and Ellisman, M. H., 1983, The varicose tubule: A direct connection between rough endoplasmic reticulum and the Golgi apparatus, *J. Cell Biol.* **97**:304a.

Maccioni, H. J., Landa, C., Arce, A., and Caputto, R., 1977, The biosynthesis of brain gangliosides—evidence for a "transient pool" and an "end product pool" of gangliosides, *Adv. Exp. Med. Biol.* **83**:267–281.

Marchase, R. B., Koro, L. A., Kelly, C. M., and McClay, D. R., 1982, Retinal ligatin recognizes glycoproteins bearing oligosaccharides terminating in phosphodiester-linked glucose, *Cell* **28**:813–820.

Marchisio, P. C., Sjostrand, J., Aglietta, M., and Karlsson, J. O., 1973, The development of axonal transport of proteins and glycoproteins in the optic pathway of chick embryos, *Brain Res.* **63**:273–284.

Marchisio, P. C., Gremo, F., and Sjostrand, J., 1975, Axonal transport in embryonic neurons. The possibility of a proximo-distal axolemmal transfer of glycoproteins, *Brain Res.* **85**:281–285.

Margolis, R. K., and Margolis, R. U., 1972, Disposition of fucose in brain, *J. Neurochem.* **19**:1023–1030.

Margolis, R. U., and Margolis, R. K., 1979, Perspectives and functional implications, in: *Complex Carbohydrates of Nervous Tissue* (R. U. Margolis and R. K. Margolis, eds.), pp. 377–386, Plenum Press, New York.

Margolis, R. K., Margolis, R. U., Preti, C., and Lai, D., 1975, Distribution and metabolism of glycoproteins and glycosaminoglycans in subcellular fractions of brain, *Biochemistry* **14**:4797–4804.

Marko, P., and Cuenod, M., 1973, Contribution of the nerve cell body to renewal of axonal and synaptic glycoproteins in the pigeon visual system, *Brain Res.* **62**:419–423.

Markov, D., Rambourg, A., and Droz, B., 1976, Smooth endoplasmic reticulum and fast axonal transport of glycoproteins, an electron microscopic radioautographic study of thick sections after heavy metals impregnation, *J. Microsc. Biol. Cell.* **25**:57–60.

Matthews, M. A., Narayanan, C. H., and Siegenthaler-Matthews, D. J., 1982a, Inhibition of axoplasmic transport in the developing visual system of the rat. III. Electron microscopy and Golgi studies of retino-fugal synapses and post-synaptic neurons in the dorsal lateral geniculate nucleus, *Neuroscience* **7**:405–422.

Matthews, M. A., West, L. C., and Clarkson, D. B., 1982b, Inhibition of axoplasmic transport in the developing visual system of the rat. II. Quantitative analysis of alterations in transport of tritiated proline or fucose, *Neuroscience* **7**:385–404.

Matthieu, J. M., Webster, H. d., DeVries, G. H., Corthay, S., and Koellreutter, B., 1978, Glial versus neuronal origin of myelin proteins and glycoproteins studied by combined intraocular and intracranial labeling, *J. Neurochem.* **31**:93–102.

McEwen, B. S., Forman, D. S., and Grafstein, B., 1971, Components of fast and slow axonal transport in the goldfish optic nerve, *J. Neurobiol.* **2**:361–377.

McLean, W. G., Frizell, M., and Sjostrand, J., 1975, Axonal transport of labeled proteins in sensory fibres of rabbit vagus nerve in vitro, *J. Neurochem.* **25**:695–698.

Monticone, R. E., and Elam, J. S., 1975, Isolation of axonally transported glycoproteins with goldfish visual system myelin, *Brain Res.* **100**:61–71.

Morre, D. J., Keenan, T. W., and Huang, C. M., 1974, Membrane flow and differentiation: Origin of Golgi apparatus membranes from endoplasmic reticulum, *Adv. Cytopharmacol.* **2**:107–125.

Neufeld, E. F., and McKusick, V. A., 1983, Disorders of lysosomal enzyme synthesis and localization. I. Cell disease and pseudo-Hurler polydystrophy, in: *The Metabolic Basis of Disease,* 5th ed. (J. B. Stanbury, J. B. Wyngaarden, D. S. Frederickson, M. S. Goldstein, and J. L. Brown, eds.), pp. 778–787, McGraw–Hill, New York.

Nichols, T. R., Smith, R. S., and Snyder, R. E., 1982, The action of puromycin and cyclohexamide on the initiation and rapid axonal transport on amphibian dorsal root neurons, *J. Physiol. (London)* **332**:441–458.

Ochs, S., 1973, Effects of maturation and aging on the rate of fast axoplasmic transport in mammalian nerve, *Prog. Brain Res.* **40**:349–362.

Ochs, S., 1982, *Axoplasmic Transport and Its Relation to Other Nerve Functions,* Wiley, New York.

Ohlson, C., and Karlsson, J. O., 1983a, Glycoproteins of axonal transport. Interaction with heparin, *Brain Res.* **274**:303–308.

Ohlson, C., and Karlsson, J. O., 1983b, Glycoproteins of axonal transport: Polypeptides interacting with the lectin from Aleuria aurantia, *Brain Res.* **264**:99–104.

Olden, K., Parent, J. B., and White, S. L., 1982, Carbohydrate moieties of glycoproteins. A reevaluation of their function, *Biochim. Biophys. Acta* **650**:209–232.

Olden, K., Bernard, B. A., Humphries, M. J., Yeo, T., Yeo, K., White, S. L., Newton, S. A., Bauer, H. C., and Parent, J. B., 1985, Function of glycoprotein glycans, *Trends Biochem. Sci.* **10**:78–82.

Padilla, S. S., and Morell, P., 1980, Axonal transport of [^{3}H]fucose-labeled glycoproteins in two intra-brain tracts of the rat, *J. Neurochem.* **35**:444–450.

Palade, G. E., 1983, Membrane biogenesis: An overview, in: *Methods in Enzymology* (S. Fleischer, and B. Fleischer, eds.), pp. xxix–xl, Academic Press, New York.

Panzetta, P., Chiarenza, A. P., and Maccioni, H. J. F., 1983, Axonal transport of gangliosides in the visual system of the developing chick embryo, *Int. J. Dev. Neurosci.* **1**:149–153.

Raghupathy, E., Ko, G. K. W., and Peterson, N. A., 1972, Glycoprotein biosynthesis in the developing rat brain. III. Are glycoprotein glycosyltransferases present in synaptosomes? *Biochim. Biophys. Acta* **286**:339–349.

Ripellino, J. A., and Elam, J. S., 1980, Differential turnover of axonally transported glycoproteins, *Neurochem. Res.* **5**:351–360.

Roger, L. J., Breese, G. R., and Morell, P., 1980, Axonal transport of proteins and glycoproteins in the rat nigro-striatal pathway and the effects of 6-hydroxydopamine, *Brain Res.* **197**:95–112.

Rosner, H., 1975, Incorporation of sialic acid into gangliosides and glycoproteins of the optic pathway following an intraocular injection of [N-^{3}H]acetylmannosamine in the chicken, *Brain Res.* **97**:107–116.

Rosner, H., and Merz, G., 1982, Uniform distribution and similar turnover rates of individual gangliosides along axons of retinal ganglion cells in the chicken, *Brain Res.* **236**:63–75.

Rosner, H., Weigandt, H., and Rahmann, H., 1973, Sialic acid incorporation into gangliosides and glycoproteins of the fish brain, *J. Neurochem.* **21**:655–665.

Sabatini, D. D., Kreibich, G., Morimoto, T., and Adesnik, M., 1982, Mechanisms for the incorporation of proteins in membranes and organelles, *J. Cell Biol.* **92**:1–22.

Salton, S. R. J., Margolis, R. U., and Margolis, R. K., 1983, Release of chromaffin granule glycoproteins and proteoglycans from potassium-stimulated PC12 pheochromocytoma cells, *J. Neurochem.* **41**:1165–1170.

Sbaschnig-Agler, M., Ledeen, R. W., Grafstein, B., and Alpert, R. M., 1984, Ganglioside changes in the regenerating goldfish optic system: Comparison with glycoproteins and phospholipids, *J. Neurosci. Res.* **12**:221–232.

Schengrund, C. L., and Rosenberg, A., 1970, Intracellular location and properties of bovine bran sialidase, *J. Biol. Chem.* **245**:6196–6200.

Schnapp, B. J., and Reese, T. S., 1986, New developments in understanding rapid axonal transport, *Trends Neurosci.* **9**:155–162.

Schwab, M. E., and Thoenen, H., 1983, Retrograde axonal transport, in: *Metabolic Turnover in the Nervous System, Handbook of Neurochemistry*, 2nd ed. (A. Lajtha, ed.), pp. 381–404, Plenum Press, New York.

Skene, J. H. P., and Willard, M., 1981a, Changes in axonally transported proteins during axon regeneration in toad retinal ganglion cells, *J. Cell Biol.* **89**:86–95.

Skene, J. H. P., and Willard, M., 1981b, Axonally transported proteins associated with axon growth in rabbit central and peripheral nervous systems, *J. Cell Biol.* **89**:96–103.

Skene, J. H. P., and Willard, M., 1981c, Characteristics of growth-associated polypeptides in regenerating toad retinal ganglion cell axons, *J. Neurosci.* **1**:419–426.

Sly, W. S., Natowicz, M., Gonzalez-Noriega, A., Grubb, J. H., and Fischer, H. D., 1981, The role of mannose-6-phosphate recognition marker and its receptor in the uptake and intracellular transport of lysosomal enzymes, in: *Lysosomes and Lysosomal Storage Disease* (J. W. Callahan and J. A. Lowden, eds.), pp. 131–146, Raven Press, New York.

Smith, R. S., and Bisby, M. A. (eds.), 1987, *Axonal Transport*, Liss, New York.

Specht, S. C., 1977, Axonal transport in the optic system of neonatal and adult hamsters, *Exp. Neurol.* **56**:252–264.

Specht, S. C., 1982, Postnatal changes in [³H]fucosyl glycopeptides of hamster optic nerve synaptosomal membranes, *Dev. Brain Res.* **4**:109–114.

Specht, S. C., and Grafstein, B., 1973, Accumulation of radioactive protein in mouse cerebral cortex after injection of ³H-fucose into the eye, *Exp. Neurol.* **41**:705–722.

Specht, S. C., and Grafstein, B., 1977, Axonal transport and transneuronal transfer in mouse visual system following injection of [³H]fucose into the eye, *Exp. Neurol.* **54**:352–368.

Stone, G. C., and Hammerschlag, R., 1981, Differential effects of cobalt on the initiation of fast axonal transport, *Cell Mol. Neurobiol.* **1**:3–17.

Stone, G. C., and Hammerschlag, R., 1983, Glycosylation as a criterion for defining subpopulations of fast-transported proteins, *J. Neurochem.* **40**:1124–1133.

Stone, G. C., and Hammerschlag, R., 1987, Molecular mechanisms involved in sorting of fast-transported proteins, in: *Axonal Transport* (R. S. Smith and M. A. Bisby, eds.), pp. 15–36, Liss, New York.

Stone, G. C., Wilson, D. L., and Hall, M. E., 1978, Two-dimensional gel electrophoresis of proteins in rapid axoplasmic transport, *Brain Res.* **144**:287–302.

Stone, G. C., Hammerschlag, R., and Bobinski, J. A., 1983, Fast-transported glycoproteins and non-glycosylated proteins contain sulfate, *J. Neurochem.* **41**:1085–1089.

Stone, G. C., Hammerschlag, R., and Bobinski, J. A., 1987, Complex compartmentation of tyrosine sulfate-containing proteins undergoing fast axonal transport, *J. Neurochem.* **48**:1736–1744.

Tartakoff, A. M., 1980, Reversible perturbation of the Golgi complex, *Cell Biol. Int. Rep.* **4**:809.

Tartakoff, A. M., Hoessli, D., and Vassalli, P., 1980, Golgi participation in intracellular transport of surface glycoproteins, *Eur. J. Cell Biol.* **22**:173.

Tedeschi, B., and Wilson, D. L., 1987, Subsets of axonally transported and periaxonal polypeptides are released from regenerating nerve, *J. Neurochem.* **48**:463–469.

Teichberg, S., and Holtzman, E., 1973, Axonal agranular reticulum and synaptic vesicles in cultured embryonic chick sympathetic neurons, *J. Cell Biol.* **57**:88–108.

Thompson, E. B., Schwartz, J. H., and Kandel, E. R., 1976, A radioautographic analysis in the light and electron microscope of identified Aplysia neurons and their processes after intrasomatic injection of L-[³H]fucose, *Brain Res.* **112**:251–281.

Toews, A. D., Saunders, B. F., and Morell, P., 1982, Axonal transport and metabolism of glycoproteins in rat sciatic nerve, *J. Neurochem.* **39**:1348–1355.

Townsend, R. R., Li, Y.-T., and Li, S.-C., 1979, Brain glycosidases, in: *Complex Carbohydrates of Nervous Tissue* (R. U. Margolis and R. K. Margolis, eds.), pp. 127–138, Plenum Press, New York.

Tytell, M., Gulley, R. L., Wenthold, R. J., and Lasek, R. J., 1980, Fast axonal transport in auditory neurons of the guinea pig: A rapidly turned-over glycoprotein, *Proc. Natl. Acad. Sci. USA* **77**:3042–3046.

Walter, P., Gilmore, R., and Blobel, G., 1985, Protein translocation across the endoplasmic reticulum, *Cell* **38**:5–8.

Weiss, P., and Hiscoe, H. B., 1948, Experiments on the mechanism of nerve growth, *J. Exp. Zool.* **107**:315–395.

Wenthold, R. J., and McGarvey, M. L., 1982, Different polypeptides are rapidly transported in auditory and optic neurons, *J. Neurochem.* **39**:27–35.

Whitnall, M. H., Currie, J. R., and Grafstein, B., 1982, Bidirectional axonal transport of glycoproteins in goldfish optic nerve, *Exp. Neurol.* **75**:191–207.

Williams, D. V., Swiedler, S. J., and Hart, G. W., 1985, Intracellular transport of membrane glycoproteins: Two closely related histocompatibility antigens differ in their rates of transit to the cell surface, *J. Cell Biol.* **101**:725–734.

Yeo, K., Parent, J. B., Yeo, T., and Olden, K., 1985a, Variability in transport rates of secretory glycoproteins through the endoplasmic reticulum and Golgi in human hepatoma cells, *J. Biol. Chem.* **260**:7896–7902.

Yeo, T., Yeo, K., Parent, J. B., and Olden, K., 1985b, Swainsonine treatment accelerates intracellular transport and secretion of glycoproteins in human hepatoma cells, *J. Biol. Chem.* **260**:2565–2569.

Zatz, M., and Barondes, S. H., 1971, Rapid transport of fucosyl glycoproteins to nerve endings in mouse brain, *J. Neurochem.* **18**:1125–1133.

Synaptic Vesicle Glycoproteins and Proteoglycans

Steven S. Carlson

1. INTRODUCTION

Immunological markers specific for synaptic vesicles have been helpful in studying synapse regeneration (Glicksman and Sanes, 1983), development (Chun and Shatz, 1983), and tracing membrane traffic in the nerve terminal (von Wedel *et al.*, 1981). Vesicle-specific antibodies have also identified a number of vesicle proteins. This chapter deals with four antigenic integral membrane proteins of the synaptic vesicle. These proteins, identified with monoclonal antibodies, are the best characterized vesicle membrane components. Three of these proteins are probably present in all vesicles of the regulated pathway of neuroendocrine cells. This includes both dense-core vesicles and small, clear synaptic vesicles. The cDNA for one of these proteins, p38, has been cloned and sequenced. The fourth protein is an integral membrane proteoglycan (SVPG) which has been found in some synaptic vesicles, but could have a wider distribution. Interestingly, this proteoglycan shares a unique antigenic site with a putative nerve terminal anchorage protein, TAP-1.

Before discussing these synaptic vesicle proteins, it is worth briefly mentioning the duties which vesicles perform. During the life of a synaptic vesicle, several different activities must be carried out successfully (Kelly and Reichardt, 1983; Kelly and Hooper, 1982; Kelly *et al.*, 1979):

1. The synaptic vesicle is filled with neurotransmitter. Classical transmitters, like acetylcholine, are thought to be concentrated in the vesicle by neurotransmitter/H^+ exchange proteins. The exchange is probably driven by a proton gradient (low pH inside) which is maintained by a proton pump (Kelly and Reichardt, 1983; Marshall and Parsons, 1987; Winkler, 1987; Winkler and Carmichael, 1982). Peptide neurotransmitter vesicles are formed containing peptides at the *trans*-Golgi network. It is

Steven S. Carlson • Department of Physiology and Biophysics, University of Washington, Seattle, Washington 98195.

thought that they are transported to the exocytotic site and used only once (Kelly and Reichardt, 1983).

2. Synaptic vesicles are axonally transported to the nerve terminal. Purified synaptic vesicles have been shown capable of being transported both anterogradely and retrogradely (Schroer *et al.*, 1985). Synaptic vesicles presumably contain receptors for both the retrograde and anterograde microtubule translocating proteins (Burgess and Kelly, 1987).

3. Synaptic vesicles are thought to interact with cytoskeletal and plasma membrane elements in the nerve terminal. At the neuromuscular junction, vesicles appear to associate with cytoplasmic presynaptic membrane specializations as a preliminary step to regulated exocytosis (Llinas and Heuser, 1977; Kelly *et al.*, 1979; Heuser and Reese, 1981). It is also hypothesized that cytoskeletal constraints involving brain spectrin and synapsin 1 can modulate the number of vesicles interacting with these presynaptic sites (Krebs *et al.*, 1987; Bahler and Greengard, 1987; Baines, 1987; Linstedt and Kelly, 1987). Presumably, the synaptic vesicle must have receptors for these nerve terminal elements.

4. A protein responsible for the fusion of the synaptic vesicle and presynaptic plasma membranes, a "fusase," may be present in the synaptic vesicle membrane (Kelly and Reichardt, 1983). Such a protein should be Ca^{2+} activated and might be similar to the viral fusion proteins which have been so well studied (White *et al.*, 1983; White and Wilson, 1987).

5. When the synaptic vesicle is formed from the *trans* Golgi network or re-formed from the plasma membrane, the synaptic vesicle components must be sorted from the components of the parent membrane (Kelly, 1985; Griffiths and Simons, 1986; Burgess and Kelly, 1987). Clathrin and the clathrin-associated proteins are most likely involved (Pearse, 1987); probably some proteins that are permanent residents of the synaptic vesicle membrane participate as well. These might be considered sorting proteins. Some vesicle proteins may also be involved in the endocytotic membrane budding process.

For some of these synaptic vesicle functions a few proteins have been identified (Stadler and Fenwick, 1983; Lee and Witzemann, 1983; Harlos *et al.*, 1984; Rephaeli and Parsons, 1982; Yamagata and Parsons, 1987; Bahr and Parsons, 1987; Walker and Agoston, 1987). Disappointingly, however, none of these functions have been definitively ascribed to any of the integral membrane synaptic vesicle proteins mentioned above. When considering activities that these identified vesicle-specific proteins might have, one possibility is worth remembering. Although some of the vesicle activities may be shared with other membranes, those vesicle proteins responsible might still be vesicle-specific. The synaptic vesicle could have its own version of a protein.

The identification and characterization of specific synaptic vesicle proteins has been greatly facilitated by the availability of purified synaptic vesicle preparations. Synaptic vesicles of mixed transmitter types can be purified from mammalian brain (Huttner *et al.*, 1983), and vesicles of only one transmitter type, acetylcholine, from elasmobranch electric organ (Carlson *et al.*, 1978; Tashiro and Stadler, 1978). The preparations from brain and electric organ involve differential centrifugation, equilibrium sedimentation or flotation on sucrose gradients, and permeation chromatography with controlled pore glass (Huttner *et al.*, 1983; Carlson *et al.*, 1978). Synaptic vesicles purified by these methods from electric organ have been demonstrated by biophysical criteria to be at least 90% pure (Carlson *et al.*, 1978). The mammalian brain vesicle preparation (Huttner *et al.*, 1983) is also relatively pure. This is evidenced by the finding that two proteins, synapsin 1 and p38, which are vesicle proteins by several criteria, are major proteins of this preparation (Jahn *et al.*, 1985).

2. SYNAPSIN 1

Synapsin 1 is a neuron-specific peripheral membrane protein associated with the cytoplasmic surface of synaptic vesicles. It is hypothesized to regulate the availability of synaptic vesicles for exocytosis (Llinas *et al.*, 1985; Bahler and Greengard, 1987; Baines, 1987). Since synapsin 1 is not an integral membrane glycoprotein, it is not a major subject of this review and will therefore be discussed only briefly.

Synapsin 1 appears to be present in all nerve terminals (De Camilli *et al.*, 1983; Volknandt *et al.*, 1987), but is associated only with small clear synaptic vesicles and not with large dense-core secretory granules (Navone *et al.*, 1984). In addition to being associated with synaptic vesicles, this protein can also interact with the cytoskeleton. It can bundle F-actin (Petrucci and Morrow, 1987; Bahler and Greengard, 1987), and is reported to interact with brain spectrin, microtubules, and microfilaments (Baines, 1987). It is thought that synapsin 1 might be part of a cytoskeletal matrix which binds synaptic vesicles in the nerve terminal (Navone *et al.*, 1984; Bahler and Greengard, 1987; Baines, 1987).

The cDNA encoding synapsin 1 has been cloned and sequenced (Kilimann and DeGennaro, 1985; McCaffery and DeGennaro, 1986). The sequence predicts a protein of 691 amino acids with a molecular weight of 73,000, in agreement with the behavior of purified synapsin 1b on SDS–PAGE. (Synapsin 1 is actually composed of two very similar polypeptides 1a and 1b.) Interestingly, the sequence also shows homology with the actin-binding proteins profilin and villin, in support of the actin bundling activity found for synapsin 1.

Synapsin 1 can be phosphorylated. The protein shows an increased phosphorylation in response to cell depolarization. It is a substrate for cAMP-dependent protein kinase, calcium/calmodulin-dependent protein kinase, and protein kinase C (Browning *et al.*, 1985). The binding of this protein to synaptic vesicles is decreased fivefold when it is phosphorylated at a specific site by calcium/calmodulin-dependent protein kinase (Schiebler *et al.*, 1986). This phosphorylation also abolishes the ability of synapsin 1 to bundle F-actin.

It is hypothesized that synapsin 1 regulates the movement of synaptic vesicles to their sites of release within the nerve terminal (Bahler and Greengard, 1987; Petrucci and Morrow, 1987; Baines, 1987; Linstedt and Kelly, 1987; Krebs *et al.*, 1987). Phosphorylation of synapsin 1 is thought to remove cytoskeletal linkages constraining synaptic vesicles and allow vesicles to move into release sites on the presynaptic plasma membrane. In support of this hypothesis, pressure injection of dephosphory-lated synapsin 1 into the squid giant synapse decreases the amount of evoked release.

Injection of calcium/calmodulin-dependent protein kinase increases the amount of evoked release (Llinas *et al.*, 1985).

3. *p38 OR SYNAPTOPHYSIN*

A synaptic vesicle protein with a molecular weight of about 38,000 has been identified in mammalian brain (Wiedemann and Franke, 1985; Jahn *et al.*, 1985). Named "p38" (Jahn *et al.*, 1985) or "synaptophysin" (Wiedemann and Franke, 1985), it is an integral membrane protein with an apparent molecular weight on SDS–PAGE of 38,000–40,000 depending on the cell type which produces it (Leube *et al.*, 1987). This vesicle protein was originally identified by monoclonal antibodies raised to both crude fractions of bovine brain coated vesicles (Wiedemann and Franke, 1985) and purified rat synaptic vesicles (Jahn *et al.*, 1985). The same protein probably has been independently identified by several other workers as well (Obata *et al.*, 1986; Devoto and Barnstable, 1987). More structural information exists for this vesicle protein than for any other, since the cDNA encoding p38 has been cloned and sequenced (Buckley *et al.*, 1987; Leube *et al.*, 1987; Sudhof *et al.*, 1987).

Several lines of evidence indicate that p38 is a synaptic vesicle protein. (1) p38 antigenicity is found to copurify with synaptic vesicles from rat brain. Jahn *et al.* (1985) isolated synaptic vesicles by the procedures of Huttner *et al.* (1983) using synapsin 1 as a vesicle marker. Synapsin 1 and p38 antigenicity showed the same enrichment throughout the purification and cochromatographed when subjected to permeation chromatography in the last step of the purification. Like synapsin 1, the p38 protein is a major constituent of these chromatographically purified synaptic vesicles (Jahn *et al.*, 1985). (2) By immunocytochemistry at the electron microscopic level, p38 appears to be localized on the cytoplasmic surface of synaptic vesicles (Wiedemann and Franke, 1985; Navone *et al.*, 1986). By immunostaining with colloidal gold, p38 was localized to the synaptic vesicles of cerebral synapses (Wiedemann and Franke, 1985), the neuromuscular junction (Wiedemann and Franke, 1985), and hypothalamic nerve endings (Navone *et al.*, 1986). Significant staining of nonvesicle membranes was not observed.

The p38 protein appears to be a transmembrane protein. It is exposed on the cytoplasmic surface of the synaptic vesicle as indicated by its protease sensitivity when intact vesicles are digested with pronase (Jahn *et al.*, 1985). However, it is protected from complete digestion unless the vesicles are lysed by detergent. Synaptophysin is glycosylated (Jahn *et al.*, 1985; Rehm *et al.*, 1986); thus, it is most likely exposed on the luminal side of the vesicle membrane as well. Further, it behaves as an integral membrane protein. The protein requires detergent to be solubilized from the synaptic vesicle membrane (Wiedemann and Franke, 1985; Jahn *et al.*, 1985). In addition, when p38 is solubilized in the two-phase detergent system of Triton X-114, this protein partitions into the detergent-rich pellet (Jahn *et al.*, 1985). Such behavior is characteristic of integral membrane proteins (Bordier, 1981).

Synaptophysin is *N*-glycosylated (Rehm *et al.*, 1986). Treatment of a rat pheochromocytoma cell line (PC12) with tunicamycin, an inhibitor of *N*-glycosylation, results in the production of p38 with a decreased molecular weight (from 40,000 to 34,000). A similar reduction in molecular weight is seen when p38 from PC12 cells is subjected to chemical deglycosylation (Navone et al., 1986).

In the synaptic vesicle membrane, p38 probably exists as a homopolymer (Rehm et al., 1986). p38, when solubilized in the presence of phosphatidylcholine by Triton X-100, behaves as a protein–detergent–phospholipid complex with a molecular weight of 220,000. The molecular weight of the protein in this complex is estimated at 120,000. Cross-linking of p38 in the synaptic vesicle membrane by -SH oxidation or glutaraldehyde yields 38,000 and 76,000 antigenic proteins on SDS–PAGE. When purified detergent-solubilized p38 is cross-linked by -SH oxidation, the monomer (38,000), dimer (76,000), trimer (114,000), and tetramer (150,000) are obtained.

Synaptophysin may be a calcium-binding protein (Rehm et al., 1986). After SDS–PAGE and transfer to nitrocellulose, p38 was able to bind $^{45}Ca^{2+}$. It appears to be one of the major $^{45}Ca^{2+}$ binding proteins in the synaptic vesicle preparation used by Rehm et al. (1986). This binding of calcium is probably not simply due to the presence of negative charge on p38 (pI 4.6–4.8). With this nitrocellulose calcium binding assay, Rehm et al. (1986) determined that on a molar basis, synaptophysin has a binding capacity for calcium about sixfold greater than ovalbumin and bovine serum albumin. These two proteins have similar acidic isoelectric points (4.6 and 5.6, respectively). However, the binding capacity is about three times less than that of calmodulin. The calcium binding region of p38 is presumably on the cytoplasmic side of the synaptic vesicle. Pronase digestion of intact vesicles yields a 27,000 synaptophysin fragment which does not bind $^{45}Ca^{2+}$. As yet the details (e.g., the rate and equilibrium constants) of calcium binding to synaptophysin have not been determined.

The p38 protein is present not only in neurons, but also in endocrine cells. By immunocytochemistry, p38 is present in the brain, spinal cord, retina, and neuromuscular junction. It has also been found in endocrine tissue: adrenal medulla, pituitary, endocrine cells of the stomach mucosa, and islet cells of the pancreas (Wiedemann and Franke, 1985; Wiedemann et al., 1986; Navone et al., 1986). However, it does not appear to be present in exocrine tissue, e.g., the exocrine pancreas (Wiedemann et al., 1986). Not surprisingly, neural and neuroendocrine tumor cell lines appear to contain p38 (Wiedemann et al., 1986). Most of the vertebrate classes have been tested and their nervous tissue found to contain p38: bony fish, amphibians, reptiles, birds, and mammals (Jahn et al., 1987).

It has been proposed that p38 is present only in small (40–60 nm diameter) synaptic vesicles without an electron-dense core, and not found in large (> 60 nm diameter) dense-core granules (Navone et al., 1986). This is a distribution similar to that found for synapsin 1 (Navone et al., 1984). Electron microscopic immunocytochemistry shows staining of small synaptic vesicles with the anti-p38 antibodies, but little or no staining of large dense-core granules (Navone et al., 1986). For example, in the hypothalamus where nerve endings contain both types of vesicles, after treatment with anti-p38 antibodies only the small vesicles show intense labeling with protein A-coated gold particles. In a tissue like the adrenal medulla, where only large secretory granules are thought to be present, these large dense-core vesicles show very low or background amounts of labeling. However, a pleiomorphic population of smooth-surfaced membranes (predominantly round or oval vesicles about the size of synaptic vesicles) are intensely labeled. In agreement with this finding, immunoblots of chromaffin granules show no p38 antigenicity (Wiedemann and Franke, 1985). Navone *et al.* (1986) suggest that presence of these labeled vesicles in adrenal medullary cells represents a heretofore unknown secretory pathway in these cells.

Recently, the opposite conclusion has been reached by Lowe *et al.* (1988), who found p38 in dense-core vesicles. These workers were able to immunoprecipitate purified chromaffin granules and dense-core vesicles from PC12 cells with a monoclonal antibody to p38. Chromaffin granules or the PC12 granules were labeled by uptake of [^{3}H]norepinephrine. As much as 80% of the vesicular [^{3}H]norepinephrine could be immunoprecipitated from fractions containing purified chromaffin granules and PC12 postnuclear supernatants. Electron microscopy of the immunoprecipitated vesicles revealed large (> 100 nm diameter) dense-core vesicles. (An essentially identical result was obtained with antibodies against two other synaptic vesicle proteins, p65 and SV2.) In addition, secretory granules containing human growth hormone isolated from PC12 cells were also precipitated (> 90%) with the p38 antibody. These granules were obtained from PC12 cells which had been transfected with DNA coding for the hormone. Lowe *et al.* (1988) suggest that the failure to observe p38 in dense-core secretory granules by Navone *et al.* (1986) with immunoelectron microscopy might be due to the higher sensitivity of the immunoprecipitation method. They further suggest that the small vesicles, seen by Navone *et al.* (1986) in adrenal medullary cells, might be membranes of the endocytotic pathway observed by Patzak and Winkler (1986), rather than representing a separate secretory pathway. The p38 protein might be more concentrated in these membranes and, thus, detected more easily by immunocytochemistry.

The cDNA for the p38 protein has been cloned and sequenced by three separate groups of investigators (Leube *et al.*, 1987; Sudhof *et al.*, 1987; Buckley *et al.*, 1987). The sequence data from the three studies are essentially in agreement. The cDNAs were identified by either screening expression libraries with antibodies to p38 (Leube *et al.*, 1987; Buckley *et al.*, 1987) or screening cDNA libraries with synthetic oligonucleotide probes made from amino acid sequence data (Leube *et al.*, 1987; Sudhof *et al.*, 1987).

Leube *et al.* (1987) and Buckley *et al.* (1987) report finding a single 2.4- to 2.5-kb mRNA species coding for p38 in a variety of human and rat neuroendocrine cells. Sudhof *et al.* (1987) report the size as 3.4 kb. Southern blots of rat genomic DNA revealed the presence of a single synaptophysin gene (Sudhof *et al.*, 1987).

The sequenced cDNAs contain an open reading frame coding for a protein of 307 amino acids and a molecular weight of 33,000 (Leube *et al.*, 1987; Sudhof *et al.*, 1987). This agrees relatively well with the molecular weight determined for deglycosylated synaptophysin of approximately 34,000 (Rehm *et al.*, 1986; Navone *et al.*, 1986). Further, the deduced amino acid sequence was confirmed by comparison with amino acid sequence data of tryptic and chymotryptic peptides not used to construct synthetic oligonucleotide probes (Leube *et al.*, 1987; Sudhof *et al.*, 1987). The most 5' AUG of the cDNA sequence is assumed to be the N-terminus. The N-terminus of the protein appears to be blocked; thus, there are no confirmatory direct amino acid sequence data. With this N-terminus, no signal sequence appears to be present (Leube *et al.*, 1987; Sudhof *et al.*, 1987). However, this has been found for several other membrane proteins such as erythrocyte band III or the Ca^{2+}-dependent ATPase (Leube *et al.*, 1987).

From the deduced amino acid sequence, it is predicted that the p38 protein spans the vesicle bilayer four times with the N-terminus and a large C-terminal domain (≈ 90 amino acids) on the cytoplasmic side of the membrane. An analysis of the sequence for hydrophobic regions indicates the presence of four potential transmembrane domains, each 20–24 residues in length (Leube *et al.*, 1987; Sudhof *et al.*, 1987; Buckley *et al.*, 1987). One or possibly two *N*-glycosylation sites are found in the sequence: between the first and second (Leube *et al.*, 1987; Sudhof *et al.*, 1987; Buckley *et al.*, 1987), and the third and fourth (Buckley *et al.*, 1987) hydrophobic domains. Assuming that these sites are on the interior side of the vesicle membrane, the N- and C-termini must be cytoplasmic. This also places the cysteine residues, thought to be involved in interchain disulfide bonds, on the luminal side of the vesicle membrane away from the reducing cytoplasmic environment (Sudhof *et al.*, 1987).

Several pieces of evidence support a model of p38 with the C-terminus on the cytoplasmic side of the membrane. A monoclonal antibody (SY38) directed against p38 can immunoprecipitate intact secretory granules (Lowe *et al.*, 1988), indicating the presence of this antigenic site on the cytoplasmic side of the vesicle membrane. A fusion protein constructed to contain about the last half of the putative cytoplasmic C-terminal domain binds SY38 (Buckley *et al.*, 1987). In addition, the p38 sequence contains four potential collagenase cleavage sites in the latter half of the putative C-terminal cytoplasmic domain. Digestion of intact vesicles with collagenase yields an approximately 28,000 fragment which lacks the SY38 antigenic site (Leube *et al.*, 1987).

The large, putative cytoplasmic domain of p38 has three additional noteworthy features. (1) Nine to ten copies of a tetrapeptide or pentapeptide repeat are present (Leube *et al.*, 1987; Sudhof *et al.*, 1987). (2) The sequence contains three potential phosphorylation sites: two tyrosines and one threonine. However, there is as yet no report that p38 is phosphorylated (Buckley *et al.*, 1987). (3) As mentioned previously, the Ca^{2+} binding activity of p38 is thought to be located in the cytoplasmic tail. No homology is found between p38 and any of the known Ca^{2+} binding proteins. Synaptophysin may contain a new Ca^{2+} binding sequence (Leube *et al.*, 1987; Buckley *et al.*, 1987).

No function for p38 is known; however, several hypotheses have been put forward. Synaptophysin might bind the synaptic vesicle membrane to the cytoskeleton or to the presynaptic plasma membrane as a preliminary step to exocytosis (Buckley *et al.*, 1987). It has been suggested that the similarity of the collagenlike sequences present in the cytoplasmic domain of p38 and synapsin 1 might indicate an interaction between the two proteins (Leube *et al.*, 1987). Synapsin 1 is thought to hold synaptic vesicles in a cytoskeletal meshwork (Navone *et al.*, 1984; Baines, 1987; Linstedt and Kelly, 1987). However, synapsin 1 is not found in adrenal medullary cells where p38 is present (Navone *et al.*, 1986).

Sudhof *et al.* (1987) have hypothesized that p38 might be a channel. Sudhof *et al.*

(1987) and Leube *et al.* (1987) point out that the structure of p38 is similar to that of the gap junction protein. Each 32-kDa gap junction monomer contains four potential membrane-spanning domains, and as a hexamer forms a channel (Unwin, 1986; Paul, 1986). The gap junction can pass molecules of about 1000 Da or less (Gilula, 1985). If p38 exists in the membrane as a trimer or a tetramer (Rehm *et al.*, 1986), then the polymer would bring 12–16 transmembrane regions together (Sudhof *et al.*, 1987). Thus, synaptophysin might be involved in packaging molecules into synaptic vesicles, such as Ca^{2+}, since p38 appears to bind Ca^{2+} (Buckley *et al.*, 1987). Another possible function might be that p38 participates in Ca^{2+}-dependent exocytosis (Buckley *et al.*, 1987). It might be a component of the transient "fusion pore," postulated to be an early step in the fusion of secretory granule and plasma membranes during exocytoisis (Breckenridge and Almers, 1987).

4. p65

A 65-kDa synaptic vesicle protein of the mammalian brain has been characterized (Matthew *et al.*, 1981b). It is likely an integral membrane protein, although that has not been unequivocally demonstrated. This protein is present in both neural and endocrine tissue, and in both small clear synaptic vesicles and large dense-core granules (Matthew *et al.*, 1981b; Lowe *et al.*, 1988). p65 protein was originally identified by two monoclonal antibodies (mAb30 and mAb48) which bind to antigenic sites on the cytoplasmic side of the vesicle membrane (Matthew *et al.*, 1981b). This same protein has probably been identified independently by other workers (Obata *et al.*, 1987). Antibodies to p65 have been very useful as immunocytochemical presynaptic markers for developmental studies (Chun and Shatz, 1983; Burry *et al.*, 1986; Sarthy and Bacon, 1985; Greif and Reichardt, 1982; Bixby and Reichardt, 1985).

The monoclonal antibodies which bind p65 were screened from a library of monoclonals made against synaptic junctional complexes (Matthew *et al.*, 1981b). Supernatants from two hybridoma cell lines, called serum 30 and 48, were found to contain antibodies against p65. The synaptic junctional complexes (Wang and Mahler, 1976) used for immunization were prepared from fractions containing synaptic membranes (Jones and Matus, 1974) by extraction with Triton X-100. The detergent extracts about 90% of the protein, presumably leaving proteins associated with insoluble cytoskeletal and extracellular matrix, including synaptic density and cleft material (Matthew *et al.*, 1981a). Whether the presence of p65 in this preparation is due just to contamination, or to a real association of this vesicle protein with an extracellular matrix or cytoskeletal component, is not clear.

Electron microscopic immunocytochemistry with the monoclonal antibodies against p65 strongly suggests that it is a synaptic vesicle protein (Matthew *et al.*, 1981b). This result might be expected from light microscopic immunocytochemistry, which showed staining of synapse-rich areas of the nervous system. For example, p65 antigenicity is found only in the synaptic layers of the frog retina (Matthew *et al.*, 1981b). At the electron microscopic level using the peroxidase–antiperoxidase method, Matthew *et al.* (1981b) showed reaction product on the outside of synaptic vesicles at cerebellar synapses. Mitochondrial and plasma membranes facing synaptic vesicles also show peroxidase staining; however, this was thought to be due to the diffusion of reaction product. Unlike the immunogold methods used for the localization of p38, the more sensitive peroxidase methods can be subject to this ambiguity. When the adrenal medulla was stained with anti-p65 antibodies, chromaffin granules also showed reaction product associated with their surface.

Biochemical evidence supports the immunocytochemical data that p65 is a secretory vesicle protein. The monoclonal antibodies to p65 are able to immunoprecipitate a variety of synaptic vesicle and secretory granules: (1) Membrane vesicles the size of synaptic vesicles were selectively bound to polyacrylamide beads by an antibody to p65 (Matthew et al., 1981b). A crude lysed synaptosome preparation was incubated with a monoclonal antibody against p65, and the resulting immune complexes removed with protein A-coated polyacrylamide beads. Electron microscopic inspection showed the beads to be covered with membrane vesicles the size of synaptic vesicles. Although mitochondria, resealed synaptosomes, and cellular debris were common in the starting material, there was a 30-fold enrichment of membrane profiles the size of synaptic vesicles. (2) Secretory granules labeled with [^{3}H]norepinephrine could be immunoprecipitated from a homogenate of PC12 cells. Using the same polyacrylamide bead procedure, Matthew et al. (1981b) were able to immunoprecipitate 95% of the vesicular [^{3}H]norepinephrine with about a 20-fold enrichment for protein (as measured by incorporated [^{14}C]leucine) over the starting material. (3) About 80% of purified chromaffin granules (labeled with [^{3}H]norepinephrine) could be precipitated by beads coated with mAb48 (Lowe et al., 1988). In addition, 90% of PC12 secretory granules containing human growth hormone could be immunoprecipitated by bead-bound mAb48. These granules were prepared from PC12 cells stably transfected with DNA encoding this hormone (Lowe et al., 1988). (4) Substance P contained in synaptic vesicles, isolated from rat brain, was immunoprecipitated by mAb48 (Floor and Leeman, 1985). After exposure of the vesicles to the monoclonal antibody, 60% of the vesicular substance P was bound by second antibody-coated polyacrylamide beads.

The two monoclonal antibodies (mAb48 and mAb30) directed against p65 identify a 65-kDa polypeptide on reducing SDS–PAGE. The antigenic sites on p65 are probably overlapping, since these antibodies compete with one another for binding. However, these sites are not identical; mAb48 binds with higher affinity than mAb30. Clearly, the ability of these antibodies to immunoprecipitate secretory vesicles indicates that antigenic sites are exposed on the cytoplasmic side of the vesicle bilayer (Matthew et al., 1981b).

The p65 protein is probably an integral membrane protein (Matthew et al., 1981a). It cannot be solubilized by extraction of membranes with high (1 M NaCl) or low salt (10 mM NaCl). It is solubilized by nondenaturing detergents, like Triton X-100 (Matthew et al., 1981b) or CHAPS (Bixby and Reichardt, 1985). However, a more rigorous test of whether p65 is an integral membrane protein has not been made. For example, there has been no attempt to reconstitute the purified protein into the bilayer of a liposome (Brunner et al., 1978).

If p65 is not an integral membrane protein, it could be a peripheral membrane protein bound tightly to an integral membrane protein. However, the latter is not likely. Immunoaffinity purification of p65 solubilized in nondenaturing detergent yields essentially one major polypeptide (plus a degradation product) and a few minor proteins on SDS-PAGE (Bixby and Reichardt, 1985). Unfortunately, additional clues suggesting integral membrane protein status are lacking. For example, it has not been reported whether p65 is glycosylated; thus, there is no evidence to suggest that the p65 polypeptide might traverse the vesicle bilayer and be exposed in the lumen of the vesicle.

The p65 protein has been found in every part of the central and peripheral nervous system that has been assayed (Matthew et al., 1981a), including the neuromuscular junction (Bixby and Reichardt, 1985). It also appears to be present in endocrine tissue (Matthew et al., 1981b), but is very low or absent in the exocrine pancreas. It is also very low or absent in cells without a regulated pathway, such as liver, erythrocytes, or the extrajunctional regions of the diaphragm. Thus, the tissue distribution of p65 is similar to p38. Also like p38, p65 is found in all classes of vertebrates that have been checked. A 65-kDa protein is detected in mammalian, chicken, frog, and shark brain extracts (Matthew et al., 1981b). Although the mAb48 antigenic site does not appear to be present on electric organ synaptic vesicles, the p65 protein may be present. An antiserum made to purified electric organ vesicles (Carlson and Kelly, 1980) blocks the binding of mAb48 to mammalian brain synaptic vesicles (Matthew, 1981; Carlson, unpublished observations).

5. SV2

SV2 is a transmembrane glycoprotein present in the secretory granules of vertebrate neurons and endocrine cells (Buckley and Kelly, 1985). It has a molecular weight of about 100,000, but this can vary among cell types from 75,000 to 110,000, presumably due to differences in glycosylation. It was identified by a monoclonal antibody (mAb10H3) generated against purified cholinergic synaptic vesicles. The vesicles were purified from the electric organ of the marine elasmobranch electric ray (Carlson et al., 1978). A synaptic vesicle protein having a similar molecular weight (86,000) has been identified in the electric organ of Torpedo marmorata with antiserum made to purified synaptic vesicles (Walker et al., 1986).

Both biochemical and immunocytochemical evidence support the hypothesis that the SV2 protein is a synaptic vesicle protein. (1) The SV2 protein copurifies with electric organ synaptic vesicles (Buckley and Kelly, 1985). In the last step of the vesicle purification, the SV2 antigenicity quantitatively elutes with synaptic vesicles during permeation chromatography on controlled pore glass. (2) Purified electric organ synaptic vesicles are immunoprecipitated by the anti-SV2 antibody (Buckley and Kelly, 1985). Intact synaptic vesicles were labeled with ^{125}I and about 30% were bound by anti-SV2-coated beads. (3) PC12 secretory granules and chromaffin granules have also been immunoprecipitated with anti-SV2-coated beads (Lowe et al., 1988). These granules were labeled by uptake of [^{3}H]norepinephrine; about 65% of this vesicular ^{3}H label could be immunoprecipitated from PC12 cell postnuclear supernatants, and about 36% from a preparation of purified chromaffin granules. In PC12 cells transfected with DNA coding for the human growth hormone, 99% of the granules containing the hormone could be immunoisolated. It is not clear whether these differences in the efficiencies of immunoprecipitation are just artifactual or reflect actual differences in granule composition. (4) SV2 antigenicity was detected on synaptic vesicles by electron microscopic immunocytochemistry of electric organ nerve terminals. Anti-SV2 antibodies were shown to be bound by synaptic vesicles in partially disrupted nerve terminals, using second antibody conjugated to horseradish peroxidase (Buckley and Kelly, 1985). In addition, about 80% of the SV2 antigenicity was found restricted to synaptic vesicles when sections of Lowicryl-embedded electric organ were stained (Carlson *et al.*, 1986; also see Figure 3). The remaining 20% was located on the nerve terminal plasma membrane. Here, antibody binding was visualized with second antibody-coated gold particles.

The molecular weight of the SV2 polypeptide is about 100,000 on SDS–PAGE (Buckley and Kelly, 1985), but it can vary from tissue to tissue (and species to species). For example, immunoblots show a molecular weight between 75,000 and 85,000 in brain extracts; with extracts of endocrine cells the molecular weight is 105,000–110,000. In purified electric organ synaptic vesicles from *Discopyge ommata*, the molecular weight is about 97,000. Presumably this variation is due to differences in glycosylation. The protein appears as a relatively wide band on SDS–PAGE.

The SV2 protein contains N-linked carbohydrate (Buckley and Kelly, 1985). When PC12 cells are labeled with [^{35}S]methionine, immunoprecipitated SV2 protein has a molecular weight of 110,000. If the cells are grown in tunicamycin, an inhibitor of N-glycosylation, the molecular weight of the SV2 protein decreases to about 62,000. Furthermore, short-term labeling of PC12 cells without tunicamycin also yields a protein of about 64,000, presumably a precursor with little oligosaccharide attached.

The SV2 protein is most likely a transmembrane protein (Buckley and Kelly, 1985). The immunoprecipitation studies clearly indicate that the SV2 antigenic site is on the cytoplasmic side of the vesicle membrane. One would expect the glycosylated region of the SV2 polypeptide to be exposed on the luminal side of the vesicle membrane. Thus, it is anticipated that the SV2 protein should span the bilayer at least once.

The SV2 protein may be a universal marker for vertebrate nerve terminals (Buckley and Kelly, 1985). In nervous tissue, it is found in synaptic areas and is not restricted to any one class of nerve terminals. This distribution of SV2 is very similar to those of p38 and p65. By immunocytochemistry at the light microscopic level, the antigen was found associated with all the synaptic areas of the rat and frog spinal cord, rat cerebellum, rat hippocampus, and frog retina. By immunocytochemical or biochemical analysis, the SV2 antigenic site has been found in all vertebrates that have been investigated: mammals, birds, amphibians, and the cartilaginous fish (Buckley and Kelly, 1985).

Although the SV2 protein is found only in cells with a regulated secretory pathway, it is not a universal marker for this pathway (Buckley and Kelly, 1985). Light microscopic immunocytochemistry shows the SV2 antigen to be present in the adrenal medulla (not the cortex), the intermediate and anterior lobes of the pituitary, and islet cells of the pancreas. The antigen is present in neuroendocrine tumor cell lines: AtT-20, PC12, GH3, and HIT cells. However, the antigen is not seen in cells lacking regulated secretion, e.g., liver cells. In addition, this antigen is absent from exocrine glands: acinar cells of the pancreas, and the submaxillary salivary gland. Thus, the tissue distribution of SV2 is very similar to those of p65 and p38. These proteins are present only in a subset of cells possessing the regulated secretory pathway: neuroendocrine, but not exocrine cells.

Both p65 and SV2 are present in clathrin-coated vesicles from brain (Pfeffer and Kelly, 1985). About 25% of purified brain clathrin-coated vesicles contain SV2 protein, 13% contain p65, and about 3% contain both proteins. This has been demonstrated by quantitative immunoprecipitation of purified coated vesicles. These subfractions of coated vesicles contain a 38-kDa protein that is quite likely the synaptic vesicle p38 protein (see discussion of p38). This is consistent with the hypothesis that clathrin-coated vesicles are involved in the recycling of synaptic vesicle proteins from the nerve terminal membrane after exocytosis (Heuser and Reese, 1973, 1981; Heuser et al., 1979), and the generation of secretory granules at the trans-Golgi network (Griffiths and Simons, 1986). Although it was believed that clathrin-coated vesicles are also involved in shuttling membrane between the Golgi stacks or from the Golgi to the cell surface, these hypotheses have been called into question (Orci et al., 1986; Griffiths and Simons, 1986; Pfeffer and Rothman, 1987).

6. SYNAPTIC VESICLE PROTEOGLYCAN

A proteoglycan of between 100,000 and 200,000 daltons has been found in elasmobranch electric organ synaptic vesicles (Carlson and Kelly, 1983; Stadler and Dowe, 1982). It is present in the lumen of the vesicle and is probably an integral membrane protein. The proteoglycan is thought to be a major component of these cholinergic vesicles, making up approximately 25% of the total protein (Carlson and Kelly, 1983). It may be in the heparin/heparan sulfate family of proteoglycans (Carlson and Kelly, 1983; Stadler and Dowe, 1982). An antiserum directed against the molecule has detected its presence at the rat neuromuscular junction (Volknandt and Zimmerman, 1986).

The synaptic vesicle proteoglycan (SVPG) was originally identified as a specific synaptic vesicle antigen which appeared to behave anomalously on SDS–PAGE (Carlson and Kelly, 1983). This vesicle component smeared through the stacking gel and remained at the top of the running gel in the standard Laemmli system. Several monoclonal antibodies as well as antisera made to synaptic vesicles identified this vesicle-specific antigen (Carlson and Kelly, 1983; Kelly et al., 1983; Stadler and Dowe, 1982; Volknandt and Zimmerman, 1986): A rabbit antisynaptic vesicle antiserum (RASVA), after a simple adsorption, identified antigens on both sides of the synaptic vesicle membrane that are specific to synaptic vesicles (Carlson and Kelly, 1980). This specificity was demonstrated by the facts that the RASVA antigens copurified with electric organ synaptic vesicles, and that these antigens segregated with synaptic vesicles when electric organ was fractionated by a method not utilized in the

Figure 1. Blocking of Tor 70 binding to synaptic vesicles by anti-synaptic vesicle antibodies. The percent inhibition of Tor 70 binding to vesicles is plotted against the blocking antibody. Monoclonal antibodies, Tor 70 and anti-SV1, as well as rabbit anti-synaptic vesicle antiserum (RASVA) show about 100% inhibition of [^{125}I]-Tor 70 antibody binding. Monoclonal antibodies directed against synaptic vesicle antigens SV2 and SV4 and preimmune rabbit serum (Prelm.) show es-

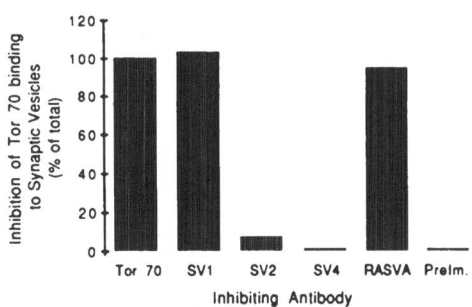

sentially no inhibition. The conditions of this solid-phase antibody binding assay are those described by Carlson and Kelly (1980, 1983). Purified synaptic vesicles are nonspecifically adsorbed to plastic 96-well microtiter plates. When vesicles bind to the plastic, they appear to lyse and expose internal antigens. The adsorbed vesicles are then exposed to a concentration of unlabeled antibody which is known to saturate the vesicular antigenic sites present in the microtiter well. The [^{125}I]-Tor 70 antibody is then added to the microtiter well. The data for RASVA are from Carlson and Kelly (1983). The data for SV2, SV1, and SV4 are unpublished observations (Carlson).

vesicle purification. At least 75% of the RASVA antigenicity appears only on synaptic vesicles and is not associated with other electric organ membranes. A high percentage of this antigenicity showed anomalous behavior on SDS–PAGE (Carlson and Kelly, 1983; Kelly *et al.*, 1983). That this anomalously behaving material might be a single component was suggested by a monoclonal antibody, Tor 70, which gave a very similar staining pattern to RASVA on Western blots (Carlson and Kelly, 1983; Kelly *et al.*, 1983). The Tor 70 antigenicity copurified with synaptic vesicles on the last step of the vesicle purification and was inhibited from binding its vesicle antigen by RASVA (Carlson and Kelly, 1983, and Figure 1). The Tor 70 monoclonal antibody was screened from a library of monoclonals made to *Torpedo californica* electric organ synaptosomes (Buckley *et al.*, 1983; Kushner and Reichardt, 1981). The Tor 70 antigenic site was reidentified and renamed SV1 when monoclonal antibodies were made to synaptic vesicles (Caroni *et al.*, 1985). The anti-SV1 monoclonal antibody stained Western blots of synaptic vesicles identically to Tor 70 and completely inhibited Tor 70 from binding to synaptic vesicles (Figure 1).

In support of the biochemical data, electron microscopic immunocytochemical evidence finds the majority of the SV1 antigen also restricted to synaptic vesicles (Figure 2). Lowicryl sections of electric organ were stained with anti-SV1 antibody and 10-nm gold particles coated with goat anti-mouse IgG. Essentially all of the SV1 antigenicity is found associated with the nerve terminal (Figure 2). Within the nerve terminal, about 80% of the gold particles are found on synaptic vesicles and 20% on the nerve terminal plasma membrane (Figure 3). Further, these data complement the electron microscopic immunocytochemical staining of intact nerve terminals with Tor 70 (Buckley *et al.*, 1983). Using HRP-coupled second antibody, Buckley *et al.* (1983) showed that Tor 70 antigenicity was restricted to synaptic junctional regions of the nerve terminal surface. Presumably, this antigenicity associated with the nerve terminal plasma membrane was due to inserted vesicle membrane awaiting endocytosis.

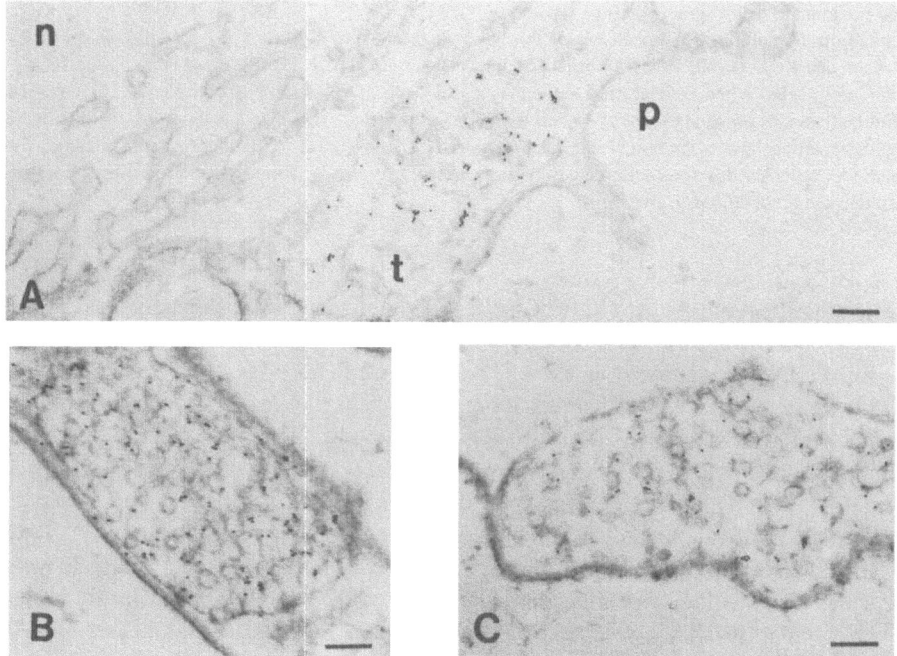

Figure 2. Electron microscopic localization of the SV1 antigen in electric organ. The monoclonal antibody was bound to ultrathin sections of electric organ embedded with Lowicryl K4M. Binding was detected using goat anti-mouse IgG coupled to 10-nm gold particles. The methods were those described in Carlson *et al.* (1986). (A) Shown is the nerve terminal (t) attached to a postsynaptic electrocyte (p), and the uninnervated face (n) of an adjacent ventral electrocyte with its canalicular network. The gold particles are restricted to the nerve terminal, primarily on synaptic vesicles, and are not found in the electrocyte. (B, C) Two other examples of nerve terminals stained for the SV1 monoclonal antibody. The gold particles are primarily associated with synaptic vesicles and a small amount is associated with the nerve terminal membrane. Bars = 200 nm. S. S. Carlson (unpublished observations).

Figure 3. The distribution of gold particles bound to the SV4, SV1, and SV2 antigen associated with synaptic vesicles or nerve terminal plasma membrane in electron micrographs. Each monoclonal antibody was bound to ultrathin sections of electric organ embedded with Lowicryl K4M. Bound antibody was detected using goat anti-mouse IgG coupled to 10-nm gold particles (as shown in Figure 2 for SV1). In 43 micrographs stained with a monoclonal antibody directed against SV4, the number of gold particles associated with either synaptic vesicles or nerve terminal plasma membrane was determined for 55 nerve terminals. For SV1, 35 micrographs

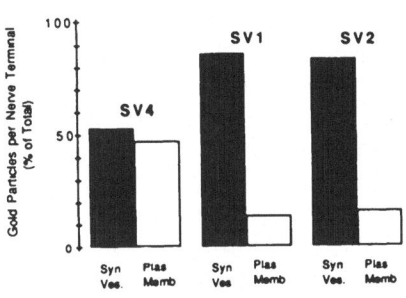

were used and gold particles bound to 42 nerve terminals were counted; for SV2, 24 micrographs were used and gold particles bound to 29 were counted. The data are presented in a bar graph as the average percentage of gold particles for each nerve terminal associated with either synaptic vesicles or nerve terminal plasma membrane. The data for SV4 and SV2 are from Carlson *et al.* (1986); the data for SV1 are unpublished observations.

The SV1 (Tor 70) antigen appears to be a proteoglycan (Carlson and Kelly, 1983). Several lines of evidence support this conclusion. (1) The chromatographic behavior of the SV1 antigen on anion exchangers suggests that it has the high negative charge density of sulfated proteoglycans. The SV1 antigen under denaturing conditions (8 M urea) in nondenaturing detergent binds to DEAE-Sephacel at pH 5 and requires ≈ 0.46 M salt to be eluted (Carlson *et al.*, 1986). These elution conditions are characteristic of proteoglycans; proteins without sulfated glycosaminoglycans elute at NaCl concentrations below 0.3 M (Yanagishita *et al.*, 1987). (2) A proteoglycan copurifies with synaptic vesicles on the last step of the vesicle purification. This is indicated by the release of a glycosaminoglycan upon pronase digestion of lysed vesicles. The amount of glycosaminoglycan released quantitatively correlates with the amount of vesicular membrane in the digested sample (Carlson and Kelly, 1983). (3) Glycosaminoglycan containing protein and Tor 70 antigenicity elute together when purified vesicles are fractionated on Sepharose 6B under denaturing and reducing conditions (Carlson and Kelly, 1983). Glycosaminoglycan is enriched over Tor 70 antigenicity in fractions of higher molecular weight, but Tor 70 antigenicity is found in all fractions containing protease-releasable glycosaminoglycan. This might reflect heterogeneity in glycosylation of the proteoglycan protein core (see later discussion on the nature of the glycosaminoglycan SV1 antigen).

Metabolic labeling studies of electric organ synaptic vesicle components with $^{35}SO_4$ also suggest that a proteoglycan is a synaptic vesicle component (Stadler and Dowe, 1982). The high SO_4 content of the glycosaminoglycan side chains makes metabolic labeling with $^{35}SO_4$ a relatively selective way of radioactively tagging proteoglycans. Injection of the electromotor nucleus with isotope leads to incorporation and axonal transport of labeled nerve terminal components to electric organ. Synaptic vesicles purified from electric organs labeled in such a manner show incorporated $^{35}SO_4$ copurifying with vesicles in the last step of the purification. By gel filtration on Ultrogel ACA 34 under denaturing conditions, the majority of the $^{35}SO_4$ appears to be incorporated into a high-molecular-weight (about 200,000–350,000) protein. When pronase-digested material was subjected to thin-layer electrophoresis, the incorporated isotope migrated with a heparan sulfate standard.

That this $^{35}SO_4$-labeled proteoglycan is a specific vesicle component was demonstrated with an antiserum made to purified electric organ synaptic vesicles (Walker *et al.*, 1983). This antiserum, after adsorption with liver membranes and an electric organ soluble fraction, appeared specific for synaptic vesicles: (1) The antigenicity copurifies with synaptic vesicles in the last step of the purification (Walker *et al.*, 1983). (2) Immunocytochemistry at the electron microscopic level with the antiserum shows HRP reaction product only on neuronal membrane and synaptic vesicles (Borroni *et al.*, 1985). Eighty-six percent of the $^{35}SO_4$-labeled material in purified detergent-lysed vesicles could be precipitated by this antiserum. None of the incorporated label was precipitated by preimmune serum (Stadler and Dowe, 1982).

Variation in the size and number of the glycosaminoglycan side chains found on each polypeptide is thought to be partially responsible for the molecular weight heterogeneity of proteoglycans (Hassell *et al.*, 1986). The heterogeneity of proteoglycans somewhat complicates estimating the molecular weights for these molecules, and may be responsible for the discrepancies in the size estimates of the SVPG by different workers. Using glycosaminoglycan content as a marker, Carlson and Kelly (1983) estimate the molecular weight of the SVPG to be about 100,000–200,000 by gel filtration in SDS. The $^{35}SO_4$-labeled material in purified vesicles was estimated by similar procedures to have an approximate molecular weight of 200,000–350,000 (Stadler and Dowe, 1982). Staining Western blots of purified synaptic vesicles with an anti-synaptic vesicle antiserum, Walker *et al.* (1983) obtained an estimate of 120,000–300,000. Western blots with the anti-SV1 monoclonal antibody give an estimate of between 100,000 and 200,000 (Figure 4).

It is useful to make a technical comment. For the Western blots shown in Figure 4, we used SDS-polyacrylamide gels with a 2.4% stacking gel and a 2.4–15% polyacrylamide gradient in the running gel. Even large proteoglycans (1.2×10^6 daltons) can enter these gels (Carlson and Wight, 1987). The anomalies seen previously with Tor 70 in higher percentage SDS gels (Carlson and Kelly, 1983) are not seen here.

The SVPG is exposed on the inside lumen of the vesicle attached to the vesicle membrane (Buckley and Kelly, 1985). As mentioned previously, intact purified synaptic vesicles can be bound to polyacrylamide beads coated with the anti-SV2 mAb. However, only very low amounts of intact synaptic vesicles are immunoprecipitated by the anti-SV1 mAb under the same conditions. If the purified vesicles are lysed by sonication, the anti-SV1-coated beads bind about as much vesicle membrane as anti-SV2 beads with intact vesicles (or sonicated vesicles). Tor 70 also binds sonicated, but not intact vesicles (Carlson and Kelly, 1983). For all these immunoprecipitations, vesicles were radioactively labeled intact on the outside surface with a membrane-impermeable reagent. Since the anti-SV1 (or Tor 70) was able to precipitate radioactive label after the vesicles were ruptured, the SVPG must be attached to the vesicle membrane (Carlson and Kelly, 1983).

The SVPG is probably a membrane protein. It is firmly attached to the synaptic vesicle membrane; conditions that remove peripheral membrane proteins (Elliot *et al.*, 1980) fail to remove the proteoglycan. That is, when vesicles are lysed by several cycles of freezing and thawing at pH 11, the SVPG is still pelleted with the vesicle membranes by centrifugation in an airfuge at 120,000*g* for 10 min (S. S. Carlson,

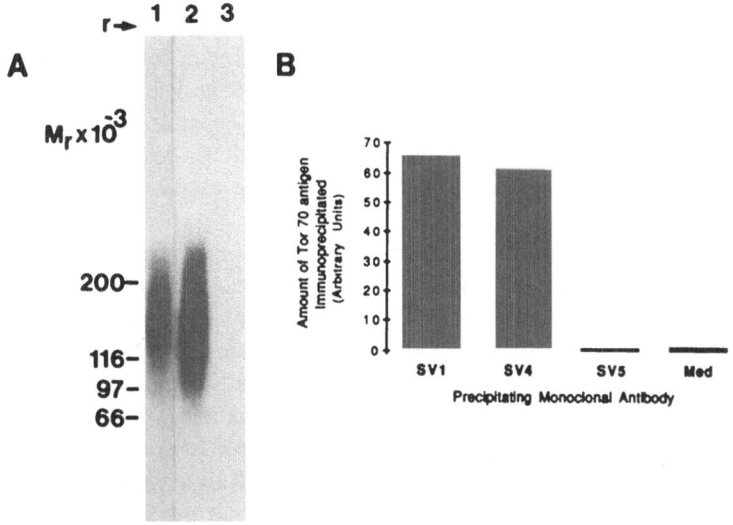

Figure 4. The SV1 and SV4 antigenic sites are associated with the same synaptic vesicle component. (A) The monoclonal antibodies to SV1 and SV4 appear to identify the same antigen on SDS–PAGE. Shown is an immunoblot autoradiogram of purified electric organ synaptic vesicles electrophoresed on SDS–PAGE and stained with monoclonal antibodies to SV4 (lane 1) and SV1 (lane 2). r indicates the beginning of the running gel. Fifteen micrograms (lane 1) and seven micrograms (lane 2) of vesicle protein were applied to the 2.4–8% polyacrylamide gel. The stacking gel was 2.4%. Lane 3 contains the proteins used for molecular weight standard exposed to the anti-SV1 antibody; nothing in this lane is stained. The standards used are myosin (200,000), β-galactosidase (116,000), phosphorylase B (97,000), and BSA (66,000). The monoclonal antibody was detected with ^{125}I-labeled goat anti-mouse IgG. (B) The monoclonal antibodies to SV1 and SV4 are equally effective in immunoprecipitating the Tor 70 antigen from detergent-solubilized synaptic vesicles. The amount of Tor 70 antigenicity immunoprecipitated by polyacrylamide beads coated with monoclonal antibodies to SV1, SV4, SV5, or monoclonal culture supernatant is shown as a bar graph. Purified electric organ synaptic vesicles were solubilized in SDS with β-mercaptoethanol, diluted under nondenaturing detergent conditions, and the vesicle antigens immunoprecipitated by polyacrylamide beads coated with the appropriate monoclonal antibody. The antigens were released from the beads by boiling in SDS under reducing conditions, and the amount of Tor 70 antigenicity was quantified by dot-blot with ^{125}I-labeled Tor 70 monoclonal antibody. These methods are described in Carlson and Kelly (1983) and Carlson *et al.* (1986). The data presented are unpublished observations.

unpublished observations). Further, the SVPG appears to be reconstituted into liposomes (Carlson *et al.*, 1986). When liposomes are formed in the presence of isolated SVPG and [³H]cholesterol, the proteoglycan and the cholesterol label coelute from DEAE-Sephacel at the salt concentration characteristic of the SVPG (about 0.47 M NaCl).

The SVPG may be a heparin/heparan sulfate proteoglycan, but some uncertainty remains. Carlson and Kelly (1983) determined that digestion of purified SVPG with pronase yields a glycosaminoglycan which migrates with heparin and keratan sulfate standards on cellulose acetate electrophoresis. Acid hydrolysis and amino acid analysis of the isolated proteoglycan gives glucosamine and no galactosamine, consistent with the heparin/heparan sulfate or keratan sulfate family (Carlson and Kelly, 1983). Stadler and Dowe (1982) found that pronase digestion of $^{35}SO_4$-labeled synaptic vesicles released a glycosaminoglycan which coelectrophoresed with a heparan sulfate standard, but not with heparin. Analysis of the entire synaptic vesicle showed the presence of glucosamine and uronic acid, consistent with a heparin/heparan sulfate proteoglycan, but not keratan sulfate (Stadler and Dowe, 1982). Clearly, the sensitivity of the proteoglycan to glycosaminoglycan lyases would be useful additional data to resolve these discrepancies.

The SVPG may comprise a significant proportion of the synaptic vesicle proteins. About 25% of the vesicle protein appears to be proteoglycan (Carlson and Kelly, 1983). From the glucosamine content of the isolated proteoglycan (0.75 nmole/μg protein), it can be estimated that there is 0.36 μg glycosaminoglycan/μg protein, or the SVPG is about 28% carbohydrate. If the molecular weight of the SVPG is 100,000–200,000, then the core protein is 70,000–140,000. Since the particle weight of the synaptic vesicle is 176×10^6 and the vesicle is about 3% protein (Carlson et al., 1978), there should be about 10–20 copies per vesicle (Carlson and Kelly, 1983).

A degradation product of the SVPG was mistakenly identified as a specific synaptic vesicle component. Given the name *vesiculin*, this component was incorrectly identified as a protein and then reidentified as a glycosaminoglycan (Stadler and Whittaker, 1978). These free glycosaminoglycan chains probably do not occur in synaptic vesicles, but arise by inadvertent proteolysis of the SVPG (Walker et al., 1983).

The SVPG appears to be present in synaptic vesicles of the rat neuromuscular junction (Volknandt and Zimmerman, 1986). An antiserum made to the SVPG (Stadler and Kiene, 1987) stains the rat neuromuscular junction by immunocytochemistry at the light microscopic level (Volknandt and Zimmerman, 1986). In addition, subcellular fractionation of rat diaphragm demonstrates that the antigenicity identified by this antiserum cofractionates with synaptic vesicle components (Volknandt and Zimmerman, 1986). Chromatography of impure rat diaphragm vesicles on Sephacryl S-1000 shows that SVPG antigenicity, vesicular ATP, vesicular acetylcholine, and the synaptic vesicle protein p65 (see earlier discussion) all elute together in the volume of the column.

In addition to SV1, another antigenic site, SV4, has been identified on the SVPG. It was found when monoclonal antibodies were raised to synaptic vesicles (Caroni et al., 1985). Unlike the anti-SV1 mAb, the anti-SV4 mAb does not prevent Tor 70 from binding to synaptic vesicles (Figure 1); thus, the SV4 site is distinct from SV1. However, by Western blot analysis both anti-SV4 and anti-SV1 monoclonal antibodies appear to bind the same antigen (Figure 4A). This is confirmed by coimmunoprecipitation experiments; the anti-SV4 mAb is about as effective in immunoprecipitating Tor 70 antigenicity as the anti-SV1 mAb (Figure 4B). The ratio of SV1/SV4 antigenicities on SVPG is about 3 (Caroni et al., 1985).

7. TERMINAL ANCHORAGE PROTEIN ONE (TAP-1)

TAP-1 is an integral membrane proteoglycan found on the nerve terminal surface of electric organ bound to the extracellular matrix (ECM). Because of these properties, this proteoglycan is hypothesized to be important in anchoring the nerve terminal to the

ECM (Carlson *et al.*, 1986; Carlson and Wight, 1987). Some members of this group of molecules, nerve terminal anchorage proteins, are quite likely important in recognizing extracellular matrix cues during nerve regeneration at the neuromuscular junction (Kelly *et al.*, 1987; Sanes and Chiu, 1983).

The identification of TAP-1 resulted from the work done on the SVPG antigens. Curiously, it was found that an abundance of the SV4 antigen is present in an electric organ ECM fraction, while other synaptic vesicle proteins are present only in low amounts. Originally it was hypothesized that the SVPG was being selectively deposited in the synaptic cleft ECM (Caroni *et al.*, 1985). However, when the SV4 antigen in the ECM fraction was characterized, it was found to be a much larger and more negatively charged molecule than the SVPG (Carlson *et al.*, 1986). By purification of this ECM-SV4 antigen, we have determined that it is a completely different proteoglycan which we now call TAP-1 (Carlson and Wight, 1987; see also Table 1). TAP-1 is not present in purified synaptic vesicles (Carlson *et al.*, 1986).

Three properties of this ECM proteoglycan indicate that it is a good candidate for a nerve terminal anchorage protein. (1) It is present on the nerve terminal plasma membrane. (2) It is tightly associated with an electric organ ECM fraction. (3) It has the properties of an integral membrane protein. Thus, TAP-1 could act to link the nerve terminal surface with the ECM (Carlson *et al.*, 1986).

By electron microscopic immunocytochemistry, the SV4 antigen is exclusively localized to the nerve terminal; it is not found on postsynaptic structures (Carlson *et al.*, 1986). Within the nerve terminal, when the SV4 antigenicity is localized with 10-nm gold particles, about 50% of the particles are on the nerve terminal plasma membrane and 50% on synaptic vesicles (Figure 3). This is consistent with the biochemically demonstrated presence of SV4 on a membrane protein (TAP-1) and a vesicle protein (SVPG). A marker which is biochemically *only* present on the SVPG (Kelly *et al.*, 1987), SV1, is found primarily (86%) in synaptic vesicles and only a small amount (14%) is found in the plasma membrane. A similar distribution is found for another synaptic vesicle protein, SV2 (Figure 3).

That the SV4 antigen on the nerve terminal surface is primarily associated with TAP-1 was demonstrated by Caroni *et al.* (1985). A preparation of postsynaptic membrane sheets with intact nerve terminals attached can be prepared from electric organ (Heuser and Salpeter, 1979). By incubating this "membrane sheet" preparation with the Tor 70 and anti-SV4 mAbs, these antigens can be visualized in the electron microscope on the surface of the nerve terminals with HRP–second antibody (Buckley *et al.*, 1983; Kelly *et al.*, 1987). When these mAb-treated "membrane sheets" are used to prepare an ECM fraction (rather than being used for electron microscopy), 75% of the anti-SV4 mAb bound to nerve terminals was found in the ECM fraction. In contrast, only 11% of nerve terminal-bound anti-SV1 mAb was carried into the ECM fraction, and only 3% for another synaptic vesicle component. Thus, most of the SV4 antigen on the nerve terminal surface is the ECM-bound TAP-1 proteoglycan.

The TAP-1 antigen is very tightly associated with an electric organ ECM fraction. This fraction is a pellet resulting from extractions with low-salt, high-salt, and detergent-containing solutions. It is enriched in collagen (Godfrey *et al.*, 1984) and is the preparation from which the acetylcholine receptor clustering factor, agrin, is obtained (Godfrey *et al.*, 1984; Reist *et al.*, 1987; Nitkin *et al.*, 1987). Denaturing conditions

(SDS or guanidine-HCl/CHAPS) are required to solubilize TAP-1 from this fraction (Carlson *et al.*, 1986).

TAP-1 behaves as an integral membrane protein (Carlson *et al.*, 1986). The TAP-1 antigen, isolated under denaturing conditions by chromatography on DEAE-Sephacel, can be reconstituted into liposomes. After liposomes are formed in the presence of TAP-1, TAP-1 sediments on sucrose gradients to the density of liposomes, not to the density of detergent-solubilized TAP-1. Further, TAP-1 is not simply trapped inside liposomes, since most of the proteoglycan is exposed on the liposome surface. This is indicated by the sensitivity of the TAP-1 to protease digestion. Because TAP-1 acts this way, it most likely contains a hydrophobic domain which allows it to intercalate into a phospholipid bilayer.

TAP-1 has been completely purified and found to be a large chondroitin sulfate proteoglycan (Carlson and Wight, 1987). Gel filtration of the intact proteoglycan gives a molecular weight estimate of 10^6; the free glycosaminoglycan side chains chromatograph on Sepharose 6B with an average molecular weight of about 42,000. Visualization of purified TAP-1 in the electron microscope reveals a "bottlebrush" structure expected for a proteoglycan. The molecule has an average total length of 345 $\pm$ 17 nm with 20 $\pm$ 2 side projections of 113 $\pm$ 5 nm. From these dimensions, we predict that TAP-1 should have a molecular weight of about 1.2×10^6 and the glycosaminoglycan side chains a weight of approximately 50,000. Thus, the electron microscopic and the biochemical estimates are in relatively good agreement.

A comparison of the size of TAP-1 and the size of the synaptic cleft suggests that this proteoglycan could make a significant structural contribution to the synaptic basal lamina (Figure 5). Since the synaptic cleft is only about 60 nm wide, the proteoglycan could easily touch both pre- and postsynaptic cells. The actual space that TAP-1 might occupy could be reduced considerably if it were complexed with other proteins. However, the large hydrodynamic volumes characteristic of proteoglycans could be important in filling the synaptic cleft and holding the nerve terminal a fixed distance from the postsynaptic cell. The large proteoglycans of cartilage are thought to act as cushions and reversibly resist compressive force (Hascall and Hascall, 1981). Proteoglycans might do this and at the same time provide an aqueous environment for the relatively unobstructed diffusion of neurotransmitters (Carlson and Wight, 1987).

Much remains to be elucidated about the nature of SV4 and SV1 antigenic sites. Both sites are probably carbohydrate. They are insensitive to pronase digestion. The SV4 antigen may be a glycosaminoglycan (S. S. Carlson, unpublished observations). Alkaline borohydride cleavage of TAP-1 yields SV4 antigenic fragments with an average molecular weight of approximately 45,000. This alkali treatment is known to cleave glycosaminoglycan chains from their protein core (Heinegard and Sommarin, 1987). Pronase digestion of TAP-1 gives similar size fragments. Like a glycosaminoglycan, this antigenic fragment binds well to DEAE-nitrocellulose, but very poorly to regular nitrocellulose.

The SV4 antigens can be detected in the neuronal cell bodies of the electromotor nucleus as well as at their nerve terminals in the electric organ (Caroni *et al.*, 1985). Ligature of the nerve connecting the nucleus to the electric organ indicates that the SV4 antigenicity is axonally transported in both the anterograde and retrograde direction.

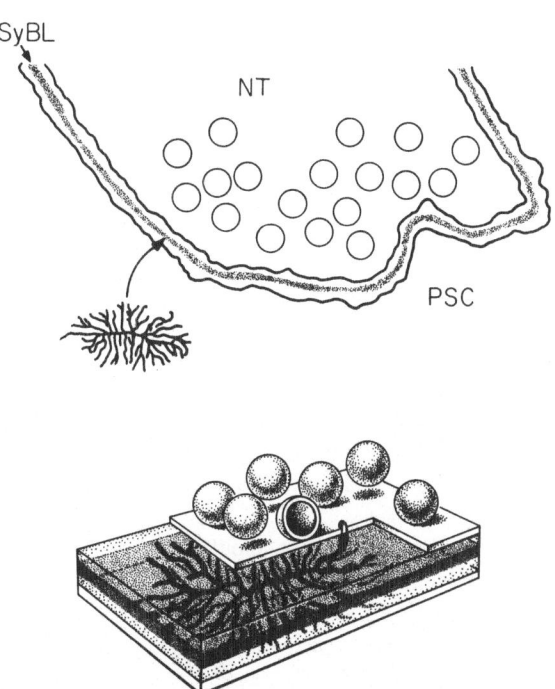

Figure 5. (Upper) Comparison of the size of TAP-1 and the synaptic cleft. The two-dimensional image of a TAP-1 molecule derived from electron microscopic visualization of the purified TAP-1 molecule (Carlson and Wight, 1987) is shown next to a cross section of the electric organ synapse; both the nerve terminal and TAP-1 are drawn to scale. TAP-1 is large enough to span the synaptic cleft. Thus, it could make a major structural contribution to the nerve terminal basal lamina. SyBL, synaptic basal lamina; NT, nerve terminal; PSC, postsynaptic cell. (Lower) Hypothetical placement of TAP-1 in the synaptic cleft. A small three-dimensional cutaway section of the synaptic cleft containing TAP-1 is shown. TAP-1 is represented as a transmembrane protein in the figure. However, it has only been shown to have the properties of an integral membrane protein (Carlson *et al.*, 1986); it has not yet been determined whether it spans the nerve terminal membrane bilayer. Reprinted with permission from Carlson and Wight (1987).

The proteins which carry the SV4 antigenic site can be metabolically labeled and axonally transported to the electric organ by injection of the electromotor nucleus with [^{35}S]methionine. These data are consistent with TAP-1 and the SVPG being products of the neuron. Presumably, the SVPG is axonally transported in synaptic vesicles and TAP-1 is transported in another vesicular compartment.

The SV1 antigenicity, like SV4, is present in the electromotor nucleus (Caroni *et al.*, 1985). Unlike the SV4 antigen, however, ligature of the nerve innervating the electric organ indicates that this exclusive SVPG antigen is only retrogradely, not anterogradely, transported. Such one-way transport suggests that the SV1 antigen is being produced at the nerve terminal in electric organ. How could this occur? One possibility might be that since the SV1 antigenic site is probably carbohydrate, it could be produced by cleavage of a sugar side chain.

Clearly, TAP-1 and the SVPG are very different proteoglycans (Table 1). Although both molecules have the SV4 antigenic site, they contain different glycosaminoglycan side chains. Further, TAP-1 is about an order of magnitude larger than the SVPG in molecular weight. The SVPG contains both the SV4 and the SV1 antigenic sites; TAP-1 contains only the SV4 site. It has been hypothesized that the SVPG might be a breakdown product of TAP-1 (Kelly *et al.*, 1987; Carlson *et al.*, 1986). Since the SV1 site appears to be generated at the nerve terminal (Caroni *et al.*, 1985), it

Table 1. Comparison of Synaptic Vesicle Proteoglycan with Terminal Anchorage Protein One (TAP-1)

Property	Synaptic vesicle proteoglycan	TAP-1
Molecular weight	100,000–200,000	1.2×10^6
Glycosaminoglycan	Heparin/heparan sulfate	Chondroitin sulfate
SV4 antigen	Present	Present
SV1 antigen	Present	Absent
Associates with liposomes	Yes	Yes

might result from the degradation of the larger proteoglycan. However, the finding that the major glycosaminoglycans of TAP-1 and the SVPG are different makes the hypothesis that the two molecules share a precursor–product relationship less likely (Carlson and Wight, 1987).

An intriguing question about TAP-1 is whether it is a glue invented by the electric organ, or whether it is used by other synapses as well. In addition, how is TAP-1 attached to the ECM? What proteins bind it? Is TAP-1 bound by synapse-specific proteins? The importance of synaptic glues holding the nerve terminal to the basal lamina has been shown by experiments on nerve regeneration at the neuromuscular junction. The interaction of the returning neurite with synaptic basal lamina components can cause the neurite to differentiate into a nerve terminal (Sanes *et al.*, 1978; Glicksman and Sanes, 1983).

8. CONCLUSIONS

The identification and characterization of the four synaptic vesicle integral membrane proteins, p38, p65, SV2, and the SVPG, should have several important future uses:

1. The monoclonal antibodies against p38, p65, and SV2 will be important markers for studying membrane traffic in the nerve terminal. Since it is clear that the nerve terminal plasma membrane contains proteins which the synaptic vesicle does not share (Miljanich *et al.*, 1982), one would like to know where the segregation of secretory vesicle proteins from nerve terminal components occurs after exocytosis (Burgess and Kelly, 1987). Is it during endocytosis or during the formation of vesicles from an endosomal membrane (Heuser and Reese, 1973)? Is receptor-mediated endocytosis confined to separate endocytotic vesicles?

2. The cloning and sequencing of p38 will be important for investigation of membrane protein sorting in the secretory pathway (Buckley *et al.*, 1987). The structural cues used to sort proteins into the regulated and constitutive secretory pathways at the *trans*-Golgi network are not understood (Kelly, 1985; Burgess and Kelly, 1987). Likewise, the structural cues involved in sorting vesicle from plasma membrane components during membrane recycling are not known (Burgess and Kelly, 1987). What

aspect of these proteins is recognized by the sorting machinery of the cell? Construction of DNAs encoding modified versions of the p38 protein and expression in neuroendocrine cell lines might identify regions of the molecule which act as these cues (Buckley and Kelly, 1985). In addition, the cloning and sequencing of p65, SV2, and SVPG cDNAs could give some insight into the nature of these sorting cues. A structural comparison of these proteins with the p38 might reveal some common features.

3. Understanding which proteins are responsible for known or proposed synaptic vesicle activities may be aided by cloning the cDNA of these integral membrane proteins. Transfection of neuroendocrine cell lines with DNA encoding antisense p38 RNA may allow an investigator to suppress the expression of the endogenous p38 gene. Assay of the transfected cells and the isolated vesicles from these cells might reveal a deficit in secretory granule function. Antisense RNA has been used successfully in several cell culture systems to suppress gene expression (Izant and Weintraub, 1984; Kim and Wold, 1985; Green *et al.*, 1986).

4. The monoclonals to p38, p65, and SV2 may be very useful additional tools for the purification of secretory granules or the granule membrane. This might be especially important from extracts where complete purification is difficult to achieve with classical purification methods alone. As affinity reagents, all of these monoclonals have been used successfully to immunoisolate secretory vesicles from cultured cell extracts (Lowe *et al.*, 1988). Such isolation methods may help identify new vesicle activities and components.

5. The antibodies directed against p38, p65, and SV2 should serve as important immunocytochemical markers for developmental studies, especially since these antibodies have broad cross-reactivity. Antibodies to p65 have already been used for detecting the accumulation of synaptic vesicles in neurites (Chun and Shatz, 1983; Burry *et al.*, 1986; Sarthy and Bacon, 1985; Greif and Reichardt, 1982). Monoclonal antibodies to p38 are commercially available (SY38, Boehringer Mannheim Biochemicals, Indianapolis, Indiana).

ACKNOWLEDGMENTS

The work described from the author's laboratory was supported by NIH Grant NS-22367. Some of the unpublished data presented in this chapter were obtained by the author as a postdoctoral fellow in the laboratory of Dr. Regis Kelly at the University of California, San Francisco. The work in Dr. Kelly's laboratory was supported by a Muscular Dystrophy Association award and NIH Grants NS-09878 and NS-16073 awarded to Dr. Kelly. In addition, I thank Connie Missimer for editorial help; Rebecca Cruz for biochemical technical assistance; and Deborah Crumrine for assistance with the immunocytochemistry.

9. REFERENCES

Bahler, M., and Greengard, P., 1987, Synapsin I bundles F-actin in a phosphorylation-dependent manner, *Nature* **326:**704–707.

Bahr, B. A., and Parsons, S. M., 1987, The Vesamicol (AH 5183) receptor in VP_1 cholinergic synaptic vesicles: Partial purification, *Soc. Neurosci. Abstr.* **13**:670.

Baines, A. J., 1987, Synapsin I and the cytoskeleton, *Nature* **326**:646.

Bixby, J. L., and Reichardt, L. F., 1985, The expression and localization of synaptic vesicle antigens at neuromuscular junctions *in vitro, J. Neurosci.* **5**:3070–3080.

Bordier, C., 1981, Phase separation of integral membrane proteins in Triton X-114 solution, *J. Biol. Chem.* **256**:1604–1607.

Borroni, E., Ferretti, P., Fiedler, W., and Fox, G. Q., 1985, The localization and rate of disappearance of a synaptic vesicle antigen following denervation, *Cell Tissue Res.* **241**:367–372.

Breckenridge, L. J., and Almers, W., 1987, Currents through the fusion pore that forms during exocytosis of a secretory vesicle, *Nature* **328**:814–817.

Browning, M. D., Huganir, R., and Greengard, P., 1985, Protein phosphorylation and neuronal function, *J. Neurochem.* **45**:11–23.

Brunner, J., Hauser, J., and Semenza, G., 1978, Single bilayer lipid–protein vesicles formed from phosphatidylcholine and small intestinal sucrase-isomaltase, *J. Biol. Chem.* **253**:7538–7546.

Buckley, K., and Kelly, R. B., 1985, Identification of a transmembrane glycoprotein specific for secretory vesicles of neural and endocrine cells, *J. Cell Biol.* **100**:1284–1294.

Buckley, K., Schweitzer, E. S., Miljanich, G. P., Clift-O'Grady, L., Kushner, P. D., Reichardt, L. F., and Kelly, R. B., 1983, A synaptic vesicle antigen is restricted to the junctional region of the presynaptic plasma membrane, *Proc. Natl. Acad. Sci. USA* **80**:7342–7346.

Buckley, K. M., Floor, E., and Kelly, R. B., 1987, Cloning and sequence analysis of cDNA encoding p38, a major synaptic vesicle protein, *J. Cell Biol.* **105**:2447–2456.

Burgess, T. L., and Kelly, R. B., 1987, Constitutive and regulated secretion of proteins, *Annu. Rev. Cell Biol.* **3**:243–293.

Burry, R. W., Ho, R. H., and Matthew, W. D., 1986, Presynaptic elements formed on polylycine-coated beads contain synaptic vesicle antigens, *J. Neurocytol.* **15**:409–419.

Carlson, S. S., and Kelly, R. B., 1980, An antiserum specific for cholinergic synaptic vesicles from electric organ, *J. Cell Biol.* **87**:98–103.

Carlson, S. S., and Kelly, R. B., 1983, A highly antigenic proteoglycan-like component of cholinergic synaptic vesicles, *J. Biol. Chem.* **258**:11082–11091.

Carlson, S. S., and Wight, T., 1987, Nerve terminal anchorage protein 1 (TAP-1) is a chondroitin sulfate proteoglycan: Biochemical and electron microscopic characterization, *J. Cell Biol.* **105**:3075–3086.

Carlson, S. S., Wagner, J. A., and Kelly, R. B., 1978, Purification of synaptic vesicles from elasmobranch electric organ and the use of biophysical criteria to demonstrate purity, *Biochemistry* **17**:1188–1199.

Carlson, S. S., Caroni, P., and Kelly, R. B., 1986, A nerve terminal anchorage protein from electric organ, *J. Cell Biol.* **103**:509–520.

Caroni, P., Carlson, S. S., Schweitzer, E., and Kelly, R. B., 1985, Presynaptic neurons may contribute a unique glycoprotein to the extracellular matrix at the synapse, *Nature* **314**:441–443.

Chun, J. J. M., and Shatz, C. J., 1983, Immunochemical localization of synaptic vesicle antigens in developing cat cortex, *Soc. Neurosci. Abstr.* **9**:692.

De Camilli, P., Cameron, R., and Greengard, P., 1983, Synapsin 1 (protein 1), a nerve terminal-specific phosphoprotein. 1. Its general distribution in synapses of the central and peripheral nervous system demonstrated by immunofluorescence in frozen and plastic sections, *J. Cell Biol.* **96**:1337–1354.

Devoto, S. H., and Barnstable, C. J., 1987, SVP38: A synaptic vesicle protein whose appearance correlates closely with synaptogenesis in the rat nervous system, *Ann. N.Y. Acad. Sci.* **493**:493–496.

Elliot, J., Blanchard, S. G., Woo, W., Miller, J., Strader, C. D., Hartig, P., Moore, H.-P., Racs, J., and Raftery, M. A., 1980, Purification of *Torpedo californica* post-synaptic membranes and fractionation of their constituent polypeptides, *Biochem. J.* **185**:667–677.

Floor, E., and Leeman, S. E., 1985, Evidence that large synaptic vesicles containing substance P and small synaptic vesicles have a surface antigen in common in rat, *Neurosci. Lett.* **60**:231–237.

Gilula, N. B., 1985, Gap junctional contact between cells, in: *The Cell in Contact* (G. M. Edelman and J.-P. Thiery, eds.), pp. 395–409, Wiley, New York.

Glicksman, M. A., and Sanes, J. R., 1983, Differentiation of motor nerve terminals formed in the absence of muscle fibres, *J. Neurocytol.* **12**:661–671.

Godfrey, E. W., Nitkin, R. M., Wallace, B. G., Rubin, L. L., and McMahan, U. J., 1984, Components of Torpedo electric organ and muscle that cause aggregation of acetylcholine receptors on cultured muscle cells, *J. Cell Biol.* **99**:615–627.

Green, P. J., Pines, O., and Inouye, M., 1986, The role of antisense RNA in gene regulation, *Annu. Rev. Biochem.* **55**:569–597.

Greif, K. F., and Reichardt, L. F., 1982, Appearance and distribution of neuronal cell surface and synaptic vesicle antigens in the developing rat superior cervical ganglion, *J. Neurosci.* **2**:843–852.

Griffiths, G., and Simons, K., 1986, The *trans* Golgi network: Sorting of the exit site of the Golgi complex, *Science* **234**:438–443.

Harlos, P., Lee, D. A., and Stadtler, H., 1984, Characterization of a Mg^{2+}-ATPase and a proton pump in cholinergic synaptic vesicles from the electric organ of *Torpedo marmorata*, *Eur. J. Biochem.* **144**: 441–446.

Hascall, V. C., and Hascall, G. K., 1981, Proteoglycans, in: *Cell Biology of the Extracellular Matrix* (E. D. Hay, ed.), pp. 39–63, Plenum Press, New York.

Hassell, J. R., Kimura, J. H., and Hascall, V. C., 1986, Proteoglycan core protein families, *Annu. Rev. Biochem.* **55**: 539–567.

Heinegård, D., and Sommarin, Y., 1987, Isolation and characterization of proteoglycans, *Methods Enzymol.* **144**:319–373.

Heuser, J. E., and Reese, T. S., 1973, Evidence for recycling of synaptic vesicle membrane during transmitter release at the frog neuromuscular junction, *J. Cell Biol.* **57**:315–344.

Heuser, J. E., and Reese, T. S., 1981, Structural changes after transmitter release at the frog neuromuscular junction, *J. Cell Biol.* **88**:564–580.

Heuser, J. E., and Salpeter, S. R., 1979, Organization of acetylcholine receptors in quick-frozen, deep-etched, and rotary-replicated *Torpedo* postsynaptic membrane, *J. Cell Biol.* **82**:150–173.

Heuser, J. E., Reese, T. S., Dennis, M. J., Jan, Y., Yan, L., and Evans, L. J., 1979, Synaptic vesicle exocytosis captured by quick freezing and correlated with quantal transmitter release, *J. Cell Biol.* **81**: 275–300.

Huttner, W. B., Schiebler, W., Greengard, P., and De Camilli, P., 1983, Synapsin I (protein I), a nerve terminal-specific phosphoprotein. III. Its association with synaptic vesicles studied in a highly purified synaptic vesicle preparation, *J. Cell Biol.* **96**:1374–1388.

Izant, G. J., and Weintraub, H., 1984, Inhibition of thymidine kinase gene expression by anti-sense RNA: A molecular approach to genetic analysis, *Cell* **36**:1007–1015.

Jahn, R., Schiebler, W., Ouimet, C., and Greengard, P., 1985, A 38,000-dalton membrane protein (p38) present in synaptic vesicles, *Proc. Natl. Acad. Sci. USA* **82**:4137–4141.

Jahn, R., Navone, F., Greengard, P., and De Camilli, P., 1987, Biochemical and immunochemical characterization of p38, an integral membrane glycoprotein of small synaptic vesicles, *Ann. N.Y. Acad. Sci.* **493**:497–498.

Jones, D. H., and Matus, A. I., 1974, Isolation of synaptic plasma membranes from brain by combined flotation–sedimentation density gradient centrifugation, *Biochim. Biophys. Acta* **356**:276–287.

Kelly, R. B., 1985, Pathways of protein secretion in eukaryotes, *Science* **230**:25–32.

Kelly, R. B., and Hooper, J. E., 1982, Cholinergic vesicles, in: *The Secretory Granule* (A. M. Poisner and J. M. Trifaro, eds.), pp. 81–118, Elsevier/North-Holland, Amsterdam.

Kelly, R. B., and Reichardt, L. R., 1983, A molecular description of nerve terminal function, *Annu. Rev. Biochem.* **52**:871–926.

Kelly, R. B., Deutsch, J. W., Carlson, S. S., and Wagner, J. A., 1979, Biochemistry of neurotransmitter release, *Annu. Rev. Neurosci.* **2**:399–446.

Kelly, R. B., Buckley, K. M., Burgess, T. L., Carlson, S. S., Caroni, P., Hooper, J. E., Katzen, A., Moore, H-P, Pfeffer, S. R., and Schroer, T. A., 1983, Membrane traffic in neurons and peptide-secreting cells, *Cold Spring Harbor Symp. Quant. Biol.* **48**:697–705.

Kelly, R. B., Carlson, S. S., and Caroni, P., 1987, Extracellular matrix components of the synapse, in: *The Biology of the Extracellular Matrix*, Vol. 2 (T. N. Wight and R. P. Mecham, eds.), pp. 247–265, Academic Press, New York.

Kilimann, M. W., and DeGennaro, L. J., 1985, Molecular cloning of cDNAs for the nerve-cell specific phosphoprotein, synapsin I, *EMBO J.* **4**:1997–2002.

Kim, S. K., and Wold, B. J., 1985, Stable reduction of thymidine kinase activity in cells expressing high levels of antisense RNA, *Cell* **42**:129–138.

Krebs, K. E., Zagon, I. S., Sihag, R., and Goodman, S. R., 1987, Brain protein 4.1 subtypes: A working hypothesis, *BioEssays* **6**:274–279.

Kushner, P., and Reichardt, L. F., 1981, Monoclonal antibodies against *Torpedo* synaptosomes, *Soc. Neurosci. Abstr.* **7**:120.

Lee, D. A., and Witzemann, V., 1983, Photoaffinity labeling of a synaptic vesicle specific nucleotide transport system from *Torpedo marmorata*, *Biochemistry* **22**:6123–6130.

Leube, R. E., Kaiser, P., Seiter, A., Zimbelmann, R., Franke, W. W., Rehm, H., Knaus, P., Prior, P., Betz, H., Reinke, H., Beyreuther, K., and Wiedenmann, B., 1987, Synaptophysin: Molecular organization and mRNA expression as determined from cloned cDNA, *EMBO J.* **6**:3261–3268.

Linstedt, A. D., and Kelly, R. B., 1987, Overcoming barriers to exocytosis, *Trends Neurosci.* **10**:446–448.

Llinas, R. R., and Heuser, J. E., 1977, Depolarization–release coupling systems in neurons, *Neurosci. Res. Prog. Bull.* **15**(4):560–687.

Llinas, R. R., McGuinness, T. L., Leonard, C. S., Sugimori, M., and Greengard, P., 1985, Intraterminal injection of synapsin I or calcium/calmodulin-dependent protein kinase II alters neurotransmitter release at the squid giant synapse, *Proc. Natl. Acad. Sci. USA* **82**:3035–3039.

Lowe, A. W., Madeddu, L, and Kelly, R. B., 1988, Endocrine secretory granules and neuronal synaptic vesicles have three integral membrane proteins in common, *J. Cell Biol.* **106**:51–59.

Marshall, I. G., and Parsons, S. M., 1987, The vesicular acetylcholine transport system, *Trends Neurosci.* **10**:174–177.

Matthew, W. D., 1981, Biochemical studies using monoclonal antibodies to neural antigens, Ph.D. thesis, University of California, San Francisco.

Matthew, W. D., Tsavaler, L., and Reichardt, L. F., 1981a, Identification of a synaptic vesicle-specific membrane protein with a wide distribution in neuronal and neurosecretory tissue, *J. Cell Biol.* **91**:257–269.

Matthew, W. D., Reichardt, L. F., and Tsavaler, L., 1981b, Monoclonal antibodies to synaptic membranes and vesicles, in: *Monoclonal Antibodies to Neural Antigens* (R. McKay, M. C. Raff, and L. F. Reichardt, eds.), pp. 163–180, Cold Spring Harbor Laboratory, Cold Spring Harbor, N.Y.

McCaffery, C. A., and DeGennaro, L. J., 1986, Determination and analysis of the primary structure of the nerve terminal specific phosphoprotein, synapsin I, *EMBO J.* **5**:3167–3173.

Miljanich, G. P., Brasier, A. R., and Kelly, R. B., 1982, Partial purification of presynaptic plasma membrane by immunoadsorption, *J. Cell Biol.* **94**:88–96.

Navone, F., Greengard, P., and DeCamilli, P., 1984, Synapsin I in nerve terminals: Selective association with small synaptic vesicles, *Science* **226**:1209–1211.

Navone, F., Jahn, R., Di Gioia, G., Stukenbrok, H., Greengard, P., and DeCamilli, P., 1986, Protein p38: An integral membrane protein specific for small vesicles of neurons and neuroendocrine cells, *J. Cell Biol.* **103**:2511–2527.

Nitkin, R. M., Smith, M. A., Magill, C., Fallon, J. R., Yao, Y.-M., Wallace, B. G., and McMahan, U. J., 1987, Identification of agrin, a synaptic organizing protein from *Torpedo* electric organ, *J. Cell Biol.* **105**:2471–2478.

Obata, K., Nishiye, H., Fujita, S., Shirao, T., Inoue, H., and Uchizono, K., 1986, Identification of a synaptic vesicle-specific 38,000 dalton protein by monoclonal antibodies, *Brain Res.* **375**:37–48.

Obata, K., Kojima, N., Nishiye, H., Inoue, H., Shirao, T., Fujita, S., and Uchizono, K., 1987, Four synaptic vesicle-specific proteins: Identification by monoclonal antibodies and distribution in the nervous tissue and the adrenal medulla, *Brain Res.* **404**:169–179.

Orci, L., Glick, B. S., and Rothman, J. E., 1986, A new type of coated vesicular carrier that appears not to contain clathrin: Its possible role in protein transport within the Golgi stack, *Cell* **46**:171–184.

Patzak, A., and Winkler, H., 1986, Exocytotic exposure and recycling of membrane antigens of chromaffin granules: Ultrastructural evaluation after immunolabeling, *J. Cell Biol.* **102**:510–515.

Paul, D. L., 1986, Molecular cloning of cDNA for rat liver gap junction protein, *J. Cell Biol.* **103**:123–134.

Pearse, B. M. F., 1987, Clathrin and coated vesicles, *EMBO J.* **6**:2507–2512.

Petrucci, T. C., and Morrow, J. S., 1987, Synapsin I: An actin-bundling protein under phosphorylation control, *J. Cell Biol.* **105**:1335–1363.

Pfeffer, S. R., and Kelly, R. B., 1985, The subpopulation of brain coated vesicles that carry synaptic vesicle proteins contains two unique polypeptides, *Cell* **40**:949–957.

Pfeffer, S. R., and Rothman, J. E., 1987, Biosynthetic protein transport and sorting by the endoplasmic reticulum and Golgi, *Annu. Rev. Biochem.* **56**:829–852.

Rehm, H., Wiedemann, B., and Betts, H., 1986, Molecular characterization of synaptophysin, a major calcium-binding protein of synaptic vesicle membrane, *EMBO J.* **5**:535–541.

Reist, N. E., Magill, C., and McMahan, U. J., 1987, Agrin-like molecules at synaptic sites in normal, denervated, and damaged skeletal muscles, *J. Cell Biol.* **105**:2457–2469.

Rephaeli, A., and Parsons, S. M., 1982, Calmodulin stimulation of $^{45}Ca^{2+}$ transport and protein phosphorylation in cholinergic synaptic vesicles, *Proc. Natl. Acad. Sci. USA* **79**:5783–5787.

Sanes, J. R., and Chiu, A. Y., 1983, The basal lamina of the neuromuscular junction, *Cold Spring Harbor Symp. Quant. Biol.* **48**;667–678.

Sanes, J. R., Marshall, L. M., and McMahan, U. J., 1978, Reinnervation of muscle fiber basal lamina after removal of myofibers, *J. Cell Biol.* **78**:176–198.

Sarthy, P. J., and Bacon, W., 1985, Developmental expression of a synaptic vesicle-specific protein in the rat retina, *Dev. Biol.* **112**:284–291.

Schiebler, W., Jahn, R., Doucet, J.-P., Rothlein, J., and Greengard, P., 1986, Characterization of synapsin I binding to small synaptic vesicles, *J. Biol. Chem.* **261**:8383–8390.

Schroer, T. A., Brady, S. T., and Kelly, R. B., 1985, Fast axonal transport of foreign synaptic vesicles in squid axoplasm, *J. Cell Biol.* **101**:568–572.

Stadler, H., and Dowe, G. H. C., 1982, Identification of a heparan sulfate-containing proteoglycan as a specific core component of cholinergic synaptic vesicles from *Torpedo marmorata*, *EMBO J.* **1**:1381–1384.

Stadler, H., and Fenwick, E. M., 1983, Cholinergic synaptic vesicles from *Torpedo marmorata* contain an atractyloside-binding protein related to the mitochondrial ADP/ATP carrier, *Eur. J. Biochem.* **136**: 377–382.

Stadler, H., and Kiene, M.-L., 1987, Synaptic vesicles in electromotoneurones. II. Heterogeneity of populations is expressed in uptake properties; exocytosis and insertion of a core proteoglycan into the extracellular matrix, *EMBO J.* **6**:2217–2221.

Stadler, H., and Whittaker, V. P., 1978, Identification of vesiculin as a glycosaminoglycan, *Brain Res.* **153**: 408–413.

Sudhof, T. C., Lottspeich, F., Greengard, P., Mehl, E., and Jahn, R., 1987, A synaptic vesicle protein with a novel cytoplasmic domain and four transmembrane regions, *Science* **238**:1142–1144.

Tashiro, T., and Stadler, H., 1978, Chemical composition of cholinergic synaptic vesicles from *Torpedo marmorata* based on improved purification, *Eur. J. Biochem.* **90**:479–487.

Unwin, N., 1986, Is there a common design for cell membrane channels? *Nature* **323**:12–13.

Volknandt, W., and Zimmerman, H., 1986, Acetylcholine, ATP, and proteoglycan are common to synaptic vesicles isolated from the electric organs of electric eel and electric catfish as well as from rat diaphragm, *J. Neurochem.* **47**:1449–1462.

Volknandt, W., Naito, S., Ueda, T., and Zimmerman, H., 1987, Synapsin 1 is associated with cholinergic nerve terminals in the electric organs of *Torpedo electrophorus*, and *Malapterurus* and copurifies with *Torpedo* synaptic vesicles, *J. Neurochem.* **49**:342–347.

von Wedel, R. J., Carlson, S. S., and Kelly, R. B., 1981, Transfer of synaptic vesicle antigens to the presynaptic plasma membrane during exocytosis, *Proc. Natl. Acad. Sci. USA* **78**:1014–1018.

Walker, J. H., and Agoston, D. V., 1987, The synaptic vesicle and the cytoskeleton, *Biochem. J.* **247**:249–258.

Walker, J. H., Obrocki, J., and Zimmerman, C. W., 1983, Identification of a proteoglycan antigen characteristic of cholinergic synaptic vesicles, *J. Neurochem.* **41**:209–216.

Walker, J. H., Kristjansson, G. I., and Stadler, H., 1986, Identification of a synaptic vesicle antigen (M_r 86,000) conserved between *Torpedo* and rat, *J. Neurochem.* **46**:875–881.

Wang, Y.-J., and Mahler, H. R., 1976, Topography of the synaptosomal membrane, *J. Cell Biol.* **71**:639–658.

White, J. M., and Wilson, I. A., 1987, Anti-peptide antibodies detect steps in a protein conformational change: Low-pH activation of the influenza virus hemagglutinin, *J. Cell Biol.* **105**:2887–2896.

White, J. M., Kielian, M., and Helenius, A., 1983, Membrane fusion proteins of enveloped animal viruses, *Q. Rev. Biophys.* **16**:151–195.

Wiedemann, B., and Franke, W. W., 1985, Identification and localization of synaptophysin, an integral membrane glycoprotein of M_r 38,000 characteristic of presynaptic vesicles, *Cell* **41**:1017–1028.

Wiedemann, B., Franke, W. W., Kuhn, C., Moll, R., and Gould, V. E., 1986, Synaptophysin: A marker protein for neuroendocrine cells in neoplasms, *Proc. Natl. Acad. Sci. USA* **83**:3500–3504.

Winkler, H., 1987, Composition and transport function of membranes of chromaffin granules, *Ann. N.Y. Acad. Sci.* **493**:252–258.

Winkler, H., and Carmichael, S. W., 1982, The chromaffin granule, in: *The Secretory Granule* (A. M. Poisner and J. M. Trifaro, eds.), pp. 3–79, Elsevier/North-Holland, Amsterdam,

Yamagata, S. K., and Parsons, S. M., 1987, Molecular weight and purification of the Ca^{2+}/Mg^{2+} ATPase of cholinergic synaptic vesicles, *Soc. Neurosci. Abstr.* **13**:671.

Yanagishita, M., Midura, R. J., and Hascall, V. C., 1987, Proteoglycans: Isolation and purification from tissue cultures, *Methods Enzymol.* **138**:279–289.

Carbohydrate Recognition, Cell Interactions, and Vertebrate Neural Development

M. A. Hynes, J. Dodd, and T. M. Jessell

1. INTRODUCTION

The formation of selective connections between distinct subsets of neurons is a critical step in the generation of functional neural circuits. During embryogenesis, several sequential steps contribute to the final pattern of neural connectivity. First, the differential adhesive properties of neural and nonneural epithelial cells result in the segregation and shaping of early neural tissues. The establishment of basic morphological features is accompanied by the migration of neuroblasts and their differentiation into distinct subsets of neurons. The axons of differentiated neurons then project to their prospective cellular targets under the influence of a series of diffusible cell surface and matrix-associated guidance cues. Axonal growth cones then appear to recognize and select appropriate cellular targets with which to form stable contacts. Each of these developmental steps is dependent on a precisely coordinated program of intercellular recognition.

Over the past few years, some of the molecules that contribute to neuronal cell recognition have been characterized and their roles as adhesive ligands have been demonstrated (Jessell, 1988; Rutishauser and Jessell, 1988). Analysis of the biochemical properties and anatomical distribution of molecules involved in neural cell recognition and adhesion has provided detailed information on two major classes of adhesion molecules. The first class comprises general adhesion molecules of which NCAM and the cadherins represent the predominant members (Edelman, 1986; Take-

M. A. Hynes • Howard Hughes Medical Institute, Center for Neurobiology and Behavior, Columbia University, College of Physicians and Surgeons, New York, New York 10032.   *J. Dodd* • Department of Physiology, Columbia University, College of Physicians and Surgeons, New York, New York 10032. *T. M. Jessell* • Howard Hughes Medical Institute, Center for Neurobiology and Behavior, and Department of Biochemistry and Molecular Biophysics, Columbia University, College of Physicians and Surgeons, New York, New York 10032.

ichi, 1988; Rutishauser and Jessell, 1988). NCAM and the cadherins appear to mediate, respectively, the major Ca^{2+}-independent and Ca^{2+}-dependent adhesive interactions of neural and many nonneural cells. Recent studies have characterized a family of cell surface glycoproteins that also mediate neuronal adhesion but that exhibit more restricted patterns of expression in the nervous system (Rutishauser and Jessell, 1988). Most of these molecules are expressed on developing axons and have been implicated in the extension of growth cones along other axonal surfaces and in fasciculation. The mechanism by which general and axonal adhesion molecules mediate cell–cell interactions is still poorly understood. However, several lines of evidence suggest that the binding function of most general and axonal adhesion molecules involves homophilic interactions between the same molecular species present on the surface of participating cells (Edelman *et al.*, 1987; Nagafuchi *et al.*, 1987; Lagenaur and Lemmon, 1987).

The identification of molecules that mediate neuronal adhesion via direct protein–protein interactions has resulted in a somewhat diminished emphasis on the potential role of protein–carbohydrate interactions in neural cell adhesion and recognition. However, a role for cell surface oligosaccharides in neural cell recognition has been invoked many times over the past 30 years. The earliest suggestions derived from the detection of a high degree of heterogeneity in oligosaccharide structures, particularly gangliosides, on neural cells, and from the discovery of cell surface glycosyltransferase enzymes that are capable of interacting with surface oligosaccharides (Roseman, 1970; Roth *et al.*, 1971). The difficulty in characterizing and purifying complex oligosaccharide structures combined with a period of uncertainty over the validity of detection methods for cell surface glycosyltransferases, however, has delayed progress in defining the role of carbohydrate recognition in embryogenesis in general and in neural development in particular.

Recently, the expression of complex oligosaccharides on early embryonic and neural cells has received renewed attention. Monoclonal antibodies have revealed that many complex oligosaccharides are restricted to subsets of vertebrate cells and are expressed as gradients in developing neural tissues. In addition, many of the complex oligosaccharides present on developing neurons are expressed on other cell types, in particular on cells of the preimplantation-stage mouse embryo. Moreover, there is now considerable evidence for a functional role of cell surface oligosaccharides in mediating the interactions of early embryonic cells. Several classes of carbohydrate-binding proteins with specificity for cell surface oligosaccharides have been detected in vertebrates. In this chapter we discuss the increasing evidence that has emerged from both neural and nonneural systems that implicates carbohydrate recognition in the mediation of cell–cell interactions.

2. COMPLEX OLIGOSACCHARIDE STRUCTURES ON NEURAL CELLS

Detailed accounts of the major structural classes of oligosaccharides in the nervous system are presented elsewhere in this volume. Of these, three major classes have been implicated in cell adhesion and recognition in mammals. These classes are defined by their polysaccharide backbone sequences (Table 1). Those that express the

Table 1. Defined Oligosaccharides Expressed by Subsets of Developing Neurons[a]

Structure	MAb	Distribution
Lactoseries		
Gal(β1-4)GlcNAc-R	A5, 1B2	DRG neurons
Gal(α1-3)Gal(β1-4)GlcNAc-R	2C5	DRG neurons
Gal(β1-4)GlcNAc-R with Fuc(α1-3)	SSEA-1, AC4	Early CNS neuronal subsets; Peripheral sensory structures (taste buds/merkel cells)
SO$_4$-3GlcUA(β1-3)Gal(β1-4)GlcNAc-R	HNK-1, Leu 7, L2, NC1	Postmitotic neurons; Neural crest
Globoseries		
R-3GalNAc(β1-3)Galα1-4R	SSEA-3	DRG neurons
NeuAc(α2-3)Gal(β1-3)GlcNAcβ1-R	SSEA-4	DRG neurons
Ganglioseries		
Gal(α1-3)Gal(β1-3)GalNAcβ1-4Galβ1-4Glcβ1-R with NeuAc(α2-3) and Fuc(α1-2)	LD2, LA4	DRG neurons
NeuAc(α2-8)NeuAc(α2-3)Galβ1-R	R24	Many differentiated neurons
Neu5,9Ac(α2-3)Gal(β1-4)Glcβ1-R	D1.1, JONES	Many differentiated neurons
NeuAc(α2-8)NeuAc(α2-3)Gal(β1-4)Glcβ1-R	18B8	Retina

[a]Structurally characterized oligosaccharides that are expressed on neuronal surfaces. Details are given in the text.

Galβ1-3(4)GlcNAc-R structure are termed lactoseries carbohydrates; those that contain the GalNAcβ1-3Galα1-4Gal-R structure are termed globoseries carbohydrates;
and those that contain Galβ1-3GalNAcβ1-R are termed ganglioseries carbohydrates
(Hakomori, 1981). In each class, the backbone sequence can be modified extensively
by the addition of terminal saccharides. The attachment of the same saccharide via
multiple linkages, combined with the existence of branched carbohydrate chains of the
same or differing structure, thus results in the potential for enormous diversity in the
complex oligosaccharide structures on vertebrate cells.

The synthesis of complex oligosaccharides is achieved by the regulated activity of
glycosyltransferase enzymes, each of which is capable of adding specific saccharides
via defined linkages to a highly restricted set of oligosaccharide substrates. Thus, the
structural diversity in cell surface oligosaccharides is likely to be defined in large part
by the cellular expression and substrate specificity of these glycosyltransferase
enzymes.

3. PROTEINS INVOLVED IN SACCHARIDE RECOGNITION

A large number of proteins that are capable of binding to surface oligosaccharides
on vertebrate cells have been characterized. Recent structural and functional information has made it possible to subdivide these proteins into several categories (see
Drickamer, 1988) (Figure 1): (1) Ca^{2+}-dependent transmembrane or soluble animal
lectins [proteins that fall into this class show structural homology to the rat hepatic
lectin (also known as the asialoglycoprotein receptor)]; (2) Ca^{2+}-independent soluble
carbohydrate binding proteins; (3) Ca^{2+}-independent transmembrane carbohydrate
binding proteins; (4) membrane-bound or soluble glycosyltransferases which, in addition to their enzymatic function, can form stable complexes with their acceptor
oligosaccharide structures and thus exhibit lectinlike properties.

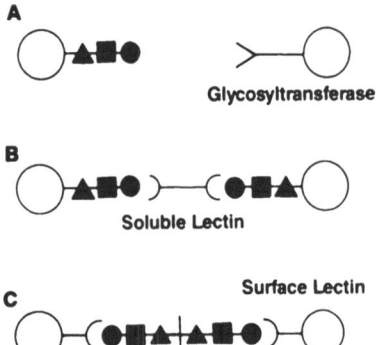

Figure 1. Potential mechanisms of carbohydrate recognition
in vertebrate cell interactions.

3.1. Ca^{2+}-Dependent Carbohydrate Binding Proteins

Molecular cloning and analysis of the rat hepatic asialoglycoprotein receptor (ASGP-R) has revealed four functional domains in this protein: an NH_2-terminal cytoplasmic domain, a membrane-spanning sequence, an extracellular neck region, and, at the COOH terminus, a carbohydrate recognition domain (CRD) (Drickamer, 1988). The CRD of the rat ASGP-R was identified by proteolytic separation of this portion of the protein and the demonstration that this fragment retains the ability to bind carbohydrate affinity matrices with the same affinity as the intact receptor (Chiacchia and Drickamer, 1984). The chick hepatic lectin also contains a structurally segregated CRD and although chick and rat hepatic lectins have nonoverlapping sugar specificities, the CRDs of these two lectins are structurally homologous. Examination of a number of other Ca^{2+}-dependent carbohydrate binding proteins has revealed a high degree of structural homology between their CRD and that of the rat hepatic lectin (Drickamer, 1988). The CRD of these proteins consists of a pattern of 18 amino acids, conserved within a 130-amino-acid domain. A salient feature of the CRD in these proteins is the conserved placement of cysteine residues which appear to be involved in the formation of disulfide bonds. Carbohydrate binding function has since been predicted for proteins that contain this domain but which had not previously been identified as lectins. These proteins include dog pulmonary surfactant (Haagsman et al., 1987) and proteoglycan core proteins (Halberg et al., 1988; Krusius et al., 1987). The structural conservation of this CRD therefore suggests that carbohydrate binding proteins of this class may be identifiable on the basis of structure alone, in the absence of functional information.

3.2. Ca^{2+}-Independent Soluble Lectins

Low-molecular-weight soluble carbohydrate binding proteins isolated from a number of species and tissues constitute a separate class of carbohydrate binding proteins with β-galactoside specificity (Barondes, 1984). The primary structure of a number of these lectins has been deduced by direct amino acid sequence determination and by molecular cloning (Gitt and Barondes, 1986; Clerch et al., 1988; Paroutaud et al., 1987). Soluble lectins isolated from a number of species and tissues show a high degree of structural homology but are not identical. In humans in particular, there is evidence for the existence of distinct genes encoding structurally heterogeneous β-galactoside lectins (Gitt and Barondes, 1986).

Although many of the Ca^{2+}-dependent and Ca^{2+}-independent lectins have similar saccharide binding specificities, the structures of their carbohydrate binding domains appear to be completely different. Most members of the Ca^{2+}-dependent lectin class contain multiple functional domains. In contrast, the majority of the soluble, Ca^{2+}-independent lectins do not appear to contain distinct functional domains. Recently, however, two Ca^{2+}-independent β-galactoside-specific lectins have been identified that appear to contain functional domains in addition to their saccharide binding site. First, a 35-kDa fibroblast lectin has been shown to contain sequences homologous to ribonuclear proteins at its NH_2 terminus (Jia and Wang, 1988). Second, preliminary biochemical characterization of the elastin receptor has suggested that this protein contains a β-galactoside binding domain (Hinek et al., 1988).

3.3. Glycosyltransferases

Glycosyltransferases located in the Golgi apparatus are responsible for the synthesis of oligosaccharide units in complex carbohydrates. In several neural systems, in particular the chick retina, glycosyltransferases have also been detected on the cell surface (Bayna et al., 1986). Interactions between glycosyltransferases and membrane glycoconjugates have been implicated in cellular recognition and adhesion (Roseman, 1970; Roth et al., 1971; Bayna et al., 1986). The glycosyltransferases may contribute to cellular recognition in two possible ways: they may be responsible for the synthesis or modification of extracellular glycoconjugates that interact with lectinlike proteins, or they may bind to external glycoconjugates directly in a receptor–ligand mode of interaction. Molecular cloning may provide information on the structural features and diversity of glycosyltransferases expressed by vertebrate cells since cDNAs encoding a β1-4 galactosyltransferase have recently been isolated (Shaper et al., 1986, Narimatsu et al., 1986).

4. CARBOHYDRATE-MEDIATED CELL–CELL INTERACTIONS

The biochemical and structural studies described above have revealed that similar or identical oligosaccharides and carbohydrate binding proteins can be expressed by developing neural and nonneural cells. The analysis of carbohydrate recognition in nonneural systems can therefore provide an important conceptual framework for assessing the functional role of similar or identical molecules within the nervous system. Nonneural systems in which the basic principles of carbohydrate-mediated recognition have emerged are discussed in the following sections.

4.1. Recognition of Circulating Glycoproteins by Hepatic Lectins

The receptor-mediated endocytosis of circulating serum glycoproteins by the hepatic ASGP-R provided one of the first physiological roles for carbohydrate recognition in vertebrates. In rat, the removal of terminal sialic acid residues from native serum glycoproteins exposes penultimate galactose residues and results in the accelerated clearance of these glycoproteins from the circulation (Ashwell and Harford, 1982) (Figure 2). The dependence of this process on a protein–carbohydrate interaction and the specificity of the ASGP-R for galactose residues was established by blocking endocytosis of glycoproteins by enzymatic alteration of their terminal residues. The ASGP-R exhibits a high degree of specificity for glycoproteins with exposed galactose residues (Ashwell and Harford, 1982). Studies using model neoglycoproteins that vary in molar ratios of galactose to protein have demonstrated a direct correlation between

Figure 2. Proposed mechanisms of carbohydrate-mediated adhesion and recognition in nonneural systems. (A) Asialoglycoprotein receptors are expressed on mammalian hepatocytes. Circulating serum glycoconjugates with exposed galactose residues are bound by the receptor, endocytosed, and targeted to the lysosome for degradation. (B) The carbohydrate moiety of the egg zona pellucida glycoprotein, ZP3, is necessary for the binding of sperm to egg. The sperm receptor has not been identified but may be a GalTase or a lectin. (C) Two phases of embryonic compaction appear to be mediated by different molecular mechanisms; E-cadherin is involved in the early Ca^{2+}-independent phase of compaction. (D) Protein receptors on subsets of lymphocytes bind to glycoconjugates on high endothelial venules to permit selective migration of circulating lymphocytes to appropriate lymphatic tissue. *Abbreviations:* GalTase, galactosyltransferase; NeuAc, neuraminic acid, sialic acid; Gal β1-4, β-galactoside-linked glycoconjugates; SSEA-1, stage-specific embryonic antigen-1; HEV, high endothelial venule.

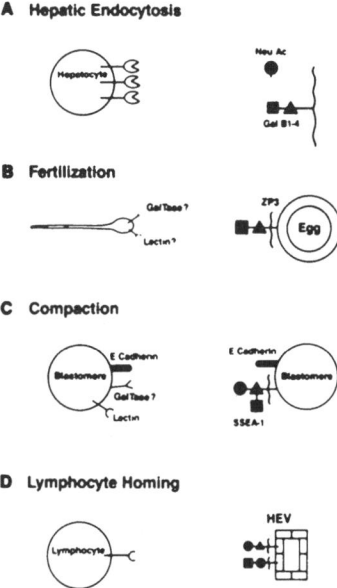

binding affinity and the extent of saccharide substitution (Krantz *et al.*, 1976). The extent of oligosaccharide branching also significantly affects ASGP-R binding affinity. For example, triantennary oligosaccharides exhibit a 30-fold greater inhibitory potency than biantennary oligosaccharides of the same linear structure (see Drickamer, 1987). Thus, the degree of oligosaccharide branching may be a major determinant in the affinity of oligosaccharide binding to their protein receptors.

Both the chick and rat hepatic lectins are transmembrane proteins and *in vivo* each of these receptors exists in the membrane as multimers. The active species of the chicken hepatic lectin is a hexamer. Thus, although a single receptor molecule is capable of recognizing carbohydrate, the multimeric nature of receptors in the membrane may contribute to the enhanced affinity of the receptor for ligands that contain clusters of terminal sugars (Kuhlenschmidt *et al.*, 1984). The oligomeric conformation of carbohydrate binding receptors may also expand the range of sugar binding capabilities of each of these proteins, either by producing a shift in sugar binding specificity and/or by increasing the ability of these proteins to bind to branched oligosaccharides with different backbone structures (Drickamer, 1987).

4.2. Mammalian Sperm–Egg Interactions

The binding of sperm to mammalian eggs exhibits a striking species and stage specificity that provides a powerful system with which to examine the selectivity of intercellular recognition in vertebrates. Initial specificity in mammalian sperm–egg interactions results from the binding of sperm to receptors present on the zona pellucida membrane that surrounds the egg (Wassarman, 1987) (Figure 2). In the mouse, the zona pellucida consists of only three major glycoproteins, termed ZP-1, ZP-2, and ZP-3 (Wassarman, 1987). The ZP-3 glycoprotein has been shown to act as the receptor for sperm and is glycosylated with both N-linked and O-linked oligosaccharides. The use of selective oligosaccharide cleavage reagents has demonstrated that the sperm receptor constitutes an O-linked oligosaccharide on the ZP-3 protein that may have a lactosaminoglycan backbone structure (see Wassarman, 1987). The structure of the ZP-3 receptor oligosaccharide has not been precisely characterized; however, addition of fucans and other fucose polymers such as fucoidin has been shown to inhibit sperm binding in a wide variety of species. Fucose structures may therefore be involved in sperm binding (Huang and Yanagimachi, 1984). More recently the use of selective glycosidases has also implicated an α-linked galactose as a crucial structural component of the ZP-3 receptor (Bleil and Wassarman, 1988).

The nature of proteins on the sperm surface that bind to the ZP-3 oligosaccharide receptor is not clear. One candidate for the ZP-3 binding protein is a galactosyltransferase (GalTase) enzyme (Lopez et al., 1985). Several lines of evidence are consistent with this possibility. Inhibitors of GalTase activity, in particular α-lactalbumin and UDP-dialdehyde, inhibit sperm binding assayed in vitro. The addition of soluble GalTase or anti-GalTase antibodies also inhibits the binding of sperm. In addition, UDP-galactose (the nucleotide sugar donor for the enzyme) but not other nucleotide sugars is able to dissociate sperm from the egg zona pellucida. Since this GalTase catalyzes the transfer of galactose to terminal N-acetylglucosamine, the ZP-3 receptor might be expected to exhibit this latter terminal saccharide. The observation that a preexisting terminal α-galactose on ZP-3 is required for sperm receptor activity (Bleil and Wassarman, 1988) is therefore not entirely consistent with the direct involvement of a GalTase.

The carbohydrate binding protein on the sperm surface in some invertebrates is a lectinlike molecule termed bindin (Glabe et al., 1982). It thus remains possible that a lectin devoid of enzymatic activity may exist on the surface of mammalian sperm and function by binding ZP-3. Further characterization of the structure of the ZP-3 oligosaccharide should provide information on the nature of the ZP-3 binding protein on the sperm surface. Collectively, these studies provide strong evidence for carbohydrate-mediated cell recognition in vertebrates.

4.3. Compaction of Preimplantation Mouse Embryos

Studies on the molecular basis of cell adhesion in preimplantation mouse embryos have provided evidence for a role of both general adhesion molecules and surface oligosaccharides. Individual blastomeres in early eight-cell mouse embryos exhibit a low degree of cell–cell contact and retain clearly defined cell boundaries (Calarco-Gillam, 1985) (Figure 2). A striking increase in blastomere adhesion and in cell–cell contact occurs at the eight-cell stage of development, a process termed compaction (Hyafil et al., 1980; Takeichi, 1985, 1988). The earliest phase of compaction appears to be Ca^{2+}-dependent, whereas at later stages, compaction becomes progressively less dependent on Ca^{2+}. It is now clear that the Ca^{2+}-dependent cell adhesion molecule E-cadherin is expressed on blastomeres at the eight-cell stage and mediates the early,

Ca^{2+}-dependent phase of compaction. In contrast, a variety of studies have suggested that carbohydrate recognition plays a role in the later, Ca^{2+}-independent phase of compaction.

Examination of the role of oligosaccharides that are involved in the adhesion of blastomeres at compaction has focused on a group of lactoseries oligosaccharides, originally defined as stage-specific embryonic antigens (SSEAs) (Solter and Knowles, 1978). These lactoseries structures, in particular the fucosylated SSEA-1 structure, are first expressed on mouse blastomeres immediately preceding compaction. The addition of soluble, multivalent neoglycoproteins expressing the SSEA-1-reactive trisaccharide inhibits the compaction of early mouse embryos (Fenderson *et al.*, 1984). This inhibition of compaction is specific to the SSEA-1 structure since analogues of this molecule do not inhibit compaction. Further, the selective cleavage of embryonal cell surface polylactosaminoglycans with the bacterial enzyme endo-β-galactosidase markedly delays the recompaction of previously decompacted embryos (Rastan *et al.*, 1985). Similar enzyme treatment does not appear to affect earlier stages of compaction, suggesting that cell surface oligosaccharides that are sensitive to endo-β-galactosidase may be involved only in the later stages of compaction.

The mechanisms by which lactoseries oligosaccharides contribute to blastomere adhesion are not understood. However, it seems likely that they interact with carbohydrate binding proteins on the blastomere surface. In support of this, mouse blastomeres and teratocarcinoma cells have been shown to express lectinlike molecules on their surface, when assayed by erythrocytic rosetting or hemagglutination (Grabel *et al.*, 1983). A 56-kDa lectin purified from mouse teratocarcinoma cells (Grabel, 1984; Grabel *et al.*, 1985) has been reported to react with fucosylated oligosaccharides similar to those present on eight-cell embryos.

There is also evidence for the involvement of cell surface GalTases in embryonic compaction (Shur, 1983; Bayna *et al.*, 1988). GalTase activity on the surface of mouse blastomeres increases markedly during the later phases of compaction. Moreover, addition of antibodies that abolish GalTase activity or of α-lactalbumin results in the decompaction of morulae. The study of preimplantation mouse embryo compaction thus constitutes an important model for analysis of the respective roles of protein- and carbohydrate-mediated adhesion in the nervous system.

4.4. Lymphocyte Homing

A critical event in the recirculation of lymphocytes is their migration from the bloodstream to lymphatic ducts in peripheral lymphoid tissues. This process has been termed lymphocyte homing and involves the adhesion of lymphocytes to a specialized set of endothelial cells called high endothelial venules (HEV) that are located in the postcapillary beds (Gallatin *et al.*, 1986) (Figure 2). There is now considerable evidence that lymphocyte homing to HEV is mediated by a set of lymphocyte cell surface proteins that interact with oligosaccharide structures located on the HEV cell surface. Different subsets of recirculating lymphocytes interact in an organ-specific manner with HEV in peripheral lymph nodes and in mucosa-associated lymphoid tissues such as Peyer's patches. The selective migration of lymphocytes appears to involve the expression of predetermined homing receptors on lymphocyte subsets since B lympho-cytes bind preferentially to Peyer's patches while T lymphocytes bind preferentially to peripheral node HEV. Lymphocyte homing therefore represents a further example of selective cell recognition that appears to involve protein–carbohydrate interactions.

A role for oligosaccharide structures in the recirculation of lymphocytes and in their homing to HEV was first suggested by the experiments of Gesner and Ginsburg (1964). More recently, the establishment of a quantitative *in vitro* assay for lympho-cyte adhesion to HEV cells in cryostat sections of lymphoid tissue (Stamper and Woodruff, 1976) has permitted a more detailed analysis of lymphocyte adhesive in-teractions. The binding of lymphocytes to HEV cells in this *in vitro* assay is specifical-ly inhibited by mannose 6-phosphate and by yeast phosphomannan polysaccharides (Stoolman *et al.*, 1984; Rosen and Yednock, 1986; Yednock *et al.*, 1987a,b). Sialidase treatment of HEV cells also perturbs lymphocyte adhesion (Rosen *et al.*, 1985), suggesting a rather complex spectrum of saccharide recognition by circulating lympho-cytes. Recently, polyacrylamide surfaces derivatized with carbohydrates have been used to define, more precisely, the carbohydrate binding specificity of lymphocyte homing receptors (Brandley *et al.*, 1987). These lymphocytes adhere specifically to polyacrylamide gels derivatized with either PPME or with fucoidan (a fucose sulfate polymer isolated from marine algae) but not to gels derivatized with other saccharides (Brandley *et al.*, 1987). The binding of lymphocytes to both saccharides suggests that distinct classes of homing receptors with differing saccharide specificities may exist on a single population of lymphocytes.

Phosphomannans have been detected in mammalian glycoconjugates and a role for phosphomannans in glycoprotein sorting (Sly and Fischer, 1982) has been pro-posed. The presence of phosphomannan residues in HEV is therefore not surprising. In contrast, fucose sulfate-containing glycoconjugates have not been widely reported in vertebrates although HEV cells are known to express high levels of sulfated glycocon-jugates (Andrews *et al.*, 1983). Further characterization of the glycoconjugates ex-pressed on the HEV surface is clearly required.

5. CARBOHYDRATE RECOGNITION IN THE NERVOUS SYSTEM

The studies on cell adhesion and recognition described above have established a physiological role for carbohydrate recognition in diverse cellular systems. Since many of the saccharide structures (Table 1), carbohydrate binding proteins, and glycosyl-transferases (Table 2) that mediate adhesion and recognition in nonneural systems are expressed in the developing nervous system, it seems likely that at least some of these molecules subserve similar functions. However, the complexity of cell interactions in the nervous system has made it substantially more difficult to determine the precise function of carbohydrate recognition than for example in sperm–egg interactions or blastomere compaction. Despite this, in three areas of neural development there is now reasonable evidence that carbohydrate–protein interactions may contribute to neural adhesion and recognition. These areas are discussed in the following sections.

Table 2. Carbohydrate Binding Proteins Located in or on Neural Cells

Carbohydrate binding protein	Tissue system	Ligand
GalTase	Sperm: fertilization	ZP-3 *O*-linked oligosaccharide
	Early embryonic cell adhesion	SSEA-1 antigen; polylactosamine glycoconjugates
	Migration on basal lamina	ECM proteoglycans
GalNAcTase	Retina: initially throughout retina; becomes restricted to synaptic layers with development	Unknown
	NMJ: adult rat	Unknown
Mannose binding	Cerebellum: found in large- and intermediate-size neurons	Phosphomannan glycoproteins
	Rat liver sinusoidal cells	Phosphomannan glycoproteins
14.5- and 29-kDa β-galactoside binding lectins	Dorsal root ganglion, motor neurons of spinal cord and brain stem	Lactoseries glycolipids and glycoproteins

*GalTase, galactosyltransferase; GalNAcTase, *N*-acetylgalactosaminyltransferase.

5.1. Cell Recognition in the Retinotectal System

Innervation of the tectum by retinal ganglion cell axons has provided a classical model system with which to study the formation of specific topographic projections in the nervous system. Early histological and electrophysiological experiments established that retinal ganglion neurons form a point-to-point representation of the retina in the tectum, and thereby form a spatially continuous map on the tectum. Cells from a subregion of the retina always project to the same, defined area of the tectum. Thus, ganglion cell axons from the dorsal part of the retina project to the ventral part of the tectum whereas axons of the ventral retina project to the dorsal tectum. Similarly, nasal (anterior) retinal axons project to the posterior regions of the tectum and temporal (posterior) retinal axons project to the anterior tectum.

Based on the behavior of regenerating retinal axons in lower vertebrates, Sperry (1963) proposed a "chemoaffinity hypothesis" to explain the precise innervation of the tectum by retinal cells axons. The basis of this hypothesis is that the precise projection of the retina onto the tectum is brought about by the selective affinities of ganglion cells for specific areas of the tectum. These affinities were proposed to derive from the existence of complementary gradients of biochemical signals on the surface of retinal ganglion axons and the tectum.

5.1.1. Functional and Biochemical Gradients

Recent biochemical and functional analyses have indicated that a number of molecules with both adhesive and inhibitory properties are distributed in a graded fashion in retina and tectum. These molecules represent potential candidates in the mediation and refinement of specific retinotectal interactions. Using a functional *in vitro* bioassay, evidence has been obtained that molecular differences along the nasotemporal axis in chick may contribute to the ordering of retinotectal projections. When retinal axons are presented with a choice of tectal membrane substrates on which to grow, axons from temporal retina adhere to (Halfter *et al.*, 1981) and extend preferentially (Walter *et al.*, 1987a) on membranes derived from the anterior tectum. The ability of temporal retinal axons to discriminate tectal membrane components was found to be a graded function of the position along the anterior–posterior axis of the tectum from which membranes were prepared (Bonhoeffer and Huf, 1982). While the molecular nature of the difference in tectal cell surfaces has not been defined, the ability of retinal axons to discriminate the positional origin of tectal membranes appears to reflect an inhibitory cell surface component on posterior tectal cells rather than an adhesive component restricted to anterior tectal cells (Walter *et al.*, 1987b).

Biochemically defined gradients along the dorsoventral axis of the retina and tectum have also been defined. A 47-kDa cell surface glycoprotein, the TOP antigen, is present in a graded distribution along the dorsoventral axis of embryonic chick retina (Trisler *et al.*, 1981) and is present along a reversed gradient in the tectum, before the arrival of retinal axons (Trisler and Collins, 1987). The TOP protein has therefore been proposed to contribute to the ordered projection of retinal axons along the dorsoventral axis. The reciprocal dorsoventral gradient of the TOP antigen in retina and tectum is at variance with the earlier models of Marchase (1977). The reversed polarity of the TOP gradient in retina and tectum suggests that TOP molecules on the axons of dorsal retina interact with TOP molecules on ventral tectal cells (Trisler *et al.*, 1981; Trisler and Collins, 1987) in a homophilic fashion, rather than in a complementary protein–carbohydrate interaction.

5.1.2. Carbohydrates and Adhesive Gradients

Additional experimental support for gradients in the retinotectal system was obtained with the demonstration of preferential adhesion of dissociated dorsal or ventral retinal cells to their physiologically matched tectal cells *in vitro* (Barbera *et al.*, 1973; Barbera, 1975). In a detailed biochemical analysis of the adhesive properties of retinal and tectal cells, Marchase (1977) established that protease treatment of ventral retinal cells or ventral tectum blocked the preferential adhesion of the retinal cells to their matching tectal halves whereas similar protease treatment of dorsal retinal cells or dorsal tectum did not alter adhesion (Marchese, 1977). In contrast, treatment of dorsal retinal or dorsal tectal cells with *N*-acetylhexosaminidase or sialidase resulted in a decrease in specific retinotectal adhesion (Marchase, 1977). From these studies a model was proposed in which two molecules, each localized on the surface of retinal cells and the tectum in a dorsal–ventral gradient, would be responsible for retinotectal adhesive specificity. A protease-insensitive molecule would be more concentrated in dorsal retina and tectum and a protease-sensitive molecule would be more concentrated in ventral retina and tectum. Marchase (1977) was able to show that binding of GM2 ganglioside to retina and tectum exhibited a ventral-to-dorsal gradient and therefore suggested that GM2 may be a substrate for the more ventrally concentrated protease-sensitive molecule. GM1 synthetase, the enzyme that converts GM2 to GM1 by the addition of a terminal galactose residue, was suggested as a candidate for the protease-sensitive molecule and was shown to be present, as detected enzymatically, in a ventral-to-dorsal gradient in the retina. To date, firm evidence that GM2 and GM1 synthetase are responsible for mediating specific retinotectal connections has not been obtained. However, the idea that a graded distribution of molecules on the surface of retinal axons interacts with a reverse gradient on the tectum to produce specific retinotectal connections is still viable.

Further evidence for the involvement of the gangliosides in generation of retinotectal specificity has been suggested on the basis of observations that retinal cells show adhesive interactions with gangliosides, and that some ganglioside species show a graded distribution in both retina and tectum. *In vitro* assays have revealed a specific adhesion between gangliosides and neural retina cells (Blackburn *et al.*, 1986). Moreover, the relative strength of adhesion of neural retinal cells differs between various ganglioside species. These adhesive interactions appear specific in that they are not detected between neural retinal cells and other lipids, or between gangliosides and other cell types such as hepatocytes. These observations suggest that retinal cells may express surface carbohydrate binding proteins that interact with gangliosides to mediate cell adhesion.

In both retina and tectum, gangliosides have been localized to subpopulations of neural cells. A monoclonal antibody (18B8) detects a number of developmentally regulated ganglioside species in chicken retina and brain (Dubois *et al.*, 1986). The major ganglioside species recognized by antibody 18B8 in retina is GT3, which is associated with the soma of most cells in the retina in early development but becomes progressively restricted to synaptic layers during later development (Grunwald *et al.*, 1985).

Another set of gangliosides, distinct from those detected with mAb 18B8, is recognized by mAb JONES. The JONES antigen is expressed in a dorsal–ventral gradient in the retina and is also present in the tectum of early postnatal rats. Biochemical characterization of the JONES antigen revealed that it is a 9-O-acetyl derivative of GD3 ganglioside and that it is similar or identical to the D1.1 ganglioside which was previously identified on developing rat neuroectoderm (Levine *et al.*, 1984). The 9-O-acetylated form of GD3 shows a different pattern of expression than nonacetylated GD3. In retina and tectum, the JONES antigen is distributed in a dorsal–ventral gradient, whereas GD3 staining appears to be uniformly distributed (Blum and Barnstable, 1987).

Biochemical and functional studies have demonstrated that N-acetylgalactosaminyltransferase (GalNAcTase) is found on the surface of embryonic chick neural retina cells. In preimplantation embryos, glycosyltransferases may function as lectins by binding to oligosaccharide substrates and forming stable molecular complexes (see above). GalNAcTase from embryonic chick neural retina can be isolated both as a particulate complex associated with its endogenous acceptor, and as a soluble protein (Balsamo *et al.*, 1986). The two GalNAcTase forms are immunologically cross-reactive but have different molecular masses. Under most conditions, the enzyme remains associated with its endogenous carbohydrate acceptor (Balsamo *et al.*, 1986), suggesting the possibility of a lectinlike function for this enzyme in retina. Immunochemical analysis using antibodies which do not distinguish the two forms of the enzyme reveals that at least one form of the enzyme is associated with cells throughout the retina in the early embryo but becomes restricted to synaptic layers and to the outer segment of the photoreceptor in the retina of adult animals. The isolation of both soluble and particulate forms of the enzyme from the retina, together with the developmental change in its anatomical localization, suggests that the soluble (enzymatic function) and particulate (lectin function) forms of the enzyme may play distinct roles in the organization or maintenance of retinotectal synaptic connections.

Taken together, these results suggest that distinct molecular mechanisms may contribute to the generation of retinotectal topography along the dorsoventral and anteroposterior axes of the retina and tectum. Glycoconjugates remain likely candidates as mediators of retinotectal recognition along the dorsoventral axis, via graded increases in adhesive affinity. In contrast, the studies of Bonhoeffer and colleagues suggest that topography along the anteroposterior axis involves tectal protein components that inhibit the migration of incoming temporal retinal axons. Functional and immunocytochemical studies have also indicated that molecular distinctions in anterior and posterior retinal axons may contribute to the generation of selective projections along this axis (Bonhoeffer and Huf, 1984).

5.2. Carbohydrate Recognition in the Spinal Cord

Primary sensory neurons in the dorsal root ganglion (DRG) transmit peripheral sensory information from cutaneous and muscle receptors to second-order neurons in the spinal cord. Analysis of the receptive properties and specialized terminal morphology of peripheral sensory endings has delineated more than a dozen functional classes of DRG neurons. Individual sets of cutaneous sensory afferents project to segregated domains of the dorsal horn (Brown, 1981) that coincide with the laminar divisions originally defined on the basis of spinal cord neuronal cytoarchitecture. Specificity in the central projections of different classes of primary afferents is apparent from the time that afferent fibers first enter the embryonic rat dorsal gray matter (Smith, 1983; Yamada, 1985). The central branches of distinct classes of DRG neurons migrate toward appropriate targets over a similar period of embryonic development (Smith, 1983; Yamada, 1985). Axonal guidance mechanisms may therefore operate independently of any temporal and positional constraints on axonal outgrowth. The existence of specific targeting mechanisms for sensory axons is supported by anatomical and functional studies on cutaneous sensory afferents that make initial errors in their ventral projections and enter the spinal cord via the central roots. Ventral root afferents appear to project to appropriate regions of the dorsal horn via aberrant spinal pathways (see Dodd and Jessell, 1986).

Several studies have now provided evidence that cell-surface oligosaccharide structures and complementary carbohydrate-binding proteins are expressed in a temporally and positionally restricted manner on subsets of DRG and spinal cord neurons during development. These molecules therefore, constitute a potential recognition system that may be involved in guiding sensory axons to spinal cord targets.

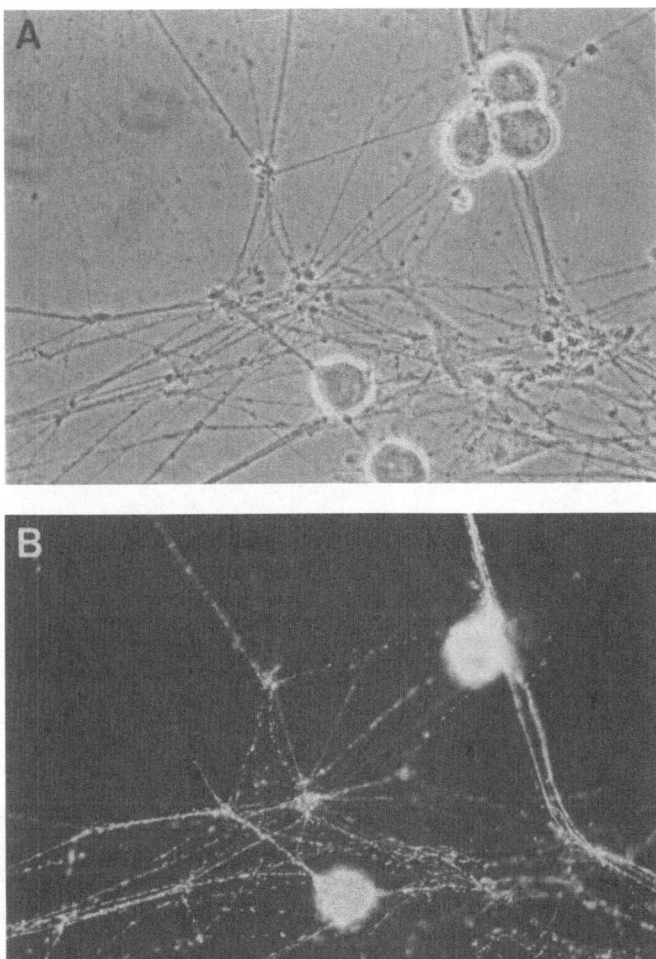

Figure 3. Expression of terminal α-galactose oligosaccharides on the surface of DRG neurons. (A) Phase-contrast micrograph of DRG neurons grown in dissociated cell culture. (B) Fluorescence micrograph of the same field, showing labeling of cell bodies and processes of a subset of the neurons by mAb LA4. Bar = 30 μm.

5.2.1. Sensory-Specific Surface Oligosaccharides

Immunocytochemical studies have identified three families of oligosaccharides, the globoseries, ganglioseries, and lactoseries carbohydrates that are expressed on subsets of DRG neurons (Dodd *et al.*, 1984; Dodd and Jessell, 1985; Jessell and Dodd, 1985) (Figure 3). The expression of these oligosaccharides by DRG neurons correlates with the established morphology and projection sites of identified subsets of DRG neurons.

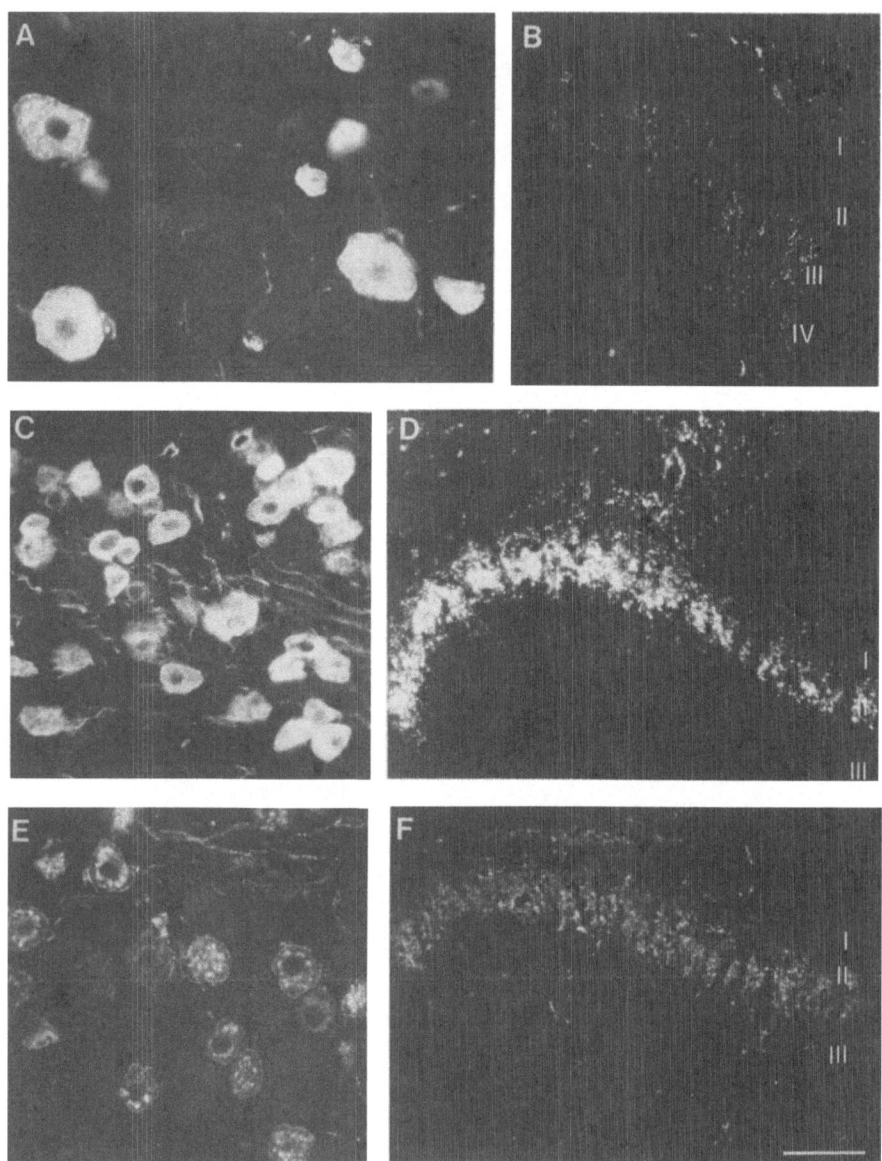

Figure 4. Expression of different classes of oligosaccharide by distinct subsets of sensory neurons. (A) mAb anti-SSEA-3 labels a subset of large- and small-diameter DRG neurons. (B) Section of the dorsal horn of the spinal cord in which the terminals of SSEA-3 $^+$ DRG neurons can be seen in laminae III and IV. A few fibers also project medially toward the central canal. SSEA-3 $^+$ fibers can also be seen in lamina I. Lamina II is devoid of staining. (C) mAb A5 labels approximately 50% of the cell bodies in the DRG. (D) Afferent terminals recognized by mAb LD2 are found in the outer region of lamina II. (E) LA4 labels a separate population of DRG cell bodies whose terminals are found in the inner part of lamina II (F). Calibrations: A, 35 μm; B, 150 μm; C, 75 μm; D, E, 40 μm; F, 50 μm.

Globoseries glycoconjugates can be recognized by the mAbs anti-SSEA-3 and anti-SSEA-4 and are associated with DRG neurons of both small and large diameter. DRG neurons that express globoseries structures do not contain neuropeptides or other cytochemical markers that define subpopulations of small-diameter DRG neurons. The central terminals of SSEA-3/4$^+$ DRG neurons are concentrated in lamina III and the medial part of lamina IV, with a sparse projection to lamina I (Dodd et al., 1984) (Figure 4). The distribution of afferent fibers that express this class of oligosaccharides suggests that globoseries determinants are restricted to myelinated primary afferents conveying low-threshold cutaneous information to deeper laminae and high-threshold mechanoreceptive or thermoreceptive information to lamina I.

Approximately 50% of adult DRG neurons express the type 2 lactoseries structures recognized by mAbs A5 and 1B2. This antigen appears during embryonic development and is present in approximately 10% of neurons in E18 DRG. During postnatal week 1, the proportion of neurons expressing this structure increases to 50%. The A5-reactive N-acetyllactosamine structure is expressed on the surface of small- and intermediate-diameter DRG neurons and on the central terminals of these neurons in laminae I and II of the dorsal horn. The A5$^+$ population of DRG neurons contains the peptides substance P and somatostatin, and the sensory neuron-specific acid phosphatase, FRAP (Dodd and Jessell, 1985; Jessell and Dodd, 1985).

The expression of α-galactose-extended ganglio- or lactoseries structures is associated with more restricted subsets of small-diameter DRG neurons (Dodd and Jessell, 1985; Jessell and Dodd, 1985). The complex-type α-galactose-extended structure identified by mAb LD2 is expressed by a subpopulation of small DRG neurons that projects to the dorsal region of lamina II. The LD2-reactive oligosaccharide cosegregates with the peptide phenotype of small-diameter sensory neurons, in that it is expressed by all SOM$^+$ but not by SP$^+$ DRG neurons. A related galactose-extended structure, recognized by mAb LA4, is found on a separate population of small- and intermediate-diameter DRG neurons (constituting 40–50% of total DRG neurons) that project predominantly to the ventral part of lamina II (Dodd and Jessell, 1985; Jessell and Dodd, 1985) (Figure 4). Other complex lactoseries oligosaccharides delineate further subsets of small-diameter DRG neurons (Dodd and Jessell, 1985). The distribution of afferent terminals expressing these sacchardies suggests that they represent C fibers and also some A-delta fibers (Gobel et al., 1981; Rethelyi, 1977).

5.2.2. Spinal Cord-Specific Oligosaccharides

Distinct oligosaccharides of the lactoseries group are also found in gradients along the dorsoventral axis of the developing spinal cord. A fucosylated lactoseries structure that includes the lactofucopentaose III trisaccharide is expressed selectively in embryonic spinal cord (Dodd and Jessell, 1986). This antigen, identified by mAb AC4, is first detected on cells in the dorsal spinal cord at E13–14. The dorsally restricted expression of this molecule does not correlate with the stage of differentiation of spinal cord cells at this stage. Moreover, the appearance of the AC4 antigen in dorsal spinal cord is transient, reaching a peak at E14–14.5. By E15.5, the antigen is almost undetectable in the dorsal spinal cord. At later stages of embryonic development, both the AC4 antigen and a closely related fucosylated lactoseries structure that is recognized by mAb anti-SSEA-1 are expressed in other regions of the spinal cord. By birth, SSEA-1 and AC4 are distributed throughout the spinal cord. This distribution is maintained in adult animals.

5.2.3. Carbohydrate Binding Proteins in Developing Sensory Neurons

Cellular interactions mediated by cell-surface oligosaccharides are likely to involve complementary receptors that function as carbohydrate binding proteins (Barondes, 1970). Several proteins that have been isolated from nonneural tissues are known to recognize β-galactoside linkages contained within the lactoseries structures described above. β-Galactoside binding proteins have been detected in embryonic brain and spinal cord (DeWaard et al., 1976; Kobiler and Barondes, 1977; Kobiler et al., 1978; Eisenbarth et al., 1978; Childs and Feizi, 1979). This class of carbohydrate binding protein exists in neural tissue in soluble form and appears to have multiple carbohydrate binding sites (Barondes, 1984). There is some evidence for the release of these lectinlike molecules from the cells that synthesize them (Beyer and Barondes, 1982; Cerra et al., 1984; Bols et al., 1986). Lectins therefore represent ligands that might mediate interactions between lactoseries structures on DRG neurons and spinal cord cells.

Antibodies raised against rat β-galactoside binding lectins originally isolated from lung (Cerra et al., 1985) have been used to identify two related or identical proteins in DRG and spinal cord (Regan et al., 1986) (Figure 5). These two lectins, termed RL 14.5 and RL 29, have subunit M_r values of 14,500 and 29,000, respectively, and are present in a subset of DRG neurons and in their central terminals in the superficial dorsal horn (Figure 5A,B). Both lectins are synthesized in the population of DRG neurons that expresses cell-surface lactoseries glycoconjugates (Regan et al., 1986; Hynes et al., 1988). Dorsal horn neurons that express lactoseries structures do not appear to synthesize these two β-galactoside binding proteins. RL 14.5 is first detectable in DRG neurons and sensory axons in the DREZ at E13–14, whereas RL 29 cannot be detected before E15 (Figure 5). From E16 both lectins are present in sensory afferent fibers as they enter the spinal cord (Figure 6).

β-Galactoside binding lectins similar to those identified in DRG neurons are known to be released from cells and can interact with membrane-bound or extracellular glycoconjugates on adjacent cells (Barondes and Haywood-Reid, 1981; Beyer and Barondes, 1982). These experiments suggest that the RL 14.5 and RL 29 may be localized extracellularly in the embryonic dorsal horn. At present, the role of these molecules in sensory neuron development has not been established. However, several possible modes of interaction can be envisaged. For example, the transient expression of lactoseries structures on embryonic dorsal horn neurons, combined with the onset of synthesis of β-galactoside binding lectins, could provide a signaling mechanism that both initiates the ingrowth and restricts the dorsoventral projection of sensory axons that express lactoseries oligosaccharides. Attempts to perturb sensory projections with antibodies or multivalent oligosaccharide ligands that block saccharide–carbohydrate binding protein interaction may provide more direct evidence about the role of these molecules.

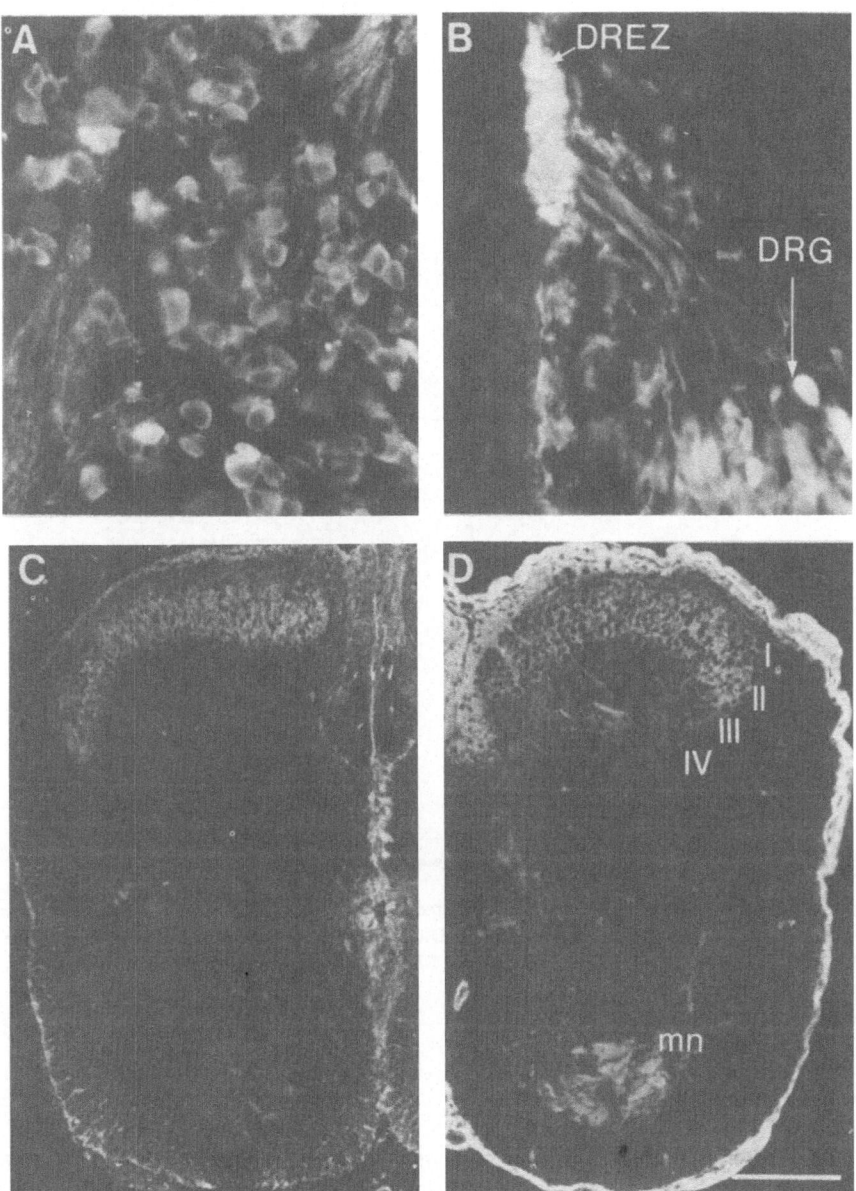

Figure 5. Immunocytochemical localization of β-galactoside binding lectins in rat DRG and spinal cord. (A) RL 29 in embryonic day (E) 20 DRG. A subpopulation of neurons expresses the lectin. (B) Localization of RL 14.5 in the DRG and in afferent fibers projecting to the spinal cord at E14. Although axons in the dorsal root entry zone (DREZ) express RL 14.5, immunoreactive sensory fibers are not detectable in the dorsal horn at this age. (C) RL 29 in the spinal cord on postnatal day 0. Immunoreactive sensory fibers are restricted to laminae I and II. (D) Intense RL 14.5 reactivity is found in sensory terminals in laminae I and II. A subset of motorneurons (mn) also appears to contain RL 14.5. Calibrations: A, 55 μm; B, 75 μm; C, D, 150 μm.

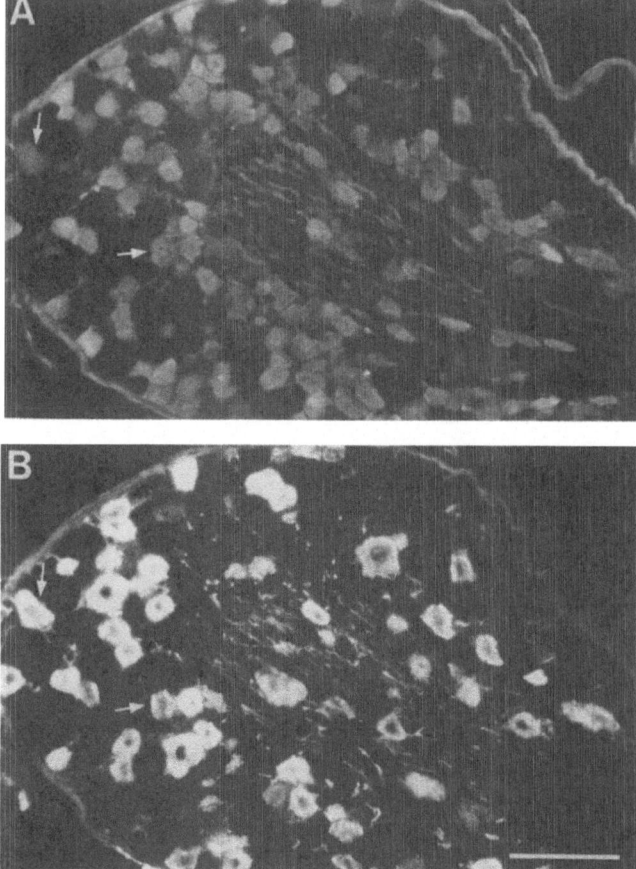

Figure 6. β-Galactoside binding lectins are synthesized by DRG neurons that express lactoseries oligosaccharides. (A) RL 29 labeling in a section of adult rat DRG. (B) LD2 immunoreactivity in the same section as that in A. Most neurons that are LD2+ also contain RL 29. Arrows indicate the same pair of cells in each section for orientation purposes. Bar = 85 μm.

5.3. Neuromuscular Junction

Carbohydrate components of glycoconjugates may be involved in the establishment or stabilization of synaptic connections. The possible role of glyconjugate interactions in synaptic development has been examined in most detail at the neuromuscular junction. Of particular interest is the finding by Sanes and Cheney (1982) that the *Dolichos biflorus* (DBA) lectin binds to a carbohydrate structure in the postsynaptic junction on rat skeletal muscle, whereas extrasynaptic sites are not labeled. Further characterization of this carbohydrate epitope has been carried out using multiple lectins, antibodies, and glycosidases as histochemical probes. The DBA lectin recognizes terminal GalNAc residues and selectively stains synaptic sites. Other lectins which bind to terminal GalNAc but that can also bind other saccharides label synaptic sites intensely, but also stain extrasynaptic sites. Lectins specific for sugars other than GalNAc stain both synaptic and extrasynaptic regions of muscle fibers equally (Scott *et al.*, 1988). From these studies it is clear that while a number of different carbohydrate structures are present on the surface of muscle fibers, molecules expressing terminal GalNAc are highly concentrated at the neuromuscular junction.

Two different synapse-specific glycoconjugates contain a carbohydrate epitope with terminal β-GalNAc residues and represent candidate molecules recognized by the DBA lectin. One molecule is the asymmetric form of the enzyme acetycholinesterase (AChE) (Scott *et al.*, 1988). The second molecule has been identified using anti-SSEA-3 antibodies which recognize glycolipids which contain β-GalNAc residues. Pretreatment of sections with chloroform–methanol to remove glycolipids abolishes reactivity in the postsynaptic densities, providing some evidence that this epitope resides on a glycolipid in the neuromuscular junction. Under the same conditions, antibody recognition of AChE, and DBA lectin binding sites are maintained, indicating that the carbohydrate epitope present on AChE is distinct from the carbohydrate epitope present on the glycolipid antigen.

The significance of two molecules with a synapse-specific localization and which contain a similar or identical carbohydrate structure is unclear at present. A synapse-specific carbohydrate may result in preferential localization of motor neurons to restricted sites on the muscle. In this light it is interesting that Obata *et al.* (1977) reported that the glycolipid globoside, which is recognized by anti-SSEA-3, perturbs normal synaptogenesis when added to nerve–muscle cocultures.

In addition to the selective localization of a carbohydrate with terminal β-GalNAc residues in the neuromuscular junction, GalNAcTase has also been found to be highly concentrated in neuromuscular junctions in normal adult muscle. Moreover, denervation results in the appearance of the enzyme at extrasynaptic sites. Two days after denervation, anti-GalNAcTase continues selectively to stain synaptic sites, suggesting that this antigen is associated with the muscle and not with the presynaptic fibers. Additional evidence that GalNAcTase is localized at synaptic sites in the muscle is the detection of a 220-kDa protein (the size of retinal GalNAcTase) on Western blots of synapse-rich but not synapse-free regions of the muscle. Furthermore, a 10- to 20-fold enrichment of GalNAcTase activity in synapse-rich, versus synapse-free portions of the muscle has been detected.

The restricted expression of the GalNAcTase may contribute to the localization of synapse-specific carbohydrates at the adult neuromuscular junction (Scott *et al.*, 1988). These observations raise the possibility that GalNAcTase might contribute to the establishment of synaptic connections, either by the regulation of glycoconjugate synthesis or by other adhesive interactions with molecules at the synaptic surface.

6. CARBOHYDRATE STRUCTURES ON NEURAL CELL ADHESION MOLECULES

Studies of sperm–egg interactions in mammalian development described above have established that the oligosaccharide structures associated with glycoproteins can serve as functional receptors in cell adhesion and recognition, independent of their

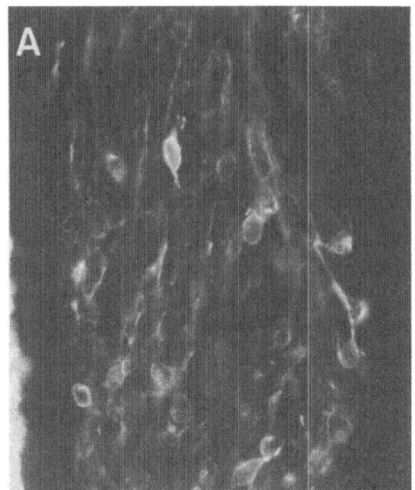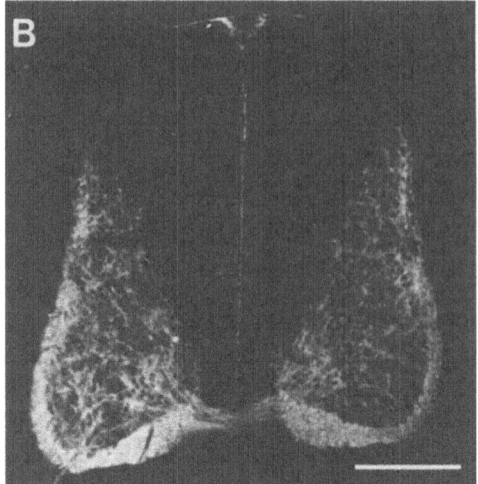

Figure 7. HNK-1 expression in embryonic spinal cord and sensory neurons. (A) A subset of E14 DRG expresses the HNK-1 epitope. (B) HNK-1 is expressed by differentiated cells in the ventral horn of E14 spinal cord. In addition, HNK-1 labels the roof plate of the spinal cord. Calibrations: A, 65 μm; B, 300 μm.

protein backbone. Analysis of the structure of several glycoproteins that have been implicated in neural cell adhesion has revealed that many of these proteins express a common carbohydrate epitope recognized by antibodies HNK-1, L2, NC1, and Leu 7 (McGarry *et al.*, 1985; Kruse *et al.*, 1984; see Kunemund *et al.*, 1988) (Figure 7). The structure of this epitope has been determined as a 3-sulfated glucuronyl derivative of a lactoseries oligosaccharide (Chou *et al.*, 1986; Ariga *et al.*, 1987). In peripheral nerve and in CNS, this structure is associated with glycolipids, glycoproteins, and proteoglycans (see Chapters 1 and 3). Defined neural cell surface glycoproteins that express the HNK-1 epitope include NCAM (Edelman, 1986; Rutishauser and Goridis, 1986), L1 (NILE) (Schachner *et al.*, 1985), TAG-1 (Dodd *et al.*, 1988), cytotactin (tenascin) (see Kunemund *et al.*, 1988), J1 (Kruse *et al.*, 1985), myelin-associated glycoproteins (Poltorak *et al.*, 1987), the fibronectin receptor α-subunit (a member of the integrin class of receptors) (Pesheva *et al.*, 1987), the myelin protein P0 (Bollensen and Schachner, 1987). It seems likely that this epitope is expressed by a much greater number of glycoproteins, since there are numerous HNK-1-reactive bands detected on Western blots that do not correspond to these defined glycoproteins.

The functional role of the HNK-1/L2 epitope is unclear. Fab fragments of L2 antibodies have been shown to perturb neuron–astrocyte and astrocyte–astrocyte adhesion *in vitro* (Kunemund *et al.*, 1988) and HNK-1 antibodies appear to inhibit process outgrowth of vertebrate neurons on conditioned medium or basement membrane glycoprotein substrates (Riopelle *et al.*, 1986). However, it has been difficult from these experiments to exclude the steric inhibition of functional protein domains as a consequence of antibody binding to the HNK-1/L2 epitope. The absence of *N*-linked oligosaccharides expressing the L2 epitope on NCAM does not appear to affect the adhesive properties of the molecule, suggesting that, at least in this case, homophilic binding function is not dependent on the presence of this carbohydrate structure (Cole and Schachner, 1987). The use of glycolipids that express the HNK-1/L2 structure has, however, provided some evidence that this oligosaccharide may influence cell–substratum adhesion and process outgrowth. Addition of HNK-1/L2-reactive glycolipids or tetrasaccharides purified from peripheral nerve has been shown to reduce the migration of cells from nerve tissue explants and to inhibit neurite extension on poly-D-lysine or laminin substrates (Kunemund et al., 1988). Of other glycolipids and saccharides examined, only sulfatide and heparin produced similar effects. The HNK-1/L2 glycolipid also appears to inhibit the adhesion of neural cells to laminin substrates. These observations provide some evidence that the HNK-1/L2 epitope can, under certain circumstances, contribute to neural cell adhesion and to neurite extension.

The mechanism by which HNK-1/L2 carbohydrates may modify neural cell function is not clear. Since the HNK-1/L2 structure is associated with lactoseries backbone structures, it may represent a ligand for β-galactoside binding lectins that are found in subsets of central and peripheral neurons (see above). Structural studies of the binding site specificity of the RL 14.5 and RL 29 lectins have indicated that both proteins can accommodate charged saccharides, such as sialic acid, linked to the terminal β-linked galactose (Leffler and Barondes, 1986). An alternative possibility is that the HNK-1/L2 structure modifies the adhesive function of glycoproteins that express this epitope. NCAM, in particular, has been shown to express a heparin binding domain that appears to regulate NCAM function. This domain may also interact with the HNK-1/L2 structure although, as discussed above, the absence of this carbohydrate epitope does not appear to modify NCAM adhesive function. Other roles for the HNK-1/L2 epitope are certainly possible. It is conceivable that the carbohydrate represents a recognition signal involved in the targeting of certain sets of glycoproteins to the cell surface. At present the diverse localization and function of proteins that express the HNK-1/L2 epitope makes it difficult to ascribe a general function to this structure. It also appears that the HNK-1/L2 epitope is not the only complex oligosaccharide associated with neural cell adhesion molecules. Recently a novel carbohydrate structure, defined as the L3 epitope, has been identified on L1 and other neural surface glycoproteins (Kucherer et al., 1987), although no function has been ascribed to this structure.

7. CONCLUSIONS

Over the past decade, the potential role of cell surface oligosaccharides as mediators of cell–cell recognition and adhesion in vertebrates has received strong experimental support in several diverse systems. It is perhaps not unexpected that the clearest functional evidence for carbohydrate recognition has emerged from the simplest of cell interactions, that of sperm–egg recognition. The complexity of cellular interactions in the nervous system, in comparison to events that occur in preimplantation embryos, suggests that an appreciation of the precise role of carbohydrate recognition in neural function is still some way off. However, several features that have emerged from a comparative analysis of neural and nonneural systems in which carbohydrate recognition has been involved provide an encouraging prognosis for future progress.

The structural and immunological characterization of oligosaccharides that has been achieved in recent years has provided strong evidence for a conservation in basic saccharide structures on functionally distinct neural and nonneural cell types. Thus, the oligosaccharide profile of the cell surface of the preimplantation mouse embryo is recapitulated by developing primary sensory neurons and at newly formed nerve–muscle synapses. The detection of cell surface glycosyltransferases at many stages of development and the availability of probes with which to modify or perturb enzyme function have also revived interest in the role of these enzymes as mediators of cell recognition. In the case of galactosyltransferases in particular, the range of molecular probes that are available has provided compelling support for the original proposal of their function as recognition molecules. Moreover, a clearer understanding of the relationship and interactions of glycosyltransferase- and cadherin-mediated adhesive systems that are coincidentally expressed on individual cells may provide important insights into the combined function of these adhesion and recognition molecules. The coincident expression of multiple adhesive systems on preimplantation embryos has, in fact, close parallels in the developing retina.

While the structural classes of oligosaccharides have been at least partially delineated, the extent of diversity in carbohydrate binding proteins that interact with cell surface oligosaccharides is still not clear. Biochemical analysis and molecular cloning have begun to reveal the existence of several structurally distinct classes of carbohydrate binding proteins. There is now extensive documentation of the developmental expression and neuronal specificity of individual glycosyltransferases and soluble β-galactoside binding proteins. However, in nonneural systems the Ca^{2+}-dependent carbohydrate binding proteins of the hepatic ASGP-R class have been shown to represent the most diverse molecular and functional class. It seems likely that members of this class of protein will be found within the nervous system. Identification of the proteins involved in carbohydrate recognition may also permit a more powerful genetic dissection of the function of carbohydrate-mediated cell interactions in the vertebrate nervous system.

ACKNOWLEDGMENTS

Research in the authors' laboratory is supported by the Howard Hughes Medical Institute (M.A.H. and T.M.J.), the NIH, The McKnight Foundation, and the National Multiple Sclerosis Society. We thank Rita Lenertz for her help and patience in preparing the manuscript.

8. REFERENCES

Andrews, P., Milsom, D. W., and Stoddart, R. W., 1983, Glycoconjugates from high endothelial cells. I. Partial characterization of a sulphated glycoconjugate from the high endothelial cells of rat lymph nodes, *J. Cell Sci.* **59**:231–244.
Ariga, T., Kohriyama, T., Freddo, L., Latov, N., Saito, M., Kon, K., Ando, S., Suzuki, M., Hemling, M. E., Rinehart, K. L., Kusunoki, S., and Yu, R. K., 1987, Characterization of sulfated glucuronic

acid containing glycolipids reacting with IgM proteins in patients with neuropathy. *J. Biol. Chem.* **262:** 848–854.

Ashwell, G., and Harford, J., 1982, Carbohydrate-specific receptors of the liver, *Annu. Rev. Biochem.* **51:** 531–554.

Balsamo, J., Pratt, R. S., and Lilien, J., 1986, Chick neural retina N-acetylgalactosaminyltransferase/acceptor complex: Catalysis involves transfer of N-acetylgalactosamine phosphate to endogenous acceptors, *Biochemistry* **25:**5402–5407.

Barbera, A. J., 1975, Adhesive recognition between developing retinal cells and the optic-tecta of the chick embryo, *Dev. Biol.* **46:**167–191.

Barbera, A. J., Marchase, R. B., and Roth, S., 1973, Adhesive recognition and retinotectal specificity, *Proc. Natl. Acad. Sci. USA* **70:**2482–2486.

Barondes, S. H., 1970, Brain glycomacromolecules and interneuronal recognition, in: *Neurosciences: Second Study Program* (F. O. Schmitt, ed.), pp. 747–760, Rockefeller University Press, New York.

Barondes, S. H., 1984, Soluble lectins: A new class of extracellular proteins, *Science* **233:**1259–1264.

Barondes, S. H., and Haywood-Reid, P. L., 1981, Externalization of an endogenous chicken muscle lectin with in vivo development, *J. Cell Biol.* **91:**568–572.

Bayna, E. M., Runyan, R. B., Scully, N. F., Reichner, J., Lopez, L. C., and Shur, B. D., 1986, Cell surface galactosyltransferase as a recognition molecule during development, *Mol. Cell. Biochem.* **72:** 141–151.

Bayna, E. M., Shaper, J. H., and Shur, B. D., 1988, Temporally specific involvement of cell surface β-1,4 galactosyltransferase during mouse embryo morula compaction, *Cell* **53:**145–157.

Beyer, E. C., and Barondes, S. H., 1982, Secretion of endogenous lectin by chicken intestinal goblet cells, *J. Cell Biol.* **92:**28–33.

Blackburn, C. C., Swank-Hill, P., and Schnaar, R. L., 1986, Gangliosides support neural retina cell adhesion, *J. Biol. Chem.* **261:**2873–2881.

Bleil, J. D., and Wassarman, P. M., 1988, Galactose at the nonreducing terminus of O-linked oligosaccharides of mouse egg ZP3 is essential for the glycoprotein's sperm receptor activity, *Proc. Natl. Acad. Sci. USA* **85**(18):6778–6782.

Blum, A. S., and Barnstable, C. J., 1987, O-acetylation of a cell-surface carbohydrate creates discrete molecular patterns during neural development, *Proc. Natl. Acad. Sci. USA* **84:**8716–8720.

Bollensen, E., and Schachner, M., 1987, The peripheral myelin glycoprotein PO expresses the L2/HNK-1 and L3 carbohydrate structures shared by neural adhesion molecules, *Neurosci. Lett.* **82:**77–82.

Bols, N. C., Roberson, M. M., Haywood-Reid, P. L., Cerra, R. F., and Barondes, S. H., 1986, Secretion of a cytoplasmic lectin from *Xenopus laevis* skin, *J. Cell Biol.* **102:**492–499.

Bonhoeffer, F., and Huf, J., 1982, *In vitro* experiments on axon guidance demonstrating an anterior-posterior gradient on the tectum, *EMBO J.* **1:**427–431.

Bonhoeffer, F., and Huf, J., 1984, Position-dependent properties of retinal axons and their growth cones, *Nature* **315:**409–410.

Brandley, B. K., Ross, T. S., and Schnaar, R. L., 1987, Multiple carbohydrate receptors on lymphocytes revealed by adhesion to immobilized polysaccharides, *J. Cell Biol.* **105:**991–997.

Brown, A. G., 1981, *Organization in the Spinal Cord*, p. 239, Springer-Verlag, Berlin.

Calarco-Gillam, P., 1985, Cell–cell interactions in mammalian preimplantation development, in: *Developmental Biology: A Comprehensive Synthesis*, Vol. 2 (L. Browder, ed.), pp. 329–371, Plenum Press, New York.

Cerra, R. F., Haywood-Reid, P. L., and Barondes, S. H., 1984, Endogenous mammalian lectin localized extracellularly in lung elastic fibers, *J. Cell Biol.* **98:**1580–1589.

Cerra, R. F., Gitt, M. A., and Barondes, S. H., 1985, Three soluble rat β-galactoside-binding lectins, *J. Biol. Chem.* **260:**10474–10477.

Chiacchia, K. B., and Drickamer, K., 1984, Direct evidence for the transmembrane orientation of the hepatic glycoprotein receptors, *J. Biol. Chem.* **259:**15440–15446.

Childs, R. A., and Feizi, T., 1979, Calf heart lectin reacts with blood group Ii antigens and other precursor chains of the major blood group antigens, *FEBS Lett.* **99:**175–179.

Chou, D. K., Ilyas, A. A., Evans, J. E., Costello, C., Quarles, R. H., and Jungalwala, F. B., 1986, Structure of sulfated glucuronyl glycolipids in the nervous system reacting with HNK-1 antibody and some IgM paraproteins in neuropathy, *J. Biol. Chem.* **261:**11717–11725.

Clerch, L. B., Whitney, P., Hass, M., Brew, K., Miller, T., Werner, R., and Massaro, D., 1988, Sequence of a full-length cDNA for rat lung β-galactoside-binding protein: Primary and secondary structure of the lectin, *Biochemistry* **27**:692–699.

Cole, G. J., and Schachner, M., 1987, Localization of the L2 monoclonal antibody binding site on N-CAM and evidence for its role in N-CAM mediated cell adhesion, *Neurosci. Lett.* **78**:227–232.

DeWaard, A., Hickman, S., and Kornfeld, S., 1976, Isolation and properties of β-galactoside-binding lectins of calf heart and lung, *J. Biol. Chem.* **251**:7581–7587.

Dodd, J., and Jessell, T. M., 1985, Lactoseries carbohydrates specify subsets of dorsal root ganglion neurons projecting to superficial dorsal horn of rat spinal cord, *J. Neurosci.* **5**:3278–3294.

Dodd, J., and Jessell, T. M., 1986, Cell surface glycoconjugates and carbohydrate-binding proteins: Possible recognition signals in sensory neurone development, *J. Exp. Biol.* **129**:225–238.

Dodd, J., Solter, D., and Jessell, T. M., 1984, Monoclonal antibodies against carbohydrate differentiation antigens identify subsets of primary sensory neurons, *Nature* **311**:469–472.

Dodd, J., Morton, S. B., Karagogeos, D., Yamamoto, M., and Jessell, T. M., 1988, Spatial regulation of axonal glycoprotein expression on subsets of embryonic spinal neurons, *Neuron* **1**(2):105–116.

Drickamer, K., 1987, Membrane receptors that mediate glycoprotein endocytosis: Structure and biosynthesis, *Kidney Int.* **32**:S167–S183.

Drickamer, K., 1988, Two distinct classes of carbohydrate-recognition domains in animal lectins, *J. Biol. Chem.* **263**:9557–9560.

Dubois, C., Magnani, J. L., Grunwald, G. B., Spitalnik, S. L., Trisler, G. D., Nirenberg, M., and Ginsburg, V., 1986, Monoclonal antibody 18B8, which detects synapse-associated antigens, binds to ganglioside G_{T3} (II^3(NeuAc)$_3$ LacCer), *J. Biol. Chem.* **261**:3826–3830.

Edelman, G. M., 1986, Cell adhesion molecules in the regulation of animal form and tissue pattern, *Annu. Rev. Cell Biol.* **2**:81–116.

Edelman, G. M., Murray, B. A., Mege, R. M., Cunningham, B. A., and Gallin, W. A., 1987, Cellular expression of liver and neural cell adhesion molecules after transfection with their cDNAs results in specific cell–cell binding, *Proc. Natl. Acad. Sci. USA* **84**:8502–8506.

Eisenbarth, G. S., Ruffolo, R. R., Walsh, F. S., and Nirenberg, M., 1978, Lactose sensitive lectin of chick retina and spinal cord, *Biochem. Biophys. Res. Commun.* **83**:1246–1252.

Fenderson, B. A., Zehavi, U., and Hakomori, S. I., 1984, A multivalent lacto-N-fucopentose III-lysyllysine conjugate decompacts preimplantation mouse embryos, while the free oligo-saccharide is ineffective, *J. Exp. Med.* **160**:1591–1596.

Gallatin, M., St. John, T. P., Siegelman, M., Reichert, R., Butcher, E. C., and Weissman, I. L., 1986, Lymphocyte homing receptors, *Cell* **44**:673–680.

Gesner, B. M., and Ginsberg, V., 1964, Effect of glycosidases on the fate of transfused lymphocytes, *Proc. Natl. Acad. Sci. USA* **52**:750–755.

Gitt, M. A., and Barondes, S. H., 1986, Evidence that a human soluble β-galactoside-binding lectin is encoded by a family of genes, *Proc. Natl. Acad. Sci. USA* **83**:7603–7607.

Glabe, C. G., Grabel, L. B., Vacquier, V. D., and Rosen, S. D., 1982, Carbohydrate specificity of sea urchin sperm bindin: A cell surface lectin mediating sperm–egg adhesion, *J. Cell Biol.* **94**:123–128.

Gobel, S., Falls, W. M., and Humphrey, E., 1981, Morphology and synaptic connections of ultrafine primary axons in lamina 1 of the spinal dorsal horn: Candidates for the terminal axonal arbors of primary neurons with unmyelinated axons, *J. Neurosci.* **1**:1163–1179.

Grabel, L. B., 1984, Isolation of a putative cell adhesion mediating lectin from teratocarcinoma stem cells and its possible role in differentiation, *Cell Differ.* **15**:121–124.

Grabel, L. B., Singer, M. S., Rosen, S. D., and Martin, G. R., 1983, The role of carbohydrates in the intercellular adhesion and differentiation of teratocarcinoma stem cells, in: *Teratocarcinoma Stem Cells*, Vol. 10 (L. M. Silver, G. R. Martin, and S. Strickland, eds.), pp. 145–163, Cold Spring Harbor Laboratory, Cold Spring Harbor, N.Y.

Grabel, L. B., Singer, M. S., Martin, G. R., and Rosen, S. D., 1985, Isolation of a teratocarcinoma stem cell lectin implicated in intercellular adhesion, *FEBS Lett.* **183**(2):228–231.

Grunwald, G. B., Fredman, P., Magnani, J. L., Trisler, D., Ginsburg, V., and Nirenberg, M., 1985, Monoclonal antibody 18B8 detects gangliosides associated with neuronal differentiation and synapse formation, *Proc. Natl. Acad. Sci. USA* **82**:4008–4012.

Haagsman, H. P., Haegood, S., Sargeant, I., Buckley, D., White, R. T., Drickamer, K., and Bernson, B. J., 1987, The major lung surfactant protein, SP 28–36, is a calcium-dependent, carbohydrate-binding protein, *J. Biol. Chem.* **262:**13877–13880.

Hakomori, S., 1981, Glycosphingolipids in cellular interaction, differentiation and oncogenesis, *Annu. Rev. Biochem.* **50:**733–764.

Halberg, D. F., Proulx, G., Doege, K., Yamada, Y., and Drickamer, K., 1988, A segment of the cartilage proteoglycan core protein has lectin-like properties, *J. Biol. Chem.* **263:**9486–9490.

Halfter, W., Claviez, M., and Schwarz, U., 1981, Preferential adhesion of tectal membranes to anterior embryonic chick retina neurites, *Nature* **292:**67–70.

Hinek, A., Wrenn, D. S., Mecham, R. P., and Barondes, S. H., 1988, The elastin receptor is a galactoside binding protein, *Science* **239:**6159.

Huang, T., and Yanagimachi, R., 1984, Fucoidin inhibits attachment of guinea pig spermatozoa to the zona pellucida through binding to the inner acrosomal membrane and equitonal domains, *Exp. Cell Res.* **153:**363–373.

Hyafil, F., Morello, D., Babinet, C., and Jacob, F., 1980, A cell surface glycoprotein involved in the compaction of embryonal carcinoma cells and cleavage stage embryos, *Cell* **21:**927–934.

Hynes, M. A., Buck, L. B., Casano, F. I., Huang, K. K., Barondes, S. H., and Jessell, T. M., 1988, Cloning and expression of a soluble 14 kDa β-galactoside binding lectin in rat nervous system, *Soc. Neurosci. Abstr.* **14:**71.

Jessell, T. M., 1988, Adhesion molecules and the hierarchy of neural development, *Neuron* **1:**3–13.

Jessell, T. M., and Dodd, J., 1985, Structure and expression of differentiation antigens on functional subclasses of primary sensory neurons, *Philos. Trans. R. Soc. London Ser. B* **308:**271–281.

Jia, S., and Wang, J. J., 1988, Carbohydrate binding protein 35: Complementary DNA sequence reveals homology with proteins of the hnRNP, *J. Biol. Chem.* **263:**6009–6011.

Kobiler, D., and Barondes, S. H., 1977, Lectin activity from embryonic chick brain, heart and liver: Changes with development, *Dev. Biol.* **60:**326–330.

Kobiler, D., Beyer, E. C., and Barondes, S. H., 1978, Developmentally regulated lectins from chick muscle, brain and liver have similar chemical and immunological properties, *Dev. Biol.* **64:**265–272.

Krantz, M. J., Holtman, N., Stowell, C., and Lee, Y. C., 1976, Attachment of thioglycosides to proteins: Enhancement of liver membrane binding, *Biochemistry* **15:**3963.

Kruse, J., Mallhammer, R., Wernecke, H., Falssner, A., Sommer, I., Goridis, C., and Schachner, M., 1984, Neural cell adhesion molecules and myeline-associated glycoprotein share a common carbohydrate moiety recognized by monoclonal antibodies L2 and HNK-1, *Nature* **311:**153–155.

Kruse, J., Keilhauer, G., Falssner, A., Timpl, R., and Schachner, M., 1985, The J1 glycoprotein—A novel nervous system cell adhesion molecule of the L2/HNK-1 family, *Nature* **316:**146–148.

Krusius, T., Gehlsen, K. R., and Ruoslahti, E., 1987, A fibroblast chondroitin sulfate proteoglycan core protein contains lectin-like and growth factor-like sequences, *J. Biol. Chem.* **262:**13120–13125.

Kucherer, A., Faissner, A., and Schachner, M., 1987, The novel carbohydrate epitope L3 is shared by some neural cell adhesion molecules, *J. Cell Biol.* **104:**1597–1602.

Kuhlenschmidt, T. B., Kuhlenschmidt, M. S., Roseman, S., and Lee, Y. C., 1984, Binding and endocytosis of glycoproteins by isolated chicken hepatocytes, *Biochemistry* **23:**6437–6444.

Kunemund, V., Jungalwala, F. B., Fischer, G., Chou, D. K. H., Keilhauer, G., and Schachner, M., 1988, The L2/HNK-1 carbohydrate of neural cell adhesion molecules is involved in cell interactions, *J. Cell Biol.* **106:**213–223.

Lagenaur, C., and Lemmon, V., 1987, An L1-like molecule, the 8D9 antigen, is a potent substrate for neurite extension, *Proc. Natl. Acad. Sci. USA* **84:**7753–7757.

Leffler, H., and Barondes, S. H., 1986, Specificity of binding of three soluble rat lung lectins to substituted and unsubstituted mammalian β-galactosides, *J. Biol. Chem.* **261:**10119–10126.

Levine, J. M., Beasley, L., and Stallcup, W. B., 1984, the D1.1 antigen: A cell surface marker for germinal cells of the central nervous system, *J. Neurosci.* **4:**820–831.

Lopez, L. C., Bayna, E. M., Litoff, D., Shaper, N. L., and Shur, B. D., 1985, Receptor function of mouse sperm surface galactosyltransferase during fertilization, *J. Cell Biol.* **101:**1501–1510.

Marchase, R. B., 1977, Biochemical investigations of retinotectal adhesive specificity, *J. Cell Biol.* **75:**237–257.

McGarry, R. C., Riopelle, R. J., and Roder, J. C., 1985, Accelerated regenerative neurite formation by a neuronal surface epitope reactive with the monoclonal antibody, Leu 7, *Neurosci. Lett.* **56**:95–100.

Nagafuchi, A., Shirayoshi, Y., Okazaki, K., Yasuda, K., and Takeichi, M., 1987, Transformation of cell adhesion properties by exogenously introduced E-cadherin cDNA, *Nature* **329**:341–343.

Narimatsu, H., Sinha, S., Brew, K., Okayama, H., and Qasba, P. K., 1986, Cloning and sequencing of cDNA of bovine N-acetylglucosamine (β1-4) galactosyltransferase, *Proc. Natl. Acad. Sci. USA* **83**: 4720–4724.

Obata, K., Oide, M., and Handa, S., 1977, Effects of glycolipids on *in vitro* development of neuromuscular junction, *Nature* **266**:369–371.

Paroutaud, P., Levi, G., Teichberg, V. I., and Strosberg, A. D., 1987, Extensive amino acid sequence homologies between animal lectins, *Proc. Natl. Acad. Sci. USA* **84**:6345–6348.

Pesheva, P., Horwitz, A. F., and Schachner, M., 1987, Integrin, the cell surface receptor for fibronectin and laminin, expresses the L2/HNK-1 and L3 carbohydrate structures shared by adhesion molecules, *Neurosci. Lett.* **83**:303–306.

Poltorak, M., Sadoul, R., Keilhauer, G., Landa, C., and Schachner, M., 1987, The myelin-associated glycoprotein (MAG), a member of the L2/HNK-1 family of neural cell adhesion molecules, is involved in neuron–oligodendrocyte and oligodendrocyte–oligodendrocyte interaction, *J. Cell Biol.* **105**:1893–1899.

Rastan, S., Thorpe, S. J., Scudder, D. P., Brown, S., Gooi, H. C., and Feizi, T., 1985, Cell interactions in preimplantation embryos: Evidence for involvement of saccharides of the poly-N-acetyllactosamine series, *J. Embryol. Exp. Morphol.* **87**:115–128.

Regan, L., Dodd, J., Barondes, S. H., and Jessell, T. M., 1986, Selective expression of endogenous lactose-binding lectins and lactoseries glycoconjugates in subsets of rat sensory neurons, *Proc. Natl. Acad. Sci. USA* **83**:2248–2252.

Rethelyi, M., 1977, Preterminal and terminal arborizations within substantia gelatinosa of cat spinal cord, *J. Comp. Neurol.* **172**:511–528.

Riopelle, R. J., McGarry, R. C., and Roder, J. C., 1986, Adhesion properties of a neuronal epitope recognized by the monoclonal antibody HNK-1, *Brain Res.* **367**:20–25.

Roseman, S., 1970, The synthesis of complex carbohydrates by multiglycosyltransferase systems and their potential function in intercellular adhesion, *Chem. Phys. Lip.* **5**:270–297.

Rosen, S. D., and Yednock, T. A., 1986, Lymphocyte attachment to high endothelial venules during recirculation: A possible role for carbohydrates as recognition determinants, *Mol. Cell. Biochem.* **72**: 153–164.

Rosen, S. D., Singer, M. S., Yednock, T. A., and Stoolman, L. M., 1985, Involvement of sialic acid on endothelial cells in organ-specific lymphocyte recirculation, *Science* **228**:1005–1007.

Roth, S., McGuire, E. J., and Roseman, J., 1971, Evidence for cell-surface glycosyltransferases: Their potential role in cellular recognition, *J. Cell Biol.* **51**:526–547.

Rutishauser, U., and Jessell, T. M., 1988, Cell adhesion molecules in vertebrate neural development, in: *Physiological Reviews*, The American Physiological Society, Bethesda, **68**(3):819–857.

Rutishauser, U., and Goridis, C., 1986, N-CAM: The molecule and its genetics, *Trends Genet.* **2**:72–76.

Sanes, J. R., and Cheney, M., 1982, Lectin binding reveals a synapse specific carbohydrate in skeletal muscle, *Nature* **300**:646–647.

Schachner, M., Faissner, A., Fischer, G., Keilhauer, G., Kruse, J., Kunemund, V., Linder, J., and Wernecke, H., 1985, Functional and structural aspects of the cell surface in mammalian nervous system development, in: *The Cell in Contact* (G. M. Edelman and J. P. Thiery, eds.), pp. 257–275, Wiley, New York.

Scott, L. J. C., Bacon, F., and Sanes, J. R., 1988, A synapse-specific carbohydrate of the neuromuscular junction: Association with both acetylcholinesterase and a glycolipid, *J. Neurosci.* **8**:932–944.

Shaper, N. L., Shaper, J. H., Meuth, J. L., Fox, J. L., Chang, H., Kirsch, I. R., and Hollis, G. F., 1986, Bovine galactosyltransferase: Identification of a clone by direct immunological screening of a cDNA expression library, *Proc. Natl. Acad. Sci. USA* **83**:1573–1577.

Shur, B. C., 1983, Embryonal carcinoma cell adhesion: The role of surface galactosyltransferase and its 90K lactosaminoglycan substrate, *Dev. Biol.* **99**:360–372.

Sly, W. S., and Fischer, H. D., 1982, The phosphomannosyl recognition system for intracellular and intercellular transport of lysosomal enzymes, *J. Cell Biochem.* **18**:67–85.

Smith, C. S., 1983, The development and postnatal organization of primary afferent projections to the rat thoracic spinal cord, *J. Comp. Neurol.* **220**:29–43.

Solter, D., and Knowles, B. B., 1978, Monoclonal antibody defining a stage-specific mouse embryonic antigen (SSEA-1), *Proc. Natl. Acad. Sci. USA* **75**:5565.

Sperry, R. W., 1963, Chemoaffinity in the orderly growth of mouse fiber patterns and connections, *Proc. Natl. Acad. Sci. USA* **50**:703–710.

Stamper, H. B., Jr., and Woodruff, J. J., 1976, Lymphocyte homing into lymph nodes: *In vitro* demonstration of the selective affinity of recirculating lymphocytes for high-endothelial venules, *J. Exp. Med.* **144**:828–833.

Stoolman, L. M., Tenforde, T. S., and Rosen, S. D., 1984, Phosphomannosyl receptors may participate in the adhesive interaction between lymphocytes and high endothelial venules, *J. Cell Biol.* **99**:1535–1540.

Takeichi, M., 1986, Molecular basis for teratocarcinoma cell–cell adhesion, in: *Developmental Biology,* Vol. 2 (L. M. Brounder, ed.), pp. 373–388, Plenum Press, New York.

Takeichi, M., 1988, The cadherins: Cell–cell adhesion molecules controlling animal morphogenesis, *Development* **102**:639–655.

Trisler, G. D., and Collins, F., 1987, Corresponding spatial gradients of TOP molecules in the developing retina and optic tectum, *Science* **237**:1208–1209.

Trisler, G. D., Schneider, M. D., and Nirenberg, M., 1981, A topographic gradient of molecules in retina can be used to identify neuron position, *Proc. Natl. Acad. Sci. USA* **78**:2145–2149.

Walter, J., Henke-Fahle, S., and Bonhoeffer, F., 1987a, Avoidance of posterior tectal membranes by temporal retina axons, *Development* **101**:909–914.

Walter, J., Kern-Veits, B., Huf, J., Stolze, B., and Bonhoeffer, F., 1987b, Recognition of position-specific properties of tectal cell membranes by retinal axons *in vitro, Development* **101**:685–696.

Wassarman, P. M., 1987, Early events in mammalian fertilization, *Annu. Rev. Cell Biol.* **3**:109–142.

Yamada, T., 1985, Development of afferent fibre projections in rat spinal cord, Thesis, pp. 1–54, Tsukuba University.

Yednock, T. A., Stoolman, L. M., and Rosen, S. D., 1987a, Phosphomannosyl-derivatized beads detect a receptor involved in lymphocyte homing, *J. Cell Biol.* **104**:713–723.

Yednock, T. A., Butcher, E. C., Stoolman, L. M., and Rosen, S. D., 1987b, Receptors involved in lymphocyte homing: Relationship between a carbohydrate-binding receptor and the MEL-14 antigen, *J. Cell Biol.* **104**:725–731.

12

Polysialic Acid as a Regulator of Cell Interactions

Urs Rutishauser

Polysialic acid, referring specifically to linear homopolymers of α2-8-linked *N*-acetyl-neuraminic acid (abbreviated here as PSA), is a remarkable carbohydrate structure. The basis for this statement begins with some surprising general observations: (1) the abundant presence of this structure in three very different biological contexts [bacterial capsules (Finne, 1985; Troy, 1979), fish eggs (Kitajima *et al.*, 1986; Inoue *et al.*, 1987), and surfaces of a variety of vertebrate cells (Finne, 1985; Margolis and Margolis, 1983; Chuong and Edelman, 1984; Finne *et al.*, 1987)], (2) the nearly complete restriction of vertebrate PSA to a single cell surface protein [the neural cell adhesion molecule (NCAM)] (Hoffman *et al.*, 1982; Finne *et al.*, 1983), and (3) the fact that in each of these situations the carbohydrate appears to form a space or barrier around a cell. In this chapter I will focus on the molecular, cell, and tissue biology of NCAM PSA, with particular emphasis on the hypothesis (Rutishauser *et al.*, 1988) that variations in this carbohydrate during development serve as an overall regulator of cell–cell and possibly cell–matrix interactions.

1. AN INTRODUCTION TO NCAM

1.1. Classification of Different Cell Adhesion Molecules

Although this chapter deals primarily with one type of CAM, it is useful to place this topic within a larger context by providing a brief summary of the spectrum of CAMs that have been identified. Although each CAM by definition is capable of forming a cell–cell bond, differences in their overall biological role suggest a division into two major groups, general and restricted CAMs (Rutishauser and Jessell, 1988).

Urs Rutishauser • Department of Genetics and Center for Neuroscience, Case Western Reserve University, School of Medicine, Cleveland, Ohio 44106.

The assignment of a particular CAM to either the general or restricted class is made primarily on the basis of its spatial and temporal distribution during development. General CAMs tend to be expressed at many sites and times during formation of the embryo, so that their individual binding functions affect several aspects of development. Restricted CAMs are found on distinct tissue subregions and are likely to mediate a more specialized type of cell–cell adhesion. There appear to be a relatively small number of general CAMs. At present these are the neural cell adhesion molecule (NCAM) and the members of the calcium-dependent CAMs called cadherins. The cadherins have been associated with a particular type of cell–cell interaction, the adherens-type junction (Gumbiner and Simons, 1986; Boller et al., 1985; Volk and Geiger, 1986), and are proposed to be involved in the selective segregation of cells into tissues (Takeichi, 1987). In contrast, NCAM appears to be involved with a very wide variety of seemingly unrelated cell membrane events, as described below.

1.2. Biochemistry of the NCAM Polypeptide

A variety of evidence is consistent with the hypothesis that NCAM participates in cell–cell adhesion via homophilic interaction between NCAM molecules present on the surface of both interacting cells (Rutishauser et al., 1982). This includes the demonstrations that (1) the binding of neurons and reconstituted NCAM vesicles, the latter containing only purified NCAM and lipids, is inhibited by exposing the cells alone to anti-NCAM Fab, (2) NCAM-mediated adhesion between two cells occurs only if both cells have NCAM on their surface, (3) antibodies directed against other membrane components or regions of the molecule not involved in binding have no effect on cell–cell adhesion, (4) purified soluble NCAM or NCAM incorporated into vesicles binds only to the surface of cells that express the molecule, (5) NCAM vesicles self-associate in suspension, and this aggregation is blocked by anti-NCAM Fab, and (6) purified NCAM, as observed by electron microscopy, exists as a multimer in which the intermolecular contact involves the putative homophilic binding region of the molecule (Rutishauser et al., 1982; Watanabe et al., 1986; Hall and Rutishauser, 1987).

All known forms of NCAM contain a single polypeptide chain, and there is considerable structural heterogeneity in both the protein and carbohydrate portions of the molecule. In this chapter, the major variants are named according to their high (H) or low (L) content of PSA (Hoffman et al., 1982; Sunshine et al., 1987) and the apparent molecular mass (in kilodaltons) of their polypeptide (Rougon et al., 1982; Hirn et al., 1983), as determined for the L form by SDS-PAGE. For example, the largest polypeptide with a high content of PSA is designated NCAM-180H.

A schematic model for the NCAM molecule is shown in Figure 1. This model is largely derived from the location of monoclonal antibody epitopes associated with known structures or activities of the molecule (Watanabe et al., 1986; Frelinger and Rutishauser, 1986), the predicted amino acid sequence from analysis of cDNA clones (Cunningham et al., 1987; Barthels et al., 1987; Small et al., 1987), and the appearance and dimensions of the metal-shadowed molecule in electron micrographs (Hall and Rutishauser, 1987). All epitopes associated with cell–cell binding are represented in a single, compact domain contained within an amino-terminal binding frag-

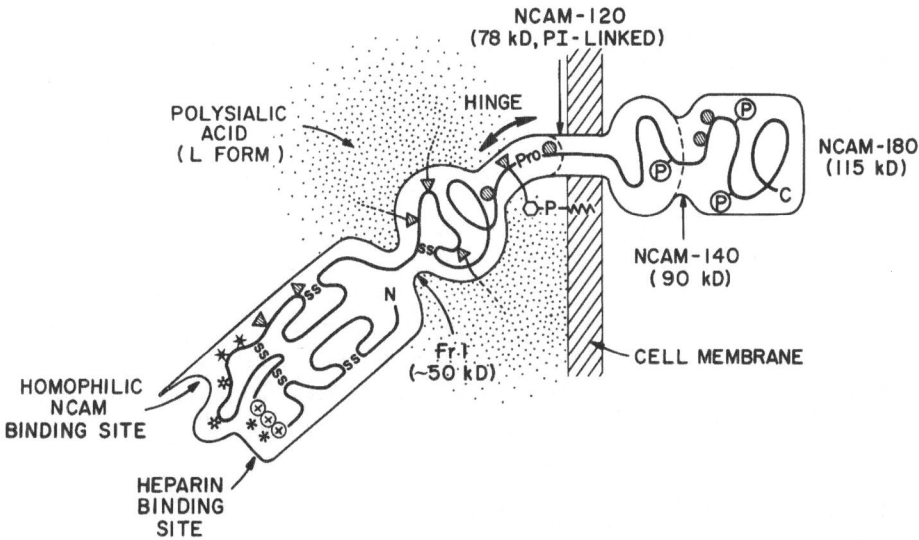

Figure 1. Schematic representation of NCAM, indicating the relative locations of the adhesion domain with two distinct binding regions, the carbohydrate-rich domain containing long polymers of sialic acid, and the cytoplasmic domains whose length represents the three major polypeptide forms. The approximate location of monoclonal antibody epitopes associated with adhesion functions (stars) and other regions of the molecule (circles and triangles) is indicated. Adapted from Frelinger et al. (1986).

ment Fr1 (Cunningham et al., 1983). These epitopes are clustered into two distinct groups, which are depicted as a heparin-binding site (Cole et al., 1986) and a homophilic binding site. The central domain is marked by both polypeptide- and carbohydrate-dependent epitopes, neither of which appear to be directly associated with binding activities. All PSA is confined to this domain (Crossin et al., 1984). The cytoplasmic domains include two compact domains representing NCAM-140 and NCAM-180, each of which contain phosphorylated threonine or serine residues (Sorkin et al., 1984; Gennarini et al., 1984).

An interesting feature of the NCAM gene is that it demonstrates a significant homology with the immunoglobulin gene superfamily, in particular with myelin-associated glycoprotein (Arquint et al., 1987) and the ICAM, LFA-3, and CD2 cell recognition glycoproteins (Seed, 1987; Simmons et al., 1988). The homology within NCAM occurs as five successive segments, each coding for about 100 amino acids and containing an intrachain disulfide bond (Cunningham et al., 1987). At present it is not clear whether this homology reflects conservation of domains that are related directly to the recognition function of the molecule.

1.3. NCAM Function in Development

The broad distribution of NCAM during development includes the primitive neural ectoderm of the neural plate, transient expression in morphogenetically active structures such as the notochord, neural crest, somites, placodes, some epidermis and mesenchyme, and the mesonephros, a uniform and persistent presence on neurons, and a spatially and temporally regulated expression by some glial and muscle cells (Thiery *et al.*, 1982; Keane *et al.*, 1988; Silver and Rutishauser, 1984; Maier *et al.*, 1986; Tosney *et al.*, 1986). Therefore, a substantial portion of embryonic tissues are composed of cells that produce, or whose precursors have produced, detectable amounts of NCAM. On this basis alone it is clear that the NCAM-mediated adhesion, with its relatively simple binding mechanism, is not well suited for identification of individual cells or specification of precise cell–cell interactions. Instead, the molecule may be more accurately described as a versatile "glue" which is used where and when a cell is required to establish adhesive contacts with an NCAM-positive neighbor. Most studies on NCAM function have focused on interactions among neurons, glia, and muscle cells, and examples are given below in which a combination of immunohistology and antibody-mediated perturbation has been used to illustrate that this general adhesion can contribute to a variety of important developmental events.

1.3.1. Growth of Optic Axons along Glial Endfeet

When the axons of retinal ganglion cells exit the eye, they follow a stereotyped route along the outer margin of the neuroepithelium, with their growth cones in close apposition to the endfest of radial glial cells. Moreover, axons originating from one region of the retina tend to remain close to each other as they grow in the optic nerve. Studies on the distribution of NCAM in the developing visual system of the chick have shown that the molecule is uniformly distributed on axons but preferentially expressed at the endfeet of the radial glia (Silver and Rutishauser, 1984). The asymmetrical pattern of expression on glia suggested that the preference of growth cones for endfeet might in part reflect NCAM-mediated adhesion. In support of this hypothesis, it was demonstrated that the presence of antibodies against NCAM can alter the route of the axons and at the same time reduces the adhesions that help stabilize fiber–fiber order (Silver and Rutishauser, 1984; Thanos *et al.*, 1984).

1.3.2. Innervation of Muscle

While the structure and function of the nerve–muscle synapse are well described, relatively little is known about the molecular events which lead to this cell–cell interaction. However, there is evidence that the initial innervation of the muscle may be regulated by NCAM expression. For example, in the case of the chick hindlimb, NCAM-positive nerves are observed to wait at the periphery of the muscle until the latter, by its own internal program, also produces large amounts of the molecule (Tosney *et al.*, 1986). At that time, the nerve–muscle association becomes more intimate, with extensive ramification of the axon into the muscle, and ultimately results in the formation of electrically active synapses. This correlation is supported by studies both *in vitro* and *in vivo* which indicate that inhibition of NCAM by antibody compromises the extent of axon–muscle contact (Rutishauser *et al.*, 1983; Landmesser *et al.*, 1988).

1.3.3. Establishment of Junctional Communication in the Neural Plate

An important question in neural induction is how the differentiation of neural tissue is contained within well-defined boundaries. One hypothesis has been that communication between cells via gap junctions serves to specify the neural phenotype. But what controls the ability of a cell on the border to communicate with cells on one side but not on the other? There is evidence that NCAM-mediated adhesion may be a prerequisite for the formation of stable gap junctions (Keane et al., 1988). NCAM is the earliest known positive marker for neural induction (Jacobson and Rutishauser, 1986), and its expression appears to precede junctional communication. In addition, there is a precise correlation between the ability of two cells to exchange a fluorescent dye and the presence of NCAM on their surfaces. Finally, whereas the block of junctional communication by the src gene product does not prevent NCAM expression, the addition of anti-NCAM Fab to histotypic cultures of neuroepithelial cells delays the establishment of extensive communication among cells that express NCAM. Since there is no reason to believe that NCAM itself contributes to the structure of the channels, it is more likely that in neuroepithelium, NCAM-mediated adhesion is required to hold cells together long enough to allow the assembly of stable junctions.

2. NCAM POLYSIALIC ACID

2.1. Content and Properties

As mentioned above, NCAM exhibits considerable heterogeneity with respect to its content of sialic acid. This variation is evident in differences in the electrophoretic mobility of the molecule, with apparent molecular masses in SDS-PAGE ranging from 120–180 kDa for the polypeptides having low levels of sialic acid (NCAM-L) to well over 400 kDa for heavily sialylated material (NCAM-H) (Hoffman et al., 1982). In terms of chemical composition, there is a change from about 10% to 30% sialic acid by weight when comparing total NCAM from adult brain with that derived from embryonic brain (Hoffman et al., 1982; Rothbard et al., 1982). Since there is a distribution of electrophoretic mobilities for NCAM in these tissues, it is likely that there are NCAM forms with even higher and lower contents of sialic acid. The large effect of PSA on NCAM electrophoretic mobility relative to its true molecular mass suggests that this saccharide has a large excluded volume. In fact, gel filtration studies on NCAM's PSA-containing peptide suggest that its hydrated volume may be much larger than the protein itself (Rutishauser et al., 1988).

Most if not all the sialic acid in NCAM appears to be in the form of linear homopolymers of the α2-8-linked saccharide. This conclusion is based both on chemical studies (Finne, 1982; Finne et al., 1983), and by examining the effects of NCAM of a phage-derived endoneuraminidase (endo N) which specifically cleaves chains of sialic acid having this linkage and a minimum polymer length of about five residues (Vimr et al., 1984; Finne and Mäkelä, 1985; Rutishauser et al., 1985). NCAM PSA is attached to the polypeptide domain through tri- or tetraantennary complex oligosaccharides (Finne, 1982; Finne and Mäkelä, 1985) bound to three asparaginyl residues located in the central domain of the molecule (Figure 1) (Crossin *et al.*, 1984). Although the exact number or length of PSA chains per core has not been established, it appears that differences in PSA content reflect at least in part a change in the number of residues per polymer. The implications of this variation may be considerably magnified by the fact that analysis of the structure of colominic acid, the bacterial counterpart of PSA, suggests that there is a major conformational change in the polymer with increasing length (Michon *et al.*, 1987).

Although sialic acid is found in many structures associated with a variety of membrane molecules, most if not all of the PSA detected in the embryo is confined to NCAM (Finne *et al.*, 1983; Vimr *et al.*, 1984). However, endo N-sensitive sialic acid has recently been found linked to the sodium channel (James and Agnew, 1987), and therefore a small fraction of PSA may be found elsewhere, particularly in adult animals where NCAM's PSA content is lower. In any case, the major effects of endo N on embryonic cell adhesion (see below) are not altered by drugs that block or stimulate sodium channels (J. Sunshine and U. Rutishauser, unpublished observations).

2.2. Effects on NCAM-Mediated Cell–Cell Aggregation

NCAM PSA has negative regulatory effect on cell–cell adhesion; i.e., cells or membrane vesicles having NCAM-L on their surfaces aggregate together faster than do those expressing NCAM-H (Sadoul *et al.*, 1983; Hoffman and Edelman, 1983). Furthermore, partial removal of PSA from the surface of embryonic brain membranes using endo N enhances the rate of membrane vesicle aggregation (Figure 2) (Rutishauser *et al.*, 1985). This change in adhesion properties is also reflected in the overall degree of contact between brain cells in a typical adhesion assay, with about a fivefold increase in areas of membrane–membrane apposition separated by less than 50 nm (Figure 3) (Rutishauser *et al.*, 1988). In all of these cases, the observed increase in adhesion can be completely and specifically inhibited by polyclonal antibody Fab fragments against the NCAM polypeptide.

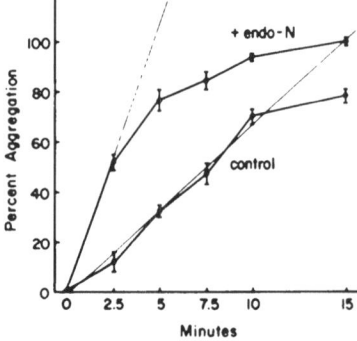

Figure 2. Effect of endo N on the rate of E14 brain membrane vesicle aggregation. The removal of NCAM PSA by this enzyme resulted in a 3.3-fold increase in the initial rate of decrease in particle number (Rutishauser *et al.*, 1985).

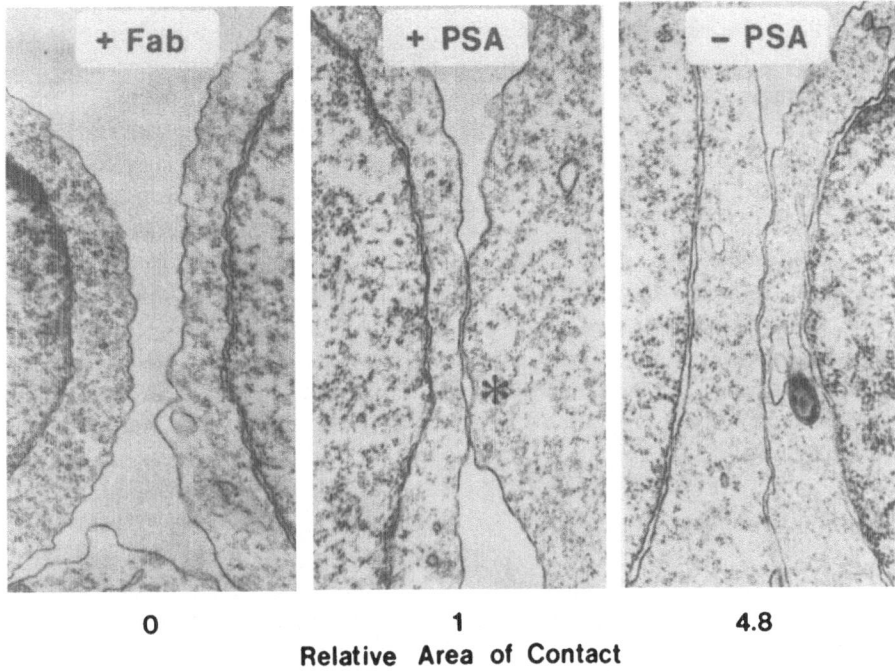

Figure 3. Apposition of cell membranes in the absence of NCAM binding and after removal of PSA from NCAM (Rutishauser *et al.*, 1988).(Left) In the presence of anti-NCAM Fab fragments, cells were rarely in contact, and their membranes were never closely apposed (within 50 nm). (Center) With NCAM having a high content of PSA, plasma membranes came into contact, but short areas of close apposition were frequently interrupted by areas of no contact (asterisk). (Right) After enzymatic removal of NCAM PSA with endo N, large regions of the plasma membrane were in close, continuous apposition. In addition, the cells were often deformed by the extensive attachment between their membranes. Bottom: Relative area of close membrane–membrane contact.

2.3. Reagents for the Analysis of NCAM PSA

Two types of probes have aided studies on the cell and developmental biology of NCAM PSA: the phage-derived endo N described above, and specific antibodies against different forms of this oligosaccharide. Endo N is a particularly fortuitous reagent in that it is highly specific for PSA and appears to convert NCAM-H into a form that is electrophoretically and functionally equivalent to the naturally occurring NCAM-L (Vimr *et al.*, 1984; Rutishauser *et al.*, 1985). Furthermore, it can be purified free of other degradative enzymes, has a pH optimum of about 7, and requires no cofactors (Hallenbeck *et al.*, 1987). Thus, endo N can be used to alter NCAM's PSA content both *in vitro* and *in vivo* (Rutishauser *et al.*, 1985).

A number of antibody preparations have been used to recognize the unique structure of PSA. Recently these have included monoclonal antibodies that appear in our preliminary studies to distinguish between different forms of the polymer. One antiPSA reagent, the 22B monoclonal (Rougon *et al.*, 1986), recognizes the full spectrum of sialylated NCAMs and stains cells both before and after treatment with endo N. In contrast, the 5A5 antibody (obtained from T. Jessell and J. Dodd) only recognizes NCAM-H. A third reagent, called 735 (Frosch *et al.*, 1985), is intermediate in its properties to the other two. Although the basis of this discrimination is not known, from the effects of endo N it is reasonable to expect that it in part reflects the known changes in conformation associated with different polymer lengths (Michon *et al.*, 1987). Thus, a combination of antibodies, together with an anti-NCAM polypeptide, may be able to assess the level of NCAM sialylation within intact tissues. This capability would be valuable since NCAM PSA is often regulated within a tissue microenvironment which is difficult to subject to a direct chemical or electrophoretic analysis.

2.4. Developmental Regulation of NCAM PSA

The original documentation of naturally occurring variation in NCAM sialylation focused on the state of the molecule isolated from the brains of embryonic and adult tissues (Rothbard *et al.*, 1982; Edelman and Chuong, 1982). This pattern of temporal regulation was found in other tissues as well, leading to the description of NCAM-H as the embryonic or E form and NCAM-L as the adult or A form. Additional studies, however, have revealed that NCAM-L also predominates very early in development (Sunshine *et al.*, 1987), that the level of PSA is regulated independently in many tissues (Rougon *et al.*, 1982; Chuong and Edelman, 1984) and can vary even within the same cell (Schlosshauer *et al.*, 1984), and that NCAM-H persists in some adult tissues such as the olfactory system (Chuong and Edelman, 1984), hypothalamus, and choroid fissure (H. Kobayashi and U. Rutishauser, unpublished observations). Thus, it is more accurate to use the simple descriptive abbreviation (H or L) in referring to NCAM glycosylation states in the embryo. It should also be mentioned that although NCAM is most abundant in neural tissues, it is also found transiently in many nonneural structures, including skeletal muscle, heart, kidney, mesenchymal tissue, the adrenal, and testis. In at least some of these cases the molecule contains PSA, indicating that the influence of this structure is not confined to the nervous system (Finne *et al.*, 1987; Roth *et al.*, 1987).

Little is known about the mechanisms that produce different levels of NCAM sialylation in tissues. In principle, differential biosynthesis, enzymatic degradation, or spontaneous hydrolysis could alter the length of these polymers. There is some evidence that PSA is synthesized by specific transferases in the Golgi (Breen *et al.*, 1987; McCoy *et al.*, 1985), and that at least some differences in NCAM PSA observed *in vitro* may be attributed to biosynthesis (Friedlander *et al.*, 1985). Developmentally regulated neuraminidases have been described (Wille and Trenkner, 1981), although none would appear to be specific for the linkages found in NCAM PSA. More than one mechanism of regulation may be involved. For example, while biosynthetic control may determine PSA levels during development, the apparent slow turnover rate of NCAM in adult neural tissue (Linnemann *et al.*, 1985; Garner *et al.*, 1986) could result in an increased impact of degradative processes.

3. EVIDENCE THAT NCAM PSA CAN REGULATE CELL INTERACTIONS

3.1. Cell Contact-Dependent Biochemical Changes

Membrane–membrane contact regulates the levels of neurotransmitter biosynthetic enzymes *in vitro* for a variety of neural crest-derived cells. With both rat and chick sympathetic neurons, cell–cell contact results in increased levels of choline acetyltransferase (ChAT). In studies on the potential role of NCAM in this process (Acheson and Rutishauser, 1988), chick sympathetic neurons were exposed, in the presence or absence of anti-NCAM Fab, to membranes containing NCAM in either its high- or low-PSA form. Under these conditions, membrane contact-mediated increases in ChAT were found to require both NCAM binding function, as indicated by inhibition with anti-NCAM Fab, and the presence of the molecule in its adhesion-promoting, low-PSA form (Figure 4). The effects of antibody were specific, in that antibodies against another abundant adhesion molecule present on these cells, the L1/G4 glycoprotein, did not alter ChAT levels. The low-PSA requirement was met either by specific enzymatic removal of PSA from NCAM on embryonic brain membranes by endo N or by the use of membranes that naturally express NCAM-L. Thus, NCAM appears to serve as a permissive regulatory factor in this system in two different ways: its binding function holds membranes together, and its PSA content independently

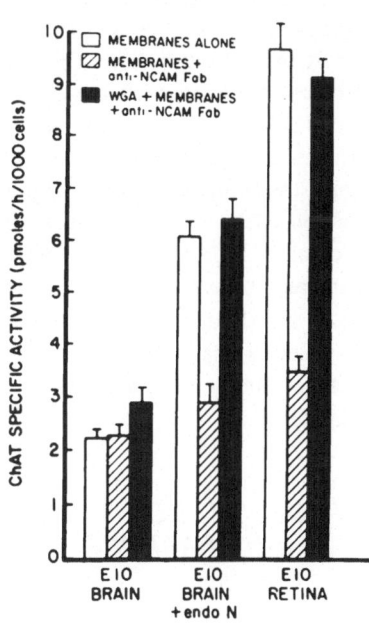

Figure 4. Membrane-mediated increase in ChAT activity in chick sympathetic neurons: role of NCAM and PSA (Acheson and Rutishauser, 1988). The increase in ChAT activity by addition of neural membranes to cells in culture (open bars) required both NCAM function, as indicated by inhibition with anti-NCAM Fab (hatched bars), and a low content of PSA. E10 retina expresses the low-PSA form of NCAM, whereas E10 brain has the high-PSA form. Selective removal of PSA from brain membranes was carried out with endo N. The effect of PSA on ChAT activity was not directly dependent on NCAM-mediated adhesion, in that the ability of an exogenous lectin, wheat germ agglutinin (WGA), to increase ChAT levels in the presence of anti-NCAM Fab (solid bars) was still inhibited by the presence of PSA.

regulates the ability of cells to transmit the relevant biochemical signal. This conclusion is also supported by the fact that the ability of anti-NCAM Fab to block increased ChAT levels can be reversed by addition of a plant lectin, but again only when the PSA content of the endogenous NCAM is low (Acheson and Rutishauser, 1988).

3.2. NCAM PSA Content Alone Can Regulate the Function of Other Cell Surface Ligands

Additional evidence for an influence of NCAM PSA on cell interactions involving molecules other than NCAM has been obtained in studies of the bundling patterns of neurites. We initially demonstrated that neurite outgrowth from dorsal root ganglia onto a collagen substrate displays increased fasciculation after treatment with endo N, suggesting that the removal of PSA can result in augmented fiber–fiber adhesion (Rutishauser et al., 1985). However, subsequent analysis revealed that the effects of endo N can also be exactly the opposite, depending on the type of neuron used (Rutishauser et al., 1988). For example, embryonic chick (E7) spinal cord neurites, whose NCAM is very heavily sialylated, grow as large fascicles on laminin or collagen and removal of PSA from these axons actually reduces fasciculation. How can these seemingly disparate results be explained? In fact, the size of bundles reflects not only neurite–neurite binding mediated by several adhesion molecules including NCAM, but also an opposing force exerted by the individual growth cones as they adhere to and migrate along the substrate. The simplest explanation of the neurite patterns obtained is therefore that with spinal ganglion cultures membrane–membrane adhesion is enhanced relative to growth cone–substrate adhesion whereas with spinal cord the opposite is the case. Consistent with this interpretation is the observation that in the spinal cord studies, the effect of endo N occurred even in the absence of NCAM-mediated adhesion, was greater than that produced by anti-NCAM Fab fragments, and could be reversed by antibodies against laminin.

A second illustration of this phenomenon was the effect of PSA on an artificial cell–cell interaction, wheat germ lectin (WGA)-mediated agglutination of embryonic chick (E10) brain membrane vesicles (Rutishauser et al., 1988). As with spinal cord neurons, the NCAM present on these membranes is heavily sialylated. In these experiments, the ability of WGA to agglutinate membranes in the absence of NCAM binding function (i.e., in the continuous presence of anti-NCAM Fab) was examined as a function of NCAM PSA content. The amount of PSA did not affect the extent of aggregate formation in the absence of WGA. However, removal of NCAM PSA caused a sixfold increase in the rate of WGA-mediated agglutination. This effect did not reflect a change in the number of lectin receptors on membranes, but rather an enhanced ability of the lectin to function when PSA was removed from the endogenous NCAM.

4. PROPOSED MECHANISM FOR NCAM-MEDIATED REGULATION OF CELL INTERACTIONS

Data from the different experimental systems presented above, together with information about the unusual properties of the molecule itself have led us to the

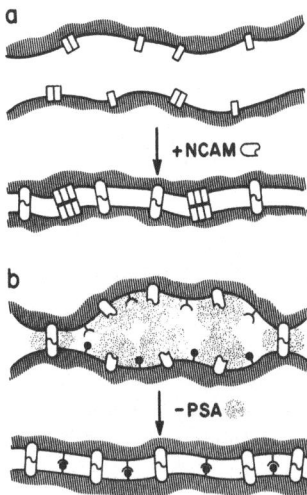

Figure 5. Two mechanisms for the regulation of cell–cell interactions by NCAM. (a) Expression of NCAM on cell surfaces and the resultant formation of NCAM–NCAM bonds enhances the probability of junction formation from interacting subunits (rectangles) by increasing the extent or duration of membrane–membrane contact. (b) Initiation of cell interactions via specific ligands (ball and socket) by a reduction in NCAM PSA content (stippled areas). The ability of the ligands to engage is enhanced by reducing the excluded volume of carbohydrate between membranes, which impedes close cell–cell contact. PSA is depicted as also compromising interaction between NCAMs, with adhesion occurring between molecules with relatively low amounts of PSA.

following conclusion: the presence of NCAM on the cell surface can have either a positive or a negative effect on its overall interaction with other cells or substrates, depending on the molecule's PSA content. To account for these phenomena, two mechanisms have been proposed (Figure 5) (Rutishauser *et al.*, 1988).

For NCAM with a relatively low PSA content, the molecule's presence would increase the extent or duration of membrane contact and thereby promote other interactions (Figure 5a). This mechanism is consistent with our findings on the relationship between NCAM and calcium-dependent adhesion (Rutishauser *et al.*, 1988), and the ability of NCAM expression to regulate the formation of gap junctions (Keane *et al.*, 1988). For example, although gap junctions do not directly utilize NCAM, *in vitro* antibody perturbation studies (see above) suggest that NCAM-mediated adhesion is required for the onset and spatial restriction of extensive junctional communication among neuroepithelial cells, as measured by dye transfer.

For molecules with high PSA content, the large volume occupied by the carbohydrate would impede membrane–membrane contact so that the function of some ligands, and probably even that of NCAM itself, is hindered (Figure 5b). In this case, interactions could be initiated either by a reduction in the amount of PSA or by removal of the entire NCAM molecule if adhesion can occur through another ligand. At present it is not possible to distinguish whether the effects of PSA involve changes in the extent, intimacy, or duration of cell–cell contact. Each could, in principle, enhance the efficiency of a particular cell interaction.

5. DISCUSSION

A variety of evidence suggests that PSA can influence the initiation of cell interactions which include but are not limited to those that are directly mediated by NCAM. The hypothesis that these diverse effects are based on properties intrinsic to this saccharide, i.e., largely independent of NCAM's own ligating activity, is a clear deviation from previous ideas (Sadoul *et al.*, 1983; Hoffman and Edelman, 1983; Rutishauser *et al.*, 1985). Given the covalent association of PSA with NCAM, it would seem more logical to propose a mechanistic link between PSA and NCAM that would lead to the observed range of PSA-related phenomena. However, unless major biochemical properties of the molecule have gone undetected (such as an affinity for laminin and collagen, as well as for receptors involved in contact-dependent increases in ChAT and WGA-mediated agglutination), *and* these activities are unaffected by polyclonal Fab against NCAM that is known to completely block NCAM's cell–cell adhesive function, it is difficult to imagine the basis for such a direct relationship. On the other hand, the more global effects of PSA in the present proposal are consistent with the fact that in the other known occurrences of PSA in nature—the coat of bacteria (Finne, 1985; Troy, 1979) and the pellucid zone of fish eggs (Kitajima *et al.*, 1986; Inoue *et al.*, 1987)—this saccharide forms a substantial coating which surrounds the cell.

Nevertheless, almost all PSA is attached to NCAM, and the indirect mechanism proposed here raises the interesting possibility that the chemical association of these polypeptide and carbohydrate moieties might represent a form of "molecular symbiosis." From the conservation of NCAM-like polypeptide structure and function in phylogeny (Hall and Rutishauser, 1985; Hoffman *et al.*, 1984) but the absence of sialic acid in invertebrates (Warren, 1963), it is reasonable to speculate that NCAM polypeptide-mediated adhesion was a relatively early event in evolution and that PSA was not added to the molecule until after the emergence of chordates. This expression of PSA may for example have provided an additional level of control of cell–cell interactions that was advantageous in accommodating the larger numbers of cells, pathways, and targets involved in complex vertebrate neural tissue. The association of PSA with the NCAM polypeptide, which is abundantly expressed on membranes of neurons and glia in these tissues, would simply have provided an effective vehicle for display of PSA on the appropriate cells.

The hypothesis that PSA can mediate global regulation of membrane events via its steric or repulsive properties has several interesting biological implications. Such alterations of the overall degree of close membrane apposition could allow differential use of several ligand–receptor systems according to their properties. A receptor that is small, relatively immobile, or present in low amounts may require a low NCAM PSA content in order to function, while an abundant and mobile extracellular protein receptor, or a soluble ligand, would be independent of this type of control. Embryonic axons, which have a particularly high PSA content during the formation of tracts and connections (Sunshine *et al.*, 1987), would particularly benefit from this type of selection. That is, as an axon grows through a complex environment, it must be able to avoid inappropriate interactions, such as formation of stable junctions, yet remain responsive to guidance and target cues.

As stated above, PSA could regulate cell–cell interactions by varying the extent or duration as well as the intimacy of membrane apposition. With respect to distance between membranes, it is interesting to speculate further that one of the most obvious structural features of NCAM, its appearance as a long rod with a flexible hinge in its extracellular domain (Hall and Rutishauser, 1987) (Figure 1), would allow a homo-philic NCAM–NCAM bond between cells to adjust to a continuum of such variations.

Finally, it is important to realize that PSA may be only one of several carbohy-drate structures that contribute to the regulation of membrane apposition, particularly in tissues that do not express NCAM. Thus, in some respects the ideas presented here are reminiscent of the "glycocalyx" proposed by Bennett 25 years ago (1963). Unfor-tunately, with the poor fixation of carbohydrates by aldehydes and the standard use of dehydration prior to embedding in expoxy, the more subtle steric properties of a glycocalyx carbohydrate at the cell surface have not been easy to detect by electron microscopy. However, techniques have recently been developed which allow for im-proved preservation of hydrated structures, and it should be possible to judge the completeness and importance of these ideas within the next few years.

6. REFERENCES

Acheson, A., and Rutishauser, U., 1988, CAM regulates cell contact-mediated changes in choline acetyl-transferase activity of embryonic chick sympathetic neurons, *J. Cell Biol.* **106**:479–486.

Arquint, M., Roder, J., Chla, L.-S., Down, J., Wilkinson, D., Bayley, H., Braun, P., and Dunn, R., 1987, Molecular cloning and primary structure of myelin-associated glycoprotein, *Proc. Natl. Acad. Sci. USA* **84**:600–604.

Barthels, D., Santoni, M.-J., Wille, W., Ruppert, C., Chaix, J.-C., Hirch, M.-R., Fontecilla-Camps, J. C., and Goridis, C., 1987, Isolation and nucleotide sequence of mouse NCAM cDNA that codes for a Mr 79,000 polypeptide without a membrane-spanning region, *EMBO J.* **6**:907–914.

Bennett, H. S., 1963, Morphological aspects of extracellular polysaccharides, *J. Histochem. Cytochem.* **11**:14–23.

Boller, K., Vestweber, D., and Kemler, R., 1985, Cell-adhesion molecule uvomorulin is localized at the intermediate junctions of adult intestinal epithelial cells, *J. Cell Biol.* **100**:327–332.

Breen, K. C., Kelly, P. G., and Regan, C. M., 1987, Postnatal D2-CAM/N-CAM sialylation state is controlled by a developmentally regulated Golgi sialyltransferase, *J. Neurochem.* **48**:1486–1493.

Chuong, C.-M., and Edelman, G. M., 1984, Alterations in neural cell adhesion molecules during develop-ment of different regions of the nervous system, *J. Neurosci.* **4**:2354–2368.

Cole, G., Loewy, A., Cross, N., Akeson, R., and Glaser, L., 1986, Topographic localization of the heparin-binding domain of the neural cell adhesion molecule N-CAM, *J. Cell Biol.* **103**:1739–1744.

Crossing, K. L., Edelman, G. M., and Cunningham, B. A., 1984, Mapping of three carbohydrate attach-ment sites in embryonic and adult forms of the neural cell adhesion molecule, *J. Cell Biol.* **99**:1848–1855.

Cunningham, B. A., Hoffman, S., Rutishauser, U., Hemperly, J. J., and Edelman, G. M., 1983, Molecular topography of the neural cell adhesion molecule N-CAM: Surface orientation and location of sialic acid-rich and binding regions, *Proc. Natl. Acad. Sci. USA* **80**:3116–3120.

Cunningham, B. A., Hemperly, J. J., Murray, B. A., Prediger, E. A., Brackenbury, R., and Edelman, G. M., 1987, Neural cell adhesion molecule: Structure, Immunoglobulin-like domains, cell surface modulation, and alternative RNA splicing, *Science* **236**:799–806.

Edelman, G. M., and Chuong, C.-M., 1982, Embryonic to adult conversion of neural cell adhesion molecules in normal and staggerer mice, *Proc. Natl. Acad. Sci. USA* **79**:7036–7040.

Finne, J., 1982, Occurrence of unique polysialosyl carbohydrate units in glycoproteins of developing brain, *J. Biol. Chem.* **257**:11966–11970.

Finne, J., 1985, Polysialic acid—A glycoprotein carbohydrate involved in neural adhesion and bacterial meningitis, *Trends Biochem. Sci.* **10**:129–132.

Finne, J., and Mäkelä, H., 1985, Cleavage of the polysialosyl units of brain glycoproteins by a bacterio-phage endosialidase, *J. Biol. Chem.* **260**:1265–1270.

Finne, J., Finne, U., Deagostini-Bazin, H., and Goridis, C., 1983, Occurrence of alpha 2-8 linked poly-sialosyl units in a neural cell adhesion molecule, *Biochem. Biophys. Res. Commun.* **112**:482–487.

Finne, J., Bitter-Suermann, D., Goridis, C., and Finne, U., 1987, An IgG monoclonal antibody to group B meningococci cross-reacts with developmentally regulated polysialic acid units of glycoproteins in neural and extraneural tissues, *J. Immunol.* **138**:4402–4407.

Frelinger, A. L., III, and Rutishauser, U., 1986, Topography of NCAM structural and functional determinants. II. Placement of monoclonal antibody epitopes, *J. Cell Biol.* **103**:1729–1737.

Friedlander, D. R., Brackenbury, R., and Edelman, G. M., 1985, Conversion of embryonic form to adult forms of N-CAM In vitro: Result from de novo synthesis of adult forms, *J. Cell Biol.* **101**:412–419.

Frosch, M., Gorgen, I., Boulnois, G. J., Timmis, D. N., and Bitter-Suermann, D., 1985, NZB mouse system for production of monoclonal antibodies to weak bacterial antigens: Isolation of an IgG antibody to the polysaccharide capsules of Escherichia coli K1 and group B meningococci, *Proc. Natl. Acad. Sci. USA* **82**:1194–1198.

Garner, J., Watanabe, M., and Rutishauser, U., 1986, Rapid axonal transport of the neural cell adhesion molecule, *J. Neurosci.* **6**:3242–3249.

Gennarini, F., Hirn, M., Deagostini-Bazin, H., and Goridis, C., 1984, Studies of the transmembrane disposition of the neural cell adhesion molecule N-CAM. The use of liposome-inserted radioiodinated N-CAM to study its transbilayer orientation, *Eur. J. Biochem.* **142**:65–73.

Gumbiner, B., and Simons, K., 1986, A functional assay for proteins involved in establishing an epithelial occluding barrier: Identification of a uvomorulin-like polypeptide, *J. Cell Biol.* **102**:457–468.

Hall, A. K., and Rutishauser, U., 1985, Phylogeny of a neural cell adhesion molecule, *Dev. Biol.* **110**:39–46.

Hall, A. K., and Rutishauser, U., 1987, Visualization of neural cell adhesion molecule by electron microscopy, *J. Cell Biol.* **104**:1579–1586.

Hallenbeck, P. C., Vimr, E. R., Yu, F., Bassier, B., and Troy, F. A., 1987, Purification and properties of a bacteriophage-induced endo-N-acetylneuraminidase specific for poly-alpha-2,8-sialosyl carbohydrate units, *J. Biol. Chem.* **262**:3553–3561.

Hirn, M., Ghandour, M. S., Deagostini-Bazin, H., and Goridis, C., 1983, Molecular heterogeneity and structural evolution during cerebellar ontogeny detected by monoclonal antibody of the mouse cell surface antigen BSP-2, *Brain Res.* **265**:87–100.

Hoffman, S., and Edelman, G. M., 1983, Kinetics of homophilic binding by embryonic and adult forms of the neural cell adhesion molecule, *Proc. Natl. Acad. Sci. USA* **80**:5762–5766.

Hoffman, S., Sorkin, B. C., White, P. C., Brackenbury, R., Mailhammer, R., Rutishauser, U., Cunningham, B. A., and Edelman, G. M., 1982, Chemical characterization of a neural cell adhesion molecule (N-CAM) purified from embryonic brain membranes, *J. Biol. Chem.* **257**:7720–7729.

Hoffman, S., Chuong, C., and Edelman, G. M., 1984, Evolutionary conservation of key structures and binding functions of neural cell adhesion molecules, *Proc. Natl. Acad. Sci. USA* **81**:6881–6885.

Inoue, S., Kitajima, K., Ionue, Y., and Kudo, S., 1987, Localization of polysialoglycoprotein as a major glycoprotein component in cortical alveoli of the unfertilized eggs of Salmo gairdneri, *Dev. Biol.* **123**:442–454.

Jacobson, M., and Rutishauser, U., 1986, Induction of neural cell adhesion molecule (NCAM) in Xenopus embryos, *Dev. Biol.* **116**:524–531.

James, W. M., and Agnew, W. S., 1987, Multiple oligosaccharide chains in the voltage-sensitive Na channel from Electrophorus electricus: Evidence for 2,8-linked polysialic acid, *Biochem. Biophys. Res. Comm.* **148**:817–826.

Keane, R. W., Parmender, P. M., Rose, B., Honig, L. S., Lowenstein, W. R., and Rutishauser, U., 1988, Neural differentiation, NCAM-mediated adhesion and gap junctional communication in neuroectoderm. A study in vitro, *J. Cell Biol.* **106**:1307–1319.

Kitajima, K., Inoue, Y., and Inoue, S., 1986, Polysialoglycoproteins of Salmonidae fish eggs, *J. Biol. Chem.* **261**:5262–5269.

Landmesser, L., Dahm, L., Schultz, K., and Rutishauser, U., 1988, Distinct roles for adhesion during innervation of embryonic chick muscle, *Dev. Biol.* **130**(2):645–670.

Linnemann, D., Lyles, J. M., and Bock, E., 1985, A developmental study of the biosynthesis of the neural cell adhesion molecule, *Dev. Neurosci.* **7**:230–238.

Maier, C. E., Watanabe, M., Singer, M., McQuarrie, I. G., Sunshine, J., and Rutishauser, U., 1986, Expression and function of neural cell adhesion molecule during limb regeneration, *Proc. Natl. Acad. Sci. USA* **83**:8395–8399.

Margolis, R. K., and Margolis, R. U., 1983, Distribution and characteristics of polysialosyl oligosaccharides in neural tissue glycoproteins, *Biochem. Biophys. Res. Commun.* **116**:889–894.

McCoy, R. D., Vimr, E. R., and Troy, F. A., 1985, CMP-NeuNAc:Poly-α-2,8-sialosyl sialyltransferase and the biosynthesis of polysialosyl units in neural cell adhesion molecules, *J. Biol. Chem.* **260**:12695–12699.

Michon, F., Brisson, J. and Jennings, H., 1987, Conformational differences between linear alpha(2-8)-linked homosialooligosaccharides and the epitope of the group B meningococcal polysaccharide, *Biochemistry* **26**:8399–8405.

Roth, J., Taatjest, D. J., Bitter-Suermann, D., and Finne, J., 1987, Polysialic acid units are spatially and temporally expressed in developing postnatal rat kidney, *Dev. Biol.* **84**:1969–1973.

Rothbard, J. B., Brackenbury, R., Cunningham, B. A., and Edelman, G. M., 1982, Differences in the carbohydrate structures of neural cell-adhesion molecules from adult and embryonic chicken brains, *J. Biol. Chem.* **257**:11064–11069.

Rougon, G., Deagostini-Bazin, H., Hirn, M., and Goridis, C., 1982, Tissue- and developmental stage-specific forms of a neural cell surface antigen linked to differences in glycosylation of a common polypeptide, *EMBO J.* **1**:1239–1244.

Rougon, G., Dubois, C., Buckley, N., Magnani, J. L., and Zollinger, W., 1986, A monoclonal antibody against meningococcus group B polysaccharides distinguishes embryonic from adult N-CAM, *J. Cell Biol.* **103**:2429–2437.

Rutishauser, U., and Jessell, T., 1988, Cell adhesion molecules in vertebrate neural development, *Phys. Rev.* **68**(3):819–857.

Rutishauser, U., Hoffman, S., and Edelman, G. M., 1982, Binding properties of a cell adhesion molecule from neural tissue, *Proc. Natl. Acad. Sci. USA* **79**:685–689.

Rutishauser, U., Grumet, M., and Edelman, G. M., 1983, N-CAM mediates initial interactions between spinal cord neurons and muscle cells in culture, *J. Cell Biol.* **97**:145–152.

Rutishauser, U., Watanabe, M., Silver, J., Troy, F. A., and Vimr, E. R., 1985, Specific alteration of NCAM-mediated cell adhesion by an endoneuraminidase, *J. Cell Biol.* **101**:1842–1849.

Rutishauser, U., Acheson, A., Hall, A. K., Mann, D., and Sunshine, J., 1988, NCAM as a regulator of cell–cell interactions, *Science* **240**:53–57.

Sadoul, R., Hirn, M., Deagostini-Bazin, H., Rougon, G., and Goridis, C., 1983, Adult and embryonic mouse neural cell adhesion molecules have different binding properties, *Nature* **304**:349.

Schlosshauer, B., Schwartz, U., and Rutishauser, U., 1984, Topological distribution of different forms of NCAM in the developing chick visual system, *Nature* **310**:141–143.

Seed, B., 1987, An LFA-3 cDNA encodes a phospholipid-linked membrane protein homologous to its receptor CD2, *Nature* **329**:840–842.

Silver, J., and Rutishauser, U., 1984, Guidance of optic axons in vivo by a preformed adhesive pathway on neuroepithelial endfeet, *Dev. Biol.* **106**:485–499.

Simmons, D., Makgoba, M. W., and Seed, B., 1988, ICAM, an adhesion ligand of LFA-1, is homologous to the neural cell adhesion molecule NCAM, *Nature* **331**:624–627.

Small, S. J., Shull, G., Santoni, M.-J., and Akeson, R., 1987, Identification of a cDNA clone that contains the complete coding sequence for a 140-kD rat NCAM polypeptide, *J. Cell Biol.* **105**:2335–2345.

Sorkin, B. C., Hoffman, S., Edelman, G. M., and Cunningham, B. A., 1984, Sulfation and phosphorylation of the neural cell adhesion molecule, N-CAM, *Science* **225**:1476–1478.

Sunshine, J., Balak, K., Rutishauser, U., and Jacobson, M., 1987, Changes in neural cell adhesion molecule (NCAM) structure during vertebrate neural development, *Proc. Natl. Acad. Sci. USA* **84**:5986–5990.

Takeichi, M., 1987, Cadherins: A molecular family essential for selective cell–cell adhesion and animal morphogenesis, *Trends Genet.* **3**:213–217.

Thanos, S., Bonhoeffer, F., and Rutishauser, U., 1984, Fiber–fiber interactions and tectal cues influence the development of the chick retinotectal projection, *Proc. Natl. Acad. Sci. USA* **81**:1906–1910.

Thiery, J.-P., Duband, J.-L., Rutishauser, U., and Edelman, G. M., 1982, Cell adhesion molecules in early chicken embryogenesis, *Proc. Natl. Acad. Sci. USA* **79**:6737–6741.

Tosney, K. W., Watanabe, M., Landmesser, L., and Rutishauser, U., 1986, The distribution of NCAM in the chick hindlimb during axon outgrowth and synaptogenesis, *Dev. Biol.* **114:**468–481.

Troy, F. A., 1979, The chemistry and biosynthesis of selected bacterial capsular polymers, *Annu. Rev. Microbiol.* **33:**519–560.

Vimr, E. R., McCoy, R. D., Voliger, H. F., Wilkison, N. C., and Troy, F. A., 1984, Use of prokaryotic-derived probes to identify poly(sialic acid) in neonatal membranes, *Proc. Natl. Acad. Sci. USA* **81:** 1971–1975.

Volk, T., and Geiger, B., 1986, A-CAM: A 135-KD receptor of intercellular adherens junctions. I. Immunoelectron microscope localization and biochemical studies, *J. Cell Biol.* **103:**1441–1450.

Warren, L., 1963, The distribution of sialic acids in nature, *Comp. Biochem. Physiol.* **10:**153–171.

Watanabe, M., Frelinger, A. L., III, and Rutishauser, U., 1986, Topology of NCAM structural and functional determinants. I. Classification of monoclonal antibody types, *J. Cell Biol.* **103:**1721–1727.

Wille, W., and Trenkner, E., 1981, Changes in particulate neuraminidase activity during normal and staggerer mutant mouse development, *J. Neurochem.* **37:**443–446.

Extracellular Matrix Adhesive Glycoproteins and Their Receptors in the Nervous System

Philippe Douville and Salvatore Carbonetto

1. INTRODUCTION

The extracellular space occupies about 40% of the immature brain (Bondareff and Pysh, 1968), decreasing to 20% in the adult (van Harreveld *et al.*, 1971; Nevis and Collins, 1967). This space is filled with a matrix (ECM) of insoluble glycoconjugates (glycoproteins, proteoglycans) through which neural cell precursors migrate to take up their final positions in the central nervous system (CNS). In the adult peripheral nervous system, regenerating axons must also make their way through the ECM to reach their targets. In these and other instances, adhesion of cells to a matrix or other cells is essential for motility, which results in large part from contractile processes in the cell. A common feature of such adhesions is that they involve interaction of glycoconjugates in the ECM with others (glycoproteins, proteoglycans, and glycolipids) at the cell surface; for example, the adhesive ECM glycoproteins laminin, fibronectin, and collagens bind to receptors on the cell surface. The ramifications of these ECM–receptor interactions in the nervous system may be profound. In some instances they rival those of hormones and growth factors (Edgar *et al.*, 1984) by modifying gene expression (Bissell *et al.*, 1982) and leading to neural cell differentiation (Reh *et al.*, 1987), proliferation (Kleinman *et al.*, 1984), synaptogenesis (Nitkin *et al.*, 1983), neuroblast migration (Newgreen and Thiery, 1980; Boucaut *et al.*, 1984; Liesi, 1985a), Schwann cell ensheathment of peripheral axons (Bunge *et al.*, 1986), and possibly formation of the blood–brain barrier (Arthur *et al.*, 1987).

In this chapter we will discuss briefly the structure and function of ECM adhesive glycoproteins with special attention to cell-surface receptors for these proteins. It

Philippe Douville and Salvatore Carbonetto • Center for Neuroscience Research, McGill University, Montreal General Hospital Research Institute, Montreal, Quebec H3G 1A4, Canada.

would be impossible to present a cohesive account of this topic without reference to work on nonneural cells. However, space limitations permit us to review this work only insofar as it bears directly on neural systems. We will conclude by discussing the special case of neural cell motility represented by nerve fiber growth in an attempt to integrate what is known about cell–matrix interactions at the cellular and molecular levels. We will not discuss glycosaminoglycans and proteoglycans, which are covered in other chapters of this volume, nor will we review cell adhesion molecules (see Chapters 11 and 12). It should be kept in mind, however, that the functions of cell adhesion molecules and ECM receptors may be complementary (Bixby et al., 1987) or overlapping (Fahrig et al., 1987; Makgoba et al., 1988).

2. ORGANIZATION OF THE ECM OF THE NERVOUS SYSTEM

A striking feature of the ECM is basement membranes which are deposited beneath epithelial cells, separating them from cells of mesenchymal origin and orienting the epithelium into basal and luminal aspects. In the nervous system, basement membranes are found over the abaxonal surfaces of Schwann cells (Bunge et al., 1986), beneath meninges lining the CNS (Peters et al., 1976), at the interface of capillary endothelium with processes of astroglia or ependymal tanycytes (Fakuda and Hashimoto, 1987), and as part of the inner limiting membrane of the retina. Basement membranes also surround muscle cells and are specialized at myoneural junctions (Chiu and Sanes, 1984). Surrounding the basement membrane in peripheral nerves is a perineural space filled with a stroma of ECM that includes collagen fibrils oriented with the long axis of the nerve (Low, 1976). Electron microscopy of basement membranes has revealed a fine structure consisting of multiple layers (Laurie et al., 1982). The layer closest to the cell is known as the lamina lucida due to its translucent appearance in electron micrographs. However, it is difficult to strictly define a boundary between the plasma membrane with its associated proteins (cell surface) and the lamina lucida which together form a continuum that extends to the next layer of the basement membrane, the lamina densa. The lamina densa is composed of a network of type IV collagen molecules in addition to laminin, entactin, fibronectin, and basement membrane proteoglycans (Laurie et al., 1982; Sanes et al., 1986). Finally, some basement membranes contain a third layer, known as the lamina rara (or reticular lamina), which connects to the underlying mesenchyme by means of fibrous collagen (type VII) projections (Keene et al., 1987).

Basement membranes are absent among neural cells in the adult mammalian CNS, although densities of extracellular material are commonly seen within synaptic clefts in electron micrographs of the CNS (Peters et al., 1976). The intercellular space of the adult CNS probably contains heparan sulfate proteoglycans but little or no chondroitin sulfate proteoglycans (Aquino et al., 1984; and Chapter 3), and may be deficient in certain ECM adhesive glycoproteins (Liesi, 1985a,b; cf. Manthorpe et al., 1988). In contrast, embryonic CNS tissues contain several adhesive glycoproteins (laminin, cytotactin, and fibronectin) which are distributed in a diffuse manner outside of basement membranes (Crossin et al., 1986; Stewart and Pearlman, 1987; Liesi,

1985a). These differences in ECM composition correlate with the ability of these tissues to support nerve fiber growth *in vivo* and in an *in vitro* culture assay (Carbonetto *et al.*, 1987). Moreover, proteoglycans or glycosaminoglycans have been shown to support little neurite growth in culture (Carbonetto *et al.*, 1982, 1983; Davis *et al.*, 1985b), and to inhibit adhesive interactions with fibronectin (Carbonetto *et al.*, 1983) and NCAM (Cole and Glaser, 1986). Thus, changes in the amounts and distribution of ECM components may alter the milieu of the CNS to influence neural development.

3. ADHESIVE GLYCOPROTEINS OF THE ECM

3.1. Laminin

Laminin is a large (900 kDa) glycoprotein (Timpl *et al.*, 1979; Chung *et al.*, 1979) that is a major constituent of the lamina densa (Laurie *et al.*, 1982). The best-studied form of laminin is from the murine Engelbreth–Holm–Swarm (EHS) sarcoma, which is a rich source of basement membranes. EHS laminin consists of three subunits: an A chain (400 kDa), a B1 chain (215 kDa), and a B2 chain (205 kDa) (reviewed in Martin and Timpl, 1987). The subunits are synthesized as the products of separate genes and are disulfide-bonded into a cruciform molecule (Figure 1A) with the two B chains forming a coiled-coil, α-helical structure along a portion of the A chain. Nucleic acid sequence data have revealed 30–40% homology between the two B chains which have likely evolved from a common ancestral gene (Sasaki and Yamada, 1987; Sasaki *et al.*, 1987). The predicted secondary amino acid structures of the two B chains suggest that they are divided into six distinct domains (Figure 1A; Barlow *et al.*, 1984; Sasaki and Yamada, 1987; Sasaki *et al.*, 1987). Domains I and II consist of α-helical regions which associate with the long arm of the A chain. Domains III and V include a repeating 50-amino-acid cysteine-rich sequence with homology to epidermal growth factor. Domains IV and VI form globular structures which bind collagen (Sasaki and Yamada, 1987). This predicted structure agrees well with rotary shadowing images of EHS laminin which show a cruciform molecule with three short arms and a long arm (Engel *et al.*, 1981). The B chains are heavily glycosylated and contain many potential sites for *N*-linked oligosaccharides (Sasaki and Yamada, 1987).

Laminin is a multifunctional molecule with sites that bind to other ECM constituents as well as to cells. For example, in addition to the collagen binding sites mentioned above (Rao *et al.*, 1982), there is a heparin binding site on the globular end of the long arm (Figure 1A; Ott *et al.*, 1982). Limited proteolytic digestion of native laminin yields a series of fragments which may contain one or more of these functional domains. One such fragment, E1, includes the short arms surrounding the center of the cross, but lacks the globular regions. E1 encompasses the major cell attachment site(s) within laminin (Rao *et al.*, 1982; Timpl *et al.*, 1983). Graf *et al.* (1987a,b) have synthesized synthetic peptides derived from the nucleotide sequence of the B1 chain. One peptide in domain III of E1 has been reported to stimulate cell attachment when immobilized on culture substrata. In solution, this same peptide inhibits attachment to laminin-coated substrata, presumably by competing with binding of native laminin to

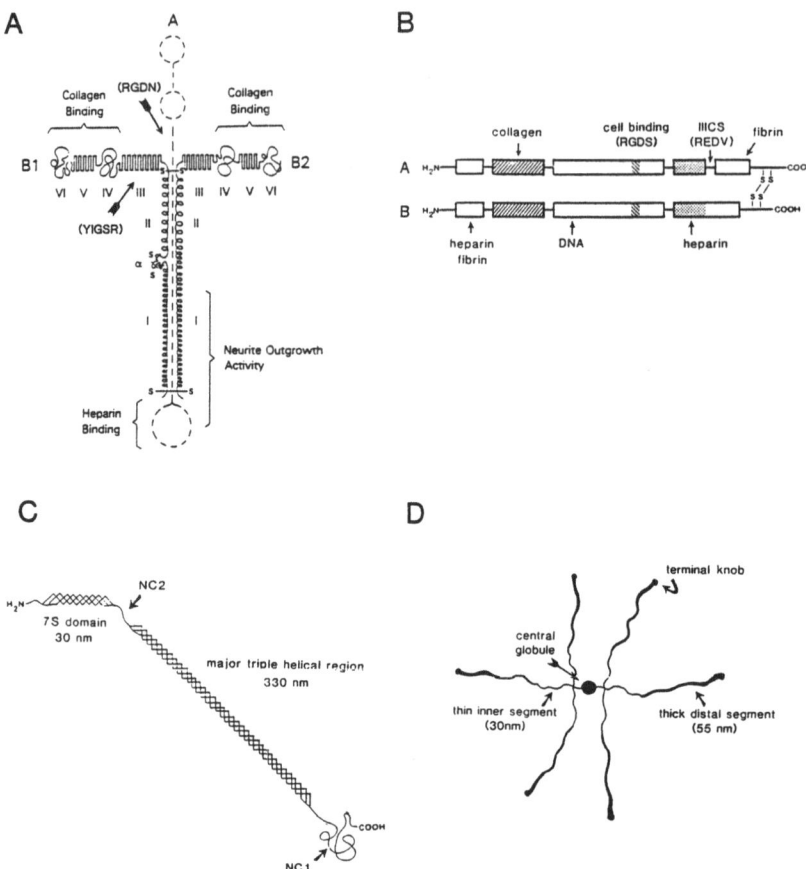

Figure 1. Schematic representations of the ECM adhesive glycoproteins laminin, fibronectin, type IV collagen, and cytotactin. (A) Cruciform structure of EHS laminin with the B1 and B2 chains forming a coiled-coil α-helix along the A chain. The globular domains in the two B chains (domains IV and VI) participate in collagen binding and the globular domain at the end of the A chain binds heparin. The main cell attachment site contains the active sequence YIGSR (arrow) and is located in domain III on the B1 chain. The neurite-promoting domain is found at the end of the long arm toward the heparin binding domain. Note the RGDN sequence (arrow) in the central cross region on the A chain. Modified from Sasaki and Yamada (1987). (B) Human plasma fibronectin is composed of two subunits (A and B chains) whose mRNAs are alternately spliced in the IIICS region. Both subunits bind heparin, fibrin, collagen, DNA, and cells. The main cell attachment site contains the active sequence RGDS and is located on both chains. Human plasma fibronectin contains an alternate cell attachment site (REDV; RGDV in the rat). (C) Structure of type IV collagen showing amino-terminal 7 S domain which associates with other collagen IV molecules in anti-parallel fashion. The 7 S domain is separated from the main triple-helical region by the NC2 (noncollagenous 2) domain. Type IV collagen also includes a carboxyl globular domain (NC1) which can also self-associate with other collagen IV molecules. (D) Rotary-shadowed image showing the hexabrachion structure of cytotactin (Erickson and Inglesias, 1984). Six arms composed of thin and thick segments can be seen emanating from the central globule.

the cell surface (Graf *et al.*, 1987a). The sequence critical for cell attachment may be YIGSR (Tyr-Ile-Gly-Ser-Arg) or, more effectively, YIGSR-NH$_2$. Substitution of any amino acid in this sequence apparently reduces its biological activity (Graf *et al.*, 1987a; Iwamoto *et al.*, 1987). Interestingly, NG108 cells (a neuroblastoma–glioma hybrid) attach to substrata coated with YIGSR peptides but do not extent processes as they do on laminin (Graf *et al.*, 1987a).

Laminin has dramatic effects on cultured cells, including stimulation of Schwann cell mitosis and migration (McCarthy *et al.*, 1984; Kleinman *et al.*, 1984), induction of tyrosine hydroxylase in adrenal chromaffin cells (Acheson *et al.*, 1986), transdifferentiation of pigmented epithelial cells into neurons (Reh *et al.*, 1987), and stimulation of neurite outgrowth from CNS as well as PNS neurons (Manthorpe *et al.*, 1983; Rogers *et al.*, 1983; Hopkins *et al.*, 1985; Tomaselli *et al.*, 1986; Cohen *et al.*, 1986; cf. Akers *et al.*, 1981; Kleitman *et al.*, 1988). The neurite-promoting activity maps to a region of the long arm near a heparin binding site (Figure 1A; Edgar *et al.*, 1984; Engvall *et al.*, 1986). This same region stimulates attachment and spreading of myoblasts and glioblastoma cells (Goodman *et al.*, 1987), as well as neurite outgrowth from dorsal root ganglion (DRG) neurons, which otherwise require NGF. This latter activity was first described as a trophic effect of conditioned medium from muscle and other cell types on DRG and sympathetic neurons (Lander *et al.*, 1985; Davis *et al.*, 1985a,b). The activity was traced to a heparan sulfate proteoglycan–laminin complex in which laminin was the active neurite-promoting factor (Lander *et al.*, 1985; Davis *et al.*, 1985a,b). These conditioned media forms of laminin appear to lack an A chain (Palm and Furcht, 1983; Davis *et al.*, 1985a). The function of the proteoglycan in these complexes is unknown, but it clearly does not interfere with the binding of laminin to neurons as heparin does with NCAM binding to itself (Cole and Glaser, 1986).

The neurite-promoting activity of laminin in culture raises obvious questions concerning its function in neural regeneration and development. Laminin has been localized immunocytochemically in the embryonic mammalian CNS (see Table 1) at about the time when neuroblasts migrate away from the periventricular region (Liesi, 1985a). At early stages of development, laminin can be found distributed diffusely within the brain (Liesi, 1985a), in the embryonic spinal cord (Carbonetto *et al.*, 1987), and has been reported on the surface of radial glial fibers (Liesi, 1985a) which guide migrating neuroblasts (Rakic, 1985). During retinal development, laminin is found on neuroepithelial cell endfeet in contact with ganglion cell axons, but once these axons have extended into the optic nerve it is found only in basement membranes (Cohen *et al.*, 1987). Laminin is also found in basement membranes surrounding the neural tube and in the mesenchymal matrix along neural crest migratory routes (Tuckett and Morriss-Kay, 1986; Rogers *et al.*, 1986; Sternberg and Kimber, 1986).

In contrast laminin may be absent from most of the adult mammalian CNS (Liesi, 1985b; cf. Manthorpe *et al.*, 1988), but is abundant in the goldfish (Hopkins *et al.*, 1985) and frog visual systems as well as in rat olfactory bulb (Liesi, 1985b), adult CNS regions which either continue to grow or have the potential to do so. In the adult PNS, laminin is found in basement membranes which persist following axotomy and axonal degeneration, apparently acting as conduits for regenerating peripheral axons (Ide *et al.*, 1983). Studies of neurite growth in culture, on tissue substrata from several of

Table 1. Localization of ECM Adhesive Glycoproteins in the Nervous System

ECM protein	Localization in CNS	Localization in PNS	References
Laminin	Radial glial fibers in developing mammalian brain; embryonic spinal cord; developing and injured adult optic nerve; basement membrane of neural tube and notochord; adult vascular and meningeal basement membrane; adult olfactory bulb; adult fish and frog optic nerves	Neural crest migratory routes; adult endoneurial and perineurial basement membranes; myoneural junction	Liesi (1985a,b), Carbonetto et al. (1987), Cohen et al. (1986, 1987), Tuckett and Morriss-Kay (1986), Sternberg and Kimber (1986), Rogers et al. (1986), Hopkins et al. (1985), Laurie et al. (1983), Palm and Furcht (1983), Sanes et al. (1986), Bunge et al. (1986), Giftochristos and David (1988)
Fibronectin	Developing cerebellum and cerebral cortex; basement membrane of neural tube and notochord; vascular and meningeal basement membrane in adult and embryonic spinal cord	Migratory routes of neural crest cells; adult endoneurial and perineurial basement membrane	Sanes et al. (1986), Newgreen and Thiery (1980), Tuckett and Morriss-Kay (1986), Sternberg and Kimber (1986), Carbonetto et al. (1987), Hatten et al. (1982), Bunge et al. (1986), Laurie et al. (1983), Palm and Furcht (1983), Stewart and Pearlman (1987)
Collagen IV	Basement membranes of spinal cord and cerebral blood vessels in postnatal rat	Adult endoneurial and perineurial basement membrane	Bunge et al. (1986), Laurie et al. (1983)
Cytotactin	Radial glial cells in developing cerebellum; developing retina; basement membrane of neural tube and notochord; adult forebrain and cerebellum	Migratory routes of neural crest cells; interstitial spaces near denervated synaptic junctions	Sanes et al. (1986), Crossin et al. (1986), Chuong et al. (1987), Tan et al. (1987)
Entactin	Basement membrane of neural tube	Adult endoneurial and perineurial basement membrane	Tuckett and Morriss-Kay (1986), Bunge et al. (1986)

these neural tissues have shown that growth is most pronounced on those tissues which support growth *in vivo,* and correlates with the immunocytochemical localization of laminin in those tissues (Carbonetto *et al.,* 1987). Furthermore, Sandrock and Matthew (1987) have demonstrated that a monoclonal antibody to a heparan sulfate pro-teoglycan (HSPG)–laminin complex, which potently inhibits neurite outgrowth in conditioned media, also inhibits nerve regeneration *in vivo*. Taken together, these observations suggest that laminin is found in regions of the nervous system where it critically influences neural development and regeneration.

3.2. Fibronectin

Fibronectin is a large, extended glycoprotein (450 kDa) found in many epithelial and endothelial basement membranes and in the ECM surrounding muscle fibers and peripheral nerves (Madri *et al.*, 1984; Sanes *et al.*, 1986). Fibronectin is also found throughout mesenchymal tissue (cellular fibronectin) and in serum as a soluble protein (plasma fibronectin). In all instances, fibronectin is a dimer of two similar subunits, each approximately 220 kDa, that are associated at their carboxy-termini via disulfide bonds (Figure 1B; Engel *et al.*, 1981; Hynes, 1985). Each subunit is about 56 nm long and 2–3 nm wide (Engel *et al.*, 1981), and has protease-resistant globular domains, consisting of tandemly repeated segments of amino acids, which are interspersed with protease-sensitive regions (Odermatt *et al.*, 1985; Petersen *et al.*, 1983). These repeated segments fall into one of three categories according to their primary and secondary structures (reviewed in Hynes, 1985). For example, regions of type I and type II homology are made up of internal repeating units with multiple disulfide bonds; type III regions are predominantly β sheets (Petersen *et al.*, 1983). Separate exons within the fibronectin gene encode the three types of homology, suggesting that fibronectin arose phylogenetically from three primordial genes by duplication and fusion (Petersen *et al.*, 1983; Odermatt *et al.*, 1985; Oldberg and Ruoslahti, 1986). cDNA sequence data suggest that the two subunits of fibronectin are synthesized from a single gene whose mRNA transcripts are alternatively spliced in the IIICS (connecting segment) region (Figure 1B; Schwarzbauer *et al.*, 1985; Kornblihtt *et al.*, 1985). The ten or more isoforms of fibronectin thus far identified result from additional patterns of alternate mRNA splicing (Kornblihtt *et al.*, 1985).

As with laminin, proteolytic fragments of fibronectin have been used to identify the functional regions of this multifunctional molecule. An amino-terminal region comprised mainly of type I repeats binds heparin and fibrin (Figure 1B; Smith and Furcht, 1982; reviewed in Hynes, 1985). Adjacent to this region and toward the carboxy side is a collagen binding domain with type I and type II repeats (Hynes, 1985). A large fragment that includes the major cell binding region (Pierschbacher *et al.*, 1981), and which consists mainly of type III homologies, is found in the central portion of the molecule (Hynes, 1985; Yamada, 1983). Finally, at the carboxy end of the molecule there are one or two additional domains with binding sites for heparin, fibrin, cell surfaces (Smith and Furcht, 1982), and possibly a fibronectin self-association site (Ehrismann *et al.*, 1982).

Proteolytic fragments of fibronectin containing the major cell binding domain support neurite outgrowth (Carbonetto *et al.*, 1983). Unlike laminin (discussed above), neurite growth by DRG neurons on fibronectin requires NGF (Carbonetto *et al.*, 1983). A sequence of three amino acids, RGD (Arg-Gly-Asp), within the major cell binding domain of fibronectin is principally responsible for cell attachment (Pierschbacher and Ruoslahti, 1984). In solution, synthetic peptides containing RGD inhibit attachment of

nonneural cells to fibronectin (Yamada and Kennedy, 1984), as well as neurite extension by PC12 cells (Akeson and Warren, 1986; Letourneau, 1988). RGD sequences are
found in other ECM molecules such as laminin, collagen, fibrinogen, von Willebrand
factor, and vitronectin, where they are often involved in cell attachment (reviewed in
Ruoslahti and Pierschbacher, 1987). However, the RGD sequence in laminin (Figure
1A; M. Sasaki, personal communication) may not be active in neurite growth since
RGD-containing peptides do not inhibit growth on laminin (Akeson and Warren, 1986;
Letourneau, 1988), or [^{125}I]laminin binding to neuronal membranes (Douville et al.,
1988). In other studies, RGD sequences have been implicated in blood clotting and
tissue repair (Pytela et al., 1986), metastasis of malignant tumors (Humphries et al.,
1986a), gastrulation (Naidet et al., 1987), and neural crest cell migration (Boucaut et
al., 1984).

   More recently, non-RGD binding sites have been reported within fibronectin. For
example, a proteolytic fragment of fibronectin from a region near the carboxy-terminal
heparin binding domain supports neurite growth by CNS neurons, an activity which is
not apparent in the intact molecule (Rogers et al., 1985). Similarly, RGD peptides
inhibit neuroblastoma attachment to and neurite extension on a 120-kDa fragment of
fibronectin containing the major cell binding site but do not inhibit on native fibronectin (Waite et al., 1987). Humphries et al. (1986b) have identified an active sequence
Arg-Glu-Asp-Val (REDV) in the IIICS region of the A chain (Figure 1B) of human
plasma fibronectin which is active in melanoma cell attachment.

   As in the case of laminin, there has been considerable interest in the question of
whether there is fibronectin in the developing nervous system (Table 1). Fibronectin
has been found by several laboratories along the migratory routes of neural crest cells
(e.g., Newgreen and Thiery, 1980), and injected RGD peptides inhibit migration
(Boucaut et al., 1984). Its localization within the developing CNS is more controversial. In early development it has been reported in basement membranes surrounding the
developing neural tube but not within the neural epithelium itself (Newgreen and
Thiery, 1980; Tuckett and Morriss-Kay, 1986; Sternberg and Kimber, 1986). Some
investigators have reported fibronectin on endothelial and leptomeningeal cells but not
on neurons or glia (Schachner et al., 1978; Hynes et al., 1986). Others have detected
fibronectin in the cerebellum during migration of the external granule layer (Hatten et
al., 1982) and more recently, among neural cells in the developing cerebral cortex
(Stewart and Pearlman, 1987). The reasons for these discrepancies are unclear, but
may reflect the transient appearance of fibronectin in different regions of the CNS.

3.3. Type IV Collagen

   Collagens are a family of matrix molecules that are found in virtually every
connective tissue in the body (reviewed in Linsenmayer, 1981). They are composed of
three subunits arranged in a triple helix. The primary amino acid structure has a
repeating GLY-X-Y motif. Glycine, a small amino acid, apparently allows the three
helices to wrap tightly around each other. Proline is often found at the X and Y
positions but is converted to hydroxyproline at the Y position by the enzyme prolyl
hydroxylase and is important for extensive hydrogen bonding which helps stabilize the

triple helix. Lysine is also frequently found in the Y position and can be hydroxylated to hydroxylysine by the enzyme lysyl hydroxylase. The hydroxylysine can be glycosylated or can be converted to aldehyde forms which participate in covalent cross-links with other aldehyde lysine or hydroxylysine residues on adjacent molecules, rendering collagens highly insoluble.

Type IV collagen (380 kDa), the major form of collagen in basement membranes (Trüeb et al., 1982), forms the backbone of the lamina densa (Laurie et al., 1982). Its three subunits, two α1(IV) chains and one α2(IV) chain (Trüeb et al., 1982), are organized into four major structural domains (Figure 1C); a globular, nonhelical domain (NC1) at the carboxy-terminus; a central 330-nm mostly triple-helical region; a shorter 30-nm triple-helical region (7 S domain) located at the amino-terminus; and a short nonhelical domain (NC2) separating the two helical regions (Timpl et al., 1984). A distinctive feature of the type IV collagen triple-helical regions are the frequent interruptions of the major triple-helical region by short nonhelical regions (Kühn et al., 1985). In contrast to interstitial forms of collagen (e.g., types I and III), type IV collagen does not associate into fibrils because it is not proteolytically processed before secretion to remove the nonhelical domains (Kefalides et al., 1979). Partial cDNA sequence data for both the α2(IV) and α1(IV) chains in the NC1 domain (Hostikka et al., 1987) reveal extensive homologies between the chains but not with the non-collagenous globular domains of interstitial procollagens (Oberbäumer et al., 1985). In addition, analysis of genomic clones to the α1(IV) chain reveals that the sizes of the exons do not correspond to the characteristic tandemly duplicated 54-bp coding unit found in fibrillar collagens (Sakurai et al., 1986). This suggests that the α1(IV) gene evolved separately from genes for fibrillar collagens (Sakurai et al., 1986).

In electron micrographs, rotary-shadowed type IV collagen is seen in supramolecular polygonal arrays (Madri et al., 1984; Kühn et al., 1985). This self-association depends upon the triple-helical 7 S domains which interact to form a "star-shaped" tetramer from which is assembled a larger polygonal array similar to that in basement membranes (Timpl et al., 1984; Madri et al., (1984). The presence of cysteines in the NC1 region (Oberbäumer et al., 1985) correlates with observed intramolecular and intermolecular disulfide bonds that participate in covalent cross-linking of adjacent tetrameric units (Timpl et al., 1985).

Type IV collagen supports attachment and neurite outgrowth of PNS neurons and PC12 cells (Carbonetto et al., 1983; Turner et al., 1987), effects which may be mediated through RGD sequences (Hostikka et al., 1987). In the adult peripheral nervous system, type IV collagen is restricted to basement membranes, where it is synthesized by Schwann cells (Table 1; Bunge et al., 1986). Type IV collagen has also been localized in vascular basement membranes in the embryonic CNS (Laurie et al., 1983).

3.4. Other Adhesive ECM Glycoproteins

Several other ECM adhesive proteins have been identified including: epinectin (Enenstein and Furcht, 1984), osteonectin (Termine et al., 1981), vitronectin (Suzuki et al., 1985), chondronectin (Hewitt et al., 1980), cytotactin (Grumet et al., 1985), and entactin (discussed below). Except for cytotactin and entactin, the others have not been reported in the nervous system. Cytotactin, which is similar to J1 (Kruse *et al.*, 1985) and tenascin or myotendinous antigen (Chiquet-Ehrismann *et al.*, 1986; Chiquet, 1989), consists of two or three polypeptides of 190–220 kDa (Crossin *et al.*, 1986). These are disulfide-bonded into a large ($\simeq$ 1000 kDa) hexabrachion with a central globule, thin and thick segments, and knobs at the ends of the arms (Figure 1D; Erickson and Inglesias, 1984; Chiquet-Ehrismann *et al.*, 1986). Cytotactin contains a carbohydrate epitope recognized by the antibodies L2 and HNK-1 and which is also found on several CAMs (NCAM, L1, or NgCAM; Kruse *et al.*, 1985; Grumet *et al.*, 1985). It is unclear whether this carbohydrate is directly involved in adhesive events. However, antibodies against cytotactin block neuron–glia but not neuron–neuron adhesion (Grumet *et al.*, 1985).

In the developing nervous system, cytotactin has been localized to peripheral and central glia, but not neurons or meningeal cells (Table 1; Grumet *et al.*, 1985; Kruse *et al.*, 1985), and in the cerebellum it is associated with radial glial fibers during granule cell migration through the molecular layer (Chuong *et al.*, 1987; Crossin *et al.*, 1986). Polyclonal and monoclonal antibodies to cytotactin block granule cell migration *in vivo* (Chuong *et al.*, 1987). Cytotactin is similarly found in the optic fiber layer during extension of ganglion cell axons and decreases in amount when growth ceases (Crossin *et al.*, 1986). At much earlier times, cytotactin is found in basement membranes surrounding the neural tube, notochord, and somites where it is temporally expressed in a rostral-to-caudal gradient (Crossin *et al.*, 1986; Tan *et al.*, 1987). Like laminin and fibronectin, cytotactin is found along the ventral migratory routes taken by neural crest cells; unlike laminin and fibronectin, it is also found in the adult CNS (Crossin *et al.*, 1986). In muscle it is concentrated at the myotendinous region but not in basement membranes. However, following denervation it accumulates both in the basement membrane and within interstitial spaces near synaptic sites, possibly to guide regenerating axons to their targets (Sanes *et al.*, 1986).

Entactin is a dumbbell-shaped 158-kDa sulfated glycoprotein similar or identical to nidogen (Dziadek *et al.*, 1985) and is typically found in basement membranes (Carlin *et al.*, 1981; Paulsson *et al.*, 1987). Entactin forms stable, noncovalent complexes ($K_d \simeq 10^{-9}$ M) in a 1 : 1 ratio with laminin (Dziadek and Timpl, 1985) through binding of one of its globular domains with the center of the cross region of laminin (Martin and Timpl, 1987). In the developing embryo, entactin is localized in basement membranes surrounding the neural epithelium and neural tube along with basement membrane laminin, but not with mesenchymal laminin (Table 1; Tuckett and Morriss-Kay, 1986). Entactin is also found in basement membranes surrounding Schwann cells in the PNS (Bunge *et al.*, 1986). Its function is unclear, although it may prevent cell binding to the central cross of laminin (Aumailley *et al.*, 1987).

4. RECEPTORS FOR ECM COMPONENTS

Several indirect observations first indicated that cells adhere to ECM glycoproteins by complementary interactions. These included: selective adhesion and differentiation of cells on culture dishes coated with collagen (Hauschka and Konigsberg,

1966) or serum proteins such as fibronectin (Grinnell, 1978); adhesion of cells to proteolytic fragments of ECM adhesive proteins (discussed above); and characterization of binding sites on cells with radiolabeled ECM proteins (reviewed in von der Mark and Kühl, 1985). More recently, several receptors for ECM adhesive proteins have been isolated. The most extensively characterized ECM receptors constitute a superfamily called integrins, but other ECM receptors for laminin and collagen also appear to be involved in cell–matrix interactions.

4.1. Integrins

By affinity chromatography, similar ECM receptors have been isolated from a variety of mammalian cell types including platelets (Shadle *et al.*, 1984), human osteosarcoma cells (Pytela *et al.*, 1985a,b), melanoma cells (Cheresh and Harper, 1987), erythroid precursors (Patel and Lodish, 1986), and other cell types (Plow *et al.*, 1986). These proteins, eluted from affinity columns by RGD-peptides, are all heterodimers consisting of one α and one β subunit, and are now known to be members of a family of proteins called integrins (Tamkun *et al.*, 1986; Hynes, 1987). Integrins were first identified in cultured cells using monoclonal antibodies (CSAT and JG22) which interfere with fibroblast and myoblast attachment to fibronectin and laminin substrata (Neff *et al.*, 1982; Greve and Gottlieb, 1982). Immunoprecipitation with these antibodies yields a glycoprotein complex from chick cells consisting of two or three subunits with an average subunit molecular mass of 140 kDa (Knudsen *et al.*, 1985; Akiyama *et al.*, 1986; Horwitz *et al.*, 1985). When electrophoresed under nonreducing conditions, the immunoprecipitates resolve into three bands of approximately 160, 120, and 110 kDa (reviewed in Buck and Horwitz, 1987). This may well reflect two heterodimers consisting of distinct α subunits (160 and 120 kDa) and a common β subunit (110 kDa; discussed below), although a unique heterotrimer in chick cells has not been ruled out (Buck and Horwitz, 1987). The α subunits of some chick and mammalian integrins electrophorese at a slower rate under nonreducing conditions due to the retention of smaller (≃ 25 kDa) polypeptides which are disulfide-bonded to the larger proteins. The β subunits display the opposite behavior, electrophoresing faster under nonreducing conditions, due to internal disulfide bonds that maintain the protein in a more compact conformation (Plow *et al.*, 1986; Buck and Horwitz, 1987). CSAT and JG22 recognize the β subunit which remains associated with the α subunits following detergent extraction and is coprecipitated (Buck *et al.*, 1986). Binding of integrins to laminin or fibronectin requires an intact oligomeric complex (Buck *et al.*, 1986) and is of relatively low affinity ($K_d \simeq 10^{-6}$ M; Horwitz *et al.*, 1985).

Molecular cloning and sequencing of the cDNA for the β subunit of chick integrin indicates that it is an 803-amino-acid transmembrane protein with a large extracellular domain including four cysteine-rich repeats of 40 amino acids and a short (47 amino acid) cytoplasmic tail (Tamkun *et al.*, 1986). As mentioned above, there is wide homology with heterodimeric proteins in other cells which have subsequently been classified into three groups (Reichardt *et al.*, 1989): those that have B_1 subunits (e.g., very late antigens; Hemler *et al.*, 1987), B_2 subunits (e.g., cytoadhesins), or B_3 subunits (e.g., glycoprotein IIb/IIIa). A fourth group of proteins, comprising the position-specific antigens in *Drosophila,* may constitute another integrin family (Ruoslahti and Pierschbacher, 1987). The chick integrin β1 subunit has 47% homology with the β3 subunit of the platelet receptor (GP IIIa; Fitzgerald *et al.,* 1987; Tamkun *et al.,* 1986) and 47% homology with the β2 subunit of the cytoadhesins (also known as LFA-1, MAC-1, CR3, and p150,95) on lymphocytes (Law *et al.,* 1987). Partial sequence data also suggest homologies among the ten or more α subunits of integrins but not between α's and β's (Suzuki *et al.,* 1986; Ginsberg *et al.,* 1987). The α's are also transmembrane proteins (Buck and Horwitz, 1987) but additional structural information awaits more extensive sequence data. Each β can be complexed with different α's to generate considerable diversity, but there is as yet no example of an α complexed to different β's (Buck and Horwitz, 1987).

Integrins mediate a spectrum of cellular functions which stem from their participation in cell–matrix and cell–cell adhesion. The relatively few β subunits identified so far suggest that the α subunits are important in generating this diversity of adhesive specificities, although both subunits are required for binding (Buck *et al.,* 1986). In some instances this specificity is high. Consider that several integrins recognize RGD in ECM glycoproteins (Ruoslahti and Pierschbacher, 1987) and that the vitronectin receptor binds well to RGD in vitronectin but only poorly, if at all, to the same sequence in fibronectin (Pytela *et al.,* 1985a,b). This specificity may result not only from variations in integrin subunits but also may be encoded by structural differences in the regions surrounding RGD within fibronectin and vitronectin (Ruoslahti and Pierschbacher, 1987; Yamada and Kennedy, 1987). Other integrins, such as the platelet glycoprotein IIb/IIIa, are much more indiscriminate, binding to fibronectin, fibrinogen, von Willebrand factor, and vitronectin (Pytela *et al.,* 1986; Hynes, 1987), and may have binding sites suited to RGD in multiple conformations. The promiscuous nature of chick integrin has also been reported in experiments demonstrating competitive binding between vitronectin, fibronectin, and laminin for the same or neighboring sites (Buck and Horwitz, 1987).

In stationary fibroblasts, fibronectin fibrils on the cell surface align with microfilaments, vinculin, talin, and α-actinin inside the cell, suggesting a transmembrane link between the ECM and the cytoskeleton (Chen *et al.,* 1985; Damsky *et al.,* 1985; Buck and Horwitz, 1987). This is consistent with more recent data concerning the predicted transmembrane structure of integrin (Tamkun *et al.,* 1986), which can form complexes with talin and vinculin through its cytoplasmic domain (Horwitz *et al.,* 1986; Burridge, 1987; Tapley *et al.,* 1989). The region to which talin binds encompasses a consensus tyrosine kinase phosphorylation site that is homologous to one on EGF and insulin receptors (Tamkun *et al.,* 1986). This site can be phosphorylated by the oncogenic tyrosine kinase pp60[src] from Rous sarcoma virus which is found, along with integrin, in areas of cell–substratum contact (Hirst *et al.,* 1986; Tapley *et al.,* 1989). Moreover, the β subunit of chick integrin is phosphorylated *in vivo* following transformation of cells by Rous sarcoma virus (Hirst *et al.,* 1986). Since transformed cells lose their substratum adhesiveness, these observations raise the possibility that phosphorylation of some integrins is involved in cell transformation (discussed in Buck and Horwitz, 1987; Tapley *et al.,* 1989).

Antibodies against integrins inhibit avian neural crest cell migration (Duband *et al.,* 1986), giving rise to ectopic aggregates of neural crest cells and malformed neural tubes *in vivo* (Bronner-Fraser, 1986). Anti-integrin antibodies also inhibit neurite growth on laminin, fibronectin, and collagen from PNS (Bozyczko and Horwitz, 1986, Tomaselli *et al.*, 1986) and CNS neurons (Hall *et al.*, 1987; Cohen *et al.*, 1986, 1987) as well as from PC12 cells (Tomaselli *et al.*, 1987). In DRG cells, CSAT immunoprecipitates two prominent polypeptides (135 and 110 kDa), and several smaller polypeptides thought to be proteolytic fragments of the two major proteins (Bozyczko and Horwitz, 1986). In chick retinal cells, anti-integrin antibodies immunoprecipitate three or four polypeptides (145, 135, 120, and 110 kDa) (Hall *et al.*, 1987), and in PC12 cells similar antibodies precipitate three polypeptides, possibly two α subunits (140 and 180 kDa) and a β_1 120-kDa subunit (Tomaselli *et al.*, 1987). The implications of this structural diversity are just beginning to be studied. For example, neurons cultured from chick retina (Cohen *et al.*, 1986; Hall *et al.*, 1987) and optic lobe (Douville *et al.*, 1987) lose their ability to extend nerve processes on laminin by about embryonic day 9 but continue to grow well on astrocytic substrata (Cohen *et al.*, 1986). There is no obvious depletion of integrins as these cells mature (Cohen *et al.*, 1986; Hall *et al.*, 1987). This has led to the suggestion that integrins may somehow become functionally uncoupled (Cohen *et al.*, 1987). Such uncoupling could, in principle, occur by dissociation of subunits within the plasma membrane or perhaps through posttranslational modifications such as phosphorylation. Alternatively, Hall *et al.* (1987) have reported that the relative abundance of integrin subunits in the retina changes somewhat with development, which could alter the responsiveness of neurons to laminin.

4.2. High-Affinity 67-kDa Laminin Receptor

Studies of radioiodinated laminin binding to a variety of nonneural (Terranova *et al.*, 1983; Malinoff and Wicha, 1983; Barsky *et al.*, 1984; Huard *et al.*, 1986) and neural cells (Harvey and Carbonetto, 1986; Luckenbill-Edds *et al.*, 1986; Douville *et al.*, 1988) cells reveal a single class of binding sites with a K_d of about 10^{-9} M. The binding is specific, and is not inhibited by fibronectin, collagen, or serum (Barsky *et al.*, 1984), but is blocked by the E1 fragment of laminin (Terranova *et al.*, 1983; Barsky *et al.*, 1984). The affinity of this laminin binding contrasts with the much lower affinity ($K_d \approx 10^{-6}$ M) of fibronectin binding to cells (Gardner and Hynes, 1985; Akiyama and Yamada, 1985; Ginsberg *et al.*, 1985), and integrin binding to laminin and fibronectin (Horwitz *et al.*, 1985).

A laminin receptor with an apparent molecular weight of 67 kDa ($\approx$ 55 kDa, unreduced) has been isolated from human breast carcinoma (Barsky *et al.*, 1984), murine melanoma cells (Rao *et al.*, 1983), muscle (Lesot *et al.*, 1983), murine fibrosarcoma cells (Malinoff and Wicha, 1983), NG108 cells and EHS tumor cells (Graf *et al.*, 1987a,b), and CNS neurons (Douville *et al.*, 1987, 1988). The affinity of this receptor for laminin is at least 1000-fold higher than that of chick integrin (Graf *et al.*, 1987a,b; Douville *et al.*, 1988), and the receptor from NG108 cells can be eluted from laminin affinity columns by the YIGSR pentapeptide (Graf *et al.*, 1987a,b). Mecham and coworkers (personal communication; Hinek *et al.*, 1988) have identified a related 67-kDa protein in chondroblasts and fibroblasts that can be eluted from both laminin and elastin affinity columns along with 3 other proteins (61 kDa, 55 kDa, and 43 kDa).

The 55-kDa and possibly 61-kDa protein are thought to be integral membrane proteins to which the 67-kDa protein binds (Figure 2). The 43-kDa protein may be actin (Brown *et al.*, 1983; R. Mecham, personal communication). Interestingly, lactose disrupts binding of this receptor to laminin and elastin and dislodges the 67-kDa receptor from the 55-kDa protein (R. Mecham, personal communication; Hinek *et al.*, 1988). Lactose apparently complexes with a site on the receptor removed from that to which the laminin and elastin bind.

Monoclonal and polycolonal antibodies which detach cells from laminin have been raised to the 67-kDa laminin receptor (Liotta *et al.*, 1985; R. Ogle and A. Albini, personal communication) and used to clone a partial cDNA for this protein (Wewer *et al.*, 1986). Monoclonal antibodies to an immunologically cross-reactive 43-kDa protein from developing rat retina have subsequently been used to clone full length cDNAs (Rabacchi *et al.*, 1988) which encode the entire partial sequence published by Wewer *et al.* (1986) and are apparently the same as cDNAs enriched in human colon carcinoma (Yow *et al.*, 1988). These data firmly establish these protein(s) as non-integrins. In addition they are distinct from a muscle-specific laminin-binding protein known as aspartactin (56 kDa in chick and 66 kDa in rat; reduced) that interacts with the heparin binding domain at the end of the long arm of laminin (Hall *et al.*, 1988; Clegg *et al.*, 1988). Although no function has yet been assigned to aspartactin it is found extracellularly in close proximity to myotube plasmalemma in the extracellular matrix where it may mediate cell-matrix interactions (Hall *et al.*, 1988).

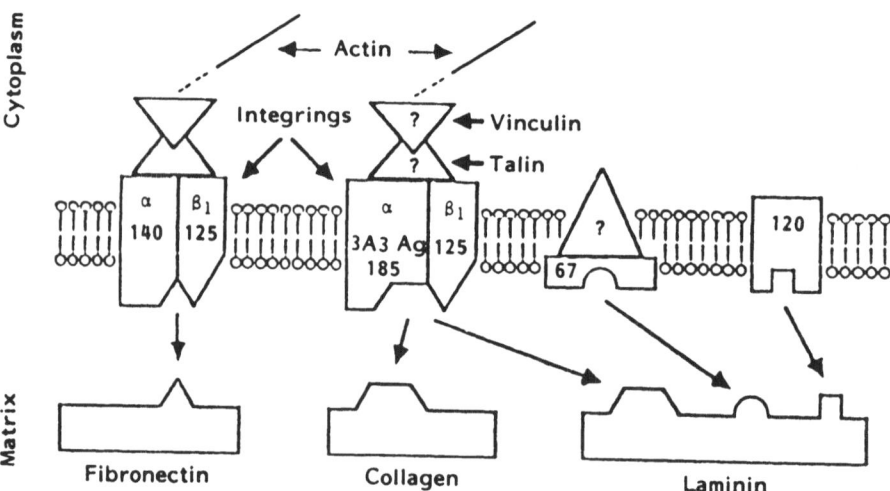

Figure 2. Possible ECM receptors on neural cells. Neural cells attach to ECM adhesive proteins through several cell surface receptors. Most notable are the integrins, which in chick neurons bind laminin, fibronectin, and collagen. Illustrated is a mammalian integrin which binds fibronectin only (left), and which may form transmembrane links with the cytoskeleton through talin and vinculin, and ultimately with actin through some unknown intermediate(s) (dashed line). Also shown is the 3A3 antigen which is an integral part of a dual laminin/collagen receptor which is an integrin. Neurons also have the 67-kDa high-affinity laminin receptor, shown associated with the lipid bilayer through docking proteins. Finally, cranin is a 120-kDa high-affinity laminin receptor on the surface of neurons.

cDNA clones to the 67-kDa laminin receptor hybridize to mRNAs of 1.1 to 1.3 kb in colon carcinoma and embryonic mouse heads (Yow *et al.*, 1988; Rabacchi *et al.*, 1988) or 1.7 kb from human breast carcinoma (Wewer *et al.*, 1986). The full-length cDNA clones contain a region corresponding to 295 amino acids that encodes a protein of only 32 kDa (Yow *et al.*, 1988; Sequi-Real *et al.*, 1988). The sequence contains no consensus sites for N-linked glycosylation, one for O-linked glycosylation, and two cysteine residues making it unlikely that the discrepancy in molecular weight (67 versus 32 kDa) results from glycosylation plus internal disulfide bonding (Yow *et al.*, 1988; Segui-Real *et al.*, 1988). The 67-kDa protein may be immunologically and functionally related to the 32-kDa protein and possibly derived by alternative splicing of a larger RNA transcript. It is noteworthy that the full-length clones lack a consensus N-terminal leader signal sequence but may contain hydrophobic regions for membrane association (Yow *et al.*, 1988).

The 67-kDa laminin receptor has been studied most extensively because of its possible involvement in tumor metastasis (reviewed in Liotta *et al.*, 1986). However the 67-kDa receptor is also found in NG-108 cells (Graf *et al.*, 1987a), and in neurons (Douville *et al.*, 1988) where its biological function, especially in relation to integrins, is unclear. Recent reports have localized the protein(s) to the dorsal third of the developing mammalian retina as well as transiently in axons of the optic nerve and in early motor neurons (Dräger *et al.*, 1988). The YIGSR sequence, to which the 67-kDa receptor binds, is found only within the central, cell-attachment region of laminin and not in the neurite-promoting region suggesting that these activities in laminin (i.e., cell attachment versus neurite outgrowth) may be mediated by separate receptors on the cell (Kleinman *et al.*, 1988). In this hypothetical scheme the 67-kDa laminin receptor, with its high affinity for laminin, may be responsible for firmly anchoring neural cell bodies to their substrata thereby offsetting tension generated by the extending axon, while integrins and/or other ECM receptors would promote neurite growth *per se*. This would be consistent with reports that substratum-bound YIGSR is capable of supporting neural cell attachment but not neurite outgrowth (Graf *et al.*, 1987a; Kleinman *et al.*, 1988).

4.3. High-Affinity 120/140-kDa Laminin Receptor

A 120-kDa laminin binding protein, called cranin, (Smalheiser and Schwartz, 1987) has been identified in chick brain, liver, and fibroblast membranes by probing electrophoresed and electroblotted membranes with [^{125}I]laminin. A related or identical protein which varies in molecular mass from 120 to 140 kDa has been similarly identified and laminin-affinity purified from embryonic chick brain, neuronal cultures (Douville *et al.*, 1988) and NG-108 cells (Kleinman *et al.*, 1988), where it can be radioiodinated on the surface of intact cells, (P. Douville and S. Carbonetto, unpublished). Despite its similar molecular mass, indirect observations suggest that the 120-kDa protein is not an integrin subunit: (1) it can be detected in electroblots with concentrations of [^{125}I]laminin (10^{-10} M) much below the K_d of integrins (Smalheiser and Schwartz, 1987; Douville *et al.*, 1988; Kleinman *et al.*, 1988); (2) it binds laminin following reduction; (3) it migrates at the same rate by SDS-PAGE under reducing or nonreducing conditions; (4) binding of [^{125}I]laminin is blocked only by unlabeled laminin (Smalheiser and Schwartz, 1987) and not by fibronectin, fibronectin-related peptides (RGDS, discussed above), or antibodies to integrin (Douville *et al.*, 1988). These findings have been complicated by recent attempts to partially sequence the protein, which have shown the presence of entactin in a laminin-affinity purified protein that appears similar to cranin (Sephel *et al.*, 1989).

In several instances, methods of isolating ECM receptors which utilize antibodies, affinity chromatography, or binding assays with ECM proteins have identified separate receptors. Cranin is a good example because it is the most obvious laminin-binding protein in electroblots of membranes probed with [^{125}I]laminin, yet, until recently (Douville *et al.*, 1988; Kleinman *et al.*, 1988), it had not been reported by investigators utilizing affinity purification schemes. Perhaps then, it is not surprising that there is much to be done to ascertain the relationships of these various ECM receptors, especially with regard to their function in neural cells. Moreover, not all ECM-binding proteins are necessarily involved in matrix adhesion. Consider that there are likely intracellular processing proteins which must bind them, or cell surface proteins which function in matrix assembly.

4.4. Other ECM Receptors

4.4.1. 3A3 Antigen

PC12 cells attach to and extend nerve fibers on laminin and collagen substrata through a Mg^{2+}-dependent mechanism (Turner *et al.*, 1987). A recently obtained monclonal antibody (3A3) directed against a PC12 surface antigen disrupts attachment and nerve fiber growth on laminin- and collagen- but not on polylysine- or wheat germ agglutinin-coated substrata (D. Turner, personal communication). The antigen is also present on rat DRG neurons where 3A3 inhibits neurite outgrowth and on rat astrocytes where it blocks cell attachment to laminin and collagen (N. Tawil and S. Carbonetto, unpublished). This antibody immunoprecipitates a heterodimeric protein from PC12 cells (Turner *et al.*, 1989) and from embryonic rat brain (N. Tawil and S. Carbonetto, unpublished) with apparent subunit molecular masses of 185 kDa and 125 kDa. The 125-kDa protein has recently been shown (after this review was submitted) to be immunologically related to a 125-kDa integrin β1 subunit in PC12 cells (L. Flier and D. Turner, unpublished). The 185-kDa polypeptide seen in 3A3 immunoprecipitates represents an α subunit previously shown to be precipitated by anti-β2-antisera (Toma-selli *et al.*, 1988; Turner *et al.*, 1989). Since 3A3 precipitates a single heterodimer from PC12 cells, it appears that it binds to a functional region of an α subunit (185 kDa) of a laminin/collagen integrin (Figure 2). Recent data (Toyota *et al.*, 1988) indicate that this receptor functions in peripheral nerve regeneration *in vivo*.

4.4.2. Collagen Receptors

Collagen affinity chromatography has been used to purify five proteins (120, 90, 56, 47, and 38 kDa) from chick fibroblasts which bind type I and IV collagen. Three of these (120, 90, and 47 kDa) can be eluted from collagen columns with RGD peptides (Ogle and Little, 1989). The 120-kDa protein is probably identical to the β subunit of chick integrin (Ogle and Little, 1989), and antibodies to this protein detach peripheral neurons from type I collagen (R. Ogle, personal communication). A collagen binding integrin with an α subunit of 160 kDa and a β1 subunit has also been reported (Santoro, 1986; Wayner and Carter, 1987). Finally, Dedhar et al. (1987) have identified and eluted three polypeptides from human osteosarcoma cells (250, 70, and 30 kDa). The 70-kDa collagen I binding protein may be related to a recently reported 67-kDa type IV binding protein (Ogle et al., 1987). The distribution of this protein and its function in neural cells is unknown.

4.4.3. Carbohydrate Receptors

The external disposition of glycoconjugates on the surfaces of cells has prompted much speculation that cell adhesion involves binding to carbohydrates. Perhaps the clearest example of cell adhesion mediated by carbohydrate receptors is in sea urchin sperm–egg recognition (reviewed in Trimmer and Vacquier, 1986). A growing body of evidence, including work by Mecham and co-workers cited previously, suggests that carbohydrates are involved in cell–matrix adhesion. For example, Rauvala and Hakomori (1981) have described a cell-surface mannosidase which, at physiological pH, binds but does not cleave its substrate and can cause cell–substratum attachment. Lectinlike molecules have been described in many tissues including developing rat cerebellum (Zanetta et al., 1985). Runyan et al. (1986) have reported that neural crest cells utilize a cell-surface galactosyltransferase to migrate along laminin, which contains GlcNAc residues. In other studies the carbohydrate epitope recognized by HNK-1 antibodies appears to interfere with cell–substratum interactions involved in neurite outgrowth (Dow et al., 1988).

4.4.4. Glycolipids

Addition of disialogangliosides or antibodies against the sialic acid portion of gangliosides partially inhibits nonneural and neuroblastoma cell attachment to fibronectin, laminin, vitronectin, collagen, and RGDS-containing peptides (Yamada et al., 1983; Cheresh et al., 1986). However, gangliosides do not appear to bind directly to ECM glycoproteins and a modulatory role for gangliosides in cell adhesion has been postulated (Cheresh et al., 1986; Ruoslahti and Pierschbacher, 1987). Apparently, gangliosides are required prosthetic groups for integrin structure and function (Ruoslahti and Pierschbacher, 1987). In addition, sulfated glycolipids, but not gangliosides, bind directly to laminin with an affinity in the nanomolar range (Roberts et al., 1985).

5. NERVE FIBER GROWTH ON EXTRACELLULAR MATRICES— CONCLUDING REMARKS

The growth cone, a motile organelle at the tip of a growing nerve fiber, has fine projections called filopodia containing bundles of microfilaments which are sent out from the growth cone, testing the substratum for its adhesiveness (Landis, 1983;

Carbonetto, 1984; Letourneau, 1983). The mechanism of filopodial extension is unknown but may involve actin polymerization similar to the extension of cellular projections during platelet activation (Pollard, 1980). The body of the growth cone appears anchored to the substratum by flattened regions called lamellipodia which are also largely composed of a microfilamentous network. From time-lapse studies, adhesion of filopedia to their substrata appears to initiate contractile forces that help extend the neurite (Letourneau, 1981, 1983; Carbonetto, 1984). From earlier discussions it should be easy to see how the ECM can direct the formation of adhesive complexes that link the matrix with the cytoskeleton. Integrins, for example, bind to ECM components on the cell surface and talin on the inside of the cell (Figure 2). Talin is in turn coupled to vinculin, which is itself somehow associated with microfilament bundles (Burridge, 1987). These interactions could organize microfilaments within cell–substratum contacts and possibly trigger contractile events needed to extend the axon. To help accomplish this, integrins, which are distributed diffusely in motile cells (Damsky et al., 1985; Duband et al., 1986), may concentrate within cell–substratum contacts (Chen et al., 1985) and compensate for the relatively low affinity of their binding to ECM adhesive proteins. Once made, adhesive contacts must be broken for the growth cone to advance. This may occur, for example, by phosphorylation of integrin transmembrane complexes (discussed above; Tapley et al., 1989) or by secretion of proteases at the growth cone (Patterson, 1985). As the "old" growth cone advances, its cytoskeleton must be integrated into the main axon cable. Simultaneously a "new" growth cone emerges, ready to repeat the cycle.

Even in this simplified model of neurite growth, on the ECM the variety of receptors present on neural cells raises many questions concerning their functions. Several of these, such as integrins, the 67-kDa receptor, and cranin, are on neurons, suggesting that growing axons have multiple adhesive systems. In the broader context of neural development and regeneration, it is not known which receptors reside on the various neural cells, how they are distributed over the cell surface, or how they are regulated. With reagents (antibodies, synthetic peptides) and approaches discussed above, we are beginning to decipher how and to what extent the expression of ECM receptors and their adhesive proteins controls the rate and direction of axonal extension during development and regeneration.

ACKNOWLEDGMENTS

We thank Drs. D. Turner, R. Ogle, and R. Mecham for generously providing us with unpublished material of their work and for helpful comments on the manuscript. We also thank Dr. M. Sasaki for allowing us to reproduce and modify the schematic shown in Figure 1A.

6. REFERENCES

Acheson, A., Edgar, D., Timpl, R., and Theonen, H., 1986, Laminin increases both levels and activity of tyrosine hydroxylase in calf adrenal chromaffin cells, J. Cell Biol. 102:151–159.

Akers, R. M., Mosher, D. F., and Lilien, J. E., 1981, Promotion of retinal neurite outgrowth by substratum-bound fibronectin, *Dev. Biol.* **86**:179–188.

Akeson, R., and Warren, S. L., 1986, PC12 adhesion and neurite formation on selected substrates are inhibited by some glycosaminoglycans and a fibronectin-derived tetrapeptide, *Exp. Cell Res.* **162**:347–362.

Akiyama, S. K., and Yamada, K. M., 1985, The interaction of plasma fibronectin with fibroblastic cells in suspension, *J. Biol. Chem.* **260**:4492–4500.

Akiyama, S. K., Yamada, S. S., and Yamada, K. M., 1986, Characterization of a 140-kD avian cell surface antigen as a fibronectin-binding molecule, *J. Cell Biol.* **102**:442–448.

Aquino, D. A., Margolis, R. U., and Margolis, R. K., 1984, Immunocytochemical localization of a chondroitin sulfate proteoglycan in nervous tissue. I. Adult brain, retina, and peripheral nerve, *J. Cell Biol.* **99**:1117–1119.

Arthur, F. E., Shivers, R. R., and Bowman, P. D., 1987, Astrocyte-mediated induction of tight junctions in brain capillary endothelium: An efficient in vitro model, *Dev. Brain Res.* **36**:155–159.

Aumailley, M., Nurcombe, V., Edgar, D., Paulsson, M., and Timpl, R., 1987, The cellular interactions of laminin fragments, *J. Biol. Chem.* **262**:11532–11538.

Barlow, D. P., Green, N. M., Kurkinen, M., and Hogan, B., 1984, Sequencing of laminin B chain cDNA, reveals C-terminal regions of coiled-coil alpha-helix, *EMBO J.* **3**:2355–2362.

Barsky, S. H., Rao, C. N., Hyams, D., and Liotta, L. A., 1984, Characterization of a laminin receptor from human breast carcinoma tissue, *Breast Cancer Res. Treat.* **4**:181–188.

Bissell, M. J., Hall, G. H., and Parry, G., 1982, How does the extracellular matrix direct gene expression? *J. Theor. Biol.* **99**:31–68.

Bixby, J. L., Pratt, R. S., Lilien, J., and Reichardt, L. F., 1987, Neurite outgrowth on muscle cell surfaces involves extracellular matrix receptors as well as Ca^{2+}-dependent and -independent cell adhesion molecules, *Proc. Natl. Acad. Sci. USA* **84**:2555–2559.

Bondareff, W., and Pysh, J. J., 1968, Distribution of the extracellular space during postnatal maturation of rat cerebral cortex, *Anat. Rec.* **160**:773–780.

Boucaut, J.-C., Darribere, T., Poole, T. J., Aoyama, H., Yamada, K. M., and Thiery, J.-P., 1984, Biologically active synthetic peptides as probes of embryonic development: A competitive peptide inhibitor of fibronectin function inhibits gastrulation in amphibian embryos and neural crest cell migration in avian embryos, *J. Cell Biol.* **99**:1822–1830.

Bozyczko, D., and Horwitz, A. F., 1986, The participation of a putative cell surface receptor for laminin and fibronectin in peripheral neurite extension, *J. Neurosci.* **65**:1241–1251.

Bronner-Fraser, M., 1986, An antibody to a receptor for fibronectin and laminin perturbs cranial neural crest cell development in vivo, *Dev. Biol.* **117**:528–536.

Brown, S. S., Malinoff, H. L., and Wicha, M. S., 1983, Connectin, a cell surface protein that binds both laminin and actin, *Proc. Natl. Acad. Sci. USA* **80**:5927–5930.

Buck, C. A., and Horwitz, A. F., 1987, Cell surface receptors for extracellular matrix molecules, *Annu. Rev. Cell Biol.* **3**:179–205.

Buck, C. A., Shea, E., Duggan, K., and Horwitz, A. F., 1986, Integrin (the CSAT antigen) functionality requires oligomeric integrity, *J. Cell Biol.* **103**:2421–2428.

Bunge, R. P., Bunge, M. B., and Eldridge, C. F., 1986, Linkage between axonal ensheathment and basal lamina production by Schwann cells, *Annu. Rev. Neurosci.* **9**:305–328.

Burridge, K., 1987, Substrate adhesions in normal and transformed fibroblasts: Organization and regulation of cytoskeletal, membrane and extracellular matrix components at focal contacts, *Cancer Rev.* **4**:18–78.

Carbonetto, S., 1984, The extracellular matrix of the nervous system, *Trends Neurosci.* **7**:382–387.

Carbonetto, S. T., Gruver, M. M., and Turner, D. C., 1982, Nerve fiber growth on defined hydrogel substrates, *Science* **216**:897–899.

Carbonetto, S., Gruver, M. M., and Turner, D. C., 1983, Nerve fiber growth in culture on fibronectin, collagen, and glycosaminoglycan substrates, *J. Neurosci.* **3**:2324–2335.

Carbonetto, S., Evans, D., and Cochard, P., 1987, Nerve fiber growth in culture on tissue substrata from central and peripheral nervous systems, *J. Neurosci.* **7**:610–620.

Carlin, B., Jaffe, R., Bender, B., and Chung, A. E., 1981, Entactin, a novel basal lamina-associated sulfated glycoprotein, *J. Biol. Chem.* **256**:5209–5214.

Chen, W.-T., Hasegawa, E., Hasegawa, T., Weinstock, C., and Yamada, K. M., 1985, Development of cell surface linkage complexes in cultured fibroblasts, *J. Cell Biol.* **100**:1103–1114.

Cheresh, D. A., and Harper, J. R., 1987, Arg-Gly-Asp recognition by a cell adhesion receptor requires its 130-kDa α subunit, *J. Biol. Chem.* **262**:1434–1437.

Cheresh, D. A., Pierschbacher, M. D., Herzig, M. A., and Mujoo, K., 1986, Disialogangliosides GD2 and GD3 are involved in the attachment of human melanoma and neuroblastoma cells to extracellular matrix proteins, *J. Cell Biol.* **102**:688–696.

Chiquet-Ehrismann, R., Mackie, E. J., Pearson, C. A., and Sakakura, T., 1986, Tenascin: An extracellular matrix protein involved in tissue interactions during fetal development and oncogenesis, *Cell* **47**:131–139.

Chiquet, M., 1989, Tenascin/J1/Cytotactin: The potential function of hexabrachion protein in neural development, *Dev. Neurosci.* **11**:in press.

Chiu, A. Y., and Sanes, J. R., 1984, Development of basal lamina in synaptic and extrasynaptic portions of embryonic rat muscle, *Dev. Biol.* **103**:456–467.

Chung, A. E., Jaffe, R., Freeman, I. L., Vergnes, J. P., Braginski, J. E., and Carlin, B., 1979, Properties of a basement membrane related glycoprotein synthesized in culture by a mouse embryonal carcinoma-derived cell line, *Cell* **16**:277–287.

Chuong, C.-M., Crossin, K. L., and Edelman, G. M., 1987, Sequential expression and differential function of multiple adhesion molecules during the formation of cerebellar cortical layers, *J. Cell Biol.* **104**:331–342.

Clegg, D. O., Helder, J. C., Hann, B. C., Hall, D. E., and Reichardt, L. F., 1988, Amino acid sequence and distribution of mRNA encoding a major skeletal muscle laminin binding protein: an extracellular matrix-associated protein with an unusual COOH-terminal polyaspartate domain, *J. Cell Biol.* **107**:699–705.

Cohen, J., Burne, J. F., Winter, J., and Bartlett, P., 1986, Retinal ganglion cells lose response to laminin with maturation, *Nature* **322**:465–467.

Cohen, J., Burne, J. F., McKinlay, C., and Winter, J., 1987, The role of laminin and the laminin/fibronectin receptor complex in the outgrowth of retinal ganglion cell axons, *Dev. Biol.* **122**:407–418.

Cole, G. C., and Glaser, L., 1986, A heparin-binding domain from N-CAM is involved in neuron-substratum adhesion, *J. Cell Biol.* **102**:403–412.

Crossin, K. L., Hoffman, S., Grumet, M., Thiery, J.-P., and Edelman, G. M., 1986, Site-restricted expression of cytotactin during development of the chicken embryo, *J. Cell Biol.* **102**:1917–1930.

Damsky, C. H., Knudsen, K. A., Bradley, D., Buck, C. A., and Horwitz, A. F., 1985, Distribution of the cell substratum (CSAT) antigen on myogenic and fibroblastic cells in culture, *J. Cell Biol.* **100**:1528–1539.

Davis, G. E., Manthorpe, M., Engvall, E., and Varon, S., 1985a, Isolation and characterization of rat Schwannoma neurite-promoting factor: Evidence that the factor contains laminin, *J. Neurosci.* **5**:2662–2671.

Davis, G. E., Varon, S., Engvall, E., and Manthorpe, M., 1985b, Substratum-binding neurite-promoting factors: Relationships to laminin, *Trends Neurosci.* **8**:528–532.

Dedhar, S., Ruoslahti, E., and Pierschbacher, M. D., 1987, A cell surface receptor complex for collagen type I recognizes the Arg-Gly-Asp sequence, *J. Cell Biol.* **104**:585–593.

Douville, P., Harvey, W., and Carbonetto, S., 1987, Identification and purification of a high affinity laminin receptor from embryonic chick brain: Evidence for developmental regulation in the CNS, *Soc. Neurosci. Abstr.* **13**:1482.

Douville, P. J., Harvey, W. J., and Carbonetto, S., 1988, Isolation and characterization of high affinity laminin receptors in neural cells, *J. Biol. Chem.* **263**:14964–14969.

Dow, K. E., Mirsky, S. E. L., Roder, J. C., Riopelle, R. G., 1988, Neuronal proteoglycans: Biosynthesis and functional interaction with neurons *in vitro*, *J. Neurosci.* **8**:3278–3289.

Dräger, U. C., and Rabacchi, S. A., 1988, A positional marker in the dorsal eye of the embryo, *Soc. Neurosci. Abs.* **14**:769.

Duband, J.-L., Rocher, S., Chen, W.-T., Yamada, K. M., and Thiery, J.-P., 1986, Cell adhesion and migration in the early vertebrate embryo: Location and possible role of the putative fibronectin receptor complex, *J. Cell Biol.* **102**:160–178.

Dziadek, M., and Timpl, R., 1985, Expression of nidogen and laminin in basement membranes during mouse embryogenesis and in teratocarcinoma cells, *Dev. Biol.* **111**:372–382.

Dziadek, M., Paulsson, M., and Timpl, R., 1985, Identification and interaction repertoire of large forms of the basement membrane protein nidogen, *EMBO J.* **4**:2513–2518.

Edgar, D., Timpl, R., and Theonen, H., 1984, The heparin-binding domain of laminin is responsible for its effects on neurite outgrowth and neuronal survival, *EMBO J.* **3**:1463–1468.

Ehrismann, R., Roth, D. E., Eppenberger, H. M., and Turner, D. C., 1982, Arrangement of attachment-promoting, self-association, and heparin-binding sites in horse serum fibronectin, *J. Biol. Chem.* **257**: 7381–7387.

Enenstein, J., and Furcht, L. T., 1984, Isolation and characterization of epinectin, a novel adhesion protein for epithelial cells, *J. Cell Biol.* **99**:464–470.

Engel, J., Odermatt, E., Engle, A., Madri, J. A., Furthmayr, H., Rohde, H., and Timpl, R., 1981, Shapes, domain organizations and flexibility of laminin and fibronectin, two multifunctional proteins of the extracellular matrix, *J. Mol. Biol.* **150**:97–120.

Engvall, E., Davis, G. E., Dickerson, K., Ruoslahti, E., Varon, S., and Manthorpe, M., 1986, Mapping of domains in human laminin using monoclonal antibodies: localization of the neurite-promoting site, *J. Cell Biol.* **103**:2457–2465.

Erickson, H. P., and Inglesias, J. L., 1984, A six-armed oligomer isolated from cell surface fibronectin preparations, *Nature* **311**:267–269.

Fahrig, T., Landa, C., Pesheva, P., Kühn, K., and Schachner, M., 1987, Characterization of binding properties of the myelin-associated glycoprotein to extracellular matrix constituents, *EMBO J.* **6**:2875–2883.

Fakuda, T., and Hashimoto, P. H., 1987, Distribution and fine structure of ependymal cells possessing intracellular cysts in the aqueductal wall of the rat brain, *Cell Tissue Res.* **247**:555–564.

Fitzgerald, L. A., Steiner, B., Rall, S. C., Jr., Lo, S., and Phillips, D. R., 1987, Protein sequence of endothelial glycoprotein IIIa derived from a cDNA clone, *J. Biol. Chem.* **262**:3936–3939.

Gardner, J. M., and Hynes, R. O., 1985, Interaction of fibronectin with its receptor on platelets, *Cell* **42**: 439–448.

Giftochristos, N., and David, S., 1988, Laminin and heparan sulfate proteoglycan in the lesioned adult mammaliam central nervous system and their possible relationship to axonal sprouting, *J. Neurocyt.* **17**:385–397.

Ginsberg, M., Pierschbacher, M. D., Ruoslahti, E., Marguerie, G., and Plow, E., 1985, Inhibition of fibronectin binding to platelets by proteolytic fragments and synthetic peptides which support fibroblast adhesion, *J. Biol. Chem.* **260**:3931–3936.

Ginsberg, M. H., Loftus, J., Ryckwaert, J., Pierschbacher, M., Pytela, R., Ruoslahti, E., and Plow, E. F., 1987, Immunochemical and amino-terminal sequence comparison of two cytoadhesins indicates they contain similar or identical β subunits and distinct α subunits, *J. Biol. Chem.* **262**:5437–5440.

Goodman, S. L., Deutzmann, R., and von der Mark, K., 1987, Two distinct cell-binding domains can independently promote nonneuronal cell adhesion and spreading, *J. Cell Biol.* **105**:589–598.

Graf, J., Iwamoto, Y., Sasaki, M., Martin, G. R., Kleinman, H. K., Robey, F. A., and Yamada, Y., 1987a, Identification of an amino acid sequence in laminin mediating cell attachment, chemotaxis, and receptor binding, *Cell* **48**:989–996.

Graf, J., Ogle, R. C., Robey, F. A., Sasaki, M., Martin, G. R., Yamada, Y., and Kleinman, H. K., 1987b, A pentapeptide from the laminin B1 chain mediates cell adhesion and binds the 67,000 laminin receptor, *Biochemistry* **26**:6896–6900.

Greve, J. M., and Gottlieb, D. I., 1982, Monoclonal antibodies which alter the morphology of cultured chick myogenic cells, *J. Cell Biochem.* **18**:221–229.

Grinnell, F., 1978, Cellular adhesiveness and extracellular substrata, *Int. Rev. Cytol.* **53**:65–144.

Grumet, M., Hoffman, S., Crossin, K. L., and Edelman, G. M., 1985, Cytotactin, an extracellular matrix protein of neural and non-neural tissues that mediates glia-neuron interaction, *Proc. Natl. Acad. Sci. USA* **82**:8075–8079.

Hall, D. E., Neugebauer, K. M., and Reichardt, L. F., 1987, Embryonic neural retinal cell response to extracellular matrix proteins: Developmental changes and effects of the cell substratum attachment antibody (CSAT), *J. Cell Biol.* **104**:623–634.

Hall, D. E., Frazer, K. A., Hann, B. C., and Reichardt, L. F., 1988, Isolation and characterization of a laminin-binding protein from rat and chick muscle, *J. Cell Biol.* **107**:687–697.

Harvey, W., and Carbonetto, S., 1986, Specific binding of ^{125}I-laminin to neural cells in culture, *Soc. Neurosci. Abstr.* **12**:1109.

Hatten, M. E., Furie, M. B., and Rifkin, D. B., 1982, Binding of developing mouse cerebellar cells to fibronectin: A possible mechanism for the formation of the external granular layer, *J. Neurosci.* **2**: 1195–1206.

Hauschka, S. D., and Konigsberg, I. R., 1966, The influence of collagen on the development of muscle clones, *Proc. Natl. Acad. Sci. USA* **55**:119–126.

Hemler, M. E., Huang, C., and Schwarz, L., 1987, The VLA protein family: Characterization of five distinct cell surface heterodimers each with a common 130,000 M_r subunit, *J. Biol. Chem.* **262**:3300–3309.

Hewitt, A. T., Kleinman, H. K., Pennypacker, J. P., and Martin, G. R., 1980, Identification of an adhesion factor for chondrocytes, *Proc. Natl. Acad. Sci. USA* **77**:385–388.

Hinek, A., Wrenn, D. S., Mecham, R. P., and Barondes, S. H., 1988, The elastin receptor: a galactoside-binding protein, *Science* **239**:1539–1541.

Hirst, R., Horwitz, A., Buck, C., and Rohrschneider, L., 1986, Phosphorylation of the fibronectin receptor complex in cells transformed by oncogenes that encode tyrosine kinases, *Proc. Natl. Acad. Sci. USA* **83**:6470–6474.

Hopkins, J. M., Ford-Holevinski, T. S., McCoy, J. P., and Agranoff, B. W., 1985, Laminin and optic nerve regeneration in the goldfish, *J. Neurosci.* **5**:3030–3038.

Horwitz, A., Duggan, K., Greggs, R., Decker, C., and Buck, C., 1985, The cell substrate attachment (CSAT) antigen has properties of a receptor for laminin and fibronectin, *J. Cell Biol.* **101**:2134–2144.

Horwitz, A., Duggan, K., Buck, C., Beckerle, M. C., and Burridge, K., 1986, Interaction of plasma membrane fibronectin receptor with talin—A transmembrane linkage, *Nature* **320**:531–533.

Hostikka, S. L., Kurkinen, M., and Trygvason, K., 1987, Nucleotide sequence coding for the human type IV collagen α_2 chain cDNA reveals extensive homology with the NC-1 domain of α_1(IV) but not with the collagenous domain or 3′-untranslated region, *FEBS Lett.* **216**:281–286.

Huard, T. M., Malinoff, H. L., and Wicha, M. S., 1986, Macrophages express a plasma membrane receptor for basement membrane laminin, *Am. J. Pathol.* **123**:365–370.

Humphries, M. J., Olden, K., and Yamada, K. M., 1986a, A synthetic peptide from fibronectin inhibits experimental metastasis of murine melanoma cells, *Science* **233**:467–470.

Humphries, M. J., Akiyama, S. K., Komoriya, A., Olden, K., and Yamada, K. M., 1986b, Identification of an alternatively spliced site in human plasma fibronectin that mediates cell type-specific adhesion, *J. Cell Biol.* **103**:2637–2647.

Hynes, R. O., 1985, Molecular biology of fibronectin, *Annu. Rev. Cell Biol.* **1**:67–90.

Hynes, R. O., 1987, Integrins: A family of cell surface receptors, *Cell* **48**:549–554.

Hynes, R. O., Patel, R., and Miller, R. H., 1986, Migration of neuroblasts along preexisting axonal tracts during prenatal cerebellar development, *J. Neurosci.* **6**:867–876.

Ide, C., Tohyama, K., Yokota, R., Nitatori, T., and Onodera, S., 1983, Schwann cell basal lamina and nerve regeneration, *Brain Res.* **288**:61–75.

Iwamoto, Y., Robey, F. A., Graf, J., Sasaki, M., Kleinman, H. K., Yamada, Y., and Martin, G., 1987, YIGSR, a synthetic laminin pentapeptide, inhibits experimental metastasis formation, *Science* **238**: 1132–1134.

Keene, D. R., Sakai, L. Y., Lunstrum, G. P., Morris, N. P., and Burgeson, R. E., 1987, Type VII collagen forms an extended network of anchoring fibrils, *J. Cell Biol.* **104**:611–621.

Kefalides, N. A., Alper, R., and Clark, C. C., 1979, Biochemistry and metabolism of basement membranes, *Int. Rev. Cytol.* **61**:167–221.

Kleinman, H. K., McGarvey, M. L., Hassell, J. R., Martin, G. R., Baron van Evercooren, A., and Dubois-Dalcq, M., 1984, The role of laminin in basement membranes and in the growth, adhesion, and differentiation of cells, in: *The Role of Extracellular Matrix in Development* (R. L. Trelstad, ed.), pp. 123–143, Liss, New York.

Kleinman, H. K., Ogle, R. C., Cannon, F. B., Little, C. C., Sweeney, T. M., and Luckenbill-Edds, L., 1988, Laminin receptors for neurite formation, *Proc. Natl. Acad. Sci. USA* **85**:1282–1286.

Kleitman, N., Wood, P., Johnson, M. I., and Bunge, R. P., 1988, Schwann cell surfaces but not extracellular matrix organized by Schwann cells support neurite outgrowth from embryonic rat retina, *J. Neurosci.* **8:**653–663.

Knudsen, K. A., Horwitz, A. F., and Buck, C. A., 1985, A monoclonal antibody identifies a glycoprotein complex involved in cell–substratum adhesion, *Exp. Cell Res.* **157:**218–226.

Kornblihtt, A. R., Umezawa, K., Vibe-Pedersen, K., and Baralle, F. E., 1985, Primary structure of human fibronectin: Differential splicing may generate at least 10 polypeptides from a single gene, *EMBO J.* **4:** 1755–1759.

Kruse, J., Keilhauer, G., Faissner, A., Timpl, R., and Schachner, M., 1985, The J1 glycoprotein—A novel nervous system cell adhesion molecule of the L2/HNK-1 family, *Nature* **316:**146–148.

Kühn, K., Glanville, R. W., Babel, W., Qian, R.-Q., Dieringer, H., Voss, T., Siebold, B., Oberbäumer, J., Schwarz, U., and Yamada, Y., 1985, The structure of type IV collagen, *Ann. N.Y. Acad. Sci.* **460:** 14–24.

Lander, A. D., Fujii, D. K., and Reichardt, L. F., 1985, Laminin is associated with the "neurite outgrowth promoting factors" found in conditioned media, *Proc. Natl. Acad. Sci. USA* **82:**2183–2187.

Landis, S. C., 1983, Neuronal growth cones, *Annu. Rev. Physiol.* **45:**567–580.

Laurie, G. W., Leblond, C. P., and Martin, G. R., 1982, Localization of type IV collagen, laminin, heparan sulfate proteoglycan, and fibronectin to the basal lamina of basement membranes, *J. Cell Biol.* **95:**340–344.

Laurie, G. W., Leblond, C. P., and Martin, G. R., 1983, Light microscopic immunolocalization of type IV collagen, laminin, heparan sulfate proteoglycan, and fibronectin in basement membranes of a variety of rat organs, *Am. J. Anat.* **167:**71–82.

Law, S. K. A., Gagnon, J., Hildreth, J. E. K., Wells, C. E., Willis, A. C., and Wong, A. J., 1987, The primary structure of the β subunit of the cell surface adhesion glycoproteins LFA-1, CR3 and p150,95 and its relationship to the fibronectin receptor, *EMBO J.* **64:**915–919.

Lesot, H., Kuhl, U., and von der Mark, K., 1983, Isolation of a laminin-binding protein from muscle cell membranes, *EMBO J.* **2:**861–865.

Letourneau, P. C., 1981, Immunocytochemical evidence for colocalization in neurite growth cones of actin and myosin and their relationship to cell–substratum adhesions, *Dev. Biol.* **85:**113–122.

Letourneau, P. C., 1983, Axonal growth and guidance, *Trends Neurosci.* **6:**451–455.

Letourneau, P. C., 1988, Interaction of growing axons with fibronectin and laminin, in: *The Current Status of Peripheral Nerve Regeneration* (V. Chan-Palay and S. L. Palay, eds.), pp. 99–110, Liss, New York.

Liesi, P., 1985a, Do neurons in the vertebrate CNS migrate on laminin? *EMBO J.* **4:**1163–1170.

Liesi, P., 1985b, Laminin-immunoreactive glia distinguish adult CNS systems from non-regenerative ones, *EMBO J.* **4:**2505–2511.

Linsenmayer, T. F., 1981, Collagen, in: *Cell Biology of Extracellular Matrix* (E. Hay, ed.), pp. 5–37, Plenum Press, New York.

Liotta, L. A., Hand, P. H., Rao, C. N., Bryant, G., Barsky, S. H., and Schlom, J., 1985, Monoclonal antibodies to the human laminin receptor recognize structurally distinct sites, *Exp. Cell Res.* **156:**117–126.

Liotta, L. A., Rao, C. N., and Wewer, U. M., 1986, Biochemical interactions of tumor cells with the basement membrane, *Annu. Rev. Biochem.* **55:**1037–1057.

Low, F. N., 1976, The perineurium and connective tissue of peripheral nerve, in: *The Peripheral Nerve* (D. N. Landon, ed.), pp. 159–187, Chapman & Hall, London.

Luckenbill-Edds, L., Ogle, R. C., and Kleinman, H. K., 1986, Laminin binds to cell membrane receptors on a neuroblastoma × glioma cell line (NG108-15), *J. Cell Biol.* **103:**261a.

Madri, J. A., Pratt, B. M., Yurchenko, P. D., and Furthmayr, H., 1984, The ultrastructural organization and architecture of basement membranes, in: *Basement Membranes and Cell Movement*, pp. 6–24, Pitman, London.

Makgoba, M. W., Sanders, M. E., Luce, G. E. G., Dustin, M. L., Springer, T. A., Clark, E. A., Mannoni, P., and Shaw, S., 1988, ICAM-1 a ligand for LFA-1-dependent adhesion of B, T and myeloid cells, *Nature* **331:**86–88.

Malinoff, H. L., and Wicha, M. S., 1983, Isolation of a cell surface receptor protein for laminin from murine fibrosarcoma cells, *J. Cell Biol.* **96:**1475–1479.

Manthorpe, M., Engvall, E., Ruoslahti, E., Longo, F. M., Davis, G. E., and Varon, S., 1983, Laminin promotes neuritic regeneration from cultured peripheral and central neurons, *J. Cell Biol.* **97**:1882–1890.

Manthorpe, M., Hagg, T., Engvall, E., and Varon, S., 1988, Luminin-like immunoreactivity in adult rat brain neurons *Soc. Neurosci. Abs.* **14**:364.

Martin, G. R., and Timpl, R., 1987, Laminin and other basement membrane components, *Annu. Rev. Cell Biol.* **3**:57–85.

McCarthy, J. B., Palm, S. L., and Furcht, L. T., 1984, Migration by haptotaxis of a Schwann cell tumor line to the basement membrane glycoprotein laminin, *J. Cell Biol.* **97**:772–777.

Naidet, C., Sémèrira, M., Yamada, K. M., and Thiery, J.-P., 1987, Peptides containing the cell-attachment recognition signal Arg-Gly-Asp prevent gastrulation in Drosophila embryos, *Nature* **325**:348–350.

Neff, N. T., Lowrey, C., Decker, C., Tovar, A., Damsky, C., Buck, C., and Horwitz, A. F., 1982, A monoclonal antibody detaches embryonic skeletal muscle from extracellular matrices, *J. Cell Biol.* **95**:654–666.

Nevis, A. H., and Collins, G. H., 1967, Electrical impedance and volume changes in brain during development, *Brain Res.* **5**:57–85.

Newgreen, D., and Thiery, J.-P., 1980, Fibronectin in early avian embryos: Synthesis and distribution along the migration pathways of neural crest cells, *Cell Tissue Res.* **211**:269–291.

Nitkin, R. M., Wallace, B. G., Spira, M. E., Godfrey, E. W., and McMahon, U. J., 1983, Molecular components of the synaptic basal lamina that direct differentiation of regenerating neuromuscular junctions, *Cold Spring Harbor Symp. Quant. Biol.* **48**:653–665.

Oberbäumer, I., Laurent, M., Schwarz, U., Sakurai, Y., Yamada, Y., Vogeli, G., Voss, T., Siebold, B., Glanville, R. W., and Kühn, K., 1985, Amino acid sequence of the non-collagenous globular domain (NC1) of the $\alpha 1$(IV) chain of basement membrane collagen derived from complementary DNA, *Eur. J. Biochem.* **14**:217–224.

Odermatt, E., Tamkun, J. W., and Hynes, R. O., 1985, Repeating modular structure of the fibronectin gene: Relationship to protein structure and subunit variation, *Proc. Natl. Acad. Sci. USA* **82**:6571–6575.

Ogle, R. C., and Little, C. D., 1989, Collagen binding proteins derived from the embryonic fibroblast cell surface recognize arginine-glycine-aspartic acid, *Biosci. Rep.*, in press.

Ogle, R. C., Laurie, G. W., Kitten, G. T., Kandel, S. L., and Bing, J. T., 1987, Isolation of a cell surface receptor for collagen type IV, *J. Cell Biol.* **105**:136a.

Oldberg, A., and Ruoslahti, E., 1986, Evolution of the fibronectin gene, *J. Biol. Chem.* **261**:2113–2116.

Ott, U., Odermatt, E., Engel, J., Furthmayr, H., and Timpl, R., 1982, Protease resistance and conformation of laminin, *Eur. J. Biochem.* **123**:63–72.

Palm, S. L., and Furcht, L. T., 1983, Production of laminin and fibronectin by Schwannoma cells: Cell–protein interactions in vitro and protein localization in peripheral nerve in vivo, *J. Cell Biol.* **96**:1218–1226.

Patel, V. P., and Lodish, H. F., 1986, The fibronectin receptor on mammalian erythroid precursor cells: Characterization and developmental regulation, *J. Cell Biol.* **86**:449–456.

Patterson, P. H., 1985, On the role of proteases, their inhibitors and the extracellular matrix in promoting neurite outgrowth, *J. Physiol. (Paris)* **80**:207–211.

Paulsson, M., Aumailley, M., Deutzmann, R., Timpl, R., Beck, K., and Engel, J., 1987, Laminin-nidogen complex: Extraction with chelating agents and structural characterization, *Eur. J. Biochem.* **166**:11–19.

Peters, A., Palay, S. L., and Webster, H., 1976, *The Fine Structure of the Nervous System: The Neurons and Supporting Cells,* Saunders, Philadelphia.

Petersen, T. E., Thogersen, H. C., Skorstengaard, K., Vibe-Pedersen, K., Sahl, P., Sottrup-Jensen, L., and Magnusson, S., 1983, Partial primary structure of bovine plasma fibronectin: Three types of internal homology, *Proc. Natl. Acad. Sci. USA* **80**:137–141.

Pierschbacher, M. D., and Ruoslahti, E., 1984, Cell attachment activity of fibronectin can be duplicated by small synthetic fragments of the molecule, *Nature* **309**:30–33.

Pierschbacher, M. D., Hayman, E. G., and Ruoslahti, E., 1981, Location of the cell-attachment site in fibronectin with monoclonal antibodies and proteolytic fragments of the molecule, *Cell* **26**:259–267.

Plow, E. F., Loftus, J. C., Levin, E. G., Fair, D. S., Dixon, D., Forsyth, J., and Ginsberg, M. H., 1986,

Immunologic relationship between platelet membrane glycoprotein GPIIb/IIIa and cell surface molecules expressed by a variety of cells, *Proc. Natl. Acad. Sci. USA* **83**:6002–6006.

Pollard, T. D., 1980, Platelet contractile proteins, *Thromb. Haemostasis* **42**:1634–1637.

Pytela, R., Pierschbacher, M. D., and Ruoslahti, E., 1985a, Identification and isolation of a 140 kd cell surface glycoprotein with properties expected of a fibronectin receptor, *Cell* **40**:191–198.

Pytela, R., Pierschbacher, M. D., and Ruoslahti, E., 1985b, A 125/115-kDa cell surface receptor for vitronectin interacts with the arginine-glycine-aspartic acid adhesion sequence from fibronectin, *Proc. Natl. Acad. Sci. USA* **82**:5766–5770.

Pytela, R., Pierschbacher, M. D., Ginsberg, M. H., Plow, E. F., and Ruoslahti, E., 1986, Platelet membrane glycoprotein IIb/IIIa: Member of a family of Arg-Gly-Asp-specific adhesion receptors, *Science* **231**:1559–1562.

Rabacchi, S. A., Neve, R. L., and Dräger, U. C., 1988, Molecular cloning of the "dorsal eye antigen": homology to the high-affinity laminin receptor, *Soc. Neurosci. Abs.* **14**:769.

Rakic, P., 1985, Contact regulation of neuronal migration, in: *The Cell in Contact: Adhesions and Junctions as Morphogenetic Determinants* (G. M. Edelman and J.-P. Thiery, eds.), pp. 67–92, Wiley, New York.

Rao, C. N., Margulies, I. M. K., Trolka, T. S., Terranova, V. P., Madri, J. A., and Liotta, L. A., 1982, Isolation of a subunit of laminin and its role in molecular structure and tumor cell attachment, *J. Biol. Chem.* **257**:9740–9744.

Rao, C. N., Barsky, S. H., Terranova, V. P., and Liotta, L. A., 1983, Isolation of a tumor cell laminin receptor, *Biochem. Biophys. Res. Commun.* **111**:804–808.

Rauvala, H., and Hakomori, S.-I., 1981, Studies on cell adhesion and recognition. III. The occurrence of α-mannosidase at the fibroblast cell surface, and its possible role in cell recognition, *J. Cell Biol.* **88**:149–159.

Reichardt, L. F., Bixby, J. L., Hall, D. E., Ignatius, M. J., Neugebauer, K. M., and Tomaselli, K. J., 1989, Integrins and cell adhesion molecules: Neuronal receptors that regulate axon growth on extracellular matrices and cell surfaces, *Dev. Neurosci.* **11**:in press.

Reh, T. A., Nagy, T., and Gretton, H., 1987, Retinal pigmented epithelial cells induced to transdifferentiate to neurons by laminin, *Nature* **330**:68–71.

Roberts, D. D., Rao, C. N., Magini, J. L., Spitalnik, S. L., Liotta, L. A., and Ginsberg, V., 1985, Laminin binds specifically to sulfated glycolipids, *Proc. Natl. Acad. Sci. USA* **82**:1306–1310.

Rogers, S. L., Letourneau, P. C., Palm, S. L., McCarthy, J. B., and Furcht, L. T., 1983, Neurite extension by peripheral and central nervous system neurons in response to substratum-bound fibronectin and laminin, *Dev. Biol.* **98**:212–220.

Rogers, S. L., McCarthy, J. B., Palm, S. L., Furcht, L. T., and Letourneau, P. C., 1985, Neuron-specific interactions with two neurite-promoting fragments of fibronectin, *J. Neurosci.* **5**:369–378.

Rogers, S. L., Edson, K. J., Letourneau, P. C., and McLoon, S. C., 1986, Distribution of laminin in the developing peripheral nervous system of the chick, *Dev. Biol.* **113**:429–435.

Runyan, R. B., Maxwell, G. D., and Shur, B. D., 1986, Evidence for a novel enzymatic mechanism of neural crest cell migration on extracellular glycoconjugate matrices, *J. Cell Biol.* **102**:432–441.

Ruoslahti, E., and Pierschbacher, M. D., 1987, New perspectives in cell adhesion: RGD and integrins, *Science* **238**:491–497.

Sakurai, Y., Sullivan, M., and Yamada, Y., 1986, α1 type IV collagen gene evolved differently from fibrillar collagen genes, *J. Biol. Chem.* **261**:6654–6657.

Sandrock, A. W., Jr., and Matthew, W. D., 1987, An in vitro neurite promoting antigen functions in axonal regeneration in vivo, *Science* **237**:1605–1608.

Sanes, J. R., Schachner, M., and Covault, J., 1986, Expression of several adhesive macromolecules (N-CAM, L1, J1, NILE, uvomorulin, laminin, fibronectin and a heparan sulfate proteoglycan) in embryonic, adult and denervated adult skeletal muscle, *J. Cell Biol.* **102**:420–431.

Santoro, S. A., 1986, Identification of a 160,000 dalton platelet membrane protein that mediates the initial divalent cation-dependent adhesion of platelets to collagen, *Cell* **46**:913–920.

Sasaki, M., and Yamada, Y., 1987, The laminin B2 chain has a multidomain structure homologous to the B1 chain, *J. Biol. Chem.* **262**:17111–17117.

Sasaki, M., Kato, S., Kohno, K., Martin, G. R., and Yamada, Y., 1987, Sequence of the cDNA encoding the laminin B1 chain reveals a multidomain protein containing cysteine-rich repeats, *Proc. Natl. Acad. Sci. USA* **84**:935–939.

Schachner, M., Schoonmaker, G., and Hynes, R. O., 1978, Cellular and subcellular localization of LETS protein in the nervous system, *Brain Res.* **158**:149–158.

Schwarzbauer, J. E., Paul, J. I., and Hynes, R. O., 1985, On the origin of species of fibronectin, *Proc. Natl. Acad. Sci. USA* **82**:1424–1428.

Segui-Real, B., Savagner, P., Ogle, R. C., Huang, T., Martin, G. R., and Yamada, Y., 1988, Unusual features of the laminin receptor predicted from cDNA clones, *Fed. Proc.* **2**:A1551.

Sephel, G. C., Burrous, B. A., and Kleinman, H. K., 1989, Laminin neural activity and binding proteins, *Dev. Neurosci.* **11**:in press.

Shadle, P. J., Ginsberg, M. H., Plow, E. F., and Barondes, S. H., 1984, Platelet–collagen adhesion: Inhibition by a monoclonal antibody that binds glycoprotein IIb, *J. Cell Biol.* **99**:2056–2060.

Smalheiser, N. R., and Schwartz, N. B., 1987, Cranin: A laminin binding protein of cell membranes, *Proc. Natl. Acad. Sci. USA* **84**:6457–6461.

Smith, D. E., and Furcht, L. T., 1982, Localization of two unique heparin binding domains of human plasma fibronectin with monoclonal antibodies, *J. Biol. Chem.* **257**:6518–6523.

Sternberg, J., and Kimber, S. J., 1986, Distribution of fibronectin, laminin and entactin in the environment of migrating neural crest cells in early mouse embryos, *J. Embryol. Exp. Morphol.* **91**:267–282.

Stewart, G. R., and Pearlman, A. L., 1987, Fibronectin-like immunoreactivity in the developing cerebral cortex, *J. Neurosci.* **7**:3325–3333.

Suzuki, S., Oldberg, A., Hayman, E. G., Pierschbacher, M. D., and Ruoslahti, E., 1985, Complete amino acid sequence of human vitronectin deduced from cDNA. Similarity of cell attachment sites in vitronectin and fibronectin, *EMBO J.* **4**:2519–2529.

Suzuki, S., Argraves, W. S., Pytela, R., Arai, H., Krusius, T., Pierschbacher, M. D., and Ruoslahti, E., 1986, cDNA and amino acid sequences of the cell adhesion protein receptor recognizing vitronectin reveal a transmembrane domain and homologies with other adhesion protein receptors, *Proc. Natl. Acad. Sci. USA* **83**:8614–8618.

Tamkun, J. W., DeSimone, D. W., Fonda, D., Patel, R. S., Buck, C., Horwitz, A. F., and Hynes, R. O., 1986, Structure of integrin, a glycoprotein involved in the transmembrane linkage between fibronectin and actin, *Cell* **46**:271–282.

Tan, S.-S., Crossin, K. L., Hoffman, S., and Edelman, G. M., 1987, Asymmetric expression in somites of cytotactin and its proteoglycan ligand is correlated with neural crest distribution, *Proc. Natl. Acad. Sci. USA* **84**:7977–7981.

Tapley, P., Horwitz, A., Buck, C., Burridge, K., Duggan, K., Hirst, R., and Rohrschneider, L., 1989, Integrins isolated from rous sarcoma virus transformed chicken embryo fibroblasts, *Oncogene,* in press.

Termine, J. D., Kleinman, H. K., William-Whitsu, S., Conn, K. M., McGarvey, M. L., and Martin, G. R., 1981, Osteonectin, a bone-specific protein linking mineral to collagen, *Cell* **26**:99–105.

Terranova, V. P., Rao, C. N., Kalebic, T., Margulies, I. M., and Liotta, L. A., 1983, Laminin receptor on human breast carcinoma cells, *Proc. Natl. Acad. Sci. USA* **80**:444–448.

Timpl, R., Rohde, H., Robey, P. G., Rennard, S. I., Foidart, J.-M., and Martin, G. R., 1979, Laminin—A glycoprotein from basement membranes, *J. Biol. Chem.* **254**:9933–9937.

Timpl, R., Johansson, S., van Delden, V., Oberbaumer, I., and Hook, M., 1983, Characterization of protease-resistant fragments of laminin mediating attachment and spreading of rat hepatocytes, *J. Biol. Chem.* **258**:8922–8927.

Timpl, R., Fujiwara, S., Dziadek, M., Aumailley, M., Weber, S., and Engel, J., 1984, Laminin, proteoglycan, nidogen and collagen IV: Structural models and molecular interactions, in: *Basement Membranes and Cell Movement,* pp. 25–43, Pitman, London.

Timpl, R., Oberbäumer, I., von der Mark, K., Bode, W., Wick, G., Weber, S., and Engel, J., 1985, Structure and biology of the globular domain of basement membrane type IV collagen, *Ann. N.Y. Acad. Sci.* **460**:58–72.

Tomaselli, K. J., Reichardt, L. F., and Bixby, J. L., 1986, Distinct molecular interactions mediate neuronal process outgrowth on nonneuronal cell surfaces and extracellular matrices, *J. Cell Biol.* **103**:2659–2672.

Tomaselli, K. J., Damsky, C. H., and Reichardt, L. F., 1987, Interaction of a neuronal cell line (PC12) with laminin, collagen IV, and fibronectin: Identification of integrin-related glycoproteins involved in attachment and process outgrowth, *J. Cell Biol.* **105**:2347–2358.

Tomaselli, K. J., Damsky, C. H., and Reichardt, L. F., 1988, Purification and characterization of mammalian integrins expressed by a rat neuronal cell line (PC12): evidence that they function as α/β heterodimeric receptors for laminin and type IV collagen, *J. Cell Biol.* **107**:1241–1252.

Toyota, B., Carbonetto, S., and David, S., 1988, Involvement of laminin in nerve regeneration in vivo, *Soc. Neurosci. Abs.* **14**:498.

Trimmer, J. S., and Vacquier, V. D., 1986, Activation of sea urchin gametes, *Annu. Rev. Cell Biol.* **2**:1–26.

Trüeb, B., Gröbli, B., Spiess, M., Odermatt, B. F., and Winterhalter, K. H., 1982, Basement membrane (type IV) collagen is a heteropolymer, *J. Biol. Chem.* **257**:5239–5245.

Tuckett, F., and Morriss-Kay, G. M., 1986, The distribution of fibronectin, laminin and entactin in the neurulating rat embryo by indirect immunofluorescence, *J. Embryol. Exp. Morphol.* **94**:95–112.

Turner, D. C., Flier, L. A., and Carbonetto, S., 1987, Magnesium-dependent attachment and neurite outgrowth by PC12 cells on collagen and laminin substrata, *Dev. Biol.* **121**:510–525.

Turner, D. C., Flier, L. A., and Carbonetto, S., 1989, Identification of a cell-surface protein involved in PC12 cell–substratum adhesion and neurite outgrowth on collagen and laminin, *J. Neurosci.*, in press.

Van Harreveld, A., Dafny, N., and Khattab, F. I., 1971, Effects of calcium on the electrical resistance and extracellular space of cerebral cortex, *Exp. Neurol.* **31**:358–367.

von der Mark, K., and Kühl, U., 1985, Laminin and its receptor, *Biochim. Biophys. Acta* **823**:147–160.

Waite, K. A., Munai, G., and Culp, L. A., 1987, A second cell-binding domain on fibronectin (RGDS-independent) for neurite extension of human neuroblastoma cells, *Exp. Cell Res.* **169**:311–327.

Wayner, E. A., and Carter, W. G., 1987, Identification of multiple cell adhesion receptors for collagen and fibronectin in human fibrosarcoma cells possessing unique α and β subunits, *J. Cell Biol.* **105**:1873–1884.

Wewer, U. M., Liotta, L. A., Jaye, M., Ricca, G. A., Drohan, W. N., Claysmith, A. P., Rao, C. N., Wirth, P., Coligan, J. E., Albrechtsen, R., Mudryj, M., and Sobel, M. E., 1986, Altered levels of laminin receptor in various carcinoma cells that have different abilities to bind laminin, *Proc. Natl. Acad. Sci. USA* **83**:7137–7141.

Yamada, K. M., 1983, Cell surface interactions with extracellular materials, *Annu. Rev. Biochem.* **52**:761–799.

Yamada, K. M., and Kennedy, D. W., 1984, Dualistic nature of adhesive protein function: Fibronectin and its biologically active peptide fragments can autoinhibit fibronectin function, *J. Cell Biol.* **99**:29–36.

Yamada, K. M., and Kennedy, D. W., 1987, Peptide inhibitors of fibronectin, laminin, and other adhesion molecules: Unique and shared features, *J. Cell Physiol.* **130**:21–28.

Yamada, K. M., Critchley, D. R., Fishman, P. H., and Moss, J., 1983, Exogenous gangliosides enhance the interaction of fibronectin with ganglioside-deficient cells, *Exp. Cell Res.* **143**:295–302.

Yow, H., Wang, J. M., Chen, H. S., Lee, C., Steele, G. D., Jr., and Chen, L. B., 1988, Increased mRNA expression of a laminin-binding protein in human colon carcinoma: complete sequence of a full-length cDNA encoding the protein, *Proc. Natl. Acad. Sci. USA* **85**:6394–6398.

Zanetta, J. P., Dontenwill, M., Meyer, A., and Roussel, G., 1985, Isolation and immunohistochemical localization of a lectin-like molecule from the rat cerebellum, *Dev. Brain Res.* **17**:233–243.

Hyaluronate and Hyaluronate-Binding Proteins of Brain

Bryan P. Toole

1. HYALURONATE SYNTHESIS DURING EMBRYONIC DEVELOPMENT

Migratory and proliferating cells in the embryo are usually surrounded by hyaluronate-enriched extracellular matrices (reviewed in Toole, 1981). For example, hyaluronate accumulates in the pathways of migration of corneal mesenchyme (Toole and Trelstad, 1971), neural crest cells (Pratt *et al.*, 1975; Derby, 1978; Pintar, 1978), endocardial cushion cells (Markwald *et al.*, 1978), and sclerotomal mesenchyme (Kvist and Finnegan, 1970; Toole, 1972; Solursh *et al.*, 1979). Frequently, subsequent differentiation is accompanied by decreases in levels of hyaluronate in these tissues.

Hyaluronate is also enriched in the developing brain relative to the adult brain (Polansky *et al.*, 1974; Krusius *et al.*, 1974; Margolis *et al.*, 1975; Werz *et al.*, 1985; Normand *et al.*, 1985; Oohira *et al.*, 1986). In the developing postnatal rat cerebellum, hyaluronate has been localized to the extracellular spaces of presumptive white matter, the granule cell layer, and the base of the molecular layer. In the adult, however, hyaluronate is mainly intracellular (Margolis *et al.*, 1987; Ripellino *et al.*, 1988; see Chapter 3). In the embryonic rat cerebral cortex, hyaluronate is present extracellularly in the marginal layer and the subplate zone (A. Delpech *et al.*, 1987). These sites correspond to areas relatively enriched in extracellular spaces and may provide hydrated conduits for the growth of nerve fibers.

2. HYALURONATE AND CELL BEHAVIOR

Hyaluronate has been shown to influence several types of cell behavior, but its effects depend greatly on at least three parameters. These are the size and concentration

Bryan P. Toole • Department of Anatomy and Cellular Biology, Tufts University Health Sciences Center, Boston, Massachusetts 02111.

of the hyaluronate and the type of cell in question. For example, several investigators have described a close positive correlation between hyaluronate synthesis and cell proliferation (Mian, 1986; Brecht *et al.*, 1986; Matuoka *et al.*, 1987). We have shown, however, that addition of high concentrations of high-molecular-weight hyaluronate to cultures of various cell types markedly inhibits proliferation, whereas addition of low-molecular-weight or low concentrations of hyaluronate is in many cases stimulatory (Goldberg and Toole, 1987).

Hyaluronate bonded to the cell culture substratum inhibits differentiation of myoblasts and supports their continued proliferation (Kujawa *et al.*, 1986a), but promotes the differentiation of chondroblasts (Kujawa and Caplan, 1986). The latter phenomenon, however, occurs efficiently only with hyaluronate of molecular weight 2–4 $\times$ 10^5 and is not at all influenced by hyaluronate of greater than 10^6 (Kujawa *et al.*, 1986b). Also, in several chondrogenic systems, various sizes of hyaluronate in solution inhibit synthesis of the proteoglycan component of cartilage (Toole, 1973; Wiebkin and Muir, 1973; Solursh *et al.*, 1980).

Hyaluronate also exhibits a concentration-dependent effect on cell aggregation. At high concentrations it blocks aggregation of several cell types whereas at low concentrations it can mediate aggregation (Fraser and Clarris, 1970; Pessac and Defendi, 1972; Wasteson *et al.*, 1973; Underhill and Dorfman, 1978; Love *et al.*, 1979; Underhill and Toole, 1981; Forrester and Lackie, 1981; Wright *et al.*, 1981). In studies of certain virally transformed cells, this apparent contradiction has been resolved. With these cells it has been shown that aggregation is due to cross bridging by hyaluronate of binding sites present in the plasma membrane of adjacent cells. However, at high concentrations of hyaluronate these receptors become saturated, thus inhibiting cross bridging (Underhill and Toole, 1981; Underhill, 1982a).

It is not at all clear which of the hyaluronate-mediated phenomena described above are relevant to cell behavior *in vivo*. However, there is a significant body of convincing evidence supporting the facilitation by hyaluronate of cell movement through tissues. In the first place, high concentrations of hyaluronate are present in the pathways of movement of cells during embryonic development (see above), tissue remodeling or regeneration (Reid and Flint, 1974; Smith *et al.*, 1975; Bertolami and Donoff, 1982; Mescher and Munaim, 1986), and tumor cell invasion (W. Knudson *et al.*, 1984; Biswas and Toole, 1987). Second, hyaluronate has been shown, in several *in vitro* and *in vivo* systems, to promote the movement of a wide variety of cell types (Turley and Roth, 1979; Bernanke and Markwald, 1979; Abatangelo *et al.*, 1983; Turley *et al.*, 1985; Doillon and Silver, 1986). Again, however, there is a dependence on cell type and the size and concentration of the hyaluronate. For example, hyaluronate appears to inhibit migration of endothelial cells (Feinberg and Beebe, 1983). In the case of leukocytes, hyaluronate inhibits movement at high concentrations (Brandt, 1970; Forrester and Wilkinson, 1981) but at low concentrations it has a marked stimulatory effect both *in vitro* and *in vivo* (Hakansson *et al.*, 1980; Hakansson and Venge, 1985).

One way by which hyaluronate seems to facilitate cellular invasion of tissues arises from the ability of hyaluronate to form meshworks which exert considerable osmotic pressures (Meyer, 1983; Meyer *et al.*, 1983). These swelling pressures may lead to the formation of hydrated pathways separating barriers to cell movement such as collagen fibers and cell layers. Examples of this are seen in the concomitance of hyaluronate accumulation with tissue hydration and separation of tissue structures at the time of migration of corneal mesenchyme (Toole and Trelstad, 1971; Hay, 1980), neural crest cells (Pratt et al., 1975), and sclerotomal cells (Solursh et al., 1979). As mentioned above, similar but smaller spaces in the developing brain may serve as hydrated conduits for axon and dendrite outgrowth. However, they are not likely to be pathways for cell migration (A. Delpech et al., 1987).

A second way that hyaluronate would influence cell movement is by reducing cell–cell (Fraser and Clarris, 1970; McBride and Bard, 1979; Underhill and Toole, 1982; Orkin et al., 1985) and cell–substratum adhesion (Culp et al., 1979; Barnhart et al., 1979; Fisher and Solursh, 1979; Mikuni-Takagaki and Toole, 1980; Abatangelo et al., 1982; Turley et al., 1985). Turley et al. (1985) have shown that addition of a hyaluronate-binding protein and hyaluronate together to fibroblasts causes an increase in their motility, reduces cell spreading, and increases cell underlapping. The hyaluronate-binding protein was localized to the leading lamellae and retraction processes of motile cells, leading to the conclusion that these phenomena are due to hyaluronate-mediated reduction in cell–substratum adhesion (Turley and Torrance, 1984; Turley et al., 1985). This is particularly interesting in light of the recent finding that invasive human bladder carcinoma cells, but not normal human fibroblasts, have hyaluronate-binding sites on their cell surface (Nemec et al., 1987).

3. HYALURONATE-BINDING PROTEINS AND HYALURONATE–CELL SURFACE INTERACTIONS

Hyaluronate is an extremely large polymer composed of the repeating disaccharide $\beta 1,4$-glucuronate-$\beta 1,3$-N-acetylglucosamine. Two classes of hyaluronate-binding proteins have been characterized that are distinguished by the number of hyaluronate disaccharide repeats necessary for binding. The first class has been mainly studied in cartilage. Two components of cartilage, a large proteoglycan species and link protein, have been shown to interact specifically with hyaluronate. For both the proteoglycan (Hardingham and Muir, 1973; Hascall and Heinegard, 1974) and the link protein (Tengblad, 1981), a minimum of five hyaluronate disaccharide repeats is required for interaction to occur with significant affinity. Similar link protein and proteoglycan species have also been found in tissues other than cartilage, including brain (Kiang et al., 1981; Crawford, 1988; see Chapter 3). Recently, extensive homology has been found in the sequences of the hyaluronate-binding regions of cartilage proteoglycan and link protein, even though the overall sequences of these two proteins are quite different (Neame et al., 1986; Doege et al., 1987; Goetinck et al., 1987).

The second class of hyaluronate-binding proteins was discovered as a result of studies of hyaluronate–cell interactions. Methods for the analysis of hyaluronate–cell surface interactions were first developed in this laboratory using various cell lines. These studies led to the detailed characterization of specific, high-affinity ($K_d \sim$ 1 nM), hyaluronate-binding sites that are present in the plasma membrane and that mediate binding of endogenously produced or exogenously added hyaluronate to the cell surface (Underhill and Toole, 1979, 1980; Underhill *et al.*, 1983; Goldberg *et al.*, 1984; Chi-Rosso and Toole, 1987; reviewed in Toole *et al.*, 1984). The cell surface glycoprotein responsible for binding of hyaluronate clearly recognizes a hyaluronate oligosaccharide comprised of three disaccharide repeats as opposed to the minimum of five repeats needed for binding to proteoglycan or link protein (Underhill *et al.*, 1983; Laurent *et al.*, 1986; Nemec *et al.*, 1987). Recently, Underhill and colleagues have identified a membrane glycoprotein of 85 kDa that has the properties of these binding sites (Underhill *et al.*, 1985, 1987a) and that is arranged in the plasma membrane in such a way that it either directly or indirectly interacts with intracellular actin filaments (Lacy and Underhill, 1987).

The methods used to characterize the interactions of hyaluronate with the surface of cell lines have recently been applied to embryonic cells and tissues, particularly those of the chick embryo limb bud (C. Knudson and Toole, 1985, 1987). A crucial stage in the differentiation of limb mesoderm is the condensation of cells which precedes final cytodifferentiation. At precisely this stage, hyaluronate-binding sites appear on the surface of the mesodermal cells. Cell aggregation and removal of hyaluronate from the matrix are fundamental to the mechanism of condensation. The hyaluronate-biding sites are presumably involved in these two phenomena since cell surface hyaluronate-binding sites have been shown to mediate both cell aggregation (Underhill and Toole, 1981; Underhill, 1982a) and endocytosis of hyaluronate (Orkin *et al.*, 1982; Smedsrod *et al.*, 1984) in other systems.

Other hyaluronate-binding proteins have been partially characterized (Underhill, 1982b; D'Souza and Datta, 1985; Turley *et al.*, 1987) but the specificity of their interaction with hyaluronate oligosaccharides has not been determined. Considerable attention, however, has been given to hyaluronate-binding proteins from brain, and these are addressed in the following section.

4. HYALURONATE-BINDING PROTEINS IN BRAIN

The initial approaches to discovery and characterization of hyaluronate-binding proteins in brain followed two very different courses, viz. the conventional methods of proteoglycan biochemistry (Kiang *et al.*, 1981; Crawford, 1988; see Chapter 3) and attempts to identify substances detected by immunochemical methods (B. Delpech and Halavent, 1981). More recently, methods used to identify cell surface binding sites for hyaluronate have been used (Underhill *et al.*, 1987b; Marks *et al.*, 1989). The tentative conclusion of these studies is that a major group of hyaluronate-binding proteins of brain have hyaluronate-binding domains whose properties are similar to those of cartilage proteoglycan and link protein. Indeed, these hyaluronic acid-binding proteins have been shown, in some cases, to be intact or partially degraded brain proteoglycans. In addition, however, brain also contains hyaluronate-binding proteins dissimilar to proteoglycans and link protein (Marks, *et al.*, 1989).

Hyaluronectin is the term given by B. Delpech and Halavent (1981) to a glycoprotein(s) prepared from acid extracts of postmortem human brain by affinity chromatography on hyaluronate-conjugated Sepharose. Hyaluronectin was detected and assayed in eluates from the affinity columns using antibodies raised against the so-called nervous system-associated antigen, NSA_3, and is assumed to be identical to this antigen(s). The major component of this preparation has a molecular mass variously estimated to be in the range 59–68 kDa (B. Delpech and Halavent, 1981; B. Delpech, 1982; D'Souza and Datta, 1986; Bignami and Dahl, 1986a). In the adult, hyaluronectin and NSA_3 are associated with nodes of Ranvier, certain neuronal cell surfaces, and the connective tissue of several normal and neoplastic tissues (A. Delpech et al., 1978, 1982; B. Delpech et al., 1979). Hyaluronectin is widely distributed in the mesenchyme of early embryonic tissues where hyaluronate is also present in high concentrations (A. Delpech and B. Delpech, 1984). In the embryonic rat cerebral cortex, hyaluronectin is codistributed with hyaluronate in the marginal layer and subplate zone (A. Delpech et al., 1987). Human hyaluronectin used as a cell culture substratum has been shown to inhibit neuronal cell attachment and neurite outgrowth (Bignami et al., 1988). Possibly, then, the ratio of hyaluronate to hyaluronectin in a given region may be a factor in regulation of neurite outgrowth in that region.

A related antigen to the above-described hyaluronectin has been studied by Bignami and Dahl. This group has prepared monoclonal antibodies to the major component present in the preparations of human hyaluronectin characterized by Delpech's group. The antigen detected by these antibodies, which has a molecular mass of 59 kDa as determined by SDS-PAGE, appears to be restricted to brain in the calf and adult human (Bignami and Dahl, 1986a), to be produced by white matter astrocytes (Bignami and Dahl, 1986b), and to appear relatively late in development (Bignami and Dahl, 1988). It has not yet been shown directly that this antigen is in fact the hyaluronate-binding component of the original hyaluronectin preparation from which the monoclonal antibodies were elicited. These and other studies (B. Delpech et al., 1979) suggest, however, that the polyclonal antibodies prepared against hyaluronectin by Delpech's group may detect more than one material. In the adult human, at least one of these materials is restricted to brain and one is present in mesenchymal tissues and possibly nervous tissues also.

Recent studies have shown: (1) that the binding specificity of hyaluronectin preparations, derived from human or mouse brain by the methods of Delpech's group, is identical to that of cartilage proteoglycan, i.e., they both require five hyaluronate disaccharides for recognition (Bertrand and Delpech, 1985; Underhill et al., 1987b), and (2) that hyaluronectin contains chondroitin sulfate chains (B. Delpech et al., 1986). It has thus been concluded that hyaluronectin is a proteoglycan or a fragment thereof.

In a separate series of studies, Margolises' group (Kiang et al., 1981; and Chapter 3) have isolated and characterized the major proteoglycan species of rat brain. Their initial studies, using conventional proteoglycan methodology, indicated that only 10% of this proteoglycan was able to bind to hyaluronate to give rise to large aggregates of proteoglycan and hyaluronate (Kiang et al., 1981). More recently, however, Ripellino et al. (1989) have shown that this brain proteoglycan is recognized by monoclonal antibodies raised against the hyaluronate-binding region of cartilage proteoglycan, inferring that these two proteoglycans have similar hyaluronate-binding properties.

Crawford (1988) has recently isolated a proteoglycan from chick embryo brain that forms large aggregates with hyaluronate. A molecule with the properties of link protein, another hyaluronate-binding protein originally isolated from cartilage, is also present in rat brain (Ripellino *et al.*, 1989; and Chapter 3).

Studies by my group (Marks and Toole, 1985; Marks *et al.*, 1989) of the hyaluronate-binding proteins of mouse brain have also revealed the presence of link protein and have shown that a proteoglycan species similar to that described by Margolises' group is a major hyaluronate-binding fraction of brain. Both my group and that of the Margolises have found that this proteoglycan preparation contains a component of approximately 65 kDa that accounts for much of the immunoreactivity with monoclonal antibodies against cartilage proteoglycan hyaluronate-binding region. Whether this protein is a separate but related protein to the proteoglycan itself or whether it is a proteolytic fragment of the proteoglycan remains to be determined, as does its relationship to hyaluronectin. Also, we have found a third major hyaluronate-binding protein that does not seem to be closely related to link protein or to the hyaluronate-binding region of proteoglycan. The specific functions of these three hyaluronate-binding macromolecules are not known but are likely to be of significant biological interest.

ACKNOWLEDGMENT

The recent work from this laboratory described herein was supported by Grant DE-05838 from the National Institutes of Health.

5. REFERENCES

Abatangelo, G., Cortivo, R., Martelli, M., and Vecchia, P., 1982, Cell detachment mediated by hyaluronic acid, *Exp. Cell Res.* **137**:73–78.

Abatangelo, G., Martelli, M., and Vecchia, P., 1983, Healing of hyaluronic acid-enriched wounds: Histological observations, *J. Surg. Res.* **35**:410–416.

Barnhart, B. J., Cox, S. H., and Kraemer, P. M., 1979, Detachment variants of Chinese hamster cells. Hyaluronic acid as a modulator of cell detachment, *Exp. Cell Res.* **119**:327–332.

Bernanke, D. H., and Markwald, R. R., 1979, Effects of hyaluronic acid on cardiac cushion tissue cells in collagen matrix cultures, *Tex. Rep. Biol. Med.* **39**:271–285.

Bertolami, C. N., and Donoff, R. B., 1982, Identification, characterization, and partial purification of mammalian skin wound hyaluronidase, *J. Invest. Dermatol.* **79**:417–421.

Bertrand, P., and Delpech, B., 1985, Interaction of hyaluronectin and hyaluronic acid oligosaccharides, *J. Neurochem.* **45**:434–439.

Bignami, A., and Dahl, D., 1986a, Brain-specific hyaluronate-binding protein: An immunohistological study with monoclonal antibodies of human and bovine central nervous system, *Proc. Natl. Acad. Sci. USA* **83**:3518–3522.

Bignami, A., and Dahl, D., 1986b, Brain-specific hyaluronate-binding protein. A product of white matter astrocytes? *J. Neurocytol.* **15**:671–679.

Bignami, A., and Dahl, D., 1988, Expression of brain-specific hyaluronectin, a hyaluronate-binding protein, in dog postnatal development, *Exp. Neurol.* **99**:107–117.

Bignami, A., Dahl, D., Gilad, V. H., and Gilad, G. M., 1988, Hyaluronate-binding protein produced by white matter astrocytes. An axonal growth repellent? *Exp. Neurol.* **100**:253–256.

Biswas, C., and Toole, B. P., 1987, Modulation of the extracellular matrix by tumor cell–fibroblast interactions, in: *Cell Membranes*, Vol. 3 (E. Elson, W. Frazier, and L. Glaser, eds.), pp. 341–363, Plenum Press, New York.

Brandt, K., 1970, Modification of chemotaxis by synovial fluid hyaluronate, *Arthritis Rheum.* **13:**308–309.

Brecht, M., Mayer, U., Schlosser, E., and Prehm, P., 1986, Increased hyaluronate synthesis is required for fibroblast detachment and mitosis, *Biochem. J.* **239:**445–450.

Chi-Rosso, G., and Toole, B. P., 1987, Hyaluronate-binding protein of simian virus 40-transformed 3T3 cells: Membrane distribution and reconstitution into lipid vesicles, *J. Cell. Biochem.* **33:**173–183.

Crawford, T., 1988, Distribution in cesium chloride gradients of proteoglycans of chick embryo brain and characterization of a large aggregating proteoglycan, *Biochim. Biophys. Acta* **964:**183–192.

Culp, L. A., Murray, B. A., and Rollins, B. J., 1979, Fibronectin and proteoglycans as determinants of cell–substratum adhesion, *J. Supramol. Struct.* **11:**401–427.

Delpech, A., and Delpech, B., 1984, Expression of hyaluronic acid-binding glycoprotein, hyaluronectin, in the developing rat embryo, *Dev. Biol.* **101:**391–400.

Delpech, A., Delpech, B., Girard, N., and Vidard, M. N., 1978, Localisation immunohistologique de trois antigenes associes au tissu nerveux (GFA, ANS₂, brain glycoprotein), *Biol. Cell.* **32:**207–214.

Delpech, A., Girard, N., and Delpech, B., 1982, Localization of hyaluronectin in the nervous system, *Brain Res.* **245:**251–257.

Delpech, A., Delpech, B., Girard, N., Bertrand, P., and Chauzy, C., 1987, Hyaluronectin and hyaluronic acid during the development of rat brain cortex, in: *Mesenchymal–Epithelial Interactions in Neural Development* (J. R. Wolff, J. Sievers, and M. Berry, eds.), pp. 77–87, Springer-Verlag, Berlin.

Delpech, B., 1982, Immunochemical characterization of the hyaluronic acid–hyaluronectin interaction, *J. Neurochem.* **38:**978–984.

Delpech, B., and Halavent, C., 1981, Characterization and purification from human brain of a hyaluronic acid-binding glycoprotein, hyaluronectin, *J. Neurochem.* **36:**855–859.

Delpech, B., Delpech, A., Girard, N., Chauzy, C., and Laumonier, R., 1979, An antigen associated with mesenchyme in human tumors that cross-reacts with brain glycoprotein, *Br. J. Cancer* **40:**123–133.

Delpech, B., Bertrand, P., Hermelin, B., Delpech, A., Girard, N., Halkin, E., and Chauzy, C., 1986, Hyaluronectin, in: *Frontiers in Matrix Biology*, Vol. 11 (L. Robert, ed.), pp. 78–89, Karger, Basel.

Derby, M. A., 1978, Analysis of glycosaminoglycans within the extracellular environment encountered by migrating neural crest cells, *Dev. Biol.* **66:**321–336.

Doege, K., Sasaki, M., Horigan, E., Hassell, J. R., and Yamada, Y., 1987, Complete primary structure of the rat cartilage proteoglycan core protein deduced from cDNA clones, *J. Biol. Chem.* **262:**17757–17767.

Doillon, C. J., and Silver, F. H., 1986, Collagen-based wound dressing: Effects of hyaluronic acid and fibronectin on wound healing, *Biomaterials* **7:**3–8.

D'Souza, M., and Datta, K., 1985, Evidence for naturally occurring hyaluronic acid binding protein in rat liver, *Biochem. Int.* **10:**43–51.

D'Souza, M., and Datta, K., 1986, A novel glycoprotein that binds to hyaluronic acid, *Biochem. Int.* **13:**79–88.

Feinberg, R. N., and Beebe, D. C., 1983, Hyaluronate in vasculogenesis, *Science* **220:**1177–1179.

Fisher, M., and Solursh, M., 1979, The influence of the substratum on mesenchyme spreading in vitro, *Exp. Cell Res.* **123:**1–14.

Forrester, J. V., and Lackie, J. M., 1981, Effect of hyaluronic acid on neutrophil adhesion, *J. Cell Sci.* **50:**329–344.

Forrester, J. V., and Wilkinson, P. C., 1981, Inhibition of leukocyte locomotion by hyaluronic acid, *J. Cell Sci.* **48:**315–331.

Fraser, J. R., and Clarris, B. J., 1970, On the reactions of human synovial cells exposed to homologous leucocytes *in vitro*, *Clin. Exp. Immunol.* **6:**211–225.

Goetinck, P. F., Stirpe, N. S., Tsonia, P. A., and Carlone, D., 1987, The tandemly repeated sequences of cartilage link protein contain the sites for interaction with hyaluronic acid, *J. Cell Biol.* **105:**2403–2408.

Goldberg, R. L., and Toole, B. P., 1987, Hyaluronate inhibition of cell proliferation, *Arthritis Rheum.* **30:**769–778.

Goldberg, R. L., Seidman, J. D., Chi-Rosso, G., and Toole, B. P., 1984, Endogenous hyaluronate–cell surface interactions in 3T3 and simian virus-transformed 3T3 cells, *J. Biol. Chem.* **259:**9440–9446.

Hakansson, L., and Venge, P., 1985, The combined action of hyaluronic acid and fibronectin stimulates neutrophil migration, *J. Immunol.* **135:**2735–2739.

Hakansson, L., Hallgren, R., and Venge, P., 1980, Regulation of granulocyte function by hyaluronic acid: In vitro and in vivo effects on phagocytosis, locomotion and metabolism, *J. Clin. Invest.* **66:**298–305.

Hardingham, T. E., and Muir, H., 1973, Binding of oligosaccharides of hyaluronic acid to proteoglycans, *Biochem. J.* **135:**905–908.

Hascall, V. C., and Heinegard, D., 1974, Aggregation of cartilage proteoglycans. II. Oligosaccharide competitors of the proteoglycan–hyaluronic acid interaction, *J. Biol. Chem.* **249:**4242–4249.

Hay, E. D., 1980, Development of the vertebrate cornea, *Int. Rev. Cytol.* **63:**263–322.

Kiang, W. L., Margolis, R. U., and Margolis, R. K., 1981, Fractionation and properties of a chondroitin sulfate proteoglycan and the soluble glycoproteins of brain, *J. Biol. Chem.* **256:**10529–10537.

Knudson, C. B., and Toole, B. P., 1985, Changes in the pericellular matrix during differentiation of limb bud mesoderm, *Dev. Biol.* **112:**308–318.

Knudson, C. B., and Toole, B. P., 1987, Hyaluronate–cell interactions during differentiation of chick embryo limb mesoderm, *Dev. Biol.* **124:**82–90.

Knudson, W., Biswas, C., and Toole, B. P., 1984, Interactions between human tumor cells and fibroblasts stimulate hyaluronate synthesis, *Proc. Natl. Acad. Sci. USA* **82:**6767–6771.

Krusius, T., Finne, J., Karkkainen, J., and Jarnefelt, J., 1974, Neutral and acidic glycopeptides in adult and developing rat brain, *Biochim. Biophys. Acta* **365:**80–92.

Kujawa, M. J., and Caplan, A. I., 1986, Hyaluronic acid bonded to cell culture surfaces stimulates chondrogenesis in stage 24 limb mesenchyme cell cultures, *Dev. Biol.* **114:**504–518.

Kujawa, M. J., Pechak, D. J., Fizman, M. Y., and Caplan, A. I., 1986a, Hyaluronic acid bonded to cell culture surfaces inhibits the program of myogenesis, *Dev. Biol.* **113:**10–16.

Kujawa, M. J., Carrino, D. A., and Caplan, A. I., 1986b, Substrate-bonded hyaluronic acid exhibits a size-dependent stimulation of chondrogenic differentiation of stage 24 limb mesenchymal cells in culture, *Dev. Biol.* **114:**519–528.

Kvist, T. N., and Finnegan, C. V., 1970, The distribution of glycosaminoglycans in the axial region of the developing chick embryo. I. Histochemical analysis, *J. Exp. Zool.* **175:**221–240.

Lacy, B. E., and Underhill, C. B., 1987, The hyaluronate receptor is associated with actin filaments, *J. Cell Biol.* **105:**1395–1404.

Laurent, T. C., Fraser, J. R. E., Pertoft, H., and Smedsrod, B., 1986, Binding of hyaluronate and chondroitin sulphate to liver endothelial cells, *Biochem. J.* **234:**653–658.

Love, S. H., Shannon, B. T., Myrvik, Q. N., and Lynn, W. S., 1979, Characterization of macrophage agglutinating factor as a hyaluronic acid–protein complex, *J. Reticuloendothelial Soc.* **25:**269–282.

Margolis, R. U., Margolis, R. K., Chang, L. B., and Preti, C., 1975, Glycosaminoglycans of brain during development, *Biochemistry* **14:**85–88.

Margolis, R. U., Ripellino, J. A., and Margolis, R. K., 1987, Cell surface and extracellular matrix glycoproteins and proteoglycans in nervous tissue, in: *Mesenchymal–Epithelial Interactions in Neural Development* (J. R. Wolff, J. Sievers, and M. Berry, eds.), pp. 65–76, Springer-Verlag, Berlin.

Marks, M. S., and Toole, B. P., 1985, Hyaluronate-binding proteins of chick and mouse brain, *J. Cell Biol.* **101:**334a.

Marks, M. S., Chi-Rosso, G., and Toole, B. P., 1989, Hyaluronate-binding proteins of murine brain, submitted for publication.

Markwald, R. R., Fitzharris, T. P., Bank, H., and Bernanke, D. H., 1978, Structural analysis on the matrical organization of glycosaminoglycans in developing endocardial cushions, *Dev. Biol.* **62:**292–316.

Matuoka, K., Namba, M., and Mitsui, Y., 1987, Hyaluronate synthetase inhibition by normal and transformed human fibroblasts during growth reduction, *J. Cell Biol.* **104:**1105–1115.

McBride, W. H., and Bard, J. B., 1979, Hyaluronidase-sensitive halos around adherent cells. Their role in blocking lymphocyte-mediated cytolysis, *J. Exp. Med.* **149:**507–515.

Mescher, A. L., and Munaim, S. I., 1986, Changes in the extracellular matrix and glycosaminoglycan synthesis during the initiation of regeneration in adult newt forelimbs, *Anat. Rec.* **214:**424–431.

Meyer, F. A., 1983, Macromolecular basis of globular protein exclusion and of swelling pressure in loose connective tissue (umbilical cord), *Biochim. Biophys. Acta* **755**:388–399.

Meyer, F. A., Laver-Rudich, Z., and Tanenbaum, R., 1983, Evidence for a mechanical coupling of glycoprotein microfibrils with collagen fibrils in Wharton's jelly, *Biochim. Biophys. Acta* **755**:376–387.

Mian, N., 1986, Analysis of cell-growth-phase-related variations in hyaluronate synthase activity of isolated plasma-membrane fractions of cultured human skin fibroblasts, *Biochem. J.* **237**:333–342.

Mikuni-Takagaki, Y., and Toole, B. P., 1980, Cell–substratum attachment and cell surface hyaluronate of Rous sarcoma virus-transformed chondrocytes, *J. Cell Biol.* **85**:481–488.

Neame, P. J., Christner, J. E., and Baker, J. R., 1986, The primary structure of link protein from rat chondrosarcoma proteoglycan aggregate, *J. Biol. Chem.* **261**:3519–3535.

Nemec, R. E., Toole, B. P., and Knudson, W., 1987, The cell surface hyaluronate binding sites of invasive human bladder carcinoma cells, *Biochem. Biophys. Res. Commun.* **149**:249–257.

Normand, G., Clos, J., Vincendon, G., and Gombos, G., 1985, Postnatal development of rat cerebellum: Glycosaminoglycan changes related to variation in water content, cell formation and organ growth, *Int. J. Dev. Neurosci.* **3**:245–256.

Oohira, A., Matsui, F., Matsuda, M., and Shoji, R., 1986, Developmental change in the glycosaminoglycan composition of the rat brain, *J. Neurochem.* **47**:588–593.

Orkin, R. W., Underhill, C. B., and Toole, B. P., 1982, Hyaluronate degradation in 3T3 and simian virus-transformed 3T3 cells, *J. Biol. Chem.* **257**:5821–5826.

Orkin, R. W., Knudson, W., and Toole, B. P., 1985, Loss of hyaluronate-dependent coat during myoblast fusion, *Dev. Biol.* **107**:527–530.

Pessac, B., and Defendi, V., 1972, Cell aggregation: Role of acid mucopolysaccharides, *Science* **175**:898–900.

Pintar, J. E., 1978, Distribution and synthesis of glycosaminoglycans during quail neural crest morphogenesis, *Dev. Biol.* **67**:444–464.

Polansky, J., Toole, B. P., and Gross, J., 1974, Brain hyaluronidase: Changes in activity during chick development, *Science* **183**:862–864.

Pratt, R. M., Larsen, M. A., and Johnston, M. C., 1975, Migration of cranial neural crest cells in a cell-free hyaluronate-rich matrix, *Dev. Biol.* **44**:298–305.

Reid, T., and Flint, M. H., 1974, Changes in glycosaminoglycan content of healing rabbit tendon, *J. Embryol. Exp. Morphol.* **31**:489–495.

Ripellino, J. A., Bailo, M., Margolis, R. U., and Margolis, R. K., 1988, Light and electron microscopic studies on the localization of hyaluronic acid in developing rat cerebellum, *J. Cell Biol.* **106**:845–855.

Ripellino, J. A., Margolis, R. U., and Margolis, R. K., 1989, Immunoelectron microscopic localization of hyaluronic acid binding region and link protein epitopes in brain, *J. Cell. Biol.* in press.

Smedsrod, B., Pertoft, H., Eriksson, S., Fraser, J. R., and Laurent, T. C., 1984, Studies in vitro on the uptake and degradation of sodium hyaluronate in rat liver endothelial cells, *Biochem. J.* **223**:617–626.

Smith, G. N., Toole, B. P., and Gross, J., 1975, Hyaluronidase activity and glycosaminoglycan synthesis in the amputated newt limb. Comparison of denervated, non-regenerating limbs with regenerates, *Dev. Biol.* **42**:221–232.

Solursh, M., Fisher, M., Meier, S., and Singley, C. T., 1979, The role of extracellular matrix in the formation of the sclerotome, *J. Embryol. Exp. Morphol.* **54**:75–98.

Solursh, M., Hardingham, T. E., Hascall, V. C., and Kimura, J. H., 1980, Separate effects of exogenous hyaluronic acid on proteoglycan synthesis and deposition in pericellular matrix by cultured chick embryo limb chondrocytes, *Dev. Biol.* **75**:121–129.

Tengblad, A., 1981, A comparative study of the binding of cartilage link protein and the hyaluronate-binding region of the cartilage proteoglycan to hyaluronate-substituted Sepharose gel, *Biochem. J.* **199**:297–305.

Toole, B. P., 1972, Hyaluronate turnover during chondrogenesis in the developing limb and axial skeleton. *Dev. Biol.* **29**:321–329.

Toole, B. P., 1973, Hyaluronate and hyaluronidase in morphogenesis and differentiation, *Am. Zool.* **13**:1061–1065.

Toole, B. P., 1981, Glycosaminoglycans in morphogenesis, in: *Cell Biology of the Extracellular Matrix* (E. D. Hay, ed.), pp. 259–294, Plenum Press, New York.

Toole, B. P., and Trelstad, R. L., 1971, Hyaluronate production and removal during corneal development in the chick, *Dev. Biol.* **26**:28–35.

Toole, B. P., Goldberg, R. L., Chi-Rosso, G., Underhill, C. B., and Orkin, R. W., 1984, Hyaluronate–cell interactions, in: *The Role of Extracellular Matrix in Development* (R. L. Trelstad, ed.), pp. 43–66, Liss, New York.

Turley, E. A., and Roth, S., 1979, Spontaneous glycosylation of glycosaminoglycan substrates by adherent fibroblasts, *Cell* **17**:109–115.

Turley, E. A., and Torrance, J., 1984, Localization of hyaluronate and hyaluronate-binding protein on motile and non-motile fibroblasts, *Exp. Cell Res.* **161**:17–28.

Turley, E. A., Bowman, P., and Kytryk, M. A., 1985, Effects of hyaluronate and hyaluronate-binding proteins on cell motile and contact behavior, *J. Cell Sci.* **78**:133–145.

Turley, E. A., Moore, D., and Hayden, L. J., 1987, Characterization of hyaluronate binding proteins isolated from 3T3 and murine sarcoma virus transformed 3T3 cells, *Biochemistry* **26**:2997–3305.

Underhill, C. B., 1982a, Interaction of hyaluronate with the surface of simian virus 40-transformed 3T3 cells: Aggregation and binding studies, *J. Cell Sci.* **56**:177–189.

Underhill, C. B., 1982b, Naturally-occurring antibodies which bind hyaluronate, *Biochem. Biophys. Res. Commun.* **108**:1488–1494.

Underhill, C. B., and Dorfman, A., 1978, The role of hyaluronic acid in intercellular adhesion of cultured mouse cells, *Exp. Cell Res.* **117**:155–164.

Underhill, C. B., and Toole, B. P., 1979, Binding of hyaluronate to the surface of cultured cells, *J. Cell Biol.* **82**:475–484.

Underhill, C. B., and Toole, B. P., 1980, Physical characteristics of hyaluronate binding to the surface of simian virus 40-transformed 3T3 cells, *J. Biol. Chem.* **255**:4544–4549.

Underhill, C. B., and Toole, B. P., 1981, Receptors for hyaluronate on the surface of parent and virus-transformed cell lines. Binding and aggregation studies, *Exp. Cell Res.* **131**:419–423.

Underhill, C. B., and Toole, B. P., 1982, Transformation-dependent loss of the hyaluronate-containing coats of cultured cells, *J. Cell. Physiol.* **110**:123–128.

Underhill, C. B., Chi-Rosso, G., and Toole, B. P., 1983, Effects of detergent solubilization on the hyaluronate-binding protein from membranes of simian virus 40-transformed 3T3 cells, *J. Biol. Chem.* **258**:8086–8091.

Underhill, C. B., Thurn, A. L., and Lacey, B. E., 1985, Characterization and identification of the hyaluronate-binding site from membranes of SV-3T3 cells, *J. Biol. Chem.* **260**:8128–8133.

Underhill, C. B., Green, S. J., Comoglio, P. M., and Tarone, G., 1987a, The hyaluronate receptor is identical to a glycoprotein of 85,000 Mr (gp85) as shown by a monoclonal antibody that interferes with binding activity, *J. Biol. Chem.* **262**:13142–13146.

Underhill, C. B., Tarone, G., and Kausz, A. T., 1987b, The hyaluronate-binding site from the plasma membrane is distinct from the binding protein present in brain, *Connect. Tissue Res.* **16**:225–235.

Wasteson, A., Estermark, B., Lindahl, U., and Ponten, J., 1973, Aggregation of feline lymphoma cells by hyaluronic acid, *Int. J. Cancer* **12**:169–178.

Werz, W., Fischer, G., and Schachner, M., 1985, Glycosaminoglycans of rat cerebellum. II. A developmental study, *J. Neurochem.* **44**:907–910.

Wiebkin, O. W., and Muir, H., 1973, The inhibition of sulphate incorporation in isolated adult chondrocytes by hyaluronic acid, *FEBS Lett.* **37**:42–46.

Wright, T. C., Underhill, C. B., Toole, B. P., and Karnovsky, M. J., 1981, Divalent cation-independent aggregation of rat-1 fibroblasts infected with a temperature-sensitive mutant of Rous sarcoma virus, *Cancer Res.* **41**:5107–5113.

Inborn Errors of Complex Carbohydrate Catabolism

Glyn Dawson and Larry W. Hancock

1. INTRODUCTION

Complex carbohydrates of the nervous system are degraded in lysosomes by the sequential action of a group of exoglycosidases known collectively as the lysosomal hydrolases. Inherited defects in the synthesis, assembly, or turnover of these hydrolases lead to storage diseases in humans (Spranger, 1987) and a variety of domestic animals. Those involving the nervous system result in spectacular neuropathology and provide the best evidence for the types of glycoconjugates synthesized by nervous tissue, as well as their rate of turnover. For example, in Tay–Sachs disease, storage material (GM2 ganglioside) predominates in nervous tissue, especially motor neurons, and is virtually absent from visceral tissue. The variable level of accumulation of GM2 in different brain regions (identified morphologically as multilamellar cytosomes) can be related to different levels of synthesis and degradation. This clearly manifests itself in patients with partial hexosaminidase (HexA) deficiencies, who exhibit symptoms of motor neuron disease, or spinocerebellar degeneration with other neuronal function (such as vision and intelligence) relatively intact. The absence of GM2 storage outside the CNS reflects the lack of GM2 synthesis in nonneural tissue. However, since lysosomal hydrolases are synthesized constitutively in *all* tissues, GM2 can be fed to fibroblasts from HexA-deficient patients, and its steady accumulation observed. Thus, storage patterns in patients with inherited enzyme defects can be used to give an accurate reflection of glycoconjugate content of the CNS versus nonneural tissue and this will be emphasized on an enzyme/disease, case-by-case basis.

Types of defects producing glycolipid/oligosaccharide storage diseases can be summarized as follows:

Glyn Dawson and Larry W. Hancock • Departments of Pediatrics and Biochemistry and Molecular Biology, University of Chicago, Chicago, Illinois 60637.

1. Gene deletion (DNA), e.g., β-Hex B in Sandhoff
2. Lack of mRNA, e.g., fucosidosis I/Sandhoff/Tay–Sachs
3. Absence of enzyme protein, e.g., β-mannosidosis
4. Absence of activator protein, e.g., β-Hex AB variant
5. Active precursor broken down, e.g., β-galactosidase–neuraminidase deficiency (lack of protective protein), fucosidosis II
6. Enzyme active but degraded, e.g., arylsulfatase A
7. Enzyme active but insoluble, e.g., HexA-deficient variant
8. Inappropriate targeting of enzyme, e.g., GlcNAc phosphotransferase deficiency
9. Transport deficiency, e.g., sialic acid storage disease

Considerable progress has been made in studying the processing of lysosomal hydrolases (Table 1) and in their cloning and chromosome localization (Table 2), so that we can understand the molecular basis for many of the defects.

Table 1. Synthesis and Processing of Some
Lysosomal Enzymes Involved
in Brain Glycoconjugate Catabolism

Enzyme	Precursor (kDa)	Product (kDa)
Neuraminidase	ND	76
β-Galactosidase	85	64
N-Acetyl-β-hexosaminidase		
α chain	67	54
β chain	63	29/25
Activator protein	23.5	21.5
α-Mannosidase	ND	100
β-Mannosidase	114	110
α-Fucosidase	53	50
α-N-Acetylglucosaminidase	86	80
α-Galactosidase A	50.5	46
β-Glucocerebrosidase	63/61	56
Arylsulfatase A	62/59.5	60.5/57
Arylsulfatase B	64	47/12;40/31
SAP-1	70	8–11
α-L-Iduronidase	76	72/66
β-Glucuronidase	ND	75
Cathepsin D	53	31
Cathepsin B	44.5/46	33/27
α-Glucosidase	110	76/70
32-kDa gp protective factor	54	32

Table 2. Chromosomal Localization and Progress in Gene Cloning of Lysosomal Hydrolases Associated with Neurodegenerative Storage Diseases

Enzyme	Chromosome	cDNA clone	Disease
Neuraminidase	6	No	Combined
32kDa protective glycoprotein	ND	No	β-Gal/neuraminidase deficiency
GlcNAc phosphotransferase	ND	No	"I-cell"
N-Acetyl-β-Hexoasaminidase			
α chain	15	Full-length	Tay–Sachs
β chain	5	Full-length	Sandhoff
Activator protein	15	No	AB variant
α-Mannosidase	19	No	α-Mannosidosis
β-Mannosidase	(4)	Partial	β-Mannosidosis
α-Fucosidase	1	Full-length	α-Fucosidosis
1-Aspartido-β-N-acetylglucosamine aminohydrolase	4	No	Aspartylglucosaminuria
α-Galactosidase A	X	Full-length	Fabry
β-Glucosidase (glucocerebrosidase)	1	Full-length	Gaucher
Arylsulfatase A	22	Partial	Metachromatic leukodystrophy
Ceramidase	ND	No	Farber
SAP-1	10	Full-length	Atypical MLD
α-L-Iduronidase	22	No	Hurler–Scheie
β-Glucuronidase		Full-length	MPS VII
Iduronate sulfate sulfatase	X	No	Hunter (MPS II)
Cathepsin B	8	Full-length	—
Cathepsin D	—	Full-length	—
Cathepsin H	15	Partial	—
Cathepsin S	—	Full-length	—
β-Galactosidase	3	Full-length	GM1 gangliosidosis
GalNAc-6-sulfate/Gal-6-sulfate sulfatase	3	—	Morquio B
GlcNAc-6-sulfatase	7	—	MPS IV
GalNAc-4-sulfatase	5	—	Maroteaux–Lamy (MPS VI)
α-Glucosidase	17	Partial	Pompe

2. INBORN ERRORS OF GLYCOLIPID AND GLYCOPROTEIN CATABOLISM

2.1. Neuraminidase Deficiency

Nervous system glycoproteins, e.g., NCAM, and glycolipids, e.g., gangliosides, are enriched in sialic acid (typically N-acetylneuraminic acid in humans) and neuraminidase deficiency always results in severe neuropathology. The catabolism of a prototypic glycoprotein is shown in Figure 1. A deficiency of glycoprotein-specific

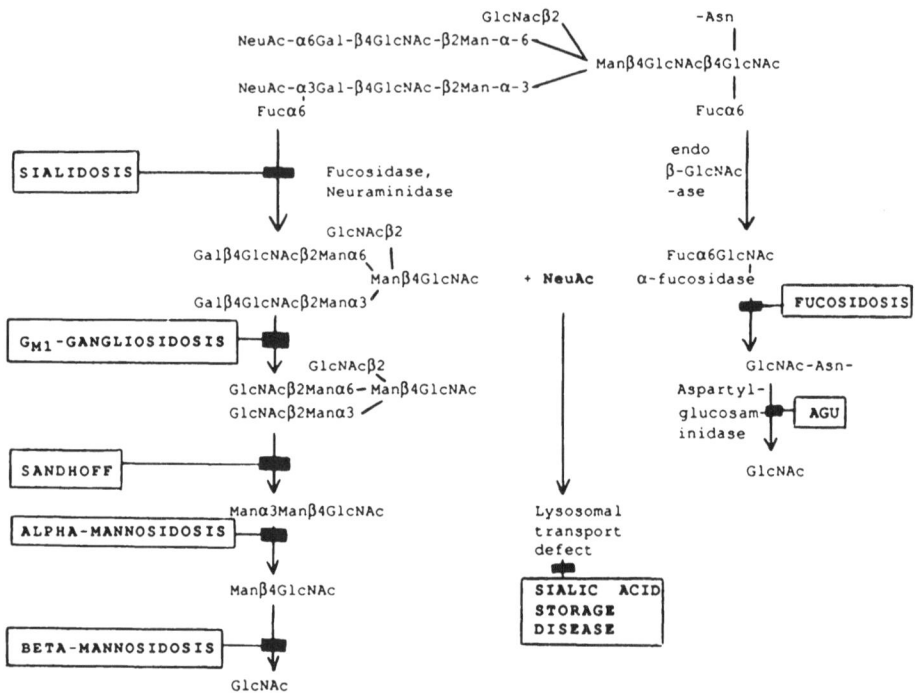

Figure 1. Proposed pathway for catabolism of a prototypic human brain glycoprotein. Diseases corresponding to enzyme deficiencies are indicated.

lysosomal neuraminidase (Verheijen *et al.*, 1987) leads to the accumulation of sialo-oligosaccharides (NeuAcα6Gal and NeuAcα3Gal terminated) in all tissues. The major oligosaccharide stored in CNS has the structure: [NeuAcα6Galβ4GlcNAc-β2Man]$_2$α3/6Manβ4GlcNAc. Four forms of inherited neuraminidase deficiency with increasing severity have been described.

- Type 1. Born with hydrops and hepatosplenomegaly, fail to thrive, suffer from recurrent infections, and die within 1 year (Laver *et al.*, 1983; Beck *et al.*, 1984). Extensive accumulation of NeuAc oligosaccharides.

- Type 2. Motor system retardation is noted at 6–12 months, accompanied by progressive neurological deterioration, skeletal abnormalities, *and* hepatosplenomegaly (Kelly and Graetz, 1977); children die in early childhood. The disease is also referred to as nephrosialidosis (Maroteaux *et al.*, 1978; Tondeur *et al.*, 1982).

- Type 3. Progressive neurological degeneration from age 6 years onwards, with ataxia, myoclonus, and cherry-red spot—the latter usually being a hallmark of neuronal lipid accumulation. Originally referred to by Spranger and associates (Cantz *et al.*, 1977; Durand *et al.*, 1977) as "mucolipidosis I," but shown by this group to be a neuraminidase deficiency.

- Type 4. The adult form, beginning in adolescence with progressive myoclonus, loss of vision (cherry-red spot—myoclonus syndrome), but little impairment of intelligence (O'Brien, 1977). Partial neuraminidase deficiency. Diagnosis is made by using either 4-methylumbelliferylαNeuAc (4MUNeuAc) synthetic substrate, neuramin-2,6-lactose, or fetuin (Kelly and Graetz, 1977).

Ganglioside catabolism is unaffected in these diseases (Figure 2), indicating that NeuAcα8NeuAc, NeuAcα3Galβ4GalNAc, and NeuAcα3Galβ4Glc linkages in glycolipids (gangliosides) are cleaved by different neuraminidases. A defect in NeuAcα3Galβ4GlcβCer (GM3) hydrolysis has been claimed in mucolipidosis IV, but never verified or confirmed by others (Crandall *et al.*, 1982; Ziegler and Bach, 1986). Sweeley and Usuki (1987) have recently reported an extralysosomal glycolipid neuraminidase which appears to be responsible for cell surface hydrolysis of GM3 to initiate cell division. Treatment with 2,3-dehydro-2-deoxy-NeuAc inhibits hydrolysis and cell division. Thus far, there is no evidence for glycolipid lysosomal neuraminidase deficiencies, but their existence is assumed.

2.2. GlcNAc Phosphotransferase Deficiency (I-Cell Disease; Mucolipidosis II)

A deficiency of UDP-GlcNAc:lysosomal enzyme precursor GlcNAc phosphotransferase results in the failure to attach the mannose 6-phosphate marker onto lysosomal hydrolases (Reitman *et al.*, 1981; Hasilik *et al.*, 1981). As a result, en-

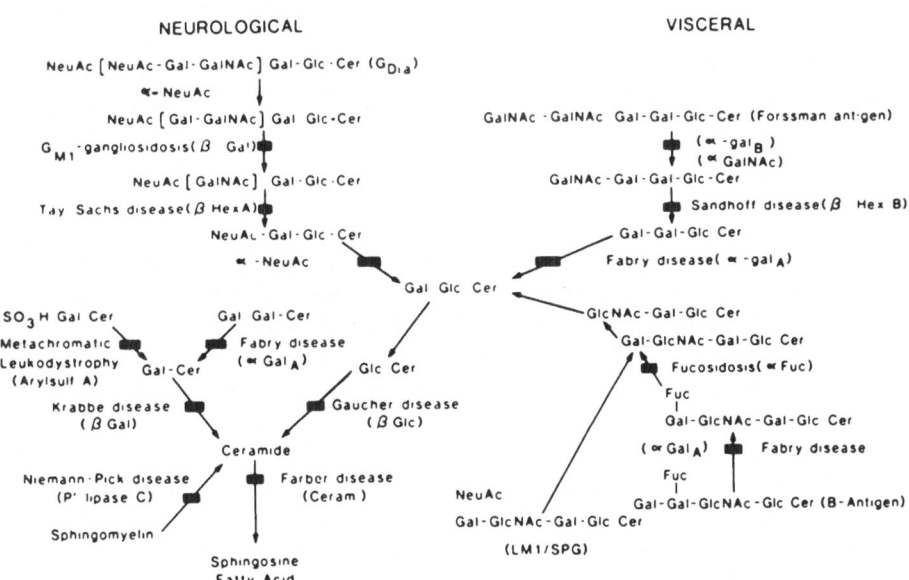

Figure 2. Catabolic pathway for the major brain glycolipids showing the site of inherited genetic defects of specific lysosomal hydrolases. Enzymes involved are indicated in parentheses, e.g., in Gaucher's disease the defective enzyme is β-glucosidase (βGlc) and the stored glycolipid is GlcCer.

zymatically active precursor forms of lysosomal hydrolases are secreted rather than being targeted to lysosomes (Kaplan et al., 1977) by Man 6-P receptors (von Figura and Hasilik, 1986; Kornfeld, 1987) for further processing and combination with activators or protective peptides, or self-association. The disease can be mimicked in fibroblasts by preventing acidification of prelysosomal vesicles via the addition of ammonium chloride or chloroquine to cultured skin fibroblasts (Weismann et al., 1975; Sando et al., 1979; Gonzalez-Noriega et al., 1980), although this mechanism involves accumulation of receptor–enzyme complexes in intracellular compartments. Although a deficiency of virtually all lysosomal hydrolases and massive storage of glycolipids and glycosaminoglycans can be demonstrated in fibroblasts (Matalon et al., 1968; Dawson et al., 1972), the patients clinically resemble those with Hurler's disease (mucopolysaccharidosis I) and excrete mainly complex sialo-oligosaccharides in urine (Strecker et al., 1977). Accumulation of storage material in brain is not striking (Gilbert et al., 1973). Oligodendrocytes have been claimed to lack Man 6-P receptors (Hill et al., 1985), and it is possible that neurons, glia, and other cell types process lysosomal hydrolases in a Man 6-P-independent manner. There is also heterogeneity among hydrolases since in I-cell disease, β-hexosaminidase, α-fucosidase, α-galactosidase, α-iduronidase, and arylsulfatase A are totally excreted, whereas α-glucosidase and cathepsin D are present at 20–50% of normal levels, and acid phosphatase and glucocerebrosidase are unaffected (Leroy et al., 1972; Gilbert et al., 1973).

Complementation studies reveal three subtypes of the phosphotransferase deficiency (Little et al., 1986; Ben-Yoseph et al., 1986). Type 1 patients are the most severely affected, with profound mental retardation, whereas Type 2 (pseudo-Hurler polydystrophy) and Type 3 show milder neurological symptoms. Types 2 and 3 may involve a gene coding for a lysosomal hydrolase binding site or a noncatalytic phosphotransferase site, rather than the phosphotransferase itself (which is deficient in Type 1).

2.3. β-Galactosidase Deficiency (GM1 Gangliosidosis)

A deficiency of lysosomal β-galactosidase (Okada and O'Brien, 1968) results in the accumulation of galacto-oligosaccharides (predominantly [Galβ4GlcNAc-α2Man]$_2$α3,6Manβ4GlcNAc; Wolfe et al., 1974) and GM1 (II^{3}NeuAcGgOse$_4$Cer; Galβ3GalNAcβ4[NeuAcα3]Galβ4GlcCer) (Ledeen et al., 1965) in brain and other tissues as shown in Figure 2. Three variants (infantile, juvenile, and adult) of GM1 gangliosidosis have been described, based on the age of appearance of neurological symptoms and hepatosplenomegaly (O'Brien, 1972b). Neurons show the massive accumulation of GM1 in multilamellar cytosomes and subsequent formation of mega-neurites (Purpura and Suzuki, 1976). The so-called Type 1 form has much more pronounced visceral storage of [Galβ4GlcNAcβ3]$_3$Galβ3[Galβ6]GalNAc-O-Ser/Thr than the Type 2 (Tsay et al., 1975; Tsay and Dawson, 1976), but no biochemical basis for this has been established (Galjaard et al., 1975). Brain autopsy studies on the milder Type 3 form have not been performed, so the nature of the storage material is unknown (Lowden et al., 1974; Koster et al., 1976; Wenger et al., 1980). Partial β-galactosidase deficiencies, reported previously, have been seen in connection with other lysosomal hydrolase deficiencies such as α-iduronidase and UDP-N-acetylglucosamine phosphotransferase (I-cell disease) (Kint *et al.*, 1973; Rushton and Dawson, 1977).

2.4. Combined Neuraminidase–Lysosomal β-Galactosidase Deficiency

Clinical symptoms similar to those seen for the neuraminidase deficiencies (devastating neurodegeneration) are caused by an inherited defect in the protective glycoprotein, leading to a deficiency of both neuraminidase and β-galactosidase. Lysosomal neuraminidase normally associates with approximately ten 64-kDa β-galactosidase molecules to form an 800-kDa complex stabilized by the 32-kDa "protective" glycoprotein (Palmeri *et al.*, 1986). Sialo-oligosaccharides are stored, together with GM1, and the variant is characterized by profound neurological degeneration (Suzuki *et al.*, 1984).

2.5. Globoid Cell Leukodystrophy (Krabbe's Disease)

A deficiency of GalCer β-galactosidase results in failure to catabolize the major myelin glycolipid, galactosylceramide (Figure 2) (Suzuki and Suzuki, 1970) resulting in its partial conversion to cytotoxic galactosylsphingosine (Svennerholm *et al.*, 1980; Shinoda *et al.*, 1987). Early death of oligodendrocytes occurs so that no overt accumulation of GalCer is observed, apart from localized cells; the so-called globoid cells. The defect can be assayed using galactosylsphingosine, galactosylceramide, galactosyldiglyceride, or lactosylceramide as the substrate (Wenger *et al.*, 1975). An authentic animal model, the Twitcher mouse, exists (Shinoda *et al.*, 1987). Developmental studies revealed no differences in the number of myelinated fibers between affected heterozygotes and normal littermates until 20 days postnatally, but the sheaths themselves were thinner at this time. Hypomyelination was therefore apparent before myelin breakdown. Engraftment of Twitcher sciatic nerves into normal mice showed that normal Schwann cells could migrate into the graft and enhance GalCer hydrolysis (Scaravilli and Suzuki, 1983), offering therapeutic prospects.

No other glycoconjugates are stored in Krabbe's disease and storage outside the nervous system is minimal.

2.6. N-Acetyl-β-hexosaminidase Deficiencies

The hydrolysis of GM2 (II^{3}NeuAcGgOse$_3$Cer; GalNAc[NeuAcα3]β4Gal-β4GlcCer) requires a functional α-chain gene (chromosome 15), β-chain gene (chromosome 5; Lalley *et al.*, 1974), and activator protein gene (chromosome 5). Defects in any of these three peptides produce a neurodegenerative disease with glycoconjugate storage (Figures 1 and 2).

2.6.1. Tay–Sachs Disease

The most common form of GM2 gangliosidosis is the infantile form, known as Tay–Sachs disease. Mutations in the α-chain peptide (synthesized as a 67-kDa precursor, which is then processed to a 56-kDa and, finally, a 54-kDa mature α-chain; Hasilik and Neufeld, 1980) typically result in failure to make any mRNA (classical Tay–Sachs; Myerowitz and Proia, 1984). The French-Canadian variant results from a major deletion in the 5′ end of the Hexα gene (Myerowitz and Hogikyan, 1986). HexA deficiencies can also result from an insoluble β-Hex protein or synthesis of an α chain which fails to combine either with activator protein or with β chain to produce the stable HexA ($\alpha\beta_2$) complex. GM2 and asialo-GM2 are the major storage glycoconjugates in Tay–Sachs brain and their accumulation swells the neuron, causing meganeurite formation (Purpura and Suzuki, 1976) and neuronal destruction. Lack of oligosaccharide storage suggests that HexB normally degrades glycoproteins and HexA, plus activator protein, degrades GM2. There is some evidence for lyso-GM2 accumulation as a potentially neurotoxic agent (Hannun and Bell, 1987). Biochemically, HexB ($\beta 4$; $2\beta_A 2\beta_B$) is overproduced so that total β-Hex activity, when assayed with synthetic substrates, appears normal (Okada and O'Brien, 1969).

2.6.2. Sandhoff Disease

Mutations in the β locus typically result in no β-chain protein synthesis, failure to form HexB (β_4), and rapid destruction of α chains through failure to combine with β chains to form HexA (O'Dowd et al., 1985, 1986). Thus, Sandhoff disease is characterized by a total β-Hex deficiency (Sandhoff et al., 1971). Hexβ chain is synthesized as a 63-kDa precursor which, in humans, is processed into β_A (29kDa) and β_B (25kDa) chains, which can then associate, via disulfide bonds, either with themselves (HexB, β_4) or with an α chain to form HexA ($\alpha\beta_2$) (Hasilik and Neufeld, 1980). α and β chains may also associate prior to processing, since some human cells (U937 monocytes) and several species (e.g., mouse) do not process the 54-kDa β-chain peptide. The deficiency of HexB results in the additional accumulation of globoside (GalNAcβ3Galα4Galβ4GlcCer) in nonneural tissues and GlcNAc-terminated oligosaccharides in all tissues (Sandhoff et al., 1971). In brain, 70% of water-soluble stored glycoconjugates were identified as a heptasaccharide, GlcNAcβ2Manα6[GlcNAcβ4][GlcNAcβ2Manα3]Manβ4GlcNAc (Warner et al., 1985). A very minor storage oligosaccharide in brain, but the major one in nonneural tissues, was the heptasaccharide GlcNAcβ2Manα6[GlcNAcβ2(GlcNAcβ4)Manα3]Manβ4GlcNAc. All tissues contained the hexasaccharide lacking the bisecting (brain) or branched (nonneural) GlcNAc residue. The storage pattern suggests that brain has an active GlcNAc transferase III, but limited GlcNAc transferase IV activity. Sandhoff patients resemble Tay–Sachs phenotypically, but milder adult variants with spinocerebellar degeneration also exist in which oligosaccharide storage is clearly seen (Bolhuis et al., 1987).

2.6.3. Activator Protein Defect

The so-called AB variant patients (Sandhoff et al., 1971) show normal HexA and HexB activity toward synthetic substrates, but lack activator protein and cannot hydrolyze GM2. Thus, the phenotype and storage pattern resemble Tay–Sachs disease. The activator protein can be substituted by the use of detergent, implying that the activator protein role is to "present" the GM2 to the active site of $\alpha\beta_2$ (HexA).

A second type (B1) synthesize normal activator protein, but a defect in the α chain [resulting from a single base substitution (G to A at 533), which codes for histidine instead of the normal arginine (Ohno and Suzuki, 1988)] results in failure to form an α-chain–GM2–activator protein complex and failure to hydrolyze GM2. The patient resembled a typical Tay–Sachs patient.

2.6.4. Juvenile and Adult β-Hex Deficiencies

Partial deficiencies of β-Hex lead to later onset of neurological diseases. Of particular interest have been a group of adult patients with progressive motor neuron disease (Kolodny and Raghavan, 1983; Hancock *et al.*, 1985; Navon *et al.*, 1986) but normal intelligence. Electron microscopic studies of Tay–Sachs fetuses have revealed storage bodies in motor neurons at 20 weeks of gestation, suggesting that GM2 is turning over faster in motor neurons than in other nerve cells. This may explain why a partial β-Hex deficiency primarily affects motor neurons. Most of the adult amyotrophic lateral sclerosis (ALS) phenocopies have primary α-chain mutations, but we described a unique primary β-chain mutation characterized by apparently normal HexA, but no HexB (Hancock *et al.*, 1985). Further cases have now been described. The mother of the patient whom we described is heterozygous for the β-chain mutation, but the father could be heterozygous for a defect in either α or β chain. Others have suggested that this group (both HexA *and* HexB deficiencies) are possibly compound heterozygotes. Rectal biopsy showed typical membranocytoplasmic (GM2) bodies in ganglia and GM2 is poorly metabolized by fibroblasts from these patients, but no autopsy studies have been carried out (Cashman *et al.*, 1986). It is, therefore, presumed that accumulation of GM2 is responsible for the motor neuron degenerative symptoms in these patients, although accumulation of unique, normally minor gangliosides cannot be excluded.

2.7. α-Mannosidosis

An inborn deficiency of lysosomal (acidic) α-mannosidase leads to accumulation of manno-oligosaccharides (predominantly Manα3Manβ4GlcNAc) (Norden *et al.*, 1974) and varying degrees of mental retardation in humans (Carroll *et al.*, 1972; Jolly *et al.*, 1981) (Figure 1). Several animal models exist (see Table 3). Meganeurite formation has been observed in the cat model, suggesting that this is *not* the hallmark of ganglioside storage, but the mental retardation in mannosidosis is much less severe than in the lipid storage diseases. The absence of predicted *branched* Manα6Man-containing storage material in mannosidosis (in contrast to the pattern when swainsonine is used to nonspecifically inhibit α-mannosidase) suggests either the existence of a second lysosomal α-mannosidase, specific for α6 linkages and unaffected in all cases of mannosidosis thus far studied, or an extralysosomal degradative pathway.

2.8. β-Mannosidosis

Originally described as a devastating neurodegenerative Manβ4GlcNAc-β4GlcNAc storage disease in goats (Jones and Dawson, 1981; Jones and Laine, 1981;

Table 3. Animal Models for Glycoconjugate Storage Diseases

Deficient enzyme	Storage	Animal model
α-Neuraminidase/β-galactosidase	Sialo-Os/GM1	Sheep
β-Galactosidase	Gal-Os	Dog, cat
β-Hexosaminidase	GlcNac-Os	Cat
α-Mannosidase	Man-Man-GlcNac-GlcNAc	Cattle, cat
β-Mannosidase	Man-GlcNAc-GlcNAc	Goat
α-Fucosidase	Fuc-GlcNAc-GlcNAc	Dog
GalNAc-4-sulfate sulfatase	[GalNac-4-SO$_3$H-IdUA/GlcUA]$_n$	Cat
α-L-Iduronidase	[IdUAGlcNAcSO$_3$H]$_n$	Dog
β-Glucosidase	GlcCer	Dog
GalCerβGalactosidase	GalCer	Dog, twitcher mouse

Dawson, 1982), β-mannosidosis has recently been described in two sets of male siblings (Cooper *et al.*, 1986) (with a storage of Manβ4GlcNAc) and in a single case in combination with a deficiency of heparan sulfate sulfamidase (Wenger *et al.*, 1986). In humans, the storage oligosaccharide is a disaccharide, Manβ4GlcNAc (Figure 1), and the mental retardation is rather mild in contrast to the dramatic hypomyelination and neuron loss associated with the goat form of the disease (Jones *et al.*, 1982). Minor storage oligosaccharides in the goat include a Manβ4GlcNAcβ4Manβ4GlcNAc-β4GlcNAc pentasaccharide, suggesting the existence of a novel type of hitherto unsuspected glycoprotein oligosaccharide unit.

2.9. α-Fucosidosis

Fucosidosis is an inherited deficiency of α-L-fucosidase (Figures 1 and 2) which results in a phenotypically heterogeneous lysosomal storage disease, characterized by dementia, paresis, tremor, spasticity, and rigidity, i.e., profound neurodegeneration (Durand *et al.*, 1969; Borrone *et al.*, 1974). Deficiencies of α-fucosidase lead to accumulation of a disaccharide (Fucα6GlcNAc) and several larger oligosaccharides (Tsay *et al.*, 1976) in brain and other tissues. Storage fucoglycolipids of the H-antigen type (Fucα2Galβ4GlcNAcβ4Galβ4GlcβCer) (Dawson *et al.*, 1972) together with stage-specific embryonic antigens SSEA 1 and 2 (Schwarting *et al.*, 1988) have been isolated from brain and other tissues. The two major clinical phenotypes (with and without angiokeratoma) cannot be distinguished genetically or biochemically (Christomanou and Beyer, 1983). However, we have recently distinguished two major types on the basis of their ability to synthesize precursor enzyme protein. Twelve of fifteen fucosidosis patients of both phenotypes were negative for 2.4-kb mRNA and enzyme protein, whereas the other three patients synthesized the normal amount of enzymatically active 53-kDa precursor. However, the cells contained no mature 50-kDa active enzyme and appeared negative for α-fucosidase activity (Johnson and Dawson, 1985) as assayed with the 4-methylumbelliferyl-α-fucopyranoside substrate. Cross-fusion studies between fucosidosis fibroblasts indicated correction between the two types, and hence a different mutation. Further, enzyme cross-feeding suggested the absence of a protective peptide as the defect in these three atypical fucosidosis patients (Dawson *et al.*, 1988).

Southern blot analysis of 20 fucosidosis patients revealed an *Eco*RI restriction fragment polymorphism in 4 patients, but all 20 patients had less than 2% of normal α-fucosidase activity and less than 6% of normal enzyme protein levels (O'Brien *et al.*, 1987). No correlation with phenotypic expression could be made.

2.10. Aspartylglucosaminuria (AGU)

A deficiency of the lysosomal enzyme 1-aspartamido-β-*N*-acetylglucosamine aminohydrolase leads to the lysosomal accumulation of the glycopeptide GlcNAc-Asn (Jenner and Pollitt, 1967; Palo *et al.*, 1972, 1973) and progressive mental retardation (Autio, 1972) (Figure 1). Electron microscopic studies of AGU brain have revealed storage bodies similar to those seen in α-mannosidosis. Several minor glycopeptides of increasing complexity up to [NeuAc-Gal-GlcNAc-Man]$_2$Man-GlcNAc-GlcNAc-Asn and [Man]$_2$[GlcNAc]$_2$Asn have been found in urine from Japanese (Sugahara *et al.*, 1976; Akasaki *et al.*, 1976) and Finnish patients.

2.11. α-Galactosidase Deficiency (Fabry's Disease)

A deficiency of α-galactosidase (αGalA) leads to this glycosphingolipid storage disease characterized by the accumulation of trihexosylceramide (Galα4Galβ4GlcCer) in extraneural tissue (Figure 2) and the additional accumulation of dihexosylceramide (Galα4Galβ4Cer) in kidney (Sweeley and Klionsky, 1963). Some peripheral nerve pathology has been observed, presumably caused by glycolipid storage. Most Fabry patients are mRNA negative. However, pulse–chase immunoprecipitation studies reveal the normal 51-kDa αGalA precursor in some Fabry patients, but the absence of the mature 46-kDa form (Lemansky *et al.*, 1987). Addition of the thiol protease inhibitor leupeptin only partially restored the 46-kDa peptide and enzyme activity in these patients.

2.12. N-Acetyl-α-galactosaminidase Deficiency

A deficiency of *N*-acetyl-α-galactosaminidase activity (αGalB) has been described (van Diggelen *et al.*, 1987) in two patients with progressive psychomotor degeneration. αGalNAc linkages have been reported in brain glycoproteins (Nakagawa *et al.*, 1986), and the disorder is believed to primarily affect astrocytes. The major storage oligosaccharide was a trisaccharide with terminal αGalNAc residue and blood group A activity (Figure 2). Plasma, leukocytes, and fibroblasts from patients showed less than 2% of normal α-*N*-acetylgalactosaminidase activity. Both parents had low normal or reduced activity, suggesting an autosomal recessive mode of inheritance.

2.13. Arylsulfatase A Deficiency (Metachromatic Leukodystrophy)

This is an inherited demyelinating disease characterized by accumulation of sul-fogalactosylceramide (sulfatide) in CNS (Figure 2) and peripheral myelin, and to a lesser extent in kidney. Ataxia and mental retardation are prominent after the age of five; adult forms of metachromatic leukodystrophy (MLD) with presenile dementia have been described. There have been reports of CNS storage of sulfo-GlcNAc-containing oligosaccharides, but these have not been fully characterized. The HNK-1 antigenic epitope (GlcUA-3-SO_3HβGal) is also a potential storage product in this disease, since it is hydrolyzed *in vitro* by arylsulfatase A.

Some forms of MLD result from a defect in an 11-kDa sphingolipid activator glycoprotein (SAP-1) which has recently been cloned (Dewji *et al.*, 1987). The 70-kDa preproSAP-1 has been shown to have extensive sequence homology with the sulfated glycoprotein-1 (SGP-1) secreted by rat Sertoli cells and a similar protein may exist in brain.

2.14. Multiple Sulfatase Deficiency (MSD)

This is a Hurler-like disorder, characterized by skeletal abnormalities and pro-found mental retardation. The storage of both glycosaminoglycans and sulfatide in CNS (Figure 2) has been attributed to multiple sulfatase deficiency. Arylsulfatase A and six other sulfatases are profoundly deficient, and complementation shows clear distinction from other individual sulfatase deficiencies (Horwitz, 1979). The molecular basis appears to be a defect at the posttranslational level (Eto *et al.*, 1983; Horwitz *et al.*, 1986), which causes instability of all seven sulfatases. Intriguingly, one of the deficient sulfatases (steroid sulfatase; arylsulfatase C) is nonlysosomal, and hybridiza-tion studies with cDNA for this enzyme (Conary *et al.*, 1987) revealed heterogeneity within MSD patients.

2.15. Ceramidase Deficiency (Farber's Lipogranulomatosis)

The accumulation of ceramide, the lipid moiety of all glycosphingolipids, occurs in all tissues of Farber patients and results in mental retardation and visceral abnor-malities (Chen *et al.*, 1981). Minor abnormalities in brain glycoconjugate composition have been observed (Moser *et al.*, 1969).

2.16. β-Glucosidase Deficiency (Gaucher's Disease)

Three forms of lysosomal β-glucosidase deficiency have been described, with the nonneurological Ashkenazi Jewish (adult or Type 1) form being the most common. Massive accumulation of GlcCer is found in spleen and other visceral tissue. An infantile (Type 2) and juvenile (Type 3) form have been described in which neu-rodegeneration is increasingly severe. All three types of Gaucher's disease contain cross-reacting immunological material representing the normal 56-kDa deglycosylated β-glucosidase protein. The reason for the neurological damage in Types 2 and 3 is unclear (Tsuji *et al.*, 1987), but probably involves macrophage invasion of the CNS. Recently, sequencing of the mutant Type 2 (acute neurological) Gaucher gene has revealed a single base change converting leucine into proline in β-glucosidase (Graves *et al.*, 1986). The new restriction site has not been found in both genes of any Type 1 patient, but does occur in both genes of most Type 3 patients. Thus, if a patient has this particular mutation, he/she has an 80–85% chance of developing neurological problems. Type 1 patients are good candidates for enzyme replacement. Recently, the enzyme activity of Type 1 Gaucher fibroblasts and lymphoblastoid cells, infected with retrovirus containing glucocerebrosidase DNA, was restored to normal levels (Sorge *et al.*, 1987).

2.17. Neuronal Ceroid Lipofuscinosis (Batten's Disease)

Three types of autosomally inherited neuronal ceroid lipofuscinosis (NCL) have been described, with age of onset at 1, 3, and 7 years of age. An adult form with presenile dementia and blindness (Kuf's disease) also exists. The storage material occurs mainly the neurons and retinal pigmentary epithelial cells (which are rapidly destroyed), and has the characteristic appearance of curvilinear or fingerprint bodies. These bodies are believed to be encapsulated by lysosomal membranes, and to be the result of lipoperoxide or aldehyde fusion of lipid and protein to form autofluorescent pigment (Dawson and Glaser, 1987). Individual components of the storage material have been hard to identify despite staining for carbohydrate (PAS-positive) and lipid (oil-red O), but recent studies have identified dolichols, dolichol phosphates, and dolichol sugars (from Dol-P-P-GlcNAc up to Dol-P-P GlcNAcMan$_9$) in abnormal amounts in NCL brain (Wolfe *et al.*, 1987; Pullarkat, 1987). This initially led to the idea of NCL as an inborn error of glycoprotein metabolism, but this has not been substantiated, and the dolichol accumulation is believed to be a secondary phenomenon. In the sheep model of NCL, the autofluorescent pigment, including that from brain, has been shown to contain small peptides 3.5 to 15 kDa in size. These may be the result of impaired lysosomal proteolysis. Interestingly, the intracerebral injection of the thiol protease inhibitor leupeptin into rats induced CNS pigment formation which was rich in dolichols (Ivy *et al.*, 1984). A cathepsin B, L, or S deficiency could produce the neuropathology observed in NCL. However, all these thiol proteases are highly susceptible to inhibition by peroxides and aldehydes, suggesting that the cathepsin deficiency could be secondary to a defect which resulted in impaired catabolism of peroxides or aldehydes. Aldehydes such as 4-hydroxynonenal could be generated chemically in CNS following oxidation of arachidonyl residues in membrane phospholipids (van Kuijk *et al.*, 1987). A primary deficiency in phospholipase A$_2$ activity could lead to aldehyde formation, cathepsin inhibition, dolichol accumulation, peptide accumulation, and lipofuscin formation. However, the fundamental biochemical defect in NCL remains unknown.

2.18. Lysosomal (Acid) α-Glucosidase Deficiency (Pompe's Disease)

Genetic deficiency of acid α-glucosidase (acid maltase; 1,4-α-D-glucan glucohydrolase) results in glycogen storage disease Type II. This encompasses a spectrum of phenotypes from a rapidly fatal, infantile, neurovisceral, degenerative form (Pompe's disease) to a slowly progressive adult-onset myopathy. Use of a partial-length cDNA fragment for α-glucosidase (Martiniuk *et al.*, 1986) revealed molecular heterogeneity ranging from lack of detectable mRNA in infantile-onset patients to reduced mRNA of smaller size in adult-onset patients.

2.19. Sialic Acid Storage Diseases

During the past several years, a number of patients have been described who exhibit as a prominent biochemical finding the accumulation of the free monosaccharide NeuAc in tissues (including brain) and cultured cells and/or NeuAc hyperexcretion in the urine. Based on biochemical and clinical findings, as discussed below, these patients may be divided into three groups—those affected with the sialuria, Salla, and infantile variants of NeuAc storage disease.

2.19.1. Sialuria

The first patient exhibiting sialuria was described in 1968 by a group of French workers (Fontaine *et al.*, 1968); he was found to excrete 5–7 g of NeuAc in his urine each day, without evidence of extensive accumulation of the monosaccharide in tissue or cultured cells. Although severely mentally retarded, and otherwise clinically affected, the patient survived at least into the second decade. More recently, a second Finnish sialuria patient (26 years old) with less profound clinical symptoms and urinary NeuAc hyperexcretion in the range of 0.5 g/day in the absence of tissue accumulation of NeuAc was described (Palo *et al.*, 1985). Although the biochemical defect has not been demonstrated in either of these patients, studies in cultured fibroblasts derived from the original French patient have suggested the oversynthesis of NeuAc, perhaps secondary to impaired feedback inhibition of UDP-*N*-acetylglucosamine-2-epimerase (Thomas *et al.*, 1985).

2.19.2. Salla Disease

This variant, which has been described in more than 60 residents of the Salla province of Finland, is marked by the urinary hyperexcretion of NeuAc (approximately 80 mg/day) and the accumulation of free NeuAc in tissues (approximately 1 μmole/g liver and 6–20 μmole/g protein in cultured fibroblasts; Renlund *et al.*, 1979, 1983a). Clinically, the patients are severely mentally retarded and show other neural and motor deficits (Renlund *et al.*, 1983b; Renlund, 1984), although their life span is apparently normal. As in the case of the sialuria and infantile variants, Salla disease is apparently inherited as an autosomal recessive trait. Recently, non-Finnish patients having a similar clinical and biochemical presentation have been described (Wolfburg-Buchholz *et al.*, 1985; Ylitalo *et al.*, 1986). Indirect studies utilizing cultured fibroblasts have suggested the lysosomal accumulation of NeuAc in Salla disease, possibly secondary to impaired transport of NeuAc out of lysosomes (Renlund *et al.*, 1986b).

2.19.3. Infantile Generalized NeuAc Storage Disease

Since our initial biochemical description of a patient with massive accumulation of NeuAc in various tissues (Hancock *et al.*, 1982), a number of patients (now numbering near 20) having similar biochemical and clinical presentations have been reported (Tondeur *et al.*, 1982; Stevenson *et al.*, 1983; Baumkotter *et al.*, 1985; Paschke *et al.*, 1986; Fois *et al.*, 1987). In the most severe cases, death has occurred before 4 years of age, and in virtually all cases the clinical presentation has been marked by mental retardation, organomegaly, and some skeletal abnormalities. NeuAc accumulation has been documented in tissue (up to 20 μmoles/g liver) and cultured fibroblasts (20–215 nmoles/mg protein), and hyperexcretion in the urine (60–200 mg/day) has been observed in all cases in which urinary hyperexcretion has been analyzed. We have directly demonstrated the lysosomal accumulation of NeuAc in cultured fibroblasts from patients affected with the infantile variant (Hildreth *et al.*, 1986), and metabolic labeling studies by us (Hancock *et al.*, 1983) and others, as well as indirect preliminary transport studies (Jonas, 1986; Mancini *et al.*, 1986; Renlund *et al.*, 1986a,b) suggest a primary lysosomal NeuAc transport defect in infantile NeuAc storage diseases. We have recently reported decreased lysosomal density (Hildreth *et al.*, 1986), and impaired proteolytic processing of lysosomal β-hexosaminidase and α-fucosidase (Hancock, 1987; Hancock *et al.*, 1988) in affected fibroblasts. While both decreased lysosomal density and impaired proteolytic processing of lysosomal enzymes appear to be secondary to intralysosomal NeuAc accumulation (Hancock, unpublished observations), it is not yet clear how these secondary effects may contribute to the pathogenesis of infantile NeuAc storage disease. It is interesting to note, however, that there appears to exist a direct correlation between the extent of NeuAc accumulation and the severity of the clinical course observed in a number of patients.

2.20. Other Diseases Involving Glycoconjugates

The accumulation of Lafora bodies (which stain positive for carbohydrate) in pathological brain may represent impaired catabolism of a glycoconjugate, but no characterization has been carried out. Organelle disorders can also disrupt glycoconjugate metabolism, e.g., the faulty regulation of LDL uptake and cholesterol storage in Niemann–Pick Type C (Pentchev *et al.*, 1987). Typical Niemann–Pick disease (sphingomyelinase deficiency) is associated with secondary ganglioside accumulation in brain, which is probably of pathological significance.

3. INBORN ERRORS OF GLYCOSAMINOGLYCAN CATABOLISM

3.1. α-L-Iduronidase Deficiency (Hurler–Scheie's Disease; Types IH and IS)

Hurler's disease is a neurovisceral storage disease that results from an autosomal recessively inherited deficiency of the lysosomal hydrolase α-L-iduronidase (Matalon and Dorfman, 1972; Dorfman *et al.*, 1972; Bach *et al.*, 1972) (Figure 3). Both heparan

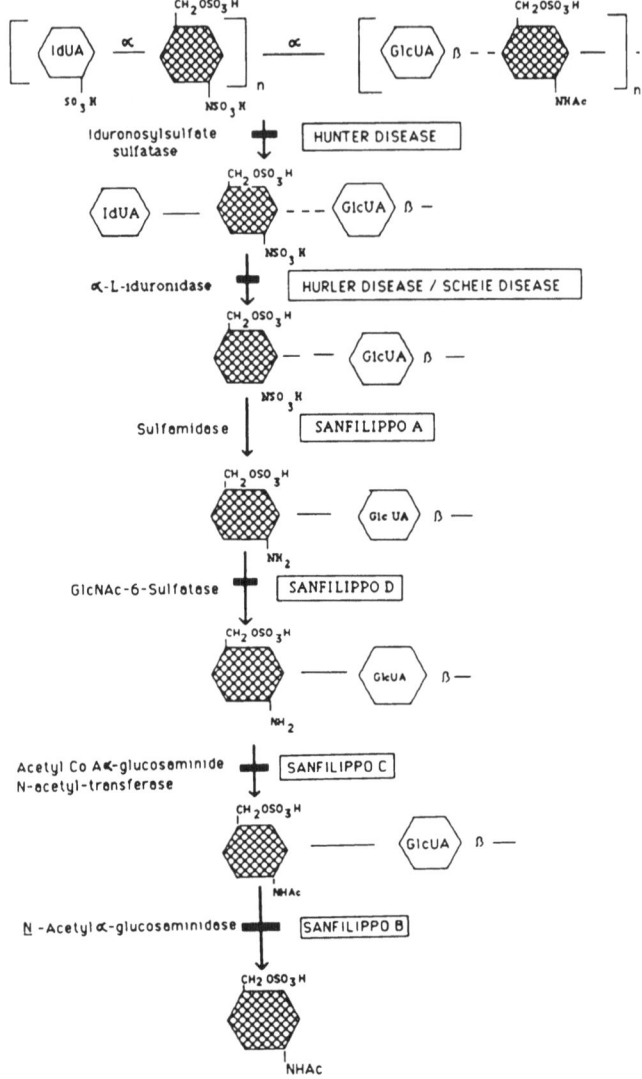

Figure 3. Schematic pathway for the catabolism of CNS heparan sulfate proteoglycan showing sites of enzyme defects leading to cerebral mucopolysaccharidoses.

sulfate and dermatan sulfate accumulate in tissues and in brain. This storage, plus secondary glycolipid GM3 and GD3 storage, manifests itself as Zebra bodies.

The Scheie syndrome (formerly Type V) is characterized by less severe mental retardation, but a similar profound deficiency of α-L-iduronidase when measured with synthetic substrates. The ratio of heparan sulfate to dermatan sulfate is less in Type IS than in IH. This may be explained by the observation that extracts of fibroblasts from IS patients release iduronic acid from desulfated heparin (IdUAα4GlcNAcα4) (Figure

3), whereas extracts of Hurler fibroblasts do not. The lack of cross-correction of [^{35}S]glycosaminoglycan accumulation following fusion of Hurler fibroblasts with different Scheie fibroblasts suggests that they are allelic mutations. Patients with probable Type IH/S, a genetic compound of the two mutations, have been observed, along with other phenotypic variants of α-L-iduronidase deficiency (Roubicek *et al.*, 1985).

3.2. Iduronate Sulfate Sulfatase Deficiency (Hunter's Disease; Type II)

Hunter's disease is a neurovisceral storage disease with additional skeletal abnormalities. Neurologically severe and mild forms of X-linked iduronate sulfate sulfatase deficiency (Sjoberg *et al.*, 1973) have been observed (Bach *et al.*, 1973). All forms show failure to hydrolyze *O*-(α-idopyranosyluronic acid-2-sulfate)4-2,5-anhydro-D-[^{3}H]mannitol-6-sulfate to a [^{3}H]monosulfated disaccharide, which is then hydrolyzed by endogenous α-L-iduronidase to [^{3}H]anhydromannitol-6-sulfate (Hall *et al.*, 1978) (Figure 3). The heparan sulfate:dermatan sulfate ratio is less than that in Hurler's disease, which may explain the milder neurological symptoms.

3.3. The Sanfilippo Syndromes (Type III)

This interesting group of inherited diseases are all characterized by the accumulation of heparan sulfate. Clinical symptoms include severe, progressive mental retardation, with some bony abnormalities, but lacking the corneal clouding, hypergingiva, abnormal facies, and hepatosplenomegaly associated with Types I and II (Dekaban and Patton, 1971; Dorfman and Matalon, 1972). The major storage material in brain is heparan sulfate (Figure 3), emphasizing the importance of this glycosaminoglycan in the nervous system.

3.3.1. Type A: Heparan *N*-Sulfate Sulfatase Deficiency

This form has the most severe CNS symptoms, implicating heparan sulfate as a major CNS proteoglycan. Diagnosis is made by using either [^{35}S]heparin fragments (Matalon and Dorfman, 1974) or a reducing terminal ^{3}H-labeled fragment from heparin (Thompson *et al.*, 1986). Of interest is the description of a combined *N*-sulfatase–β-mannosidase deficiency with excretion of both heparan sulfate and a Manβ4GlcNAc disaccharide in urine (Wenger *et al.*, 1986). This suggests the possibility of a common protective peptide which could be nonfunctional in the disorder.

3.3.2. Type B: *N*-Acetyl-α-glucosaminidase Deficiency

Accumulation of heparan sulfate leads to the early and rather severe involvement of the cortical neurons in the second and fourth layers in Sanfilippo B patients. This has been correlated with mental deterioration; the relative intactness of the third and fifth layers is consistent with the retention of motor function by Sanfilippo B patients. Diagnosis can be made with the synthetic substrate 4-methylumbelliferyl-/*p*-nitrophenyl-*N*-acetyl-α-glucosaminide (O'Brien, 1972a).

3.3.3. Type C: Acetyl CoA: α-Glucosaminide-*N*-acetyltransferase Deficiency

The transferase is a lysosomal hydrolase which catalyzes the transfer of an acetyl group from cytoplasmic acetyl CoA to terminal α-glucosamine residues of heparan sulfate within the organelle. Five patients were able to carry out acetyl CoA/CoA exchange (to form the acetyl-enzyme intermediate), but unable to transfer this bound acetyl group to glucosamine. One patient was unable to catalyze acetyl CoA/CoA exchange. Both types of patient were characterized by severe, progressive mental retardation, behavioral problems, and early death (15–20 years of age), with storage of heparan sulfate-like fragments in all organs, including brain. Diagnosis is made on the basis of failure to convert $[^3H]$-GlcNH$_2$ into $[^3H]$-GlcNAc (no longer binds to Dowex ion exchange resin) in the presence of acetyl CoA (Baum and Rome, 1986).

3.3.4. Type D: *N*-Acetylglucosamine-6-sulfate Sulfatase Deficiency

Clinical symptoms, including severe neurodegeneration, are comparable to other Sanfilippo patients (Kresse *et al.*, 1980; Gatti *et al.*, 1982). Diagnosis is made at pH 6.0 with the ^3H-labeled GlcNAc-6-SO$_3$Hβ4$[^3H]$galactitol disaccharide substrate derived from keratan sulfate or a ^3H-labeled trisaccharide from heparan sulfate.

3.4. GalNAc-6-sulfate Sulfatase Deficiency (Morquio's Disease; Type IV)

Morquio's disease results from a GalNAc-6-sulfate (or Gal-6-sulfate) sulfatase deficiency (Matalon *et al.*, 1974a; Horwitz and Dorfman, 1978). The disease is quite distinct both phenotypically and genotypically from Sanfilippo D (GlcNAc-6-sulfate sulfatase deficiency). Skeletal abnormalities are pronounced, but fragments of keratan sulfate [GalNAc-6-sulfate-β4Gal-6-sulfate]$_n$ and chondroitin-6-sulfate [GalNAc-6-sulfate-β4GlcUA]$_n$ are the major storage material. Very little storage occurs in brain, so that neural development is normal. Morquio patients have been found, on neuropathological examination, to have swollen glycoconjugate (PAS)-positive neurons and coarse, globular inclusions in the cerebral cortex (Gilles and Devel, 1971; Koto *et al.*, 1978)—similar to those seen in Hurler and Sanfilippo syndromes. Diagnosis is made by assay with the GalNAc-6-SO$_3$Hβ4GlcUAβ3-*N*-acetyl$[^3H]$galactitol trisaccharide derived from chondroitin-6-sulfate.

3.5. Proteoglycan β-Galactosidase Deficiency (Type V)

A specific β-galactosidase deficiency can result in a Morquio-like syndrome. A third, nonallelic form of the disease shows normal GalNAc-6-sulfate sulfatase and β-galactosidase activity, suggesting an activator protein defect (Horwitz and Dorfman, 1978).

3.6. GalNAc-4-sulfate Sulfatase Deficiency (Maroteaux–Lamy Syndrome; Type VI)

The Maroteaux–Lamy syndrome results from dermatan sulfate storage, a deficiency of GalNAc-4-sulfate sulfatase (Matalon *et al.*, 1974b) (arylsulfatase B), as measured with the synthetic substrate *O*-nitrocatechol sulfate. Mental retardation is rare. Such a patient has been described (Vestermark *et al.*, 1987), but multiple sulfatase deficiency was not ruled out.

3.7. β-Glucuronidase Deficiency (Type VII)

β-Glucuronidase deficiency results in a Hurler-like syndrome with delayed mental development and short stature (Sly *et al.*, 1973; Hall *et al.*, 1973). Storage material has been identified as proteoglycan fragments of the type $[GlcUA\beta4GalNAc4/6\text{-}SO_3H]_n$. A milder form with normal intelligence has been observed. Diagnosis is made by measuring the hydrolysis of 4-methylumbelliferyl-β-glucuronide. Fibroblasts from four patients contained antibody cross-reacting material, suggesting a defect in the β-glucuronidase structural gene (Bell *et al.*, 1977).

4. FUTURE DIRECTIONS

It seems likely that detailed carbohydrate analysis of autopsied brain from patients with inherited lysosomal storage diseases will continue to yield important insights into complex carbohydrates of the nervous system. Recent examples include the detection of lysoglycolipids (Hannun and Bell, 1987), stage-specific antigens in fucosidosis (Schwarting *et al.*, 1988), and increasingly complex oligosaccharides. Such glycoconjugates are normally extremely minor components of the CNS, but may be tremendously important from the point of view of neuronal–glial recognition, migration, and synapse formation. The combination of structural studies and monoclonal antibodies will be a potent one for unraveling the role of glycoconjugates in the CNS.

Because of success in cloning some of the genes for lysosomal hydrolases, and the availability of animal models, it is tempting to think of gene replacement therapy. In the past, bone marrow transplants have had very limited effect in reversing neurological symptoms in the lysosomal storage diseases, and this has been attributed to the failure of glycoproteins to cross the blood–brain barrier. Recent studies on a springer spaniel model for α-fucosidase deficiency indicate that cells of bone marrow origin can infiltrate the CNS and provide α-fucosidase to genetically deficient neurons. If the bone marrow transplant was done before the age of 6 months, neurological degeneration was essentially prevented. After 6 months, the transplants were a biochemical success, but enough neurons had degenerated to make recovery of function impossible. These studies offer a real prospect for clinical application of the immense body of knowledge which has been accumulated on the subject of inborn errors of glycoconjugate catabolism.

ACKNOWLEDGMENTS

  The work summarized in this chapter represents the sum of more than 20 years of work in many laboratories. Inevitably, because of space limitations, many fine pieces of work are not referenced directly, but can be found in other references. I would like to thank Norah McCabe and Karl Johnson for their comments and preliminary data, Sylvia Dawson for the artistic efforts, and LaJoyce Safford for typing the manuscript. Particular thanks go to the Mental Retardation Branch of the NICHD for supporting my work in this area since 1971 through USPHS Grants HD-06426 and HD-09402. L.W.H. is supported by a USPHS FIRST Award, DK-38593.

5. REFERENCES

Akasaki, M., Sugahara, K., Funakoshi, I., Aula, P., and Yamashina, I., 1976, Characterization of a mannose-containing glycoasparagine isolated from urine of a patient with aspartylglycosaminuria, *FEBS Lett.* **69**:191.

Autio, S., 1972, Aspartylglucosaminuria: Analysis of thirty-four patients, *J. Ment. Def. Res. Monogr. Ser.* **1**:1–93.

Bach, G., Friedman, R., Weismann, B., and Neufeld, E. F., 1972, The defect in the Hurler and Scheie syndrome: Deficiency of α-L-iduronidase, *Proc. Natl. Acad. Sci. USA* **69**:2048–2051.

Bach, G., Eisenberg, F., Cantz, M., and Neufeld, E. F., 1973, The defect in Hunter's disease: Deficiency of sulfoiduronate sulfatase, *Proc. Natl. Acad. Sci. USA* **70**:2134–2138.

Baum, K. J., and Rome, L. H., 1986, Genetic evidence for transmembrane acetylation by lysosomes, *Science* **233**:1087–1089.

Baumkotter, J., Cantz, M., Mendla, K., Baumann, W., Friebolin, H., Gehler, J., and Spranger, J., 1985, N-acetylneuraminic acid storage disease, *Hum. Genet.* **71**:155–159.

Beck, M., Bender, S. W., Reiter, H. L., Otto, W., Bassler, R., Dancygier, H., and Gehler, J., 1984, Neuraminidase deficiency presenting as non-immune hydrops, *Eur. J. Pediat.* **143**:135–139.

Bell, C. E., Jr., Sly, W. S., and Brot, F. E., 1977, Human β-glucuronidase deficiency mucopolysaccharidosis, *J. Clin. Invest.* **598**:97–105.

Ben-Yoseph, Y., Pack, B. A., Mitchell, D. A., Elwell, D. G., Potior, M., Melancon, S. B., and Nadler, H. L., 1986, Characterization of the mutant N-acetylglucosaminyl-phosphotransferase in I-cell disease and pseudohurler polydystrophy: Complementation analysis and kinetic studies, *Enzymology* **35**:106–116.

Bolhuis, P. A., Oonk, J. G. W., Kamp, P. E., Ris, A. J., Michalski, J. C., Overdijk, B., and Reuser, A. J. J., 1987, Ganglioside storage, hexosaminidase lability and urinary oligosaccharides in adult Sandhoff's disease, *Neurology* **37**:75–81.

Borrone, C., Gatti, R., Trias, X., and Durand, P., 1974, Fucosidosis: Clinical, biochemical, immunologic and genetic studies in two new cases, *J. Pediatr.* **84**:727–730.

Cantz, M., Gehler, J., and Spranger, J., 1977, Mucolipidosis I: Increased sialic acid content and deficiency of an α-N-acetylneuraminidase in cultured fibroblasts, *Biochem. Biophys. Res. Commun.* **74**:732–738.

Carroll, M., Dance, N., Masson, P. K., Robinson, D., and Winchester, B. G., 1972, Human mannosidosis—The enzymic defect, *Biochem. Biophys. Res. Commun.* **49**:579–583.

Cashman, N. R., Antel, J. P., Hancock, L. W., Dawson, G., Horwitz, A. L., Johnson, W. G., Huttenlocher, P. R., and Wollman, R. L., 1986, N-acetyl-β-hexosaminidase β-locus defect and juvenile motor neuron disease: A case study, *Ann. Neurol.* **19**:568–572.

Chen, W. W., Moser, A. B., and Moser, H. W., 1981, Role of lysosomal acid ceramidase in the metabolism of ceramide in human fibroblasts, *Arch. Biochem. Biophys.* **208**:444–455.

Christomanou, H., and Beyer, D., 1983, Absence of alpha fucosidase activity in two sisters showing a different phenotype, *Eur. J. Pediatr.* **140**:27–29.

Conary, J. P., Lorkowski, G., Schmidt, B., Pohlmann, R., Nagel, G., Meyer, H. E., Krentler, C., Cully, J., Hasilik, A., and von Figura, K., 1987, Genetic heterogeneity of steroid sulfatase deficiency revealed with cDNA for human steroid sulfatase, *Biochem. Biophys. Res. Commun.* **144**:1010–1017.

Cooper, A., Sardharwalla, I. B., and Roberts, M. M., 1986, Human β-mannosidosis, *N. Engl. J. Med.* **315**:1231.

Crandall, B. F., Philippart, M., Brown, W. J., and Bluestone, D. A., 1982, Mucolipidosis IV, *Am. J. Med. Genet.* **12**:301–308.

Dawson, G., 1982, Evidence for two distinct forms of mammalian β-mannosidase, *J. Biol. Chem.* **257**:3369–3371.

Dawson, G., and Glaser, P., 1987, Abnormal cathepsin B deficiency in neuronal ceroid lipofuscinosis can be explained by peroxide inhibition, *Biochem. Biophys. Res. Commun.* **147**:267–274.

Dawson, G., Matalon, R., and Dorfman, A., 1972, Glycosphingolipids in cultured human skin fibroblasts from patients with inborn errors of glycosphingolipid and mucopolysaccharide metabolism, *J. Biol. Chem.* **247**:5951–5958.

Dawson, G., McCabe, N., Hancock, L. W., and Johnson, K., 1988, Molecular biology of oligosaccharide storage disease, *Trans. Am. Soc. Neurochem.* **19**:296.

Dekaban, A. S., and Patton, V. M., 1971, Hurler's and Sanfilippo's variants of mucopolysaccharidosis; cerebral pathology and lipid chemistry, *Arch. Pathol.* **91**:434–449.

Dewji, N. N., Wenger, D. A., and O'Brien, J. S., 1987, Nucleotide sequence of cloned cDNA for human sphingolipid activator protein 1 precursor, *Proc. Natl. Acad. Sci. USA* **84**:8652–8656.

Dorfman, A., and Matalon, R., 1972, The mucopolysaccharidoses, in: *The Metabolic Basis of Inherited Disease* (J. B. Stanbury, J. B. Wyngaarden, and D. S. Fredrickson, eds.), pp. 1218–1272, McGraw-Hill, New York.

Dorfman, A., Matalon, R., Cifonelli, J. A., Thompson, J., and Dawson, G., 1972, The degradation of acid mucopolysaccharides and the mucopolysaccharidoses, in: *Sphingolipidoses and Allied Disorders* (B. W. Volk and S. M. Aronson, eds.), pp. 195–210, Plenum Press, New York.

Durand, P., Borrone, C., and Della Cella, G., 1969, Fucosidosis, *J. Pediatr.* **75**:665–674.

Durand, P., Gatti, R., Cavalieri, S., Borrone, C., Tondeur, M., Michalski, J.-C., and Strecker, G., 1977, Sialidosis (mucolipidoses I), *Helv. Paediatr. Acta* **32**:391–400.

Eto, Y., Tahara, T., Tokoro, T., and Maekawa, K., 1983, Various sulfatase activities in leukocytes and cultured skin fibroblasts from heterozygotes for the multiple sulfatase deficiency (mucosulfatidosis), *Pediatr. Res.* **17**:97–100.

Fois, A., Balestri, P., Farnetani, G. M. S., Mancini, P., Borgogni, M. A., Margollicci, M. A., Molinelli, M., Alessandrini, C., and Gerli, R., 1987, Free sialic acid storage disease: A new Italian case, *Eur. J. Pediatr.* **146**:195–198.

Fontaine, G., Biserte, G., Montreuil, J., DuPont, A., and Farriaux, J. P., 1968, La sialurie: un trouble metabolique original, *Helv. Paediatr. Acta* **23**(Suppl. XVII):3–32.

Galjaard, H., Hoogeveen, A., Keijzer, W., deWit-Verbeek, H. A., Reusor, A. J. J., Ho, M. W., and Robinson, D., 1975, Genetic heterogeneity in G_{M1}-gangliosidosis, *Nature* **257**:60–62.

Gatti, R., Borrone, C., Durand, P., DeVirtilis, S., Sanna, G., Cao, A., von Figura, K., Kresse, H., and Paschke, E., 1982, Sanfilippo Type D disease, *Eur. J. Pediatr.* **138**:168–171.

Gilbert, E. F., Dawson, G., zu Rhein, G. M., Opitz, J. M., and Spranger, J. W., 1973, I-cell disease: Mucolipidosis II, *Z. Kinderheilkd.* **114**:259–292.

Gilles, F. H., and Deuel, R. K., 1971, Neuronal cytoplasmic globules in the brain in Morquio's syndrome, *Arch. Neurol.* **25**:393–403.

Gonzalex-Noreiga, A., Grubb, J. H., Talkud, V., and Sly, W. S., 1980, Chloroquine inhibits lysosomal enzyme pinocytosis and enhances lysosomal enzyme secretion by impairing receptor recycling, *J. Cell Biol.* **85**:839–852.

Graves, P. N., Grabowski, G. A., Ludman, M. D., Palese, P., and Smith, F. I., 1986, Human acid β-glucosidase: Northern blot and S1 nuclease analysis of mRNA from HeLa cells and normal and Gaucher disease fibroblasts, *Am. J. Hum. Genet.* **39**:763–774.

Hall, C. W., Cantz, M., and Neufeld, E. F., 1973, A β-glucuronidase deficiency mucopolysaccharidosis: Studies in cultured fibroblasts, *Arch. Biochem. Biophys.* **155**:32–38.

Hall, C. W., Liebaers, I., DiNatale, P., and Neufeld, E. F., 1978, Enzymatic diagnosis of the genetic mucopolysaccharide storage diseases, *Methods Enzymol.* **50**:439–455.

Hancock, L. W., 1987, Impaired processing of lysosomal enzymes in generalized N-acetylneuraminic acid storage disease, *J. Cell Biol.* **105**:249a.

Hancock, L. W., Thaler, M. M., Horwitz, A. L., and Dawson, G., 1982, Generalized *N*-acetylneuraminic acid storage disease, *J. Neurochem.* **38**:803–809.

Hancock, L. W., Horwitz, A. L., and Dawson, G., 1983, N-acetylneuraminic acid and sialoglycoconjugate metabolism in cultured fibroblasts from a patient with generalized N-acetylneuraminic acid storage disease, *Biochim. Biophys. Acta* **760**:42–52.

Hancock, L. W., Horwitz, A. L., Cashman, N. R., Antel, J. P., and Dawson, G., 1985, N-acetyl-β-hexosaminidase B deficiency in cultured fibroblasts from a patient with progressive motor neuron disease, *Biochem. Biophys. Res. Commun.* **130**:1185–1192.

Hancock, L. W., Ricketts, J. P., and Hildreth, J., 1988, Impaired proteolytic processing of lysosomal N-acetyl-β-hexosaminidase in cultured fibroblasts from patients with infantile generalized N-acetylneuraminic acid storage disease, *Biochem. Biophys. Res. Commun.* **152**:83–92.

Hannun, Y. A., and Bell, R. A., 1987, Lysoglycosphingolipids in lipid storage diseases, *Science* **235**:670–674.

Hasilik, A., and Neufeld, E. F., 1980, Biosynthesis of lysosomal enzymes in fibroblasts: Synthesis as precursors of higher molecular weights, *J. Biol. Chem.* **255**:4937–4945.

Hasilik, A., Waheed, A., and von Figura, K., 1981, Lysosomal hydrolase synthesis and processing, *Biochem. Biophys. Res. Commun.* **98**:761–767.

Hildreth, J., Sachs, L., and Hancock, L. W., 1986, *N*-Acetylneuraminic acid accumulation in a buoyant lysosomal fraction of cultured fibroblasts from patients with infantile generalized *N*-acetylneuraminic acid storage disease, *Biochem. Biophys. Res. Commun.* **139**:838–844.

Hill, D. F., Bullock, P. N., Chiappelli, F., and Rome, L. H., 1985, Binding and internalization of lysosomal enzymes by primary cultures of rat glia, *J. Neurosci. Res.* **14**:35–47.

Horwitz, A. L., 1979, Genetic complementation studies of multiple sulfatase deficiency, *Proc. Natl. Acad. Sci. USA* **76**:6496–6498.

Horwitz, A. L., and Dorfman, A., 1978, The enzymic defect in Morquio's disease: The specificities of N-acetylhexosamine sulfatases, *Biochem. Biophys. Res. Commun.* **80**:819–825.

Horwitz, A. L., Warshawsky, L., King, J., and Burns, G., 1986, Rapid degradation of steroid sulfatase in multiple sulfatase deficiency, *Biochem. Biophys. Res. Commun.* **135**:389–396.

Ivy, G. O., Schotter, F., Wenzel, J., Baudry, M., and Lynch, G., 1984, Inhibition of lysosomal enzymes produces a rapid and massive accumulation of lipofuscin-like dense bodies in the CNS, *Science* **226**:985–987.

Jenner, F. A., and Pollitt, R. J., 1967, Large quantities of 2-acetamido-1-(β-L-aspartamido)-1,2-dideoxy-glucose in the urine of mentally retarded siblings, *Biochem. J.* **103**:48–49.

Johnson, K., and Dawson, G., 1985, Molecular defect in processing alpha-fucosidase in fucosidosis, *Biochem. Biophys. Res. Commun.* **133**:90–97.

Jolly, R. D., Winchester, B. G., Gehler, J., Dorling, P. R., and Dawson, G., 1981, Mannosidosis: A comparative review of biochemical and related clinicopathological aspects of three forms of the disease, *J. Appl. Biochem.* **3**:273–291.

Jonas, A. J., 1986, Studies of lysosomal sialic acid metabolism: Retention of sialic acid by Salla disease lysosomes, *Biochem. Biophys. Res. Commun.* **137**:175–181.

Jones, M. Z., and Dawson, G., 1981, Caprine β-mannosidosis: Inherited deficiency of β-D-mannosidase, *J. Biol. Chem.* **256**:5185–5188.

Jones, M. Z., and Laine, R. A., 1981, Caprine β-mannosidosis: Identification of the trisaccharide storage material, *J. Biol. Chem.* **256**:5181–5184.

Jones, M. Z., Cunningham, J. G., Dade, A. W., Dawson, G., Laine, R. A., Williams, C. S. F., Alessi, D. M., Mostoskey, U. V., and Vorro, J. R., 1982, Caprine β-mannosidosis, in: *Animal Models of Inherited Metabolic Diseases* (R. F. Desnick, D. F. Patterson, and D. G. Scarpelli, eds.), pp. 165–176, Liss, New York.

Kaplan, A., Fischer, D., Achord, D., and Sly, W., 1977, Phosphohexosyl recognition is a general characteristic of pinocytosis of lysosomal glycosidases by human fibroblasts, *J. Clin. Invest.* **60**:1088–1093.

Kelly, T. E., and Graetz, G., 1977, Isolated acid neuraminidase deficiency: A distinct lysosomal storage disease, *Am. J. Med. Genet.* 1:31–46.

Kint, J. A., Dacrement, G., Carton, D., Orye, E., and Hooft, C., 1973, Mucopolysaccharidosis: Secondarily induced abnormal distribution of lysosomal isoenzymes, *Science* 181:352–354.

Kolodny, E. H., and Raghavan, S. S., 1983, G_{M2}-gangliosidosis–β-hexosaminidase mutants, *Trends Neurosci.* 6:16–20.

Kornfeld, S., 1987, Trafficking of lysosomal enzymes, *FASEB J.* 1:462–468.

Koster, J. F., Nieremeijer, M. F., Loonen, M. C. B., and Galjaard, H., 1976, β-Galactosidase deficiency in an adult: A biochemical and somatic cell genetic study on a variant of G_{M1}-gangliosidosis, *Clin. Genet.* 9:427.

Koto, A., Horwitz, A. L., Suzuki, K., Tiffany, C. W., and Suzuki, K., 1978, The Morquio syndrome: Neuropathology and biochemistry, *Ann. Neurol.* 4:26–36.

Kresse, H., Paschke, E., von Figura, K., Gilberg, W., and Fuchs, W., 1980, Biochemical defect in Sanfilippo D, *Proc. Natl. Acad. Sci. USA* 77:6822–6826.

Lalley, P. A., Rattazzi, M. C., and Shows, T. B., 1974, Human β-D-N-acetylhexosaminidases A and B: Expression and linkage relationships in somatic cell hybrids, *Proc. Natl. Acad. Sci. USA* 71:1569–1573.

Laver, J., Fried, K., Beer, S. I., Ianci, T. C., Heyman, E., and Bach, G., 1983, Infantile lethal neuraminidase deficiency (sialidosis), *Clin. Genet.* 23:97–101.

Ledeen, R., Salsman, K., Gonatas, J., and Taghavy, A., 1965, Structure comparisons of the major monosialogangliosides from brains of normal humans, gargoylism and late infantile systemic lipidosis, Part 1, *J. Neuropathol. Exp. Neurol.* 24:341–351.

Lemansky, P., Bishop, D. F., Desnick, R. J., Hasilik, A., and von Figura, K., 1987, Synthesis and processing of α-galactosidase A in human fibroblasts: Evidence for different mutations in Fabry disease, *J. Biol. Chem.* 262:2062–2065.

Leroy, J. G., Ho, M. W., MacBrinn, M. C., Zielke, K., Jacob, J., and O'Brien, J. S., 1972, I-cell disease: Biochemical studies, *Pediatr. Res.* 6:752–757.

Little, L. E., Mueller, Q. T., Honey, N. K., Shows, T. B., and Miller, A. L., 1986, Heterogeneity of N-acetyl-glucosamine-1-phosphotransferase within mucolipidosis III, *J. Biol. Chem.* 261:733–738.

Lowden, J. A., Callahan, J. W., Norman, M. G., Thain, M., and Pritchard, J. S., 1974, Juvenile G_{M1} gangliosidosis, *Arch. Neurol.* 31:20–24.

Mancini, G. M. S., Verheijen, F. W., and Galjaard, H., 1986, Free N-acetylneuraminic acid (NANA) storage disorders: Evidence for defective NANA transport across the lysosomal membrane, *Hum. Genet.* 73:214–217.

Maroteaux, P., Humbel, R., Strecker, G., Michalski, J. C., and Mande, R., 1978, Un nouveau type de sialidose avec attiente renale: La nephrosialidose. *Arch. Fr. Pediatr.* 35:819–829.

Martiniuk, F., Mehler, M., Pellicer, A., Tzall, S., LaBadie, G., Hobart, C., Ellenbogen, A., and Hirschhorn, R., 1986, Isolation of a cDNA for human acid α-glucosidase and detection of genetic heterogeneity for mRNA in three α-glucosidase-deficient patients, *Proc. Natl. Acad. Sci. USA* 83:9641–9644.

Matalon, R., and Dorfman, A., 1972, Hurler's syndrome: An α-L-iduronidase deficiency, *Biochem. Biophys. Res. Commun.* 47:959–966.

Matalon, R., and Dorfman, A., 1974, Sanfilippo A syndrome: Sulfamidase deficiency in cultured skin fibroblasts and liver, *J. Clin. Invest.* 54:907–912.

Matalon, R., Cifonelli, J. A., Zellweger, H., and Dorfman, A., 1968, Lipid abnormalities in a variant of the Hurler syndrome, *Proc. Natl. Acad. Sci. USA*, 59:1097–1102.

Matalon, R., Arbogast, B., Justice, P., Brandt, I. K., and Dorfman, A., 1974a, Morquio's syndrome: Deficiency of a chondroitin sulfate N-acetyl-hexosamine sulfate sulfatase, *Biochem. Biophys. Res. Commun.* 61:759.

Matalon, R., Arbogast, B., and Dorfman, A., 1974b, Deficiency of chondroitin sulfate N-acetyl-galactosamine-4-sulfate sulfatase in Maroteaux–Lamy syndrome, *Biochem. Biophys. Res. Commun.* 61:1450–1457.

Moser, H. W., Prensky, A. L., Wolfe, H. J., and Rosman, N. P., 1969, Farber's lipogranulomatosis: Report of a case and demonstration of an excess of free ceramide and ganglioside, *Am. J. Med.* 47:869–890.

Myerowitz, R., and Hogikyan, N. D., 1986, Different mutations in Ashkenazi Jewish and non-Jewish French Canadians with Tay–Sachs disease, *Science* 232:1646–1648.

Myerowitz, R., and Proia, R. L., 1984, A cDNA clone for the α-chain of human β-hexosaminidase; deficiency of the α-chain in mRNA in Ashkenazi Tay–Sachs fibroblasts, *Proc. Natl. Acad. Sci. USA* 81:5396–5398.

Nakagawa, F., Schulte, B. A., and Spicer, S. S., 1986, Selective cytochemical demonstration of glycoconjugate containing terminal N-acetylgalactosamine on some brain neurons, *J. Comp. Neurol.* 243:280–290.

Navon, R., Argov, Z., and Frisch, A., 1986, Hexosaminidase A deficiency in adults, *Am. J. Med. Genet.* 24:179–196.

Norden, N. E., Lundblad, A., Svensson, S., and Autio, S., 1974, Characterization of two mannose-containing oligosaccharides isolated from the urine of patients with mannosidosis, *Biochemistry* 13:871–874.

O'Brien, J. S., 1972a, Sanfilippo syndrome: Profound deficiency of alpha-acetylglucosaminidase activity in organs and skin from type B patients, *Proc. Natl. Acad. Sci. USA* 69:1720–1722.

O'Brien, J. S., 1972b, G_{m1}-gangliosidosis, in: *The Metabolic Basis of Inherited Disease* (J. B. Stanbury, J. B. Wyngaarden, and D. S. Fredrickson, eds.), pp. 639–662, McGraw-Hill, New York.

O'Brien, J. S., 1977, Neuraminidase deficiency in the cherry red spot-myoclonus syndrome, *Biochem. Biophys. Res. Commun.* 79:1136–1141.

O'Brien, J. S., Willems, P. J., Fukushima, H., deWet, J. R., Darby, J. K., DiCioccio, R., Fowler, M. L., and Shows, T. B., 1987, Molecular biology of the alpha-L-fucosidase gene and fucosidosis, *Enzyme* 38:45–53.

O'Dowd, B., Quan, F., Willard, H. F., Lamhonwah, A. M., Korneluk, R. G., Lowden, J. A., Gravel, R. A., and Mahuran, D., 1985, Isolation of cDNA clones encoding the β-hexosaminidase gene, *Proc. Natl. Acad. Sci. USA* 82:1184–1188.

O'Dowd, B. F., Klavins, M. H., Willard, H. F., Gravel, R., Lowden, J. A., and Mahuran, D. J., 1986, Molecular heterogeneity in the infantile and juvenile forms of Sandhoff disease (O-variant G_{M2} gangliosidosis), *J. Biol. Chem.* 261:12680–12685.

Ohno, K., and Suzuki, K., 1988, Mutation in G_{M2}-gangliosidosis B1 variant, *J. Neurochem.* 50:316–318.

Okada, S., and O'Brien, J. S., 1968, Generalized gangliosidosis (β-galactosidase deficiency), *Science* 160:1002–1004.

Okada, S., and O'Brien, J. S., 1969, Tay–Sachs disease: Generalized absence of a β-D-acetylhexosaminidase component, *Science* 165:698–700.

Palmeri, S., Hoogeveen, A. T., Verheijen, F. W., and Galjaard, H., 1986, Galactosidosis: Molecular heterogeneity among distinct clinical phenotypes, *Am. J. Hum. Genet.* 38:137–148.

Palo, J., Riekkinen, P., Arstila, A. Y., Autio, S., and Kivimaki, T., 1972, Aspartylglucosaminuria II: biochemical studies on brain, liver, kidney, and spleen, *Acta Neuropathol.* 20:217–224.

Palo, J., Pollitt, R. J., Pretty, K. M., and Savolainen, H., 1973, Glycoasparagine metabolites in patients with aspartylglycosaminuria: Comparison between English and Finnish patients with special reference to storage materials, *Clin. Chim. Acta* 47:69–74.

Palo, J., Rauvala, H., Finne, J., Haltia, M., and Palmgren, K., 1985, Hyperexcretion of free N-acetyl-neuraminic acid—A novel type of sialuria, *Clin. Chim. Acta* 145:237–242.

Paschke, E., Trinkl, G., Erwa, W., Pavelka, M., Mutz, I., and Roscher, A., 1986, Infantile type of sialic acid storage disease with sialuria, *Clin. Genet.* 29:417–424.

Pentchev, P. G., Comly, M. E., Kruth, M. S., Tokoro, T., Butler, J., Sokol, J., Filling-Katz, M., Quirk, J. H., Marshall, D. C., Patel, S., Vanier, M. T., and Brady, R. O., 1987, Group C Niemann–Pick disease, *FASEB J.* 1:40–45.

Pullarkat, R. K., 1987, Dolichols and phosphodolichols in aging and in neurological disorders, *Chem. Scr.* 27:85–88.

Purpura, D., and Suzuki, K., 1976, Distortion of neuronal geometry and formation of aberrant synapses in neuronal storage disease, *Brain Res.* 116:1–12.

Reitman, M. L., Varki, A. P., and Kornfeld, S., 1981, Lysosomal enzyme targetting, *J. Clin. Invest.* 67:1574–1579.

Renlund, M., 1984, Clinical and laboratory diagnosis of Salla disease in infancy and childhood, *J. Pediatr.* 104:232–236.

Renlund, M., Chester, M. A., Lundblad, A., Aula, P., Raivio, K. O., Autio, S., and Koskela, S.-L., 1979, Increased urinary excretion of free N-acetylneuraminic acid in thirteen patients with Salla disease, *Eur. J. Biochem.* **101**:245–250.

Renlund, M., Chester, M. A., Lundblad, A., Parkkinen, J., and Krusius, T., 1983a, Free N-acetylneuraminic acid in tissues in Salla disease and the enzymes involved in its metabolism, *Eur. J. Biochem.* **130**:39–45.

Renlund, M., Aula, P., Raivio, K. O., Autio, S., Sainio, K., Rapola, J., and Koskela, S. I., 1983b, Salla disease: A new lysosomal storage disorder with distributed sialic acid metabolism, *Neurology* **33**:57–66.

Renlund, M., Kovanen, P. T., Raivio, K. O., Aula, P., Gahmberg, C. G., and Ehnholm, C., 1986a, Studies on the defect underlying the lysosomal storage of sialic acid in Salla disease, *J. Clin. Invest.* **77**:568–574.

Renlund, M., Tietze, F., and Gahl, W. A., 1986b, Defective sialic acid egress from isolated fibroblast lysosomes of patients with Salla disease, *Science* **232**:759–762.

Roubicek, M., Gehler, J., and Spranger, J., 1985, The clinical spectrum of alpha-iduronidase deficiency, *Am. J. Med. Genet.* **20**:471–481.

Rushton, A. R., and Dawson, G., 1977, The effect of glycosaminoglycans on the *in vivo* activity of human skin fibroblast glycosphingolipid β-galactosidases and neuraminidases, *Clin. Chim. Acta* **80**:133–139.

Sandhoff, K., Harzer, K., Wassle, W., and Jatzkewitz, H., 1971, Enzyme alterations and lipid storage in three variants of Tay–Sachs disease, *J. Neurochem.* **18**:2469–2489.

Sando, G. N., Titus-Dillon, P., Hall, C. W., and Neufeld, E. F., 1979, Inhibition of receptor-mediated uptake of a lysosomal enzyme into fibroblasts by chloroquine, procaine and ammonia, *Exp. Cell Res.* **119**:359–364.

Scaravilli, F., and Suzuki, K., 1983, Enzyme replacement in grafted nerve of *twitcher* mouse, *Nature* **305**:713–715.

Schwarting, G. A., Williams, M. A., Evans, J. E., and McCluer, R. H., 1988, Characterization of SSEA-1 glycolipids in human fucosidosis brain, *Trans. Am. Soc. Neurochem.* **19**:349.

Shinoda, H., Kobayashi, T., Katayama, M., Goto, I., and Nagara, H., 1987, Accumulation of galactosylsphingosine (psychosine) in the twitcher mouse: Determination by HPLC, *J. Neurochem.* **49**:92–99.

Sjoberg, I., Fransson, L.-A., Matalon, R., and Dorfman, A., 1973, Hunter's syndrome: A deficiency of L-iduronosulfate sulfatase, *Biochem. Biophys. Res. Commun.* **54**:1125–1132.

Sly, W. S., Quinton, B. A., McAllister, W. H., and Rimoin, D. L., 1973, β-Glucuronidase deficiency: Report of clinical, radiological and biochemical features of a new mucopolysaccharidosis, *J. Pediatr.* **82**:249–257.

Sorge, J., Kuhl, W., West, C., and Beutler, E., 1987, Complete correction of the enzymatic defect of type 1 Gaucher disease fibroblasts by retroviral-mediated gene transfer, *Proc. Natl. Acad. Sci. USA* **84**:906–909.

Spranger, J., 1987, Inborn errors of complex carbohydrates metabolism, *Am. J. Med. Genet.* **28**:489–499.

Stevenson, R. E., Lubinsky, M., Taylor, H. A., Wenger, D. A., Schroer, R. J., and Olmstead, P. M., 1983, Sialic acid storage disease with sialuria: Clinical and biochemical features in the severe infantile type, *Pediatrics* **72**:441–449.

Strecker, G., Peers, M.-C., Michalski, J.-C., Hondi-Assah, T., Fournet, B., Spik, G., Montreuil, J., Farriaux, J.-P., Maroteaux, P., and Durand, P., 1977a, Structure of nine sialyl-oligosaccharides accumulated in urine of eleven patients with three different types of sialidosis, *Eur. J. Biochem.* **75**:391–403.

Strecker, G., Michalski, J. C., Herlant-Peers, M. C., Fournet, B., and Montreuil, J., 1977b, Structure of 40 oligosaccharides and glycopeptides accumulating in the urine from patients with catabolism defect of glycoconjugates (sialidosis, fucosidosis, mannosidosis and Sandhoff's disease), *Proc. 4th Intl. Symp. Glycoconjugates*, Woods Hole, Mass.

Sugahara, K., Funakoshi, S., Funakoshi, I., Aula, P., and Yamashina, I., 1976, Characterization of one neutral and two acidic glycoasparagines isolated from the urine of patients with aspartylglucosaminuria (AGU), *J. Biochem.* **80**:195–201.

Suzuki, Y., and Suzuki, K., 1970, Krabbe's globoid cell leukodystrophy: Deficiency of galactocerebrosidase in serum leukocytes and fibroblasts, *Science* **171**:73–75.

Suzuki, Y., Sakubara, H., and Yamanaka, T., 1984, Galactosialidosis: A comparative study of clinical and biochemical data on 22 patients, in: *The Developing Brain and Its Disorders* (M. Arima, ed.), pp. 173–188, Tokyo University Press, Tokyo.

Svennerholm, L., Vanier, M.-T., and Mansson, J. E., 1980, Krabbe disease: A galactosylsphingosine (psychosine) lipidosis, *J. Lipid Res.* **21**:53–64.

Sweeley, C. C., and Klionsky, B., 1963, Fabry's disease: Classification as a sphingolipidosis and partial characterization of a novel glycolipid, *J. Biol. Chem.* **238**:3148–3150.

Sweeley, C. C., and Usuki, S., 1987, The effect of a sialidase inhibitor on the cell cycle of cultured human fibroblasts, *J. Cell Biol.* **105**:101a.

Thomas, G. H., Reynolds, L. W., and Miller, C. S., 1985, Overproduction of N-acetylneuraminic acid (sialic acid) by sialuria fibroblasts, *Pediatr. Res.* **19**:451–455.

Thompson, J. N., Roden, L., and Reynertson, R., 1986, Oligosaccharide substrates for heparin sulfamidase, *Anal. Biochem.* **152**:412–422.

Tondeur, M., Libert, J., Vamos, E., Van Hoff, F., Thomas, G. H., and Strecker, G., 1982, Infantile forms of sialic acid storage disorder, *Eur. J. Pediatr.* **139**:142–147.

Tsay, G. C., and Dawson, G., 1976, Oligosaccharide storage in brains from patients with fucosidosis, G_{M1}-gangliosidosis and G_{M2}-gangliosidosis (Sandhoff's disease), *J. Neurochem.* **27**:733–740.

Tsay, G. C., Dawson, G., and Li, Y.-T., 1975, Structure of the glycopeptide storage material in G_{M1}-gangliosidosis: Sequence determination with specific endo and exoglycosidases, *Biochim. Biophys. Acta* **385**:305–311.

Tsay, G. C., Dawson, G., and Sung, J. S.-S., 1976, Structure of the accumulating oligosaccharide in fucosidosis, *J. Biol. Chem.* **251**:5852–5859.

Tsuji, S., Choudary, P. V., Martin, B. H., Stubblefield, B. K., Mayor, J. A., Barranger, J. A., and Ginns, E. I., 1986, A mutation in the human glucocerebrosidase gene in neuronopathic Gaucher's disease, *N. Engl. J. Med.* **316**:510–575.

van Diggelen, O. P., Galjaard, H., Egge, H., Dabrowski, U., and Cantz, M., 1987, Lysosomal α-N-acetylgalactosaminidase deficiency: A new inherited metabolic disease, *Lancet* **2**:804.

van Kuijk, F. J. G. M., Sevanian, A., Handelman, G. J., and Dratz, E. A., 1987, A new role for phospholipase A_2: Protection of membranes from lipid peroxidation damage, *Trends Biochem. Sci.* **12**: 31–34.

Verheijen, F. W., Palmeri, S., and Galjaard, H., 1987, Purification and partial characterization of lysosomal neuraminidase from human placenta, *Eur. J. Biochem.* **162**:63–67.

Vestermark, S., Tonnesen, T., Schulz-Andersen, M., and Guttler, F., 1987, Neurological symptoms in a Maroteaux–Lamy patient, *Clin. Genet.* **31**:114–117.

von Figura, K., and Hasilik, A., 1986, Lysosomal enzymes and their receptors, *Annu. Rev. Biochem.* **55**: 167–193.

Warner, T. G., DeKremer, R. D., Sjoberg, E. R., and Mock, A. K., 1985, Characterization and analysis of branched chain N-acetylglucosaminyl oligosaccharides accumulating in Sandhoff disease tissue: Evidence that biantennary-bisected oligosaccharides of glycoproteins are abundant substrates for lysosomes, *J. Biol. Chem.* **260**:6194–6199.

Weismann, U., DiDonata, S., and Hershkowitz, N. N., 1975, Effect of chloroquine on cultured fibroblasts release of lysosomal hydrolases and inhibition of their uptake, *Biochem. Biophys. Res. Commun.* **66**: 1338–1343.

Wenger, D. A., Sattler, M., Clark, C., Tanaka, H., Suzuki, K., and Dawson, G., 1975, Lactosyl ceramidosis: Normal activity for two lactosyl ceramide β-galactosidases, *Science* **188**:1310–1312.

Wenger, D. A., Sattler, M., Mueller, T., Myers, G. G., Schneiman, R. S., and Nixon, G. W., 1980, Adult G_{M1} gangliosidosis, *Clin. Genet.* **17**:323–334.

Wenger, D. A., Sujansky, E., Fennessey, P. V., and Thompson, J. N., 1986, Combined heparan sulfate sulfamidase and β-mannosidase deficiency in a Sanfilippo phenotype, *N. Engl. J. Med.* **315**:1203–1206.

Wolfburg-Buchholz, K., Schlote, W., Baumkotter, J., Cantz, M., Holder, H., and Harzer, K., 1985, Familial lysosomal storage disease with generalized vacuolization and sialic aciduria, Sporadic Salla disease, *Neuropaediatrie* **16**:67–75.

Wolfe, L. S., Senior, R. G., and Ng Ying Kin, N. M. K., 1974, The structure of oligosaccharides accumulating in the liver of G_{M1}-gangliosidosis type 1, *J. Biol. Chem.* **249**:1828–1838.

Wolfe, L. S., Ivy, G. O., and Witkop, C. J., 1987, Dolichols: Lysosomal membrane turnover and relationships to the accumulation of ceroid and lipofuscin in inherited diseases, Alzheimer's disease and aging, *Chem. Scr.* **27**:79–84.

Ylitalo, V., Hagberg, B., Rapola, J., Mansson, J. E., Svennerholm, L., Sanner, G., and Tonnby, B., 1986, Salla disease variants, *Neuropediatr.* **17**:44–47.

Ziegler, M., and Bach, G., 1986, Internalization of exogenous gangliosides in cultured skin fibroblasts for the diagnosis of mucolipidosis IV, *Clin. Chim. Acta* **157**:183–190.

Index